# Sustainable Waste Management Towards Circular Bioeconomy

Vijai Singh
Editor

# Sustainable Waste Management Towards Circular Bioeconomy

## Components, Design Innovation and Impact

Springer

*Editor*
Vijai Singh
Department of Biosciences
Indrashil University
Mehsana, Gujarat, India

ISBN 978-981-95-2226-2    ISBN 978-981-95-2227-9 (eBook)
https://doi.org/10.1007/978-981-95-2227-9



This Springer imprint is published by the registered company Springer Nature Singapore Pte Ltd.
The registered company address is: 152 Beach Road, #21-01/04 Gateway East, Singapore 189721, Singapore

# Foreword

It is a great pleasure to write this foreword for the book titled *Sustainable Waste Management Towards Circular Bioeconomy*, a timely contribution in the rapidly growing field of engineering biology focusing on approaches for a fully sustainable economic and societal progress. A sustainable circular bioeconomy (CBE) is at the core of the technologies essential to reach the United Nation's Sustainable Development Goal 12 (Responsible Consumption and Production).

Recently, owing to increased pollution, climate change and global warming, as well as increasing human population, it has become evident that there is an imperative need for sustainable foods, chemicals and energy, leading to clean environments and an improved quality of human life across the globe. Part of the solution consists of abandoning our linear production to waste strategy and adopting circular approaches in which waste becomes a resource that can be converted into value-added products.

This book offers a cutting-edge overview of the circular bioeconomy and the mechanism required for its implementation, including technological advances in strain engineering and circular design. The book covers advances in biomanufacturing and resource recovery, delving into the use of waste as a feedstock and exploring the concept of the biorefinery, focusing on the conversion of biomass to biofuels, their integration and optimization. Other examples of more recent sustainable biorefineries reviewed are those using algal biomass as substrate, which open different opportunities as well as challenges. Finally, a series of chapters discusses materials including the very popular subject of recycling and upcycling of synthetic plastics into value-added products as well as the scale-up production of bio-based chemicals and biomaterials.

This book would not exist without the valuable efforts of Prof. Vijai Singh, who tirelessly worked to compile this excellent set of chapters for Springer Nature. In my opinion, this book is important to set foundational concepts enabling the application of a circular bioeconomy. I am certain that it will be useful to a broad core of students, researchers, scientists, practitioners, policymakers and stakeholders interested in unleashing the potential of the circular bioeconomy in diverse areas.

Department of Life Sciences
Imperial College London
London, UK

Jose Jimenez

# Preface

Sustainable circular bioeconomy (CBE) plays an important role in realizing the objective of Sustainable Development Goal 12 (Responsible Consumption and Production). Currently, due to rising pollution, climate changes and global warming, the human population is also increasing, leading to growing demand foods, chemicals, energy, and other essential resources. Meanwhile, massive amount of waste is being generated, which poses significant challenges for effective management and conversion into value-added products. The circular bioeconomy is a model of production and consumption. It involves in the use, reuse, repair and recycling of existing products. Therefore, it helps to minimize or reduce waste. Waste can contribute in circular bioeconomy by prioritizing reuse and recycling in a closed-loop system. It contributes to reduce carbon footprints, enhanced biodegradability and ability to be recycled without losing structural integrity.

This book covers a wide range of topics on the circular bioeconomy including introduction, mechanism of circular bioeconomy, technological advances, strain engineering, circular design and biomanufacturing, resource recovery from waste to biorefinery, biomass to biofuels, integration and optimization of biorefinery, recycling of synthetic and plastic for value-added products, production of biomaterials, algal biomass-based biorefinery and innovate, conversion of biomass into biofuels, chemicals, materials, and plastic waste to chemicals and challenges, opportunities and future of sustainable biorefineries

This book serves as a valuable resources for students, researchers, scientists, clinicians, stakeholders, policymakers and practitioners to understand sustainable circular bioeconomy. It presents well-structured and insightful content with excellent depth, clarity, and coverage, making it a good source for advancing, expanding, and accelerating knowledge in the field of circular bioeconomy.

Mehsana, Gujarat, India — Vijai Singh

# Acknowledgment

I am delighted to thank all the authors for their excellent contributions to this book. I would like to thank Dr. Soumya Basu (Associate Editor—Clinical Medicine, Biomedicine and Life Sciences) and Ms Vijayalakshmi Dayalan (Production Editor) from Springer for their excellent management of this project.

I greatly appreciate the support of my students Mr. Khushal Khambhati, Mr. Karan Murjani, Ms Khushbu Panchal, Ms Dharmisha Solanki, whose discussion and comments helped to shape this book. My sincere appreciation goes to Prof. Rakesh Rawal, Prof Sudhir P. Singh, Prof. Bharat Maitreya. Dr. Poonam Bhargava, Dr. Ravindra P. Singh, Prof. Bhabatosh Das, Dr Indra Mani, Dr. Pablo Carbonell, Dr. Satya Prakash, Prof. Suresh Ramakrishna, Dr. Dinh-Toi Chu, Prof. Antonia Sagona, Dr. Rodrigo Amaro and all other- named or unnamed-whose direct or indirect contributions have played a key role for shaping this book.

I wish to express my gratitude to my beloved wife Pritee Singh for her endless support, patience, and inspiration. Lots of affection for my kids Aaradhya and Ayush who missed me during this project. I would like to warmly thank the faculty and staff of Indrashil University for providing a great working environment.

I am aware that even despite our best efforts, the first version always comes with some error that may have crept in the compilation. I would be delighted to receive feedback from readers to further improve the future book. Last but not least, my sincere thanks to GOD for his supreme POWER for endowing me to live with joy and victory in the shape of this book.

Vijai Singh

# Contents

## About the Editor

**Vijai Singh** is a Professor and Head of the Department of Biosciences, School of Science, and Dean (Research and Innovation) at Indrashil University, Rajpur, Mehsana, India. Prior to this, he held the position of Associate Professor in the same department at Indrashil University. He has also served as Assistant Professor in the Department of Biological Sciences and Biotechnology at the Institute of Advanced Research, Gandhinagar, and in the Department of Biotechnology at the Invertis University, Bareilly, India. Earlier of his career, he worked as a Postdoctoral Fellow in the Synthetic Biology Group at the Institute of Systems and Synthetic Biology, Paris, France and at the School of Energy & Chemical Engineering at the Ulsan National Institute of Science and Technology, Ulsan, South Korea. He earned his Ph.D. in Biotechnology in 2009 from the ICAR-National Bureau of Fish Genetic Resources, Uttar Pradesh Technical University, Lucknow, India. His doctoral research focused on the development of molecular and immunoassays for the diagnosis of Aeromonas hydrophila. His research interests lies in the design and construction of novel biosynthetic pathways for production of medically and industrially important biomolecules. His laboratory is also actively engaged in developing CRISPR-Cas systems for gene knock out, knock in, gene regulation and diagnostic assay. With over 15 years of research and teaching experience, his areas of expertise include synthetic biology, metabolic engineering, bioinformatics, microbiology, and industrial microbiology. He has an extensive academic record, with over 110 articles, 89 book chapters, 26 books and 3 patents to his credits. He serves as an associate editor, editorial board member, and reviewer of several peer-reviewed journals. He is also a member of the Board of Study and Academic Council of Indrashil University and is the Member Secretary of the Institutional Biosafety Committee (IBSC) at the same university.

# An Introduction to Sustainable Circular Bioeconomy

1

Khushbu Panchal and Vijai Singh

**Abstract**

Currently, the biodiversity loss, climate change, pollution, and depletion of resources are the biggest challenges. These have prompted a move away from the conventional linear economy toward more sustainable and regenerative economic systems. The sustainable circular bioeconomy (SCB) combines the principles of both the circular economy and the bioeconomy. It encourages the utilization of renewable biological resources, waste valorization, and the implementation of closed-loop systems to reduce the consumption of fossil based fuels and decrease the environmental footprint. This chapter focuses on the chronology of the evolution of economic models from the linear economy to circular economy, then to the bioeconomy, towards sustainable circular bioeconomy. It also discusses the important principles of SCB, such as renewable resource management, circularity of bio-based materials, restoration of ecosystems, innovation-led growth, and social inclusion. With a multidisciplinary approach, the chapter points out the major technological, policy, economic, and sociocultural impediments to the large-scale adoption of SCB practices. In addition, the chapter highlights the critical need to overcome systemic and cross-sectoral challenges to realize the sustainable development goals (SDGs). It further focuses on diverse global strategies and case studies from the European Union, United States, and India, highlighting their unique approaches to fostering bio-based innovation and addressing national challenges.

K. Panchal · V. Singh (✉)
Department of Biosciences, School of Science, Indrashil University,
Rajpur, Mehsana-382715, Gujarat, India
e-mail: vijai.singh@indrashiluniversity.edu.in



V. Singh (ed.), *Sustainable Waste Management Towards Circular Bioeconomy*,
https://doi.org/10.1007/978-981-95-2227-9_1

**Keywords**

Sustainable circular bioeconomy · Renewable resources · Sustainable development goals (SDGs) · Large-scale adoption · Global strategies · Bio-based innovation

## 1.1 Introduction

Across the globe, environmental crises, social injustice, and economic turmoil, citizens of the world are beginning to raise fundamental questions that shape common future: Why do we continue to use finite resources in the name of limitless growth? Why is waste useful by-product of development? Can we support an expanding population with our present patterns of production and consumption without endangering the rights of future generations and the health of the planet? Food, air, purchases, and the communities in which we live are all influenced by these personal, intimate issues. For many decades, economies have followed a linear pattern, which adheres to a "take-make-dispose" cycle: raw material extraction, goods production, and use and disposal of it (MacArthur 2013). It is mainly focused on short-term profitability to the detriment of long-term sustainability, with the consequence of depleting resources, pollution, and damaging nature due to a lack of recycling or reuse (Sariatli 2017; Upadhayay and Alqassimi 2018). Yet it is also the cause of many of the most significant problems of our times, including the depletion of resources, climate change, pollution, loss of biological diversity, and rising social inequities.

Increased concern regarding such impacts has come up with the alternatives such as circular economy to rethink consumption and production systems that are more restorative and regenerative in character (Geissdoerfer et al. 2017). The circular economy promotes reuse, recycling, and closed-loop technologies to make resources last longer. It does not necessarily equate to social justice or environmental sustainability, even though it holds the potential for waste reduction and efficiency gains (Korhonen et al. 2018; Velenturf and Purnell 2021). Recycling in energy-intensive systems remains emissions-intensive, for example, and circular business models are not necessarily inclusive of disadvantage groups or compensating for global inequality. This awareness has led to the development of a more inclusive idea, the sustainable circular economy that integrates the principles of circularity with the higher-order goals of sustainable development (D'Amato et al. 2020). Furthermore, the sustainable circular economy is not just about the reduction of waste and the optimization of the value of resources but also about climate resilience, renewable energy, social inclusion, ethical supply chains, biodiversity, long-term ecological balance, and value maximization to reduce waste. It requires a transformation of economic models in which value creation is unbundled from resource utilization and prosperity depends on social well-being, ecosystem health, and quality of life instead of quantity of output (Holden et al. 2023; Kirchherr et al. 2017). It forces companies to grow responsibly, compels individuals to consume responsibly, and pressures governments to design frameworks that are consistent with the UN's Sustainable Development Goals (Lee et al. 2016). By problematizing conventional definitions of growth and challenging circularity into planetary boundaries and social justice, the sustainable circular economy is a potent solution to the call for change of a transformative nature. This chapter maps the origin, evolution, and core features of the sustainable circular economy, tracing its lineage from the linear, circular, bioeconomy models and emphasizing its significance as a template for a regenerative, and future-fit global society.

## 1.2 Evolution and Emergence of Sustainable Circular Bioeconomy (SCB)

Sustainable circular bioeconomy (SCB) is an amalgam of different economy models that emphasize sustainability and have been developed gradually in response to increased concern for environmental degradation, scarcity of resources, and requirements for more adaptive, inclusive, and ecologically restorative economies. It is based on the principles of circular economy, bioeconomy, and sustainable development which involves renewable biological resources, circular material flows, and socio-ecological well-being first. Initially, in early 2000s, Agenda 21 is a global plan of action was adopted by the United Nations System, governments, and major groups to address human impacts on the environment. It was based on the Rio Declaration on Environment and Development and the Statement of Principles for Sustainable Forest Management. The Commission on Sustainable Development was established in 1992 to monitor and report on the implementation of these agreements. A 5-year review of progress was agreed upon in 1997. The full implementation of Agenda 21 and the Rio principles was reaffirmed at the World Summit on Sustainable Development in 2002 (United Nations 1992). Moreover, the evolution of SCB can be comprehended by tracing the chronological development of its influential models: linear economy, circular economy, bioeconomy, and circular bioeconomy. Figure 1.1 illustrates the evolution of SCB from linear economy.

### 1.2.1 Linear Economy: The Industrial Evolution

Since the eighteenth century, the linear economy model that drove global industrial evolution has been based on take, make, use, and dispose principles. This model was relied on raw material extraction on a large scale, production, mass consumption, and waste disposal with minimal regard for environmental or social impact. It was responsible for tremendous technological progress and economic growth, particularly in the Global North, but also led to unsustainable pollution, loss of biodiversity, and greenhouse gas emissions (Stahel 2016). The Ellen MacArthur Foundation argued that the present economic design is rooted in the uneven distribution of wealth by geographic region, with industrial nations experiencing

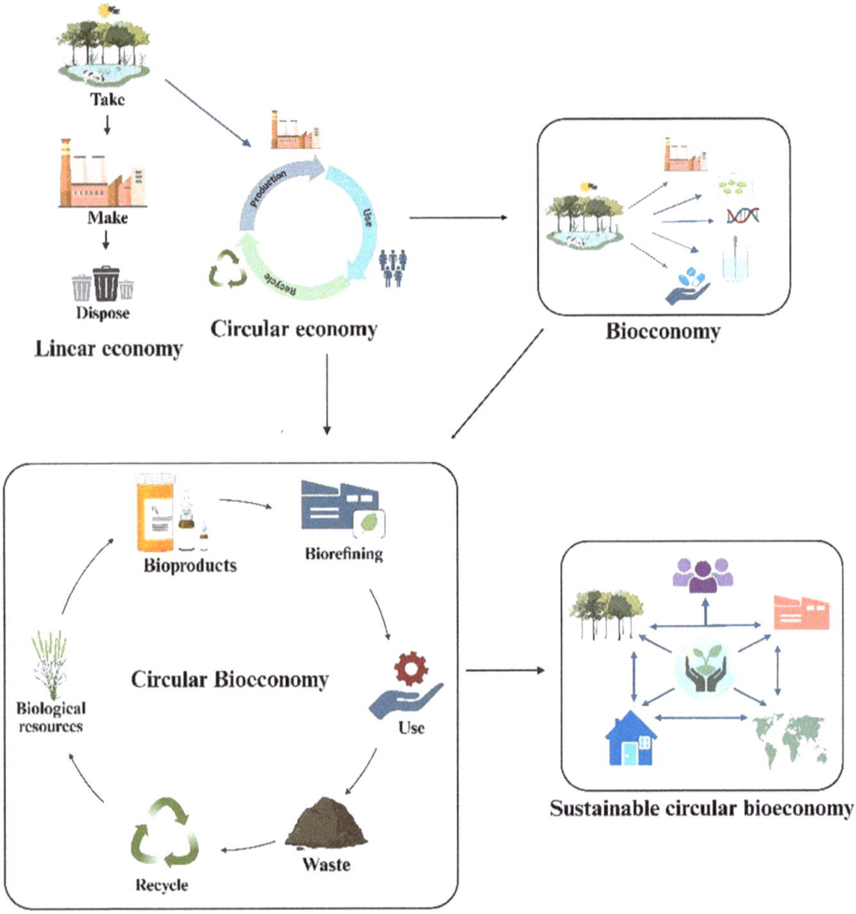


**Fig. 1.1** Schematic illustration of progression of economic models, highlighting the evolution from a traditional linear approach to the more integrated and holistic concept of a sustainable circular economy. (Created from Biorender.com)

abundant material resources and energy (MacArthur 2013). This has led to producers adopting business models that rely on extensive use of materials and economizing on human labor, which has resulted in a lack of recycling, reusing, and waste management. The regulatory, accounting, and fiscal rules have also supported this scheme, as they did not issue a protocol to charge producers with externalities, resulting in less encouragement to consider external costs of their operations (Gubeladze and Pavliashvili 2020; Morseletto 2023).

### 1.2.2 Circular Economy: Closing the Loops

By the end of the twentieth century, heightened consciousness of environmental degradation and exhaustible natural resources led to the development of the concept of circular economy. Initially developed in the 1970s and becoming more recognized in the early 2000s, the circular economy focuses on closing material loops, reducing waste, and maximizing resource value through reuse, repair, recycling, and regenerative design (Geissdoerfer et al. 2017). Some organizations such as the Ellen MacArthur Foundation and the European Union have been instrumental in bringing the concept mainstream as a sustainable means as compared to the linear economy (MacArthur 2013; Velenturf and Purnell 2021). However, the CE brings many challenges with some solutions that can be costly and inefficient or produce unintended negative impacts such as rebound effects and market distortions. Besides this, lack of universal framework hindrances the integration of CE into business models, and also there is no clarity on its social implications, such as inequality or labor exploitation. Due to these limitations, careful applications must be needed to ensure CE genuinely promotes sustainability (Morseletto 2023).

### 1.2.3 Bioeconomy: Utilizing Renewable Biological Resources

Simultaneously with the circular economy, the bioeconomy emerged as a knowledge-based economic concept revolving around sustainable biological resource utilization and biotechnology. Developed around the late 1990s and early 2000s, the bioeconomy attempts to substitute fossil-derived inputs with renewable bio-derived substitutes for food, energy, materials, and chemicals (Ferraz and Pyka 2023). Some nations and territories, such as the European Union (EU), United States, and China, implemented national bioeconomy strategies aiming to encourage new ideas, climate resilience, and rural development (Holden et al. 2023). In the twenty-first century, the bioeconomy achieved global acknowledgment through major policy initiatives, and a number of nations and institutions adopted it as a component of their frameworks for strategic growth. As per the report published by OECD (Organisation for Economic Co-operation and Development) entitled "The Bioeconomy to 2030: Designing a Policy Agenda," the bioeconomy is a key pillar of future economic and sustainable development (OECD 2009). In 2012, the European Union introduced its bioeconomy strategy, which was updated in 2018 and promotes a circular, innovation-oriented strategies (European Commission 2012). The United States published its National Bioeconomy Blueprint in 2012, which highlighted the importance of biotechnology for environmental sustainability, energy security, and public health (The White House Washington 2012). Additionally, there are many countries including Germany, Finland, Brazil, China, and South Africa that have established national bioeconomy strategies that are tailored to address their respective resources, strengths, and development goals. These strategies frequently support the larger international commitments including the Paris Agreement on Climate Change, the Convention on Biological Diversity

(CBD), and the Sustainable Development Goals (SDGs) of the UN, which place an emphasis on sustainability, inclusivity, and climate action.

### 1.2.4 Circular Bioeconomy: Merging Biology and Circularity

Seeing the confines of the standalone bioeconomy and circular economy models, the circular bioeconomy gained ground in the 2010s as a more holistically integrated concept. A circular bioeconomy involves recycling waste streams derived from renewable biological resources back into the industrial system, by using open-loop or closed-loop system or by transforming materials into energy (Venkatesh 2022). It is mainly focused on cascading biomass use, bio-based product design, valorization of biowaste, and restoration of ecosystem services (Carus and Dammer 2018). It aims to reduce fossil resource dependency and close biological loops while keeping renewable inputs and limiting environmental footprint. Yet, various key challenges, such as sustainable resource use, measuring environmental impacts, decoupling growth from resource depletion, valuing ecosystems, designing at multiple scales, securing renewable energy, and overcoming adoption barriers, are faced in its implementation, adaptation, and acceptance globally (Ferraz and Pyka 2023).

### 1.2.5 Sustainable Circular Bioeconomy: A Holistic and Transition Framework

Nonetheless, the emerging paradigm the sustainable circular bioeconomy (SCB) has been gaining popularity which shifts toward more regenerative and sustainable economic paradigms by integrating the concepts of circular economy and the bio-based economy (Carus and Dammer 2018; Rojas-Serrano and Garcia-Garcia 2024). This holistic concept closes the gaps among earlier models by integrating the three pillars of sustainability environmental, economic, and social into biological and circular processes. SCB supports a regenerative utilization of biological resources, circular design thinking, and fair sharing of benefits. It closely corresponds with the United Nations Sustainable Development Goals (SDGs), especially climate action, responsible consumption and production, clean energy, zero hunger, and life below water and on land. It mostly relies on various renewable biological resources that include microorganisms, organic waste, crops, and forests to produce energy, chemicals, materials, and foods. It aims to reduce the environmental footprints, increase economic resilience, and promote biodiversity by focusing on waste valorization, product life extension, resource efficiency, and biological cycles. It also tries to replace fossil fuels with bio-based alternatives and makes sure that biological resources are used in a sustainable and circular way (D'Amato et al. 2017). As global societies seek post-pandemic recovery, the SCB offers a strategic roadmap to recover better through local bio-based economies, climate-resilient policies, and green industrial transitions.

## 1.3 Principles of Sustainable Circular Bioeconomy

The sustainable circular bioeconomy (SCB) is mainly focused on balance economic growth and ecological resilience, resource efficiency, and social inclusion. It combines the regenerative aspect of the circular economy and the renewable resource emphasis of bioeconomy, supported by sustainability objectives (Holden et al. 2023). Globally, a number of fundamental ideas or pillars have been defined for this new integrated concept, which serves as a guide for its development and use. These ideas have been extensively covered in literature and policy frameworks (D'Amato et al. 2017; Gomez San Juan and Bogdanski 2021; OECD 2009; European Union 2018) as the cornerstone for achieving a circular bioeconomy that is inclusive and sustainable.

### 1.3.1 Sustainable and Renewable Biomass Utilization

The crucial principle of SCB is the dedication to the use of renewable biological resources in an ecologically sustainable and non-extractive fashion. These involve agricultural and forestry residues, aquatic biomass, and organic municipal solid waste. The use of biomass sustainably is regulated by land stewardship principles, biodiversity conservation, and renewal of resources. Notably, biomass feedstocks should not compete with food security nor contribute to land degradation (D'Amato et al. 2017; Scarlat et al. 2015). The European Commission (2018) underlines regional biomass sourcing, and the practice of sustainable harvesting (European Union 2018).

### 1.3.2 Circularity of Biological Flows

To enable cascading reuse, recycling, use, and energy recovery, the SCB is committed strongly to the circular flow of biological material. Products and by-products are engineered under the concept to be valorized at the end of life through biorefineries, anaerobic digestion, or composting. The concept extends the principles of circular economy to biological products such that minimal waste is produced and maximum value is retained (Carus and Dammer 2018; Kirchherr et al. 2017). Recycling of biological products from one step as feedstocks in another step, referred to as industrial symbiosis, is being used in circular bioeconomy (Kirchherr et al. 2017).

### 1.3.3 Climate Neutrality and Environmental Regeneration

The SCB works towards climate change mitigation and environmental resilience by replacing fossil-based material with bio-based substitutes, capturing carbon, and promoting sustainable land use. Interventions in climate-smart agriculture, sustainable forestry, and the production of biochar are essential interventions (Hetemäki et al. 2022). In addition, SCB interventions need to take into account the life cycle effects of bio-based products so as to ensure that they provide true environmental advantages as compared to their fossil-based equivalents.

### 1.3.4 Ecosystem Restoration and Biodiversity Conservation

In contrast to conventional models, the SCB not only aims to reduce harm but also actively restore ecosystems and promote biodiversity. This is accomplished through the regenerative agriculture, agroecology, reforestation, and rehabilitation of degraded lands. Conservation of ecosystem services pollination, nutrient cycling, and water filtration were taken as a non-negotiable part of developing a sustainable bioeconomy. Harmonization of these efforts with the Sustainable Development Goals (SDGs) enhances international environmental governance (Gomez San Juan and Bogdanski 2021; Prasad 2024).

### 1.3.5 Economic Resilience and Innovation

SCB promotes inclusive, circular, and innovation-led economies, especially in rural and disadvantage communities. It encourages the development of green jobs, circular business models, and decentralized bio-based industries. Digital technologies, synthetic biology, and bioengineering are key drivers of process efficiency and product performance. The resilience of bio-based value chains is increased through diversification, localization, and investment in circular research and development (R&D) ecosystems (D'Amato et al. 2020; OECD 2009).

### 1.3.6 Systemic Thinking and Integrated Policy Frameworks

The SCB is based on systems thinking, acknowledging the interdependence of food, energy, water, land, and waste sectors. It needs integrated policy frameworks that harmonizes circular economy strategies with bioeconomy roadmaps and sustainability goals (Gawel et al. 2019). Strong governance, cross-sectoral cooperation, and adaptive policymaking are essential to manage trade-offs and maximizing synergies in the SCB landscape (European Commission 2012).

## 1.4 Challenges and Barriers in the Sustainable Circular Bioeconomy

The sustainable circular economy (SCB) has enormous potential and growing global consensus in transition of economies, though implementation and expansion are complex and face many problems. Since these obstacles tend to be interconnected and takes place across numerous technological, economic, regulatory, social, and environmental dimensions, systemic and cross-cutting approaches are required. It has various theoretical challenges and barriers that impact its feasibility and extensive application, even though it holds promise theoretically. It is crucial to understand these hurdles for the development of bio-based scalable, inclusive, and resilient circular systems.

### 1.4.1 Technological Limitations and Infrastructure Gaps

The development and widespread adoption of SCB is limited by the persistent gap in advanced, scalable, and cost-effective technologies. Although scientific progress has led to promising technologies in bio-based biomanufacturing and circular processing systems, most of these technologies are stuck in research labs or pilot-scale demonstrations. Due to the technological limitations, key elements of SCB such as biomass conversion, waste valorization, and decentralized waste-to-energy systems are still far from reaching the level of maturity required for large-scale, economically feasible implementation (Salvador et al. 2022). Moreover, in developing and rural locations, physical infrastructure does not exist at an acceptable level which is required for the collection, sorting, transport, storage, and processing for circular bio-based system. This poor availability of good supply chains and logistics networks limits the transport of raw materials to biorefineries and the marketplace (D'Amato et al. 2020). As emphasized by the OECD (2009), conditions such as innovation finance, the creation of infrastructure, and open technology transfer practices must be instituted in order to bridge these gaps (OECD 2009).

### 1.4.2 Policy Incoherence and Regulatory Uncertainty

The sustainable circular bioeconomy is hindered by policy incoherence and regulatory uncertainty to a large extent. Though of international interest, policy architectures are usually disconnected from each other with distinct strategies for bioeconomy, circular economy, waste management, and environmental protection existing in silos (European Commission 2012). This creates redundant regulations, inconsistent incentives, and inefficiencies that discourage innovation and investment in sustainable bio-based systems. Misaligned waste classification regulations, fossil fuel subsidies, and inadequate carbon pricing mechanisms induce an unlevel playing field, where it is not easy for competing renewable, bio-based alternatives to compete. Inadequate harmonized international sustainability assessment standards, biomass sourcing, and circularity measurement create uncertainty and limit global trade and cooperation in SCB-related sectors (D'Amato et al. 2017). This policy uncertainty may slow project clearances, add compliance costs, and deter investors, particularly for industries with long-term infrastructure and R&D spending. Irrespective of policies encouraging circularity on the words, continuing support for linear economic activities from governments makes things all the more complicated. This situation needs to be addressed through crafting consistent, coherent, integrated, and cross-cutting policy guidelines that combine environmental, economic, and social objectives. Coordinated regulation, multilevel stakeholder participation, and open regulatory systems are critical to underpinning innovation, mitigating risk, and facilitating a fair transition to a sustainable circular bioeconomy (Salvador et al. 2022).

### 1.4.3 Economic Viability and Market Constraints

Sustainable circular bioeconomy (SCB) models are economically challenged, with high capital expenditures at the start, long return-on-investment cycles, and unpredictable market demand. These risks pose financial threats to public and private investors, particularly in contrast to fossil-based linear models (Angouria-Tsorochidou et al. 2021). One of the most important limitation is the lack of robust market pull mechanisms, which restricts consumer demand for bio-based products because of increased prices, unawareness, and variable quality standards. SCB technologies such as biorefineries, precision fermentation, and advanced bioconversion platforms demand high initial capital, discouraging small and medium enterprises (SMEs) and startups. Inadequate environmental externalities pricing further distorts the market, as synthetic materials and fossil fuels are artificially cheap because of subsidies, yet the ecological advantages of bio-based products that are not rewarded or monetized through public procurement policies or market mechanisms (D'Amato et al. 2017). In order to surmount these limitations, facilitating financial instruments, consumer awareness, eco-labeling, and policy-based market creation is able to reinforce demand for bio-based and circular products so that the economic systems favor the SCB transition (Rojas-Serrano and Garcia-Garcia 2024).

### 1.4.4 Knowledge Gaps and Limited Awareness

The sustainable circular bioeconomy is confronted with extensive knowledge gaps and weak awareness among diverse stakeholders, including policymakers, business leaders, academics, and the publics. The fundamental principles and practices of SCB are intangible and lack clear understanding, which inhibits implementation and pays dividends (Donner et al. 2021). The absence of integration in education systems, vocational training, and professional training programs of SCB-related subjects leads to shortages of properly qualified professionals with interdisciplinary competence in biotechnology, circular economy approaches, sustainable resource management, and socioeconomic policy. In industry, many companies lack an understanding of the possible worth of waste streams or by-products because of insufficient technical knowledge or accessibility of valorization routes (Salvador et al. 2022). The view that bio-based circular changes are difficult, risky, or expensive detains experimentation and innovation. Policymakers also struggle to create evidence-based policy because there is limited data regarding biomass availability, life cycle analyses, and socioeconomic effects. Public awareness is also low because consumers are generally unaware of the advantages of bio-based products and their contribution to sustainability outcomes.

## 1.5 Global Strategies and Case Studies in the Sustainable Circular Bioeconomy

Across the globe, to achieve the goal of sustainable circular bioeconomy, many countries and regions have implemented strategic frameworks that promote the sustainable use of biological resources, improve resource efficiency, and stimulate green economic growth. These strategies mainly focus on recognizing the potential of bio-based innovation to address critical challenges such as climate change, biodiversity loss, and socioeconomic inequalities. This section highlights notable strategies and case studies from the European Union (EU), the United States, Brazil, and India, illustrating the diversity in approaches and the tangible impacts of bioeconomy initiatives.

The European Union's bioeconomy strategy is crucial for achieving its 2030 climate and energy goals and 2050 climate neutrality while also combating biodiversity loss and pollution. Building on the progress from the 2012 and 2018 strategies, the present initiative aims to fully leverage the biotechnology revolution to foster a circular economy that benefits nature. This economic growth strategy aligns with the competitiveness compass, the clean industrial deal, and the vision for agriculture and foods. The primary goal of the bioeconomy strategy is to boost innovation and support European companies, especially SMEs and startups in rural and coastal areas, in their transition to green practices. This entails encouraging sustainable and circular production and use of biological resources for enhanced efficiency as well as mitigating pressure on natural resources. The approach targets various core areas like ensuring long-term competitiveness that is designed to attract investments and developing existing and new bio-based products and biotechnology innovations. The second core area is to improve resource efficiency and circularity in order to change the nature of the use and value of biomass resources, prioritizing high-value uses and stimulating industries and consumers to become more circular. By obtaining sustainable biomass supply, the strategy is also focused on delivering a competitive and sustainable source of biomass from domestic and international sources by obtaining sustainable biomass supply. It is also aimed at enhancing the welfare of primary producers and creating wealth and diversification of income for farmers and foresters while incentivizing them for ecosystem conservation. Lastly, the EU will be placed firmly in the expanding world market for bio-based products, biomanufacturing, biochemicals, and the agri-food and biotechnology sectors, through the utilization of available foreign policy instruments (European Commission 2012).

In the United States, initiative for the bioeconomy has been carried out which is focused on leveraging biotechnology and biomass resources. Various key initiatives have been implemented which includes National Bioeconomy Blueprint (The White House Washington 2012); this policy document highlights the role of biotechnology in advancing public health, energy security, environmental sustainability, and economic development. It lays out actions for advancing bio-based solutions, including biomanufacturing, biofuels, and green chemistry (The White House Washington 2012). Apart from this, BioPreferred Program, established by the U.S. Department of Agriculture (USDA), encourages the use of bio-based products through certification and labeling, offering a federal preference for bio-based alternatives in government procurement (USDA 2002). Furthermore, Billion-Ton Report, a comprehensive analysis of the potential for US biomass to be used as a renewable energy source and for bio-based products, helps to identify opportunities for expanding bioenergy and bioeconomy development. The latest version, the "2023 Billion-Ton Report: An Assessment of U.S. Renewable Carbon Resources," released in March 2024, establishes a sustainable production capacity of 1.1–1.5 billion tons of biomass per year, potentially tripling US bioenergy production today and satisfying about 15% of future US energy demand while leaving adequate resources available for food, feed, and fiber demands (US Department of Energy 2024).

India's bioeconomy strategy realizes economic potential from its rich biological assets and tackle domestic issues such as food security, poverty reduction, and climate change. The National Bioeconomy Strategy (2021) and the BioE3 Policy (Biotechnology for Economy, Environment, and Employment) are to make India a world biomanufacturing hub (Department of Biotechnology, Ministry of Science & Technology, Government of India 2024). Priorities include maximizing India's agricultural resources, biodiversity, and biotechnology strength to facilitate sustainable economic development. India has investigated biomass-based energy solutions, including bioelectricity and biogas generation. Programs such as the GOBARdhan Scheme promote a circular bioeconomy by converting organic waste into valuable resources like biogas and biofertilizers, with over 1200 biogas plants established (Government of India 2018). A significant aspect of India's strategy is its focus on rural development programs, empowering farmers, micro-entrepreneurs, and SMEs through the adoption of bio-based technologies, particularly in areas including biofertilizers, agro-waste recycling, and bioplastics. India's bioeconomy has expanded rapidly, accounting for 4.25% of GDP and having ambitious goals of $300 billion by 2030 and $1 trillion by 2047. India's initiatives are strategically designed to bridge the gap between industrial growth and environmental sustainability while improving the rural livelihoods through bioeconomy pathways (PIB Delhi 2025).

## 1.6 Conclusion and Future Remarks

In conclusion, the sustainable circular bioeconomy connects economy and biological processes. Different nations and regions are actively embracing SCB values. However, large-scale adoption and expansion of SCB are confronted with daunting challenges. Key challenges include continuing technology and infrastructure shortcomings, inconsistent policies that undermine achievements, economic viability concerns emanating from high initial costs and risks of markets, and pervasive knowledge gaps among actors. Globally, other countries have also launched different strategies to meet the objective of SCB. Most remarkable among these is the European Union's comprehensive strategy, constructing a robust structure for a circular bioeconomy. The United States emphasizes market-oriented innovation, leveraging its biotechnological prowess and abundant biomass resources. India, on the other hand, is rapidly emerging as a global biomanufacturing hub, leveraging its agricultural strength and biotechnological expertise. The realization of the full potential of SCB heavily depends on continued scientific and technological innovation and innovative adaptive, harmonized, and forward-looking policy regimes. By collaborative management of these challenges, the world can accelerate its shift towards a more equitable, regenerative, and sustainable world.

**Acknowledgments** K.P. and V.S. acknowledge the Science and Engineering Research Board, New Delhi (Project ID: SUR/2022/001937) for financial support.

## References

Angouria-Tsorochidou E, Teigiserova DA, Thomsen M (2021) Limits to circular bioeconomy in the transition towards decentralized biowaste management systems. Resour Conserv Recycl 164:105207

Carus M, Dammer L (2018) The circular bioeconomy—concepts, opportunities, and limitations. Ind Biotechnol 14(2):83–91

D'Amato D, Droste N, Allen B, Kettunen M, Lähtinen K, Korhonen J, Leskinen P, Matthies BD, Toppinen A (2017) Green, circular, bio economy: a comparative analysis of sustainability avenues. J Clean Prod 168:716–734

D'Amato D, Veijonaho S, Toppinen A (2020) Towards sustainability? Forest-based circular bioeconomy business models in Finnish SMEs. Forest Policy Econ 110:101848

Department of Biotechnology Ministry of Science & Technology Government of India (2024) BioE3 (Biotechnology for economy, environment and employment) policy. https://bmi.dbtindia.gov.in/pdf/folder.pdf

Donner M, Verniquet A, Broeze J, Kayser K, De Vries H (2021) Critical success and risk factors for circular business models valorising agricultural waste and by-products. Resour Conserv Recycl 165(105):236

European Commission (2012) Bioeconomy strategy. Retrieved 24th June 2025 from https://environment.ec.europa.eu/strategy/bioeconomy-strategy_en

European Union (2018) A sustainable bioeconomy for Europe strengthening the connection between economy, society and the environment: updated bioeconomy strategy. Retrieved 22nd June 2025 from https://op.europa.eu/en/publication-detail/-/publication/edace3e3-e189-11e8-b690-01aa75ed71a1/language-en

Ferraz D, Pyka A (2023) Circular economy, bioeconomy, and sustainable development goals: a systematic literature review. Environ Sci Pollut Res:1–22

Gawel E, Pannicke N, Hagemann N (2019) A path transition towards a bioeconomy—The crucial role of sustainability. Sustainability 11(11):3005

Geissdoerfer M, Savaget P, Bocken NM, Hultink EJ (2017) The circular economy–a new sustainability paradigm? J Clean Prod 143:757–768

Gomez San Juan M, Bogdanski A (2021) How to mainstream sustainability and circularity into the bioeconomy? Food and Agriculture Organization of the United Nations, Rome

Government of India (2018). Galvanizing organic bio-agro resources dhan (GOBARdhan). Retrieved 24th June 2025 from https://www.india.gov.in/spotlight/gobardhan-galvanizing-organic-bio-agro-resources-dhan
Gubeladze D, Pavliashvili S (2020) Linear economy and circular economy-current state assessment and future vision. Int J Innov Technol Econ 5:411176
Hetemäki L, Kangas J, Peltola H (2022) Forest bioeconomy and climate change. Springer Nature, Cham
Holden NM, Neill AM, Stout JC, O'Brien D, Morris MA (2023) Biocircularity: a framework to define sustainable, circular bioeconomy. Circ Econ Sustain 3(1):77–91
The White House Washington (2012) National Bioeconomy Blueprint. https://obamawhitehouse.archives.gov/sites/default/files/microsites/ostp/national_bioeconomy_blueprint_april_2012.pdf
PIB Delhi Ministry of Science and Technology (2025) The rise of India's bioeconomy from $10bn to $165.75bn in a decade. Retrieved 23rd June 2025 from https://www.pib.gov.in/PressReleasePage.aspx?PRID=2115882
Kirchherr J, Reike D, Hekkert M (2017) Conceptualizing the circular economy: an analysis of 114 definitions. Resour Conserv Recycl 127:221–232
Korhonen J, Honkasalo A, Seppälä J (2018) Circular economy: the concept and its limitations. Ecol Econ 143:37–46
Lee BX, Kjaerulf F, Turner S, Cohen L, Donnelly PD, Muggah R, Davis R, Realini A, Kieselbach B, MacGregor LS (2016) Transforming our world: implementing the 2030 agenda through sustainable development goal indicators. J Public Health Policy 37:13–31
MacArthur E (2013) Towards the circular economy, economic and business rationale for an accelerated transition. Ellen MacArthur Foundation. https://content.ellenmacarthurfoundation.org/m/4384c08da576329c/original/Towards-a-circular-economy-Business-rationale-for-an-accelerated-transition.pdf
Morseletto P (2023) Sometimes linear, sometimes circular: States of the economy and transitions to the future. J Clean Prod 390(136):138
OECD (2009) The bioeconomy to 2030: designing a policy agenda. Organization for Economic Cooperation and Development. Retrieved 21st June 2025 from https://www.oecd.org/en/publications/2009/04/the-bioeconomy-to-2030_g1gha07e.html
Prasad MNV (2024) Bioremediation, bioeconomy, circular economy, and circular bioeconomy—Strategies for sustainability. In: Bioremediation and bioeconomy. Elsevier, Amsterdam, pp 3–32
Rojas-Serrano F, Garcia-Garcia G (2024) Sustainability, circular economy and bioeconomy: a conceptual review and integration into the notion of sustainable circular bioeconomy. New Medit 23(2):3–22
Salvador R, Barros MV, Donner M, Brito P, Halog A, De Francisco AC (2022) How to advance regional circular bioeconomy systems? Identifying barriers, challenges, drivers, and opportunities. Sustain Prod Consum 32:248–269
Sariatli F (2017) Linear economy versus circular economy: a comparative and analyzer study for optimization of economy for sustainability. Visegrad J Bioecon Sustain Dev 6(1):31–34
Scarlat N, Dallemand J-F, Monforti-Ferrario F, Nita V (2015) The role of biomass and bioenergy in a future bioeconomy: Policies and facts. Environ Dev 15:3–34
Stahel WR (2016) The circular economy. Nature *531*:435–438. https://doi.org/10.1038/531435a
United Nations (1992) Agenda 21. The Sustainable Development Knowledge Platform, United Nations. Retrieved 23rd June 2025 from https://sustainabledevelopment.un.org/outcomedocuments/agenda21
United States Department of Agriculture (USDA) (2002) The BioPreferred program. Retrieved 24th June 2025 from https://www.biopreferred.gov/BioPreferred/faces/pages/AboutBioPreferred.xhtml
Upadhayay S, Alqassimi O (2018) Transition from linear to circular economy. Westcliff Int J Appl Res 2(2):62–74

US Department of Energy (2024) 2023 billion-ton report: an assessment of U.S. renewable carbon resources. Retrieved 24th June 2025 from https://www.energy.gov/eere/bioenergy/2023-billion-ton-report-assessment-us-renewable-carbon-resources

Velenturf AP, Purnell P (2021) Principles for a sustainable circular economy. Sustain Prod Consum 27:1437–1457

Venkatesh G (2022) Circular bio-economy—paradigm for the future: systematic review of scientific journal publications from 2015 to 2021. Circ Econ Sustain 2(1):231–279

# 2 Mechanism of Circular Bioeconomy

Ingita Dey Munshi and Indra Mani

**Abstract**

The circular bioeconomy (CBE) represents a transformative paradigm that synergizes circular economy principles with bio-based resource utilization to foster sustainable production and consumption systems. It is underpinned by a suite of interconnected mechanisms that transform biological resources into sustainable, closed-loop systems. Central to this paradigm are sustainable biomass sourcing (e.g., regenerative agriculture, precision forestry), cascading resource utilization (sequential extraction of high-value products from biomass), and waste valorization (conversion of organic residues into biofuels, bioplastics, and biochemicals). Integrated biorefinery systems exemplify CBE mechanisms by processing lignocellulosic biomass into multiple outputs (bioethanol, lignin-derived chemicals), while *nutrient recycling* closes agricultural loops by recovering phosphorus and nitrogen from waste streams. Cross-sector industrial symbiosis further enhances resource efficiency, linking agriculture, energy, and manufacturing. Despite technological advancements, challenges persist in scalability, economic feasibility, and regulatory alignment. Policy frameworks and stakeholder collaboration are critical to overcoming barriers and scaling these mechanisms. By harmonizing ecological stewardship with economic innovation, the CBE offers a blueprint for achieving zero-waste economies and climate-resilient systems.

I. D. Munshi
School of Computational and Integrative Sciences, Jawaharlal Nehru University, New Delhi, India

I. Mani (✉)
Department of Microbiology, Gargi College, University of Delhi, New Delhi, India
e-mail: indra.mani@gargi.du.ac.in



V. Singh (ed.), *Sustainable Waste Management Towards Circular Bioeconomy*,
https://doi.org/10.1007/978-981-95-2227-9_2

**Keywords**

Circular bioeconomy · Biomass cascading · Biorefinery · Waste valorization · Nutrient recycling · Sustainable agriculture · Industrial symbiosis · Bio-based products

## 2.1 Introduction

### 2.1.1 Understanding Bioeconomy

The bioeconomy represents a transformative economic model that harnesses biological resources—including plants, animals, and microorganisms—to sustainably produce food, energy, materials, and chemicals (Muscat et al. 2021). This paradigm shift from fossil-based to bio-based production systems spans multiple sectors including agriculture, forestry, marine ecosystems, biotechnology, and pharmaceuticals, leveraging biological processes to create sustainable industrial outputs (Leong et al. 2021). At its core, the bioeconomy framework integrates biotechnology, bio-based innovation, and enhanced resource efficiency to develop renewable alternatives to conventional petroleum-derived products (Tan and Lamers 2021).

The transition toward bioeconomic systems has gained momentum in response to pressing global challenges including climate change mitigation, resource scarcity, and the need for energy and food security. By substituting fossil fuel inputs with renewable biological resources, nations can significantly reduce greenhouse gas emissions while simultaneously stimulating rural economic development through new value chains and employment opportunities in bio-based industries (Kershaw et al. 2021). The European Union's Updated Bioeconomy Strategy (2018) and subsequent policy instruments demonstrate how regional blocs are institutionalizing this transition, with particular emphasis on creating circular flows of biological materials (Kershaw et al. 2021).

However, it is crucial to recognize that not all bioeconomy models inherently embody circular principles. Many contemporary bio-based production systems still operate within linear "take-make-dispose" frameworks, resulting in inefficient resource utilization and waste accumulation (Khanna et al. 2024). This limitation has spurred the evolution toward more integrated circular bioeconomy approaches that systematically address these shortcomings through closed-loop systems design.

### 2.1.2 The Emergence of Circular Bioeconomy

The circular bioeconomy represents the convergence of two complementary sustainability paradigms: the circular economy's emphasis on resource recirculation and the bioeconomy's focus on biological resource utilization (Kershaw et al. 2021).

This synthesis creates a robust framework where bio-based materials are continuously cycled through reuse, recycling, or biodegradation, thereby minimizing environmental impacts while maximizing resource productivity (Muscat et al. 2021) (Fig. 2.1).

Contemporary scholarship identifies three primary interpretations of circular bioeconomy: as a subset of circular economy principles applied to biological systems, as the intersection between circular economy and bioeconomy concepts, and as a novel paradigm that transcends both original frameworks. The European Commission's operational definition emphasizes "more efficient resource management of bio-based renewable resources by integrating circular economy principles into the bioeconomy" (Kershaw et al. 2021). This conceptual integration has gained substantial policy traction, evidenced by initiatives like the European Circular Bioeconomy Fund which aims to mobilize €250 million for circular bioeconomy development (Kershaw et al. 2021).

The technological foundations of circular bioeconomy are rapidly advancing, particularly in the domain of industrial biotechnology. Modern bioprocessing techniques enable the valorization of agricultural residues, food waste streams, and other organic by-products into high-value outputs including biofuels, biochemicals, and biomaterials (Khanna et al. 2024). These innovations are complemented by digital agriculture technologies that optimize input use efficiency and minimize waste generation at the production stage (Khanna et al. 2024).

### 2.1.3 Principles of Circular Bioeconomy

The circular bioeconomy operates through several interdependent principles that collectively ensure sustainable resource management:

1. *Sustainable Biomass Sourcing*

   Responsible biomass procurement forms the foundation of circular bioeconomy systems. This involves implementing regenerative agricultural practices, sustainable forest management, and marine stewardship protocols that maintain ecosystem health while meeting resource demands (Venkatesh 2022). Precision farming technologies, including sensor networks and AI-driven analytics, enable more efficient use of inputs like water and fertilizers, reducing environmental impacts while maintaining productivity (Khanna et al. 2024).
2. *Cascading Resource Utilization*

   The cascading principle maximizes value extraction from biomass by prioritizing high-value applications before energy recovery. For instance, wood resources might first yield construction materials, then fiber products, before finally being converted to bioenergy (Muscat et al. 2021). This sequential utilization dramatically improves resource efficiency compared to single-use applications.

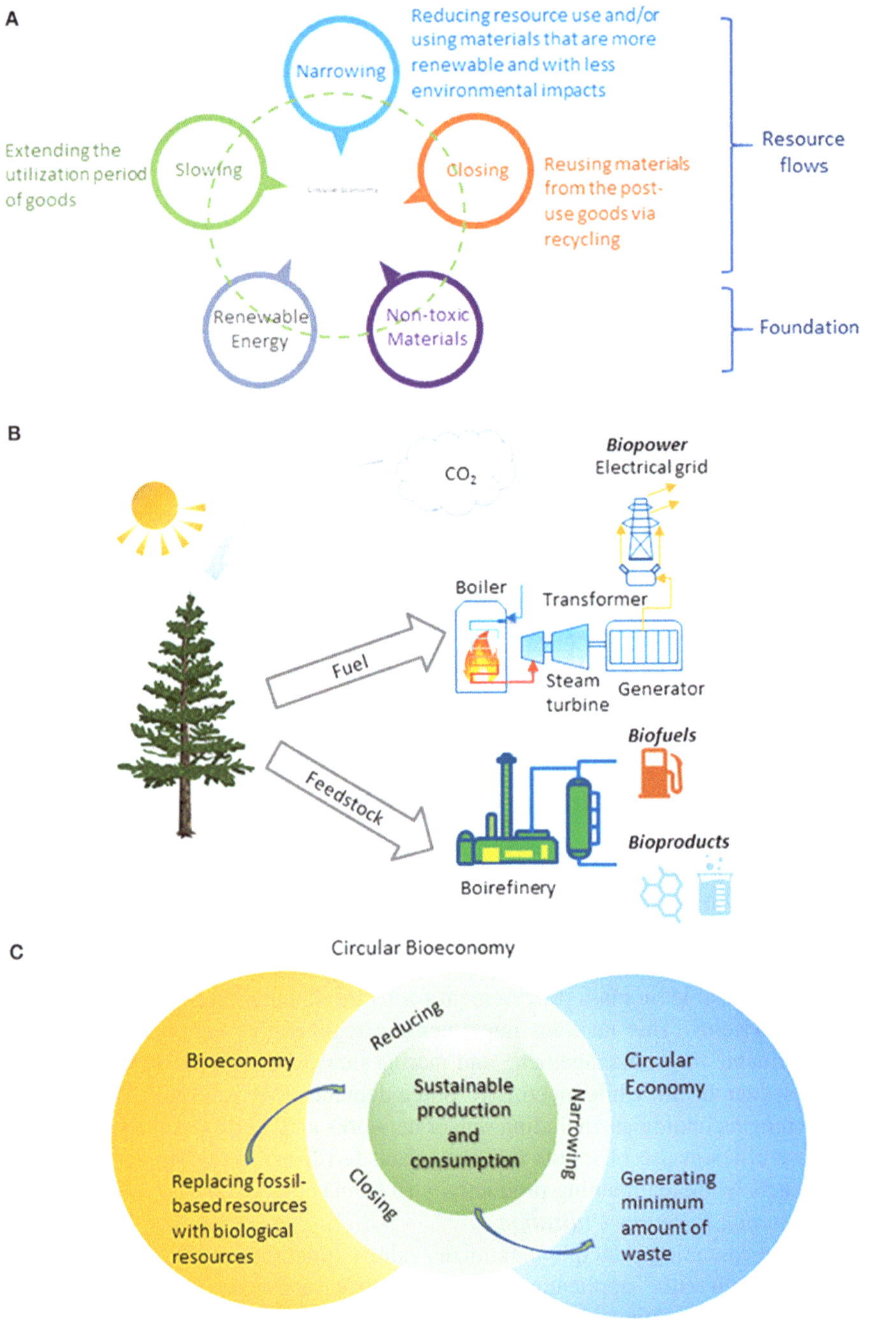


**Fig. 2.1** (**a**) Key elements contributing to an operational definition of a circular economy framework. (**b**) Visual overview of a sustainable bioeconomy: utilizing renewable biological resources for generating bioenergy, biofuels, and bio-based materials through sustainable processes. (**c**) Conceptual integration of circular economy and bioeconomy principles, illustrating their synergy in transforming biological resources into reusable, recycled, or regenerative outputs. (Adapted from Tan and Lamers 2021)

3. *Waste Valorization Technologies*
   Advanced bioconversion processes transform organic waste streams into valuable products. Anaerobic digestion converts agricultural residues to biogas and digestate fertilizers, while enzymatic hydrolysis can break down lignocellulosic biomass into fermentable sugars for biochemical production (Tan and Lamers 2021). Emerging technologies in synthetic biology and metabolic engineering are expanding the range of products that can be derived from waste biomass (Khanna et al. 2024).
4. *Circular Product Design*
   Bio-based products are increasingly designed for circularity through features like biodegradability, easy disassembly, and material purity. This includes developing bio-composites with single-material composition, bio-based adhesives that allow component separation, and polymers with controlled degradation profiles (Muscat et al. 2021).
5. *Systemic Cross-Sector Integration*
   Effective circular bioeconomies require tight coupling between traditionally separate sectors. Agricultural systems provide feedstocks for biorefineries, which supply materials to manufacturing, with by-products cycling back to agricultural soils as organic amendments (Khanna et al. 2024). Such industrial symbiosis creates resilient value networks that minimize systemic waste.

### 2.1.4 Sectoral Applications and Benefits

The implementation of circular bioeconomy principles yields transformative impacts across multiple industries:

**Agricultural Systems**
Modern agriculture stands to benefit significantly from circular bioeconomy approaches. Nutrient recovery technologies can capture and recycle phosphorus and nitrogen from livestock waste and food processing effluents, reducing synthetic fertilizer dependence while mitigating water pollution (Khanna et al. 2024). Integrated farming systems that combine crops, livestock, and agroforestry demonstrate how diversified operations can optimize biomass flows and minimize external inputs (Venkatesh 2022).

**Forestry and Wood Products**
The forest bioeconomy exemplifies circular principles through cascading wood utilization. Advanced wood engineering produces high-performance construction materials that sequester carbon for decades, while by-products furnish raw materials for bio-based chemicals and energy (Muscat et al. 2021). Innovations in lignin valorization are unlocking new pathways for converting this abundant by-product into aromatic chemicals traditionally derived from petroleum (Tan and Lamers 2021).

**Food Systems**
Circular approaches are revolutionizing food production and consumption patterns. Upcycling food processing by-products into nutritious ingredients, converting food waste into animal feed or insect protein, and developing biodegradable food packaging from agricultural residues all contribute to more circular food systems (Khanna et al. 2024). Digital platforms that connect surplus food with redistribution networks further enhance resource efficiency (Kershaw et al. 2021).

**Energy and Chemicals**
The substitution of fossil-derived fuels and chemicals with bio-based alternatives represents a major circular bioeconomy opportunity. Advanced biofuels from lignocellulosic feedstocks, renewable natural gas from organic waste, and bio-based plastics and solvents all demonstrate how biological resources can displace petroleum in key industrial applications (Tan and Lamers 2021). Emerging biorefinery concepts aim to fully utilize biomass components through integrated processing schemes that yield multiple product streams (Khanna et al. 2024).

The socioeconomic benefits of circular bioeconomy development are substantial. Rural regions gain new economic opportunities through diversified agricultural markets and decentralized biorefining operations (Kershaw et al. 2021). Employment creation spans primary production, processing, and advanced manufacturing sectors, often in areas experiencing traditional industry decline (Muscat et al. 2021). Perhaps most significantly, the circular bioeconomy enhances resource security by reducing dependence on imported fossil fuels and creating more localized, resilient production systems (Venkatesh 2022).

### 2.1.5 Policy and Governance Frameworks

Effective circular bioeconomy transitions require supportive policy environments. The European Union has emerged as a leader in this regard, with its bioeconomy strategy explicitly linking circular economy and bioeconomy objectives (Kershaw et al. 2021). Policy instruments range from research funding and innovation grants to market-pull mechanisms like bio-based product mandates and green public procurement.

Emerging governance approaches emphasize the need for multi-stakeholder engagement in circular bioeconomy development. Kershaw et al. (2021) highlight the importance of inclusive decision-making processes that incorporate diverse knowledge systems and value perspectives. This aligns with the responsible research and innovation (RRI) framework, which seeks to align technological development with societal needs and values (Kershaw et al. 2021).

At the implementation level, effective policies must address the full value chain from biomass production to end-use applications. This includes supporting farmers in adopting sustainable practices, incentivizing industry investment in biorefining capacity, and stimulating consumer demand for bio-based circular products (Khanna et al. 2024). The integration of digital tools for traceability and impact assessment further enhances the governance of circular bioeconomy systems (Khanna et al. 2024).

## 2.2 Biomass Cascading

Biomass cascading is a systematic process that maximizes resource recovery from organic waste by sequentially extracting valuable components before residual matter is used for energy or disposed of. This approach aligns with circular economy principles by ensuring that materials undergo multiple stages of utilization, minimizing waste and environmental impact. In both citrus and seafood processing industries, cascading has emerged as a key strategy for transforming low-value by-products into high-value products while reducing the ecological footprint of waste disposal (Teigiserova et al. 2022; Venugopal 2022).

The cascading process typically follows a hierarchical structure, beginning with the extraction of the most valuable compounds before progressing to secondary and tertiary applications. For citrus waste, such as orange peels, the first stage involves recovering essential oils like limonene through cold pressing or microwave-assisted extraction. Subsequent stages focus on producing pectin and citric acid via solid-state fermentation, with the remaining biomass converted into animal feed or biogas (Teigiserova et al. 2022). Similarly, seafood waste undergoes initial processing to extract proteins and omega-3 fatty acids, followed by chitin and chitosan recovery from crustacean shells and, finally, mineral extraction from fish bones (Venugopal 2022). This tiered approach ensures that each component is optimally utilized before the residue is repurposed or discarded.

One of the primary advantages of biomass cascading is its ability to enhance resource efficiency while generating multiple revenue streams. For instance, processing 1 kilogram of orange peel can yield 0.038 kg of limonene, 0.403 kg of citric acid, and 0.8 kg of animal feed, demonstrating the economic potential of this method (Teigiserova et al. 2022). In the seafood industry, cascading reduces processing waste by 50–70% compared to traditional disposal methods, significantly lowering environmental burdens (Venugopal 2022). Additionally, the integration of renewable energy sources, such as wind or biogas, can further improve the sustainability of cascading processes by reducing reliance on fossil fuels (Teigiserova et al. 2022).

Despite its benefits, implementing biomass cascading presents several challenges. Technical barriers include maintaining the stability of extracted compounds across multiple processing stages and managing the energy demands of operations like solid-state fermentation, which can consume between 7518 and 9129 kWh per ton of waste (Teigiserova et al. 2022). Economically, the high capital costs of multistage biorefineries and competition in secondary product markets can hinder widespread adoption. However, both studies highlight that these challenges can be mitigated through technological advancements, policy support, and market incentives, ensuring that cascading remains a viable and scalable solution for sustainable waste management.

Therefore, biomass cascading offers a transformative approach to waste valorization by systematically recovering valuable materials while minimizing environmental impact. Its application in industries ranging from citrus processing to seafood production demonstrates its versatility and potential to contribute to a circular bioeconomy. By addressing existing challenges and leveraging renewable energy integration, cascading can play a pivotal role in achieving sustainability goals across multiple sectors.

## 2.3 Biorefinery Approach

The biorefinery approach is a sustainable and integrated framework that processes biomass into multiple value-added products, including biofuels, biochemicals, and biomaterials, while minimizing waste. Unlike conventional single-output biofuel production, biorefineries maximize resource efficiency by utilizing all components of lignocellulosic biomass—cellulose, hemicellulose, and lignin—through a combination of biochemical, thermochemical, and mechanical conversion techniques (Devi et al. 2022). This model aligns with the principles of the circular bioeconomy, where waste streams are repurposed into high-value applications, enhancing economic viability and reducing environmental impact.

### 2.3.1 Concept and Classification of Biorefineries

Biorefineries can be categorized into three stages based on their complexity and output diversity (Devi et al. 2022). Stage I biorefineries involve a single feedstock and produce one primary product, such as biodiesel from vegetable oil. Stage II biorefineries use a single feedstock but employ multiple processes to generate several products, such as bioethanol and animal feed from corn. Stage III biorefineries, the most advanced, utilize multiple feedstocks and processes to produce a wide range of biofuels, chemicals, and energy. Lignocellulosic biorefineries, a key focus in second-generation biofuel production, fall under stage III due to their ability to fractionate biomass into fermentable sugars, lignin, and hemicellulose-derived products, ensuring no biomass component is wasted.

### 2.3.2 Key Processes in Lignocellulosic Biorefineries

The biorefinery approach involves several critical steps to convert lignocellulosic biomass into valuable products. Pretreatment is essential to break down the recalcitrant structure of biomass, increasing cellulose accessibility. Methods include physical (milling, extrusion), chemical (acid/alkali treatment, organosolv), physicochemical (steam explosion, liquid hot water), and biological (fungal/enzymatic delignification) techniques (Devi et al. 2022). For instance, steam explosion at 1–1.5 MPa has been shown to achieve 61% glucan conversion in rice straw, while ionic liquids can remove up to 91% lignin.

Following pretreatment, enzymatic hydrolysis converts cellulose into glucose, which is then fermented into ethanol. Advanced strategies such as separate hydrolysis and fermentation (SHF), simultaneous saccharification and fermentation (SSF), and consolidated bioprocessing (CBP) are employed to optimize yield and reduce costs. CBP, which combines enzyme production, hydrolysis, and fermentation in a single step, has the potential to lower production costs by 40% (Devi et al. 2022).

Lignin, often underutilized in traditional processes, can be valorized into high-value products such as vanillin, bio-oil, and carbon fibers. Companies like Lignol Innovations have demonstrated the commercial potential of lignin in producing high-purity chemicals, thereby improving the overall profitability of biorefineries (Devi et al. 2022).

### 2.3.3 Circular Bioeconomy Integration

The biorefinery model exemplifies the principles of the circular bioeconomy by co-producing multiple outputs and utilizing waste streams. For example, sugarcane bagasse can yield bioethanol with 77–95% efficiency, furfural with a 69% yield, and biogas from wastewater. Anaerobic digestion of pretreatment wastewater further offsets energy demands, enhancing the sustainability of the process. Additionally, co-products such as xylo-oligosaccharides (used as prebiotics) and lignin-based adhesives diversify revenue streams, making biorefineries economically resilient (Devi et al. 2022).

The integrated nature of biorefineries offers significant techno-economic and environmental advantages. By co-producing multiple products, the minimum selling price (MSP) of bioethanol can be substantially reduced. For instance, coupling ethanol production with biomethane synthesis has been shown to lower costs to $0.31 per liter (Devi et al. 2022). From an environmental perspective, lignocellulosic biomass-derived ethanol reduces greenhouse gas emissions by 60–80% compared to fossil fuels, contributing to climate change mitigation (Devi et al. 2022).

Despite its potential, the biorefinery approach faces several challenges. High enzyme costs, inhibitor formation during pretreatment, and the scalability of advanced processes like CBP remain significant barriers. Future research is expected to focus on metabolic engineering of robust fermentative strains, AI-driven process optimization, and policy incentives to commercialize biorefineries. Overcoming these hurdles will be critical to realizing the full potential of biorefineries in the global bioeconomy (Devi et al. 2022).

## 2.4 Waste Valorization

The valorization of waste streams is a cornerstone strategy for advancing circular bioeconomy principles in biorefineries, transforming underutilized organic residues into high-value products. Recent research has expanded beyond traditional lignocellulosic feedstocks to include mixed organic waste streams such as the organic fraction of municipal solid waste (OFMSW) and invasive aquatic biomass like water hyacinth (WH), demonstrating their potential for sustainable resource recovery (Vea et al. 2018; Shruti and Nemade 2023). This approach aligns with the broader framework of CBE, which seeks to eliminate waste by recycling and reusing components through biological and thermochemical processes (Mishra et al. 2023).

### 2.4.1 Diversified Feedstock Base for Waste Valorization

Waste valorization strategies now encompass both homogeneous agricultural residues (e.g., sugarcane bagasse, rice straw) and heterogeneous waste streams (e.g., OFMSW, water hyacinth), each requiring tailored processing methods. Lignocellulosic biomass, comprising cellulose (30–50%), hemicellulose (20–40%), and lignin (15–25%), necessitates fractionation via pretreatment (Devi et al. 2022). Meanwhile, OFMSW contains starch, proteins, lipids, and carbohydrates, posing challenges for standardization due to variability (Vea et al. 2018). WH, rich in cellulose (18–57%) and hemicellulose (17–34%) but low in lignin (4–9%), is particularly suitable for biofuel and biopolymer production (Shruti and Nemade 2023).

Agro-industrial biowaste, including paper industry sludge, food processing residues, and animal by-products, also presents significant valorization opportunities. For instance, paper industry waste, despite its high organic content, requires specialized treatment due to pollutants like sulfides and chlorinated compounds (Kamali and Khodaparast 2015). Similarly, food industry waste, such as fruit peels and used cooking oil, can be converted into biodiesel, enzymes, and nutraceuticals (Mishra et al. 2023). Animal processing waste, including collagen-rich fish scales and shrimp shells, is a valuable source of biomaterials like chitosan and hydroxyapatite (Gómez-Guillén et al. 2011; Kerton et al. 2013).

### 2.4.2 Technological Pathways and Product Spectrum

#### 2.4.2.1 Bioplastic Production

Lignocellulosic systems convert lignin into bio-based polymers (e.g., vanillin, carbon fibers), while cellulose and hemicellulose yield ethanol and organic acids (Devi et al. 2022). OFMSW-derived volatile fatty acids (VFAs) can be transformed into polyhydroxyalkanoates (PHAs), with yields ranging from 22.4 to 115 g PHA/kg OFMSW (Vea et al. 2018). Similarly, water hyacinth hydrolysate has been used to produce polyhydroxybutyrate (PHB) at 7.3 g/L, demonstrating its viability as a biopolymer feedstock (Shruti and Nemade 2023).

Municipal solid waste (MSW) and agro-industrial residues also contribute to bioplastic production. For example, collagen from fish skin and animal bones is processed into gelatin for food and biomedical applications (Gómez-Guillén et al. 2011). Additionally, chitosan derived from shrimp shells exhibits antimicrobial properties and is used in wound dressings and drug delivery systems (Peniche et al. 2008).

#### 2.4.2.2 Biopesticides and Enzymes

OFMSW serves as a substrate for *Bacillus thuringiensis* (Bt) production via semi-solid fermentation, achieving $9.3 \times 10^{14}$ CFU/kg feedstock, offering a cost-effective biopesticide alternative (Vea et al. 2018). Water hyacinth, through solid-state fermentation, yields enzymes like cellulase (46.3 FPU/g) and xylanase (57.2 IU/g), which have applications in biofuel and textile industries (Shruti and Nemade 2023).

#### 2.4.2.3 Integrated Energy and Chemical Recovery

Anaerobic digestion of OFMSW generates biogas (methane, hydrogen) alongside VFAs, while WH biomass yields 237–812 mL $CH_4$/g VS, making it a viable energy feedstock (Vea et al. 2018; Shruti and Nemade 2023). Thermochemical processes such as pyrolysis and gasification convert biowaste into bio-oil (31–58.5% yield) and syngas, which can be further upgraded into fuels and chemicals (Shruti and Nemade 2023). Similarly, agro-industrial wastes like rice husks and eggshells are processed into silica and hydroxyapatite, respectively, for industrial and biomedical applications (Adam et al. 2012; Kumar and Girija 2013).

#### 2.4.2.4 Economic and Environmental Advantages

Waste valorization offers significant economic benefits. Enzymes and biopesticides from OFMSW yield higher revenues (€1.2 M–88 M/ton) than bioplastics (€95–580/ton), while WH-derived nanocellulose and antioxidants open additional high-value markets (Vea et al. 2018; Shruti and Nemade 2023). The conversion of biowaste into biomaterials also reduces landfill dependency, lowering disposal costs and mitigating environmental pollution (Mishra et al. 2023).

From an environmental perspective, lignocellulosic ethanol reduces greenhouse gas (GHG) emissions by 60–80% compared to fossil fuels, while WH phytoremediation sequesters $CO_2$ and combats eutrophication (Devi et al. 2022; Shruti and Nemade 2023). The circular bioeconomy model further minimizes waste by repurposing agro-industrial by-products into functional materials, such as collagen from fish waste and chitosan from crustacean shells (Gómez-Guillén et al. 2011; Kerton et al. 2013).

Despite its potential, waste valorization faces several hurdles. Feedstock heterogeneity, particularly in OFMSW, complicates process optimization, while seasonal variations in WH growth necessitate efficient harvesting strategies (Vea et al. 2018; Shruti and Nemade 2023). Scaling up biorefinery technologies from lab to industrial levels remains a barrier, requiring pilot-scale demonstrations for commercial validation.

Policy integration is crucial to advancing waste valorization. The European Union's circular economy directives, which mandate 70% recycling by 2030, should incentivize cross-sector collaboration to align waste management with industrial production (Vea et al. 2018). Additionally, life cycle assessments (LCAs) must be conducted to evaluate the environmental impact of biowaste conversion processes, ensuring sustainability across the supply chain (Mishra et al. 2023).

### 2.4.3 Synergies Between Agricultural and Municipal Waste Systems

Emerging hybrid models propose cascading utilization of waste streams:

- Lignocellulosic residues (e.g., sugarcane tops) for ethanol production, supplemented by OFMSW-derived enzymes to enhance hydrolysis efficiency (Devi et al. 2022).
- WH biomass for phytoremediation followed by biofuel production, while OFMSW-derived VFAs support bioplastic manufacturing (Shruti and Nemade 2023).
- Co-locating facilities to share infrastructure (e.g., anaerobic digesters) and optimize logistics (Vea et al. 2018).

Waste valorization is a transformative approach that aligns with circular bioeconomy principles, converting underutilized organic waste into high-value biomaterials, energy, and chemicals. By leveraging diverse feedstocks—from agricultural residues to municipal waste—biorefineries can achieve economic viability while reducing environmental impact. However, overcoming technical, logistical, and policy barriers is essential for large-scale implementation. Future research should focus on optimizing pretreatment methods, scaling up technologies, and fostering policy frameworks that support sustainable waste management. Through integrated efforts, waste valorization can drive the transition toward a zero-waste, resource-efficient bioeconomy.

## 2.5 Nutrient Recycling

Nutrient recycling is a cornerstone of the circular bioeconomy, aiming to minimize waste and maximize resource efficiency by recovering and reusing nutrients from organic residuals. This process is critical for addressing global challenges such as nutrient pollution, soil degradation, and the unsustainable reliance on synthetic fertilizers. The intensification and specialization of modern agriculture have led to significant nutrient imbalances, with surpluses in some regions causing environmental harm (e.g., eutrophication) and deficits in others depleting soil fertility (Wichern et al. 2024). Nutrient recycling seeks to restore these imbalances by closing nutrient loops at local, regional, and global scales.

### 2.5.1 The Need for Nutrient Circularity

Traditional linear agricultural systems extract nutrients from the soil, transport them via crops to consumers, and discard the residuals as waste, often leading to nutrient accumulation in urban areas and depletion in rural ones (Harder et al. 2021). This linearity is exacerbated by global trade, where feed and food imports disconnect nutrient sources from sinks. For example, livestock production regions dependent on imported feed accumulate nutrients in manure, while crop-exporting regions rely on synthetic fertilizers to replenish lost nutrients (Wichern et al. 2024). Nutrient circularity addresses this by recovering nutrients from organic waste streams—such as crop residues, animal manure, food waste, and human excreta—and reintroducing them into agricultural production (Harder et al. 2021).

### 2.5.2 Frameworks for Analyzing Nutrient Circularity

Harder et al. (2021) propose a novel analytical framework to assess nutrient circularity by considering both internal and external flows. Their approach distinguishes four circularity indicators:

1. *Internal input circularity*: The fraction of nutrient inputs to local agriculture derived from recycled organic residuals.
2. *External input circularity*: The nutrient inputs required outside the region to produce imported feed/food.
3. *Internal output circularity*: The fraction of locally generated organic residuals recycled back into agriculture.
4. *External output circularity*: The fate of nutrients in exported feed/food, such as whether they are recycled in the importing region.

This framework highlights trade-offs; for instance, high internal circularity may mask external linearity if a region relies on imported feed without returning nutrients to the source (Harder et al. 2021). Similarly, Wichern et al. (2024) emphasize the importance of spatial scales, noting that circularity strategies must align with local conditions (e.g., farm-level recycling) while addressing broader systemic imbalances (e.g., regional nutrient exchanges).

### 2.5.3 Stakeholder Engagement and Innovation Platforms

Achieving nutrient circularity requires collaboration among diverse stakeholders, including farmers, waste managers, policymakers, and researchers. Van der Wiel et al. (2023) highlight the role of innovation platforms (IPs) in facilitating stakeholder interaction and systemic change. In their study of the Dutch-German Rhine-Waal region, IPs brought together actors from crop/animal production, food processing, and waste management to identify challenges and opportunities for circular nutrient management. Key findings include:

- *Challenges*: Variability in biomass quality/quantity, legal barriers (e.g., cross-border regulations), and lack of economic incentives hinder nutrient recycling (Van der Wiel et al. 2023).

- *Opportunities*: Diverse biomass streams (e.g., food waste, manure) and local valorization (e.g., composting, anaerobic digestion) can replace imported fertilizers if supported by transparent information networks (Van der Wiel et al. 2023).

The study underscores the need for participatory approaches to build trust and align stakeholder interests, particularly in nutrient-saturated regions where systemic lock-ins (e.g., reliance on imported feed) persist (Van der Wiel et al. 2023).

### 2.5.4 Technological and Practical Solutions

Advancements in nutrient recovery technologies are enabling the transformation of waste into valuable fertilizers. Examples include:

- *Anaerobic digestion*: Produces nutrient-rich digestate from organic waste, which can replace synthetic fertilizers (Wichern et al. 2024).
- *Struvite precipitation*: Recovers phosphorus from wastewater, reducing reliance on mined phosphate (Harder et al. 2021).
- *Biochar production*: Converts agricultural residues into stable carbon sinks while retaining nutrients for soil amendment (Wichern et al. 2024).

However, technical solutions alone are insufficient. Practical barriers include the economic costs of transporting low-nutrient-density waste, social resistance to waste reuse (e.g., "yuck factor" for human excreta), and regulatory hurdles (Harder et al. 2021). For example, sewage sludge application is declining in the EU due to contamination concerns, despite its potential to supply 7–34% of agricultural phosphorus demand if fully recovered (Wichern et al. 2024).

### 2.5.5 Policy Integration and Systemic Change

Successful nutrient recycling requires supportive policies and systemic shifts:

- *Economic instruments*: Subsidies for bio-based fertilizers or taxes on synthetic inputs (Wichern et al. 2024).
- *Regulatory standards*: Quality control for recycled fertilizers to ensure safety and efficacy (Van der Wiel et al. 2023).
- *Behavioral change*: Public awareness campaigns to promote waste separation and acceptance of circular products (Van der Wiel et al. 2023).

Spiller et al. (2024) demonstrated that combining circularity, efficiency, and sufficiency (e.g., dietary shifts toward plant-based foods) can halve nitrogen flows and losses in agro-food systems.

Nutrient recycling is a linchpin of the circular bioeconomy, offering a pathway to sustainable agriculture by mitigating pollution, enhancing soil health, and reducing dependency on finite resources. Achieving this requires integrated frameworks, innovative technologies, stakeholder collaboration, and supportive policies. Future efforts must prioritize scalability, equity, and holistic system redesign to ensure nutrient circularity benefits both ecosystems and societies.

## 2.6 Challenges of a Circular Bioeconomy

The transition to a CBE presents significant opportunities for sustainability but is fraught with multifaceted challenges. Drawing on the systematic review by Hamam et al. (2024) and incorporating insights from Huarachi et al. (2023), this section synthesizes key obstacles across technological, economic, regulatory, social, and methodological dimensions.

### 2.6.1 Technological and Operational Barriers

#### 2.6.1.1 Scalability and Efficiency

Insect-based bioconversion—a cornerstone of CBE—faces hurdles in scaling production. While insects like *Hermetia illucens* (black soldier fly) efficiently convert organic waste into protein, large-scale operations require advanced automation and standardized processes (Hamam et al. 2024). Similarly, biorefinery systems for biofuel production (e.g., anaerobic digestion) often struggle with process optimization and energy efficiency (Huarachi et al. 2023).

#### 2.6.1.2 Substrate Suitability and Safety

Not all organic waste is viable for insect rearing or biorefining. Contaminants (e.g., heavy metals, pesticides) in substrates can accumulate in insects or disrupt bioconversion processes (Mannaa et al. 2023). Additionally, regional variability in biomass availability complicates feedstock standardization (Huarachi et al. 2023).

#### 2.6.1.3 By-Product Valorization

While insect frass (excrement) and biorefinery residues (e.g., digestate) show promise as fertilizers, their nutrient content varies with feedstock quality, and improper application risks environmental harm (Beesigamukama et al. 2022). Extracting high-value compounds (e.g., chitin, bio-based chemicals) remains technologically immature (Huarachi et al. 2023).

### 2.6.2 Economic and Market Challenges

#### 2.6.2.1 High Production Costs

Insect-derived products (e.g., meal, oil) and biofuels (e.g., biodiesel) are often priced higher than conventional alternatives due to small-scale operations and energy-intensive processing (Hamam et al. 2024). For example, poultry feed incorporating insect meal increases production costs by 93–98% compared to traditional diets (Hamam et al. 2024). Similarly, biofuel production faces economic feasibility issues unless supported by subsidies (Huarachi et al. 2023).

#### 2.6.2.2 Logistical Constraints

Supply chain fragmentation—such as inadequate waste collection systems and distant processing facilities—inflates costs. In Nairobi, organic waste is often unsuitable for insect farming due to inorganic contamination and poor collection infrastructure (Hamam et al. 2024). Likewise, transporting biomass for biorefineries increases carbon footprints and operational expenses (Huarachi et al. 2023).

#### 2.6.2.3 Market Volatility

Dependence on subsidies and niche markets undermines long-term viability. The EU's insect protein market remains vulnerable to price fluctuations in soybean and fishmeal (Hamam et al. 2024). Similarly, biofuel markets are influenced by fossil fuel prices and policy shifts (Huarachi et al. 2023).

### 2.6.3 Regulatory and Policy Hurdles

#### 2.6.3.1 Legislative Fragmentation

EU regulations classify insects as "farmed animals," restricting substrate use (e.g., banning manure in feed) (Hamam et al. 2024). Meanwhile, biofuel policies vary widely, with some regions lacking standardized sustainability criteria (Huarachi et al. 2023).

#### 2.6.3.2 Slow Approval Processes

Despite recent approvals (e.g., insect protein in poultry feed), regulatory delays persist for novel substrates like fish waste (Hamam et al. 2024). Similarly, certification bottlenecks hinder bio-based product commercialization (Huarachi et al. 2023).

#### 2.6.3.3 Invasive Species Risks

Regulation (EU) No. 1143/2014 limits insect species for farming to prevent ecosystem disruption (Hamam et al. 2024). Likewise, introducing non-native biomass for biorefineries risks biodiversity loss (Huarachi et al. 2023).

### 2.6.4 Social and Cultural Resistance

#### 2.6.4.1 Consumer Acceptance

Western consumers exhibit "entomophagy neophobia," associating insects with disease or disgust (Sogari et al. 2022). Similarly, bio-based products (e.g., algae protein) face skepticism due to unfamiliarity (Huarachi et al. 2023).

#### 2.6.4.2 Information Asymmetry

Limited public awareness of CBE benefits exacerbates resistance. Educational campaigns highlighting environmental advantages (e.g., 100× lower GHG emissions than livestock) can mitigate this (Hamam et al. 2024).

#### 2.6.4.3 Stakeholder Misalignment

Farmers and feed producers often prioritize short-term profits over circular models (Hamam et al. 2024). Similarly, industry reluctance to adopt LCA-driven sustainability metrics slows CBE progress (Huarachi et al. 2023).

## 2.7 Conclusion and Future Perspectives

The transition to a CBE represents a transformative pathway to address global sustainability challenges by integrating biological resource efficiency with circular economy principles. By prioritizing sustainable biomass sourcing, cascading utilization, waste valorization, and nutrient recycling, the CBE framework offers systemic solutions across agriculture, forestry, energy, and food systems. These approaches not only reduce reliance on finite fossil resources but also mitigate greenhouse gas emissions, enhance resource security, and stimulate rural economies through diversified value chains (Muscat et al. 2021; Kershaw et al. 2021). Policy frameworks, such as the EU's Bioeconomy Strategy, have catalyzed progress by aligning technological innovation with market incentives, while industrial symbiosis models demonstrate the potential for cross-sectoral integration (Khanna et al. 2024; Devi et al. 2022). However, persistent challenges—including technological scalability, economic viability, regulatory fragmentation, and social acceptance—underscore the complexity of transitioning from linear to circular systems (Hamam et al. 2024; Huarachi et al. 2023). Despite these hurdles, the CBE paradigm provides a robust foundation for reconciling economic growth with planetary boundaries.

Future advancements in the circular bioeconomy will hinge on addressing existing gaps through interdisciplinary innovation and inclusive governance. Technologically, scaling biorefinery processes, optimizing enzymatic efficiency, and advancing synthetic biology could enhance resource recovery rates and reduce costs (Devi et al. 2022; Shruti and Nemade 2023). Policy reforms must prioritize harmonized regulations, subsidies for bio-based products, and mandates for waste valorization to create enabling environments (Kershaw et al. 2021; Van der Wiel et al. 2023). Concurrently, public engagement campaigns and educational initiatives are critical to overcoming cultural resistance, particularly in promoting insect-based products and bio-derived materials (Sogari et al. 2022; Huarachi et al. 2023). Methodologically, expanding life cycle assessments to incorporate socioeconomic metrics and localized data will ensure holistic sustainability evaluations (Huarachi et al. 2023). Cross-sector collaboration, global knowledge sharing, and investment in digital infrastructure for traceability will further drive systemic change. By integrating these efforts, the circular bioeconomy can evolve into a mainstream model, fostering resilient, zero-waste economies that align human prosperity with ecological integrity.

## References

Adam F, Appaturi JN, Iqbal A (2012) The utilization of rice husk silica as a catalyst: review and recent progress. Catal Today 190:2–14. https://doi.org/10.1016/j.cattod.2012.02.042

Beesigamukama D, Subramanian S, Tanga CM (2022) Nutrient quality and maturity status of frass fertilizer from nine edible insects. Sci Rep 12(1):7182. https://doi.org/10.1038/s41598-022-11336-z. PubMed PMID: 35505193; PubMed Central PMCID: PMC9064968

Devi A, Bajar S, Kour H, Kothari R, Pant D, Singh A (2022) Lignocellulosic biomass valorization for bioethanol production: a circular bioeconomy approach. Bioenergy Res 15(4):1820–1841. https://doi.org/10.1007/s12155-022-10401-9

Gómez-Guillén MC, Giménez B, López-Caballero ME, Montero MP (2011) Functional and bioactive properties of collagen and gelatin from alternative sources: a review. Food Hydrocoll 25(8):1813–1827. https://doi.org/10.1016/j.foodhyd.2011.02.007

Hamam M, D'Amico M, Di Vita G (2024) Advances in the insect industry within a circular bioeconomy context: aresearch agenda. Environ Sci Eur 36:29. https://doi.org/10.1186/s12302-024-00861-5

Harder R, Giampietro M, Smukler S (2021) Towards a circular nutrient economy: a novel way to analyze nutrient flows in food systems. Resour Conserv Recycl 172:105693

Huarachi DAR et al (2023) Life cycle thinking for a circular bioeconomy: current development, challenges, and future perspectives. Sustainability 15(11):8543. https://doi.org/10.3390/su15118543

Kamali M, Khodaparast Z (2015) Review on recent developments on pulp and paper mill wastewater treatment. Ecotoxicol Environ Saf 114:326–342. https://doi.org/10.1016/j.ecoenv.2014.05.005

Kershaw EH, Hartley S, McLeod C, Polson P (2021) The sustainable path to a circular bioeconomy. Trends Biotechnol 39(6):542–545. https://doi.org/10.1016/j.tibtech.2020.10.015

Kerton FM, Liu Y, Omari KW, Hawboldt K (2013) Green chemistry and the ocean-based biorefinery. Green Chem 15(4):860–871. https://doi.org/10.1039/C3GC36994C

Khanna M, Zilberman D, Hochman G et al (2024) An economic perspective of the circular bioeconomy in the food and agricultural sector. Commun Earth Environ 5:507. https://doi.org/10.1038/s43247-024-01663-6

Kumar GS, Girija EK (2013) Flower-like hydroxyapatite nanostructure obtained from eggshell: a candidate for biomedical applications. Ceram Int 39(7):8293–8299. https://doi.org/10.1016/j.ceramint.2013.04.003

Leong HY, Chang CK, Khoo KS et al (2021) Waste biorefinery towards a sustainable circular bioeconomy: a solution to global issues. Biotechnol Biofuels 14:87. https://doi.org/10.1186/s13068-021-01939-5

Mannaa M, Mansour A, Park I, Lee DW, Seo YS (2023) Insect-based agri-food waste valorization: agricultural applications and roles of insect gut microbiota. Environ Sci Ecotechnol 17:100287. https://doi.org/10.1016/j.ese.2023.100287. PubMed PMID: 37333762; PubMed Central PMCID: PMC10275724

Mishra B, Mohanta YK, Reddy CN, Reddy SDM, Mandal SK, Yadavalli R, Sarma H (2023) Valorization of agro-industrial biowaste to biomaterials: an innovative circular bioeconomy approach. Circ Econ 2:100050. https://doi.org/10.1016/j.cec.2023.100050

Muscat A, de Olde EM, Ripoll-Bosch R et al (2021) Principles, drivers and opportunities of a circular bioeconomy. Nat Food 2:561–566. https://doi.org/10.1038/s43016-021-00340-7

Peniche C, Argüelles-Monal W, Goycoolea FM (2008) Chitin and chitosan: major sources, properties and applications. In: Belgacem MN, Gandini A (eds) Monomers, polymers and composites from renewable resources. Elsevier, pp 517–542. https://doi.org/10.1016/B978-0-08-045316-3.00025-9
Shruti B, Nemade PR (2023) An integrated biorefinery approach for the valorization of water hyacinth towards circular bioeconomy: a review. Environ Sci Pollut Res 30:39494–39536
Sogari G, Riccioli F, Moruzzo R, Menozzi D, Sosa DAT, Li J et al (2022) Engaging in entomophagy: the role of food neophobia and disgust between insect and non-insect eaters. Food Qual Prefer 104:104764. https://doi.org/10.1016/j.foodqual.2022.104764
Spiller M, Vingerhoets R, Vlaeminck SE, Wichern F, Papangelou A (2024) Beyond circularity! Integration of circularity, efficiency, and sufficiency for nutrient management in agri-food systems. Nutr Cycl Agroecosyst 129:287
Tan ECD, Lamers P (2021) Circular bioeconomy concepts—a perspective. Front Sustain 2:701509. https://doi.org/10.3389/frsus.2021.701509
Teigiserova DA, Hamelin L, Tiruta-Barna L, Ahmadi A, Thomsen M (2022) Circular bioeconomy: life cycle assessment of scaled-up cascading production from orange peel waste under current and future electricity mixes. Sci Total Environ 812:152574
Van der Wiel BZ, Neuberger S, Darr D, Wichern F (2023) Challenges and opportunities for nutrient circularity: an innovation platform approach. Nutr Cycl Agroecosyst 129:375–391
Vea EB, Romeo D, Thomsen M (2018) Biowaste valorization in a circular economy: recent advances and future challenges. Bioresour Technol 268:732–742
Venkatesh G (2022) Circular bio-economy—paradigm for the future: systematic review of scientific journal publications from 2015 to 2021. Circ Econ Sustain 2:231–279. https://doi.org/10.1007/s43615-021-00084-3
Venugopal V (2022) Green processing of seafood waste biomass towards blue economy. Curr Res Environ Sustain 4:100164
Wichern F, García-González MC, Moral R et al (2024) Editorial: regional nutrient management in the circular economy. Nutr Cycl Agroecosyst 129:277–286

# Advances in Biorefinery Technologies

3

Mani Jayakumar, Melkiyas Diriba Muleta, Hirpha Adugna Areti, Bulcha Jifara Daba, Abas Siraj Hamda, and Lata Deso Abo

**Abstract**

Advanced biorefinery technology represents a cutting-edge approach focused on maximizing the value derived from biomass. This innovative technology employs a range of processes including biomass pretreatment, enzymatic hydrolysis, fermentation, and separation techniques to transform biomass into valuable products such as biochemicals, biofuels, and other biomaterials. By integrating advanced biorefinery technology with existing industrial processes, there is a significant likely to diminish our support on fossil fuels and lessen the environmental impact. The advancements in biorefinery technology have marked a substantial progression toward more efficient and eco-friendly operations in recent centuries. This chapter offers a concise summary of the latest developments in advanced biorefinery technologies, showcasing different types of advanced biorefineries such as lignocellulosic biomass-derived biofuel biorefineries and in addition microalgae-derived biorefineries, each at varying levels of technological readiness. While only a handful of advanced biorefineries have scaled up to commercial production, the primary objective remains achieving profitability. The chapter discusses key examples of biorefineries, emphasizing both thermochemi-

M. Jayakumar (✉)
Department of Chemical Engineering, Haramaya Institute of Technology, Haramaya University, Dire Dawa, Ethiopia

Department of Biotechnology, Faculty of Engineering, Karpagam Academy of Higher Education, Coimbatore, Tamilnadu, India

Centre for Natural Products and Functional Foods, Karpagam Academy of Higher Education, Coimbatore, Tamilnadu, India
e-mail: drjayakumarmani@haramaya.edu.et

M. D. Muleta · H. A. Areti · B. J. Daba · A. S. Hamda · L. D. Abo
Department of Chemical Engineering, Haramaya Institute of Technology, Haramaya University, Dire Dawa, Ethiopia



V. Singh (ed.), *Sustainable Waste Management Towards Circular Bioeconomy*,
https://doi.org/10.1007/978-981-95-2227-9_3

cal and biochemical processes utilized in producing advanced biofuels, alongside economic evaluations. Furthermore, it touches upon product-focused biorefineries like biomass biorefineries and genetic enhancement biorefineries, acknowledging that critical technologies are still in development, limiting the number of operational facilities. Ultimately, the chapter aims to highlight the newest technical advancements within the biorefinery sector.

**Keywords**
Biofuels · Biomass · Biorefinery · Fermentation · Gasification · Hydrogenation · Pyrolysis

## 3.1 Introduction

In modern ages, substantial advancements in technology have remained made in the field of biorefineries. These developments represent a shift toward more efficient and sustainable practices. One notable progression is the incorporation of advanced raw materials into biorefinery processes, such as forestry waste, algae, and various other waste streams (Hamda et al. 2023; Yogalakshmi et al. 2023). This shift not only reduces competition with food production but also enhances the overall sustainability of biorefinery operations by moving away from food chain-derived feedstocks like wheat, maize, barley, sugarcane, and sorghum (Raj et al. 2022). Additionally, there has been a significant focus on optimizing and intensifying processes to increase yields and decrease energy consumption (Yogalakshmi et al. 2023). Innovative practices, such as consolidated bioprocessing (CBP) and simultaneous saccharification and fermentation (SSF), have streamlined the manufacture of biochemicals and biofuels, improving together the environmental and economic viability of biorefineries (Shahid et al. 2021).

Moreover, the variety of goods made from biomass feedstocks has increased, thanks to catalytic conversion technology. By converting biomass into useful products like bio-oils, charcoal, and syngas, processes like hydrothermal liquefaction (HTL), pyrolysis, and gasification support the circular economy by maximizing the use of waste resources (Giuliano 2023; Jayakumar et al. 2023e; Gundupalli et al. 2022). Furthermore, the variety of goods made from biomass feedstocks has increased, thanks to catalytic conversion technology. By converting biomass into useful products like bio-oils, charcoal, and syngas, processes like hydrothermal liquefaction (HTL), pyrolysis, and gasification support the circular economy by maximizing the usage of waste possessions (Amoah et al. 2019).

The effective transformation of natural and waste biomass has been made possible by the development of synthetic biology and biocatalysis. Designed specifically bio-production pathways can be designed with the use of synthetic biology technologies, and complex biomass feedstocks can be converted into target molecules more easily with the help of improved catalytic activity in microbes and enzymes (Uzoejinwa et al. 2023). Furthermore, by process optimization, enabling real-time monitoring, and predictive maintenance, the adoption of digitization and

Industry 4.0 technologies has completely changed the way biorefineries operate (Dussán et al. 2022). Proactive decision-making and continual improvement are made easier by the insights into process performance provided by advanced sensors, data analytics, and machine learning algorithms.

In general, the developments in biorefinery technology are changing the face of sustainable production and present encouraging answers to the problems posed by resource depletion and climate change. Key developments propelling this change include the integration of enhanced feedstocks, process intensification, catalytic conversion technologies, and co-location of biorefineries, biocatalysis, synthetic biology, and digitalization. The efficiency, adaptability, and environmental sustainability of biorefinery operations could be further improved with ongoing study and innovation in these fields, opening the door to a more sustainable future.

## 3.2 Biomass Structure and Property Analysis

Agricultural residues are the most potential raw material for producing ecologically friendly, carbon-free alternatives to synthesized resources like biofuels and construction materials (Hamda et al. 2023; Jayakumar et al. 2023c). The quest for efficient lignin biomass utilization has resulted in extensive chemical structural investigations and the establishment of analytical techniques (Ghaffar and Fan 2013). Biomass is mostly made up of biological matter, but it also contains a trace amount of inorganic molecules that come in a range of closely related phases or minerals from diverse sources. Biofuels are composed of various physical forms of biomass, with plant species primarily containing hemicellulose, cellulose, lignin, and other extractives as key structural cell wall components (Jayakumar et al. 2023a). To characterize a specific fuel, it is common to conduct proximate and ultimate analyses as standardized methods of investigation (de Jong 2014).

Nature has given a diverse range of sources of biomass, with worldwide biomass output projected to exceed 100 billion tons per year (Ong et al. 2021). Biomass may be utilized to produce not just food but also renewable energy, carbon sequestration, and fertilizers (Jayakumar et al. 2023c). In general, biomass source is a highly renewable carbonaceous reserve that is able to be employed to produce power; chemicals (biohydrogen and biomethane); heat; biofuels (bioethanol and biodiesel); convectional fuels like fossil fuels, natural gas, and coal; and numerous other bioproducts (Jayakumar et al. 2023d). Biomass has also been shown to lower anthropogenic carbon dioxide emissions (Sulaiman et al. 2020). Lignocellulosic biomass may be converted into biodiesel, ethanol, and biogas for energy production. Nevertheless, the process of efficiently transforming second-generation biomass into biomaterial can pose challenges, often requiring a pretreatment step prior to utilization (Jayakumar et al. 2022a,b; Chozhavendhan et al. 2023). In this regard, numerous classes of investigative devices such as Brunauer-Emmett-Teller (BET), Fourier transform infrared (FTIR), scanning electron microscopy (SEM), nuclear magnetic resonance (NMR), X-ray diffraction (XRD), X-ray photoelectron spectroscopy (XPS), and thermogravimetric analysis (TGA) are employed to investigate

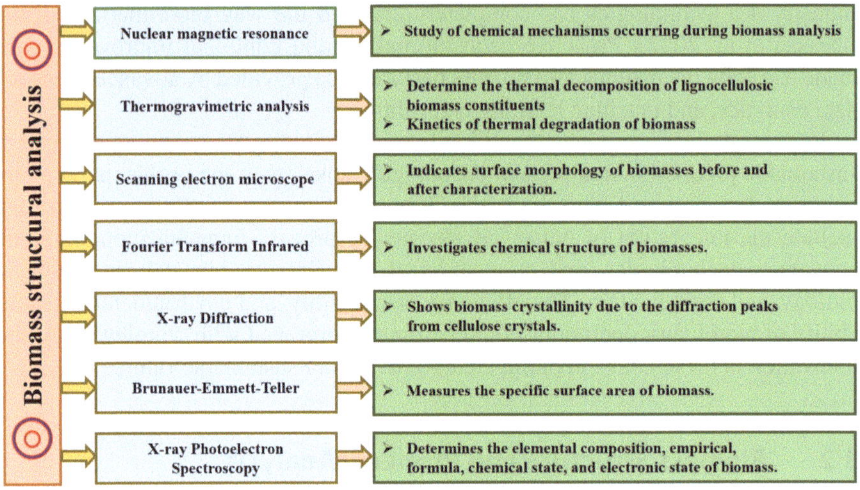


**Fig. 3.1** Biomass and its structural analysis

and determine the structural and chemical properties of second-generation lignocellulosic biomass and biomaterials (Ong et al. 2021). Fig. 3.1 summarizes the composition of biomass through structural analysis, indicating that it consists of approximately 40–50% cellulose, 10–25% lignin, and 20–30% hemicelluloses (Stoilova et al. 2010). This information can serve as a valuable resource for academics and researchers studying second-generation biomass.

### 3.2.1 Lignocellulosic Biomass Structure

The second generation of lignocellulosic biomass represents a sustainable and plentiful energy reserve that is found globally (Alonso et al. 2010). This particular type of biomass is deemed nonedible source and is most commonly found in agricultural and forest residues. Lignocellulosic biomass, also referred to as nonedible, is commonly found in agriculture and forest biomass. Besides, the bulk of second-generation biomass is discarded as garbage. In poor nations, rural residents mostly rely on conventional energy sources, including agricultural waste and firewood. Conventional methods of generating energy are not merely inefficient and expensive; however, they are also slow and dangerous to the ecosystem. By the year 2050, it is projected that lignocellulosic biomass will have the potential to contribute around 38% of the total straight fuel source globally, as well as 17% of the global energy demand (Bentsen et al. 2014; Popp et al. 2014).

Generally, all plant materials consist of three key constituents: cellulose, lignin, and hemicelluloses, as illustrated in Fig. 3.2. These essential elements are distributed unevenly throughout the cell wall and serve as the plant's framework (Vorwerk et al. 2004). The bonding between hemicelluloses, lignin, and cellulose primarily

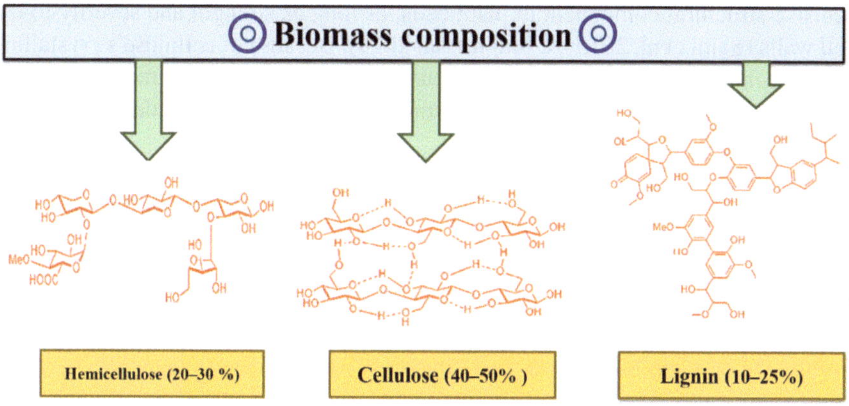


**Fig. 3.2** Biomass composition of lignocellulosic materials

arises over $H_2$ bonds, while lignin and hemicelluloses are connected by mutually covalent and hydrogen bonds.

Cellulose is a polymer of glucan composed of D-glucopyranose units linked by β-(1-4)-glycosidic bonds (Mansora et al. 2019). On the other hand, hemicelluloses are carbohydrates with a lesser unit of polymerization matched to cellulose. Hemicellulose comprises predominantly of sugars such as D-galactopyranose, D-xylopyranose, D-galactopyranosyl uronic acid, D-glucopyranosyl uronic acid, L-arabino-furanose, D-glucopyranose, and D-mannopyranose acid (Scheller and Ulvskov 2010). Moreover, lignin exhibits exceptional properties like amorphousness, a complex branched-phenolic structure, and three-dimensionality. Lignin is fabricated over an uneven biological practice involving three key phenylpropanoid monomers consisting of guaiacyl (G), p-hydroxyphenyl (H), and syringyl (S) (Alonso et al. 2010; Ong et al. 2021).

### 3.2.2 Cellulose

Cellulose can be considered a key element in the composition of plant cell walls, especially in the circumstance of second-generation biomass, where it forms a well-organized fibrous network. It is made up of anhydrous glucose components ranging from 500 to 15,000, linked together by β-1,4-glycosidic linkages to form a linear homopolysaccharide (Jayakumar et al. 2024). The smallest repeating unit within cellulose is cellobiose. The arrangement of β-1,4-glycosidic linkages results in strong intramolecular and intermolecular hydrogen bonding, leading to a highly crystalline and insoluble structure that is resistant to enzyme degradation. Researchers and scholars have identified two distinct regions within natural cellulose such as a less amorphous region and a highly crystalline region. Cellulose microfibrils create a well-organized crystalline structure that is highly resistant to attacks from microbes and enzymes. This unique composition makes cellulose a decisive structural component in plant cells, as long as strength and severity to the cell walls (Saini et al. 2015; A. Singh et al. 2019). Because of cellulose's crystalline structure, hydrolysis of a substantial amount of cellulose to glucose requires considerably lower pH, higher reaction temperatures, and/or longer residence durations than starch (Oh et al. 2015).

### 3.2.3 Hemicellulose

Hemicellulose is a complex polymer comprised of pentoses (L-arabinose and D-xylose) and hexoses (D-mannose, D-galactose, and D-glucose) with 50–200 units, forming a short and highly branched structure. The acetate groups within hemicelluloses are linked randomly to the hydroxyl groups in sugar rings through ester bonds. This unique structure allows hemicellulose to connect lignin to cellulose (Saini et al. 2015). The composition and arrangement of hemicelluloses can vary significantly among different species and cell types. These variations show a vital part in increasing the strength of the cell wall through exchanges with both cellulose and, in some instances, lignin (Scheller and Ulvskov 2010).

### 3.2.4 Lignin

Lignin is a disordered, cross-linked aromatic polymer made up of trans-sinapyl, trans-coniferyl, and trans-p-coumaryl alcohols. It entwines around the cellulose microfibril, creating an intricate structure that is chemically connected to adjacent chains on various hemicelluloses. The amount of lignin present can range from 2% to 40%. The carbon-carbon and ether links in lignin help fortify and safeguard the plant cell wall from microbial harm (A. Singh et al. 2019).

## 3.3 Sugar/Starch Crops and Waste Biomass for Biofuel Synthesis

Over the recent years, many nations such as developed and underdeveloped countries have been actively supporting efforts to produce sustainable energy, with biofuels playing a crucial role in their energy portfolios (Abo and Hailegiorgis 2021; Muleta et al. 2023). A variety of biomass sources have been tapped into for biofuel production, including crop plants, lignocellulosic materials, and organic wastes such as municipal waste, agricultural by-products, and industrial waste (Jane et al. 2021; Jifara et al. 2022; Woiciechowski et al. 2016). Sweet sorghum, sugar beet, and sugarcane are examples of sugar-rich crops commonly utilized for biofuel manufacture (Weselake et al. 2018).

The production of biofuels involves utilizing starchy feedstocks such as corn, cassava, wheat, and barley (illustrated in Fig. 3.3). Among these, corn is the utmost commonly used raw material for manufacture of bioethanol. Prior to the production

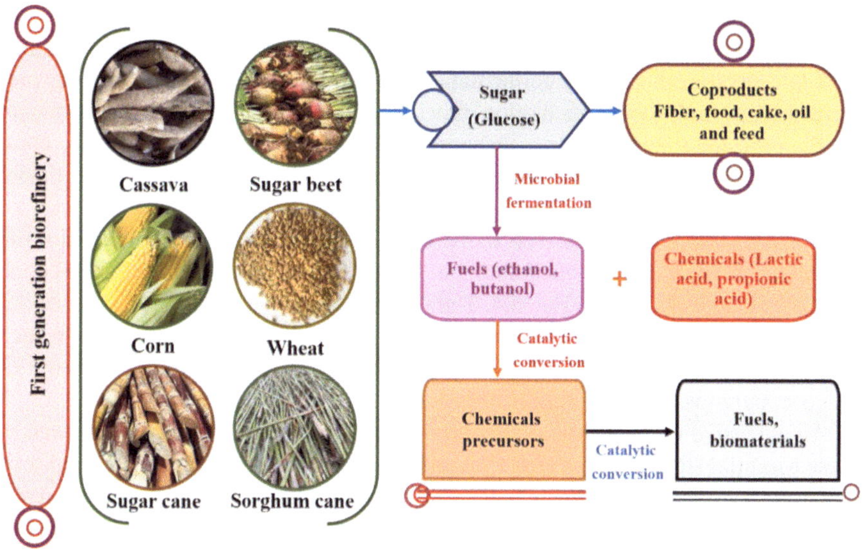


**Fig. 3.3** Biofuel synthesis from sugar/starch crops

process, the raw materials containing these feedstocks must undergo either an acidic or enzymatic pretreatment to increase the sugar content for biofuel production (Quevedo et al. 2024; Woiciechowski et al. 2016). The utilization of biofuels resulting from waste biomass has garnered significant consideration as a sustainable substitute to traditional fossil fuels. The sustainability of biofuel production is heavily influenced by factors such as feedstock selection, reagents, processing methods, and technologies, each of which carries various environmental, economic, political, and social implications (Neto et al. 2016).

Present research and improvements have concentrated on enhancing biofuel production techniques and increasing waste biomass conversion efficiency. Breakthroughs in biochemical transformation pathways, such as fermentation, and thermochemical conversion methods like direct combustion, pyrolysis (slow pyrolysis, flash pyrolysis, and fast pyrolysis), and gasification (downdraft gasifier, updraft gasifier, and cross flow gasifier) (Deso Abo et al. 2023) are expected to enhance the economic viability and expand the scope of biofuel such as biochar and bio-oil manufacturing (Daba and Hailegiorgis 2023; Quevedo et al. 2024).

## 3.4 Fermentation of Lignocellulosic Biomass

Second-generation lignocellulosic biomaterials are an extensive variety of dry plant resources that form a significant portion of sustainable carbon materials. They are considered a valuable renewable resource for producing various biofuels and biochemical substances (Østby et al. 2020). These biomaterials (natural biomass and biowaste) undergo four main unit operations such as pretreatment (acid and alkaline pretreatment), hydrolysis, fermentation, and product purification before being converted into biofuel like bioethanol through biochemical conversion processes like acid hydrolysis followed by fermentation (Jayakumar et al. 2023b). During pre-fermentation and post-fermentation processes, microorganisms break down into simple molecules such as monosaccharides, which are obtained by degrading complex polymer compounds (polysaccharides) that results in the formation bioethanol (Kucharska et al. 2020).

Numerous methods exist for transforming second-generation lignocellulosic resources into biofuel like bioethanol, biodiesel, biohydrogen, biomethane, biochar, and biogas, as illustrated in Fig. 3.4. The practice comprises three distinctive bacterial activities that are able to potentially be joined into a single reaction stage such as hydrolysis of lignin, cellulose, and hemicellulose, as well as fermentation of glucose, pyranose, arabinose, and fructose.

An extensive range of microbial species have the capability to perform each of these essential steps in converting lignocellulosic biomass such as agricultural residue, forestry residue, and municipal solid waste into bioethanol. Efforts have been made to achieve lignin hydrolysis under anaerobic conditions (Balan 2014). Additionally, there are challenges in effectively managing the anaerobic fermentation of glucose into bioethanol using mixed cultures. The rates of microbial practices described in present literature tend to be slower compared to those achieved through physicochemical pretreatments (Dionisi et al. 2014).

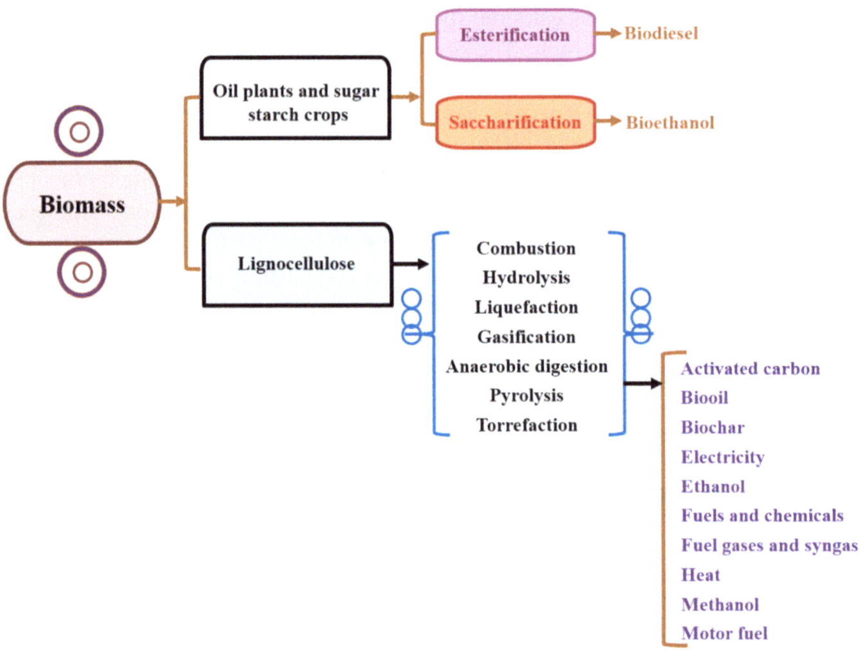


**Fig. 3.4** Various biomass transformation practices

Presently, there are various approaches for pretreating second-generation lignocellulose biomass, with technologies existing either in the laboratory or at pilot scale. Among these methods, biological pretreatments have gained popularity due to their environmentally friendly nature, low operating temperatures, and minimal release of inhibitory substances during the process (Beluhan et al. 2023). The implementation of industrial-scale biofuel manufacture through hydrolysis and biological delignification processes is quite at the laboratory stage, requiring further research. The manufacturing cost of ethanol from lignocellulose biomass remains high (Tsegaye et al. 2019), highlighting the need for advancements in this field.

Recently, it is estimated that the conversion of second-generation biomass into biofuels is costlier compared to conventional crude oil production. However, experts believe that as technology advances, the expenses will decrease over time (Balan 2014). The preprocessing of lignocellulosic biomass is essential to adjust the chemical and physical properties of the feedstock, but it is acknowledged as an expensive practice. The pretreatment of biomass can account for over 40% of the overall processing cost, and without this step, the degradation of lignocellulosic biomaterials through solid-state saccharification fermentation may only yield under 20% of total sugars. In contrast, utilizing certain pretreatment methods can result in a sugar yield ranging from 70% to 90% (Rahmati et al. 2020).

Second-generation ethanol is made from nonfood biomass, primarily lignocellulose, including switchgrass, sawdust, and black liquor from the paper sector, corn stover, and municipal garbage. Feedstock costs continue to rise, owing to processing (shredding, densifying, pulverizing, and handling) and transportation. However, they offer the potential for larger GHG emission reductions than first-generation biofuels (Demirel 2018).

## 3.5 Gasification to Syngas

The making of synthesis gas from biomass offers a variety of versatile applications. Biomass can be converted into synthesis gas using entrained flow, fluidized bed, and fixed bed reactors (Deso Abo et al. 2023). Various agents such as oxygen, steam, or their combinations are utilized in the gasification process (Molino et al. 2018). Historically, biomass gasification was primarily used for generating heat and electricity. However, recently, there has been a shift toward producing transport fuels from renewable sources. This transition is motivated by environmental concerns and the rising costs of fossil fuels (Rauch et al. 2020).

Study by Tamošiūnas et al. (2023) conducted research on converting biomass into syngas using biomass gasification in thermal water vapor arc discharge plasma. They studied the process of converting second-generation biomass like wood pellets into synthetic gas (syngas) utilizing (DC) direct current thermal arc plasma at atmospheric pressure. Water vapor was utilized as the plasma-forming gas and primary gasifying agent in the study. The study assessed the biomass gasification system based on carbon conversion efficiency, $H_2$/CO proportion, tar content, specific energy requirements, energy conversion efficiency, and producer gas composition.

The study established that the gasification enactment was most efficient at a water vapor-biomass proportion of 0.97. Synthetic gas (syngas) produced through this process is crucial for various synthesis processes, such as Fischer-Tropsch (FT) diesel or SNG production. Thermochemical gasification of solid fuels allows for the manufacture of gas mixtures suitable for industrial-scale synthesis. The DFB steam gasification technique yields nitrogen-free, hydrogen-rich product gas, while fluidized bed practice can process a wide range of fuel sources, including woody biomass, biogenic wastes, and other waste materials (Cerone and Zimbardi 2018; Schmid et al. 2021).

The utilization of biomass to value-added fuel is attained over physicochemical, thermochemical, and biochemical conversion processes (Chojnacki et al. 2020). Thermochemical transformation processes such as direct combustion, gasification, and pyrolysis stand out as one of the encouraging practices that can convert various biomass sources into valuable products like solid fuel, heat, power, liquid fuel, and gaseous fuel. These products find application across different industries due to their versatility and utility. Thermochemical conversion technology offers significant advantages by efficiently converting lignocellulosic biomass sources into syngas through gasification, direct combustion, and pyrolysis like slow, flash, and fast pyrolysis methods (Guruviah et al. 2019). This process is characterized by high conversion rates, the formation of char, rapid reaction times, cost-effective transportation, compatibility with diverse raw materials, and operation without the need for additional heating or high pressures (Deso Abo et al. 2023).

The gasification of almond and hazelnut shells in updraft mode was found to be highly efficient due to their low levels of fine particles and the effective penetration of oxidant gases through the reactive bed (Cerone and Zimbardi 2018). This investigation revealed that the extreme energy conversion efficiency from solids to gases and oils was 76% and 28%, correspondingly. In alternative research reported by Chojnacki et al. (2020), the thermal gasification of wood granules utilizing steam in a thermal reactor equipped with electric heaters was studied. The aim was to analyze how the temperature and steam flow rate inside the reactor impacted the composition of synthesis gas, including $H_2$, $CH_4$, CO, and $CO_2$, as well as the calorific value. The findings revealed that temperature played a significant role in determining the composition of synthesis gas, while the steam flow rate particularly influenced the levels of $H_2$ and $CO_2$ in the syngas.

## 3.6 Fast Pyrolysis

Pyrolysis is a method of breaking down biomass through heat, with average temperatures between 400 and 650 °C, done in the nonappearance of $O_2$ (Aboelela et al. 2023). During this process, volatile substances are released, while solid nonvolatiles are turned into biochar. Fast pyrolysis is an innovative technology that boasts high liquid yields, thanks to its fast warming rates of 10 to >1000 °C/s, shorter residence times of less than 2 s, temperatures between 400 and 650 °C, and quick cooling of vapors. For a more in-depth look at the various reactors, methods, and resulting liquid products (Dickerson and Soria 2013).

Fast pyrolysis remains an immature technique with numerous unknowns. Although several sites were built, they all suffer from a lack of working hours, and no procedure has truly been "demonstrated." Interested sectors, such as heat and electricity producers, oil corporations, and food/feed firms, are waiting for full-scale demonstrations that include continuous operation (>7000 hours per year) of several units on a scale of 5–10 tons biomass feedstock per hour (Venderbosch and Prins 2010).

The primary objective of the study was to examine the lipid-extracted residues of *Desmodesmus* sp. grown in anaerobically digested effluents. This was done using a pyrolysis-gas chromatography mass spectrometry, and thermogravimetric analyzer technology in a temperature lies between 300 and 800 °C. It was observed that quick pyrolysis at 800 °C resulted in the maximum bio-oil produce of 36.6%, whereas at 700 °C the yield remained 29% (Li et al. 2020).

The volatile products obtained from quick pyrolysis were found to be considerably obstructed by the source of lignin and the method of isolation. Additional investigations are necessary to explore the potential of dissimilar commercial lignins, considering factors such as heating rates, temperatures, and holding times under both atmospheric and vacuum pressures. Various plant species offer an abundance of lignin, each with its unique characteristics and isolation methods (Lin et al. 2015).

Carlson et al. conducted a research on the production of aromatics through catalytic fast pyrolysis (CFP) of biomass-derived materials. According to the study, solid biomass resources can be quickly pyrolyzed using a catalyst in a single reactor with brief residence durations to produce high-quality aromatic fuel additives. This practice involves the consistent thermal disintegration of biomass into lesser oxygenated compounds (Conner and Huber 2009).

## 3.7 Hydrogenation

Ethanol synthesis utilizing biomass residue emits a significant amount of $CO_2$ while incurring substantial operating costs due to expensive apparatus and fuel. The catalytic hydrogenation of $CO_2$ promises a new direction in ethanol production while also reducing environmental pollution problems (Latif et al. 2023; Ruan et al. 2023). Hydrogenation is a lignin depolymerization strategy that uses hydrogen as a reductant to produce aromatic compounds with lower oxygen concentration and greater stability than previous methods (Latif et al. 2023).

The bio-oil created from the practice of co-pyrolysis with polypropylene and corncobs underwent a hydrogenation process to create biodiesel. The co-pyrolysis took place in a stirred tank reactor (STR) with a gradual warming level of 5°Celsius per minute, reaching an extreme temperature of 500°Celsius to enhance the harvest of the nonpolar division synergistically. Polypropylene played a dual role as a $O_2$ remover and $H_2$ donor, decreasing the oxygenate content in the resulting biofuel

(Supramono et al. 2018). Various factors influence the total cost of producing biofuels, with a key factor being the efficiency of post-conversion pathways such as hydrogenation and deoxygenation processes (Hansen et al. 2020).

## 3.8 Aerobic Digestion

Aerobic digestion refers to a natural biological method employed to manage organic waste substances like sewage sludge, food scraps, or agricultural leftovers in the existence of $O_2$ (Demirbas et al. 2017). This procedure relies on aerobic microorganisms, mainly bacteria and fungi, to decompose organic materials into more straightforward compounds like carbon dioxide, water, and biomass via a sequence of chemical reactions (Adetunji et al. 2023). Aerobic digestion occurs under aerobic conditions, meaning that oxygen is readily available to support the metabolic activities of the microorganisms involved.

The process of aerobic digestion begins with the introduction of organic waste material into a reactor or tank containing a suspension of aerobic microorganisms. These microorganisms utilize the organic compounds present in the waste as a foundation of carbon and energy for their metabolism and growth. As they metabolize the organic matter, they release enzymes that break down complex molecules into smaller, soluble compounds that are easily utilized and absorbed by the microorganisms (Demirbas et al. 2017). In the practice of aerobic digestion, organic compounds undergo a sequence of biochemical changes, including acetogenesis, acidogenesis, hydrolysis, and methanogenesis. During the hydrolysis stage, intricate organic molecules like carbohydrates, fats, and proteins are cracked down into simpler compounds like fatty acids, amino acids, and sugars. This breakdown is facilitated by enzymes produced by microorganisms. These simpler compounds then go through additional metabolic processes in the following stages of aerobic digestion (Muchowska et al. 2020).

In the acidogenesis stage, the waste products produced during the breakdown of organic matter remain transformed into a range of organic acids, alcohols, and volatile fatty acids (VFA) via fermentation practice. These organic acids act as fuel for acetogenic bacteria, which formerly transform them into carbon dioxide, acetate, and hydrogen in the acetogenesis phase. In the final step, methanogenic archaea make use of these intermediary products to generate methane and carbon dioxide through methanogenesis, thereby finishing the digestion process. Aerobic digestion offers several advantages over anaerobic digestion, another common method of organic waste treatment. It typically operates at higher temperatures and pH levels, which facilitate faster reaction rates and more complete degradation of organic matter. Additionally, aerobic digestion produces fewer odors and less greenhouse gas emissions compared to anaerobic digestion, making it a more environmentally friendly option for organic waste treatment. However, aerobic digestion may require more energy input to maintain oxygen levels and mixing within the reactor, which can increase operational costs.

## 3.9 Catalytic Thermochemical Conversion

The process of catalytic thermochemical conversion utilizes catalysts to enhance chemical reactions at elevated temperatures, typically between 300 and 600 °C, deprived of the occurrence of $O_2$ (Solarte-Toro et al. 2021). The foremost goal of this technique is to transform biomass sources like agricultural residue, forestry waste, or energy crops into useful end products like biofuels, biochemicals, and biochar (Areti et al. 2024) through a series of thermochemical reactions.

The first step in catalytic thermochemical conversion is the pretreatment of biomass feedstock to prepare it for further processing (Mussatto 2016; Nanda and Vo 2021). Pretreatment methods may include size reduction, grinding, and drying to increase the surface area and accessibility of biomass particles to the catalyst. This step aims to improve the efficiency and effectiveness of subsequent thermochemical reactions. Once the biomass feedstock is prepared, it undergoes thermochemical conversion in the existence of a catalyst. In general, catalysts promote the decomposition and rearrangement of biomass molecules into simpler compounds, such as gases, liquids, and solids, under high-temperature conditions. Thermochemical reactions involved in catalytic conversion may include pyrolysis, gasification, and liquefaction, depending on the desired end products.

Practice of pyrolysis is a chemical process with higher temperatures that occurs when biomass is heated without oxygen present, resulting in a combination of gases, liquids, and solids being produced (Bhaskar et al. 2011; Dhyani and Bhaskar 2019). Utilizing catalysts in pyrolysis, also referred to as catalytic pyrolysis, enhances both the quantity and quality of the end products. This enhancement is achieved through the facilitation of supplementary chemical reactions, including cracking, reforming, and hydrogenation.

Gasification is a process that converts biomass into syngas, generally known as synthesis gas, through a thermochemical process. This synthesis gas is a combination of carbon monoxide, various other gases, and hydrogen achieved by the reaction of biomass with a specific quantity of controlled steam or oxygen. The efficiency and precision of syngas generation can be enhanced by employing catalytic gasification. This method boosts water-gas shift reactions and reforming, resulting in improved syngas production.

Liquefaction is a chemical procedure that transforms biomass into liquid fuels, like bio-oil, using solvents or catalysts at higher pressure and temperature. According to the study reported by Huang et al., catalytic liquefaction boosts the process of converting biomass into liquid fuels by facilitating hydrogenation, deoxygenation, and hydrodeoxygenation reactions (Huang et al. 2017).

In contrast, catalytic thermochemical conversion utilizes catalysts to facilitate the transformation of biomass into desired products at lower temperatures. This enables faster reaction rates, higher product selectivity, and improved energy efficiency compared to traditional thermochemical conversion processes. Table 3.1 provides a concise comparison of key aspects between traditional thermochemical conversion processes and catalytic thermochemical conversion processes (Garver

**Table 3.1** Comparison between catalytic thermochemical conversion and catalytic thermochemical conversion

| S. No | Aspect | Thermochemical conversion | Catalytic thermochemical conversion |
|---|---|---|---|
| 1. | Process type | Non-catalytic | Catalytic |
| 2. | Presence of catalyst | No catalyst required | Catalyst required |
| 3. | Reaction rate | Slower reaction rates | Faster reaction rates |
| 4. | Temperature requirements | Higher temperatures required | Lower temperatures required |
| 5. | Product selectivity | Lower product selectivity | Higher product selectivity |
| 6. | Energy efficiency | Lower energy efficiency | Higher energy efficiency |
| 7. | Reaction mechanism | Homogeneous or heterogeneous | Heterogeneous |
| 8. | Reaction control | Less control over reactions | Better control over reactions |
| 9. | By-product formation | More potential for by-products | Reduced by-product formation |

and Liu 2014; Kuila and Mukhopadhyay 2020; Sharara et al. 2012; P. K. Singh et al. 2022).

Overall, catalytic thermochemical conversion offers several advantages for biomass utilization, including higher product yields, improved product quality, and increased process efficiency. However, challenges such as catalyst deactivation, product selectivity, and process scalability need to be addressed to commercialize catalytic thermochemical conversion technologies for widespread application in the bioenergy sector.

## 3.10 Genetic Enhancement of Harvests for Manufacture of Biochemicals and Biofuels

Enhancing the genetic traits of crops to optimize their potential for producing biofuels and biochemicals requires the implementation of diverse strategies. These strategies are focused on increasing both the quantity and quality of biomass harvested from energy harvests. Three key approaches in this area include cellulose and hemicellulose modification, lignin modification, and targeted genome engineering.

### 3.10.1 Hemicellulose and Cellulose Modification

Hemicellulose and cellulose are key elements found in plant cell walls that assist as cherished resources for the manufacture of biochemicals and biofuels. These components are plentiful and offer a sustainable source of biomass for various industrial processes (Kariim et al. 2022; Kuila and Mukhopadhyay 2020; Lopes et al. 2022; Mussatto 2016).

Genetic modification techniques can be employed to alter the composition and structure of cellulose and hemicellulose, thereby improving their accessibility and digestibility for conversion into fermentable sugars (Huang et al. 2017). One approach to improving the efficiency of biomass conversion is by genetically modifying plants to naturally produce cellulases and hemicellulases. These enzymes are responsible for breaking down complex structured hemicellulose and cellulose into simpler sugars that can easily undergo fermentation practice. Furthermore, adjusting the balance between cellulose and hemicellulose in plants can enhance the biomass properties suitable for various conversion methods, including enzymatic hydrolysis or thermochemical processes (Lopes et al. 2022).

### 3.10.2 Lignin Modification

Lignin is a sophisticated polymer that plays a crucial role in providing basic sustenance to plant cell walls (PCW). However, it also acts as a hindrance to the breakdown and conversion of biomass. By utilizing genetic engineering techniques, it is possible to alter the composition and structure of lignin, ultimately minimizing its inhibitory impacts on the processing of biomass (Huang et al. 2017). This may involve downregulating genes involved in lignin biosynthesis to decrease lignin content or engineering lignin with altered chemical properties to improve its compatibility with conversion processes. By reducing lignin content or modifying its structure, biomass becomes more amenable to enzymatic hydrolysis, thermal degradation, or other conversion methods, resulting in higher yields of fermentable sugars or other valuable products.

### 3.10.3 Targeted Genome Engineering

Targeted genome engineering techniques, like the clustered regularly interspaced short palindromic repeats-associated nuclease 9 (CRISPR)/Cas9 technology, provide extremely accurate and operational means of incorporating genetic alterations into energy-producing crops (Javed et al. 2019; Saxena et al. 2020); (Hemalatha et al. 2023). These techniques enable researchers to precisely edit plant genomes to introduce desirable traits related to biomass yield, composition, and processing characteristics. For instance, precision genome editing can be employed to introduce or amplify the activity of genes related to cellulose and hemicellulose production or to alter lignin synthesis pathways. By precisely manipulating plant genomes, targeted genome engineering holds promise for developing energy crops with improved biomass traits tailored for biofuel and biochemical production. Genetic improvement of crops for biofuel and biochemical production involves a multifaceted approach that encompasses cellulose and hemicellulose modification, lignin modification, and targeted genome engineering, and their comparison is provided in Table 3.2 (D. Kumar et al. 2024; Moodley et al. 2023; Purkait and Haldar 2021). The major objective of these tactics is to enhance the features of biomass, increase

**Table 3.2** Comparison of targeted genome engineering, lignin modification, and cellulose and hemicellulose modification

| S. No | Aspect | Targeted genome engineering | Lignin modification | Cellulose and hemicellulose modification |
|---|---|---|---|---|
| 1. | Objective | Modify specific genes | Alter lignin structure | Modify cellulose/ hemicellulose structure |
| 2. | Approach | Direct genetic manipulation | Chemical or enzymatic methods | Chemical or enzymatic methods |
| 3. | Target | Lignin polymer | Lignin polymer | Cellulose/ hemicellulose polymers |
| 4. | Techniques | Chemical treatments, enzyme catalysis | Chemical treatments, enzyme catalysis | Chemical treatments, enzyme catalysis |
| 5. | Impact | Structural alterations | Structural alterations | Structural alterations |
| 6. | Applications | Biomass processing, pulp and paper industry | Biomass processing, pulp and paper manufacturing sectors | Biomass processing, biofuel manufacture |
| 7. | Challenges | Specificity, scalability | Specificity, scalability | Specificity, scalability |

its quantity and quality, and boost the effectiveness of converting biomass into chemicals and renewable fuels. This will ultimately support the sustainable production of eco-friendly energy sources.

## 3.11 Bio-Based Economy for Biorefineries

The bio-based economy signifies a transition toward the eco-friendly exploitation of renewable biological resources, like biomass, for creating a diverse array of goods like fuels, chemicals, materials, and energy. Biorefineries play a vital part in the bio-based economy, aiding as the primary hubs for transforming biomass into value-added products through a combination of biochemical, thermochemical, and other conversion techniques (Kariim et al. 2022; R. P. Kumar et al. 2020).

Biorefineries show a crucial part in driving the conversion toward an ecological bio-based economy by integrating advanced technologies to efficiently harness a diverse variety of biomass resources such as energy crops, food scraps, forest waste, and agricultural residues (Awasthi et al. 2020). These establishments utilize a range of transformation processes, including fermentation, pyrolysis, hydrolysis, and gasification, to convert biomass into valued harvests such as biofuels (such as biodiesel and bioethanol), biochemicals (such as enzymes and organic acids), biomaterials (such as bio-derived polymers and bioplastics), and bioenergy (such as biogas and bioelectricity).

A crucial factor propelling the development of the bio-based economy (BBE) is the increasing recognition of the importance of environmental sustainability and the urgency to lessen reliance on fossil fuels while addressing the impact of greenhouse gas emissions (GHGs). Biorefineries offer a sustainable alternative by harnessing renewable biomass resources, which are carbon-neutral and can help mitigate climate change by sequestering carbon dioxide through biomass growth and utilization (Singhal and Prashant. 2020).

Furthermore, the bio-based economy promotes regional development and economic growth by creating new markets for agricultural and forestry products, fostering rural employment opportunities, and stimulating investment and innovation in bio-based industries. Biorefineries show an important part in improving resource competence and utilizing waste as a valuable resource. By transforming organic wastes and by-products into bio-based products, they not only reduce environmental pollution but also fund the principles of a circular economy (Duan et al. 2022). Bio-based economy presents significant opportunities for transforming traditional industries, reducing dependence on finite fossil resources, and promoting sustainable development. Biorefineries show a crucial part in facilitating the shift toward a more sustainable, resilient, and circular economy by offering the essential infrastructure and cutting-edge technologies needed to tap into the complete potential of biomass resources.

## 3.12 Conclusion

The biorefinery industry is evolving, with a shift toward utilizing a greater variety of feedstocks beyond traditional wood sources. This expansion in biomass options has led to the development of an extensive variety of products and applications within the industry. An in-depth analysis of biorefinery plants on a global scale has been conducted, highlighting their strengths, limitations, and upcoming challenges. As the world transitions toward renewable energy sources and away from fossil fuels, expertise in managing advanced biorefinery processes has become essential. The introduction of new product lines for the chief biomass constituents, such as lignin, hemicellulose, and cellulose, signals a promising future for biorefineries. By focusing on maximizing the conversion of biomass feedstock into specialized chemicals while minimizing energy by-products, the industry has achieved success and is poised for further growth. In summary, advanced biorefinery technology offers significant potential for sustainable expansion by maximizing the value of biomass possessions and decreasing environmental impact. Continuing investigation and expansion efforts in this ground will continue to improve the efficiency and sustainability of biorefinery processes, shaping a brighter future for the industry.

## References

Abo LD, Hailegiorgis SM (2021) Investigation of scale deposits on Finchaa sugar factory ethanol plant heat transfer units. Therm Sci Eng Prog 25(July):101022. https://doi.org/10.1016/j.tsep.2021.101022

Aboelela D, Saleh H, Attia AM, Elhenawy Y, Majozi T, Bassyouni M (2023) Recent advances in biomass pyrolysis processes for bioenergy production: optimization of operating conditions. Sustainability 15(14):11238. https://doi.org/10.3390/su151411238

Adetunji AI, Oberholster PJ, Erasmus M (2023) From garbage to treasure: a review on biorefinery of organic solid wastes into valuable biobased products. Bioresour Technol Rep 24:101610. https://doi.org/10.1016/j.biteb.2023.101610

Alonso DM, Bond JQ, Dumesic JA (2010) Catalytic conversion of biomass to biofuels. Green Chem 12(9):1493. https://doi.org/10.1039/c004654j

Amoah J, Kahar P, Ogino C, Kondo A (2019) Bioenergy and biorefinery: feedstock, biotechnological conversion, and products. Biotechnol J 14(6):1800494. https://doi.org/10.1002/biot.201800494

Areti HA, Jabesa A, Daba BJ, Jibril D (2024) Response surface method based parametric optimization of Cr(VI) removal from tannery wastewater using a mixed banana peel and corn cob activated carbon: kinetic and isotherm modeling studies. J Water Process Eng 59. https://doi.org/10.1016/j.jwpe.2024.104977

Awasthi MK, Sarsaiya S, Patel A, Juneja A, Singh RP, Yan B, Awasthi SK, Jain A, Liu T, Duan Y, Pandey A, Zhang Z, Taherzadeh MJ (2020) Refining biomass residues for sustainable energy and bio-products: an assessment of technology, its importance, and strategic applications in circular bio-economy. Renew Sust Energ Rev 127:109876. https://doi.org/10.1016/j.rser.2020.109876

Balan V (2014) Current challenges in commercially producing biofuels from lignocellulosic biomass. ISRN Biotechnol 2014:1–31. https://doi.org/10.1155/2014/463074

Beluhan S, Mihajlovski K, Šantek B, Ivančić Šantek M (2023) The production of bioethanol from lignocellulosic biomass: pretreatment methods, fermentation, and downstream processing. Energies 16(19):7003. https://doi.org/10.3390/en16197003

Bentsen NS, Felby C, Thorsen BJ (2014) Agricultural residue production and potentials for energy and materials services. Prog Energy Combust Sci 40(1):59–73. https://doi.org/10.1016/j.pecs.2013.09.003

Bhaskar T, Bhavya B, Singh R, Naik DV, Kumar A, Goyal HB (2011) Chapter 3 - Thermochemical conversion of biomass to biofuels. In: Pandey A, Larroche C, Ricke SC, Dussap C-G, Gnansounou EBT-B (eds) Biofuels: alternative feedstocks and conversion processes. Academic Press, pp 51–77. https://doi.org/10.1016/B978-0-12-385099-7.00003-6

Cerone N, Zimbardi F (2018) Gasification of agroresidues for syngas production. Energies. https://doi.org/10.3390/en11051280

Chojnacki J, Najser J, Rokosz K, Peer V, Kielar J, Berner B (2020) Syngas composition: gasification of wood pellet with water steam through a reactor with continuous biomass feed system. Energies 13(17):4376. https://doi.org/10.3390/en13174376

Chozhavendhan S, Karthigadevi G, Bharathiraja B, Kumar RP, Abo LD, Prabhu SV, Balachandar R, Jayakumar M (2023) Current and prognostic overview on the strategic exploitation of anaerobic digestion and digestate: a review. Environ Res 216:114526

Conner WC, Huber ÆGW (2009) Aromatic production from catalytic fast pyrolysis of biomass-derived feedstocks. In: Top Catal, pp 241–252. https://doi.org/10.1007/s11244-008-9160-6

Daba BJ, Hailegiorgis SM (2023) Bioresource Technology Reports Torrefaction of corncob and khat stem biomass to enhance the energy content of the solid biomass and parametric optimization. Bioresour Technol Rep 21(February):101381. https://doi.org/10.1016/j.biteb.2023.101381

de Jong W (2014) Biomass composition, properties, and characterization. In: Biomass as a sustainable energy source for the future. Wiley, pp 36–68. https://doi.org/10.1002/9781118916643.ch2

Demirbas A, Coban V, Taylan O, Kabli M (2017) Aerobic digestion of sewage sludge for waste treatment. Energy Sources Part A Recovery Util Environ Eff 39(10):1056–1062. https://doi.org/10.1080/15567036.2017.1289282

Demirel Y (2018) Sugar versus lipid for sustainable biofuels. Int J Energy Res 42(3):881–884. https://doi.org/10.1002/er.3914

Deso Abo L, Arumugasamy SK, Venkatesa Prabhu S, Jayakumar M (2023) Waste biomass utilization for production of bioenergy through gasification practice. In: Encyclopedia of green materials. Springer, pp 1–12

Dhyani V, Bhaskar T (2019) Chapter 9 - Pyrolysis of biomass. In: Pandey A, Larroche C, Dussap C-G, Gnansounou E, Khanal SK, Ricke S (eds) Biomass, biofuels, biochemicals: biofuels: alternative feedstocks and conversion processes for the production of liquid and gaseous biofuels, 2nd edn. Academic Press, pp 217–244. https://doi.org/10.1016/B978-0-12-816856-1.00009-9

Dickerson T, Soria J (2013) Catalytic fast pyrolysis: a review. In: Energies, pp 514–538. https://doi.org/10.3390/en6010514

Dionisi D, Anderson JA, Aulenta F, Paton G (2014) The potential of microbial processes for lignocellulosic biomass conversion to ethanol: a review. J Chem Technol Biotechnol. https://doi.org/10.1002/jctb.4544

Duan Y, Tarafdar A, Kumar V, Ganeshan P, Rajendran K, Shekhar Giri B, Gómez-García R, Li H, Zhang Z, Sindhu R, Binod P, Pandey A, Taherzadeh MJ, Sarsaiya S, Jain A, Kumar Awasthi M (2022) Sustainable biorefinery approaches towards circular economy for conversion of biowaste to value added materials and future perspectives. Fuel 325:124846. https://doi.org/10.1016/j.fuel.2022.124846

Dussán KJ, Silva DDV, Costa AFM, Grangeiro LC, Giese EC (2022) Downstream processing in lignocellulose conversion. In: Lignocellulose bioconversion through white biotechnology. Wiley, pp 261–288. https://doi.org/10.1002/9781119735984.ch11

Garver MP, Liu S (2014) Development of thermochemical and biochemical technologies for biorefineries. In: Bioenergy research: advances and applications. Elsevier, pp 457–488. https://doi.org/10.1016/B978-0-444-59561-4.00027-9

Ghaffar SH, Fan M (2013) Structural analysis for lignin characteristics in biomass straw. Biomass Bioenergy 57:264–279. https://doi.org/10.1016/j.biombioe.2013.07.015

Giuliano A (2023) The transition of scientific research from biomass-to-energy/biofuels to biomass-to-biochemicals in a biorefinery systems framework. Energies 16(5):2261. https://doi.org/10.3390/en16052261

Gundupalli MP, Gundupalli SP, Thomas ASS, Jayakumar M, Bhattacharyya D, Gurunathan B (2022) Hydrothermal liquefaction of lignocellulosic biomass for production of biooil and by-products: current state of the art and challenges. In: Biofuels and bioenergy, pp 61–84

Guruviah KD, Sivasankaran C, Bharathiraja B (2019) Thermochemical conversion: bio-oil and syngas production. In: Prospects of renewable bioprocessing in future energy sdystems, pp 251–267

Hamda AS, Abo LD, Jayakumar M, Selvakumar KV, Periyasamy S, Emana AN (2023) Introduction: growth of agricultural waste, its disposal, and related environmental issues. In: Agricultural waste to value-added products: technical, economic and sustainable aspects. Springer, pp 1–19

Hansen S, Mirkouei A, Diaz LA (2020) A comprehensive state-of-technology review for upgrading bio-oil to renewable or blended hydrocarbon fuels. Renew Sust Energ Rev 118(November):109548. https://doi.org/10.1016/j.rser.2019.109548

Hemalatha P, Abda EM, Shah S, Prabhu SV, Jayakumar M, Karmegam N, Kim W, Govarthanan M (2023) Multi-faceted CRISPR-Cas9 strategy to reduce plant based food loss and waste for sustainable bio-economy–a review. J Environ Manag 332:117382

Huang H, Yuan X, Wu G (2017) In: Singh L, Kalia VC (eds) Liquefaction of biomass for bio-oil products BT—waste biomass management—a holistic approach. Springer, pp 231–250. https://doi.org/10.1007/978-3-319-49595-8_11

Jane S, Andy T, Palma AD (2021) The impacts of biofuel crops on local biodiversity: a global synthesis. Biodivers Conserv 30(11):2863–2883. https://doi.org/10.1007/s10531-021-02232-5

Javed MR, Noman M, Shahid M, Ahmed T, Khurshid M, Rashid MH, Ismail M, Sadaf M, Khan F (2019) Current situation of biofuel production and its enhancement by CRISPR/Cas9-mediated genome engineering of microbial cells. Microbiol Res 219:1–11. https://doi.org/10.1016/j.micres.2018.10.010

Jayakumar M, Asrat L, Feseha L, Kasirajan R, Periyasamy S (2023d) Agricultural waste as a source of fine chemicals. In: Agricultural waste to value-added products: technical, economic and sustainable aspects. Springer, pp 355–375

Jayakumar M, Gebeyehu KB, Abo LD, Tadesse AW, Vivekanandan B, Sundramurthy VP, Bacha W, Ashokkumar V, Baskar G (2023a) A comprehensive outlook on topical processing methods for biofuel production and its thermal applications: current advances, sustainability and challenges. Fuel 349:128690

Jayakumar M, Gindaba GT, Gebeyehu KB, Periyasamy S, Jabesa A, Baskar G, John BI, Pugazhendhi A (2023b) Bioethanol production from agricultural residues as lignocellulosic biomass feedstock's waste valorization approach: a comprehensive review. Sci Total Environ:163158

Jayakumar M, Hamda AS, Abo LD, Daba BJ, Prabhu SV, Rangaraju M, Jabesa A, Periyasamy S, Suresh S, Baskar G (2023e) Comprehensive review on lignocellulosic biomass derived biochar production, characterization, utilization and applications. Chemosphere:140515

Jayakumar M, Kuppusamy Vaithilingam S, Karmegam N, Gebeyehu KB, Boobalan MS, Gurunathan B (2022a) Fermentation technology for ethanol production: current trends and challenges. In: Gurunathan B, Sahadevan B (eds) Biofuels and bioenergy. Elsevier, pp 105–131. https://doi.org/10.1016/b978-0-323-90040-9.00015-1

Jayakumar M, Kuppusamy Vaithilingam S, Karmegam N, Jabesa A (2022b) Bioethanol production from green biomass resources: emerging technologies. In: Encyclopedia of green materials, pp 1–12

Jayakumar M, Prabhu SV, Abo LD, Daba BJ, Periyasamy S, Jabesa A (2023c) Microbial conversion of agricultural residues into organic fertilizers. In: Agricultural waste to value-added products: technical, economic and sustainable aspects. Springer, pp 113–138

Jayakumar M, Thiyagar T, Abo LD, Arumugasamy SK, Jabesa A (2024) Paddy straw as a biomass feedstock for the manufacturing of bioethanol using acid hydrolysis and parametric optimization through response surface methodology and an artificial neural network. Biomass Convers Biorefinery 0123456789. https://doi.org/10.1007/s13399-024-05371-1

Jifara B, Diriba M, Mengesha A (2022) Pelletization of mixed torrefied corn cob and khat stem to enhance the physicochemical and thermal properties of solid biofuel and parametric optimization. Biomass Convers Biorefinery 0123456789. https://doi.org/10.1007/s13399-022-02627-6

Kariim I, Swai H, Kivevele T (2022) Recent advances in thermochemical conversion of biomass into drop-in fuel: a review. Sci Afr 17:e01352. https://doi.org/10.1016/j.sciaf.2022.e01352

Kucharska K, Słupek E, Cieśliński H, Kamiński M (2020) Advantageous conditions of saccharification of lignocellulosic biomass for biofuels generation via fermentation processes. Chem Pap 74(4):1199–1209. https://doi.org/10.1007/s11696-019-00960-1

Kuila A, Mukhopadhyay M (2020) Biorefinery production technologies for chemicals and energy. In: Kuila A, Mukhopadhyay M (eds) Biorefinery production technologies for chemicals and energy. Wiley. https://doi.org/10.1002/9781119593065

Kumar D, Kumar S, Rajendran K (2024) Sustainable biorefining of woody biomass to biofuels and biochemicals. Elsevier. https://doi.org/10.1016/C2020-0-04152-6e

Kumar RP, Gnansounou E, Raman JK, Baskar G (2020) Refining biomass residues for sustainable energy and bioproducts: technology, advances, life cycle assessment, and economics. Elsevier. https://doi.org/10.1016/C2018-0-05005-7

Latif MN, Wan Isahak WNR, Samsuri A, Hasan SZ, Manan WN, Yaakob Z (2023) Recent advances in the technologies and catalytic processes of ethanol production. Catalysts 13(7):1093. https://doi.org/10.3390/catal13071093

Li G, Bai X, Huo S, Huang Z (2020) Fast pyrolysis of LERDADEs for renewable biofuels. IET Renew Power Gener 14:959–967. https://doi.org/10.1049/iet-rpg.2019.0852

Lin X, Sui S, Tan S, Pittman CU, Sun J, Zhang Z (2015) Fast pyrolysis of four Lignins from different isolation processes using Py-GC/MS. Energies 5107–5121. https://doi.org/10.3390/en8065107

Lopes TF, Duarte LC, Carvalheiro F, Cardona CA, Gírio F (2022) Hemicellulose biorefinery: a sustainable solution for value addition to bio-based products and bioenergy. Springer

Mansora AM, Lima JS, Anib FN, Hashima H, Hoa WS (2019) Characteristics of cellulose, hemicellulose and lignin of MD2 pineapple biomass. Chem Eng 72(1):79–84. https://doi.org/10.3303/CET1972014

Molino A, Larocca V, Chianese S, Musmarra D (2018) Biofuels production by biomass gasification: a review. Energies 11(4):811. https://doi.org/10.3390/en11040811

Moodley P, Ray RC, Kana EBG (2023) Advances in lignocellulosic biofuel production systems. Elsevier. https://doi.org/10.1016/C2020-0-04155-1

Muchowska KB, Varma SJ, Moran J (2020) Nonenzymatic metabolic reactions and life's origins. Chem Rev 120(15):7708–7744. https://doi.org/10.1021/acs.chemrev.0c00191

Muleta MD, Ramesh R, Mengstie HB, Murali A, Reddy RP (2023) Methanolysis of avocado peel oil using heterogeneous catalyst enhanced by microwave heating and parametric optimization. Energy Sources Part A Recovery Util Environ Eff 45(2):4414–4431. https://doi.org/10.1080/15567036.2023.2202623

Mussatto SI (2016) Biomass fractionation technologies for a lignocellulosic feedstock based biorefinery. Elsevier. https://doi.org/10.1016/C2014-0-01890-4

Nanda S, Vo D-VN (2021) Progressive thermochemical biorefining technologies. CRC Press. https://doi.org/10.1201/9781003098591

Neto CJD, Sydney EB, de Souza Vandenberghe LP, Soccol CR (2016) Green fuels technology, pp 387–406. https://doi.org/10.1007/978-3-319-30205-8

Oh YH, Eom IY, Joo JC, Yu JH, Song BK, Lee SH, Hong SH, Park SJ (2015) Recent advances in development of biomass pretreatment technologies used in biorefinery for the production of bio-based fuels, chemicals and polymers. Korean J Chem Eng 32(10):1945–1959. https://doi.org/10.1007/s11814-015-0191-y

Ong HC, Yu KL, Chen W, Pillejera MK, Bi X, Tran K, Pétrissans A, Pétrissans M (2021) Variation of lignocellulosic biomass structure from torrefaction: a critical review. Renew Sust Energ Rev 152(March):111698. https://doi.org/10.1016/j.rser.2021.111698

Østby H, Degn L, Svein H, Vincent JH, Anikó GHE (2020) Enzymatic processing of lignocellulosic biomass: principles, recent advances and perspectives. J Ind Microbiol Biotechnol 47(9):623–657. https://doi.org/10.1007/s10295-020-02301-8

Popp J, Lakner Z, Harangi-Rákos M, Fári M (2014) The effect of bioenergy expansion: food, energy, and environment. Renew Sust Energ Rev 32:559–578. https://doi.org/10.1016/j.rser.2014.01.056

Purkait MK, Haldar D (2021) Lignocellulosic biomass to value-added products: fundamental strategies and technological advancements. Elsevier

Quevedo RA, Blanca A, Escalera P, Aida V, Arias MR, Reynel HE, Moreno JC, Liliana P, Bonilla GA (2024) Application of waste biomass for the production of biofuels and catalysts: a review. In: Clean technologies and environmental policy (Issue 0123456789). Springer, Berlin Heidelberg. https://doi.org/10.1007/s10098-023-02728-4

Rahmati S, Doherty W, Dubal D, Atanda L, Moghaddam L, Sonar P, Hessel V, Ostrikov KK (2020) Pretreatment and fermentation of lignocellulosic biomass: reaction mechanisms and process engineering. React Chem Eng 5(11):2017–2047. https://doi.org/10.1039/D0RE00241K

Raj T, Chandrasekhar K, Kumar AN, Banu JR, Yoon JJ, Bhatia SK, Yang YH, Varjani S, Kim SH (2022) Recent advances in commercial biorefineries for lignocellulosic ethanol production: Current status, challenges and future perspectives. Bioresource technology, 344, p.126292. https://doi.org/10.1016/j.biortech.2021.126292

Rauch R, Hrbek J, Hofbauer H (2020) Biomass gasification for synthesis gas production and applications of the syngas. WIREs Energy Environ. https://doi.org/10.1002/wene.97

Ruan C, Akutsu R, Yang K, Zayan NM, Dou J, Liu J, Bose A, Brody L, Lamb HH, Li F (2023) Hydrogenation of bio-oil-derived oxygenates at ambient conditions via a two-step redox cycle. Cell Rep Phys Sci 4(7):101506. https://doi.org/10.1016/j.xcrp.2023.101506

Saini JK, Saini R, Tewari L (2015) Lignocellulosic agriculture wastes as biomass feedstocks for second-generation bioethanol production: concepts and recent developments. 3 Biotech 5:337–353

Saxena P, Singh NK, Harish, Singh AK, Pandey S, Thanki A, Yadav TC (2020) 5 - Recent advances in phytoremediation using genome engineering CRISPR–Cas9 technology. In: Pandey VC, Singh V (eds) Bioremediation of pollutants: from genetic engineering to genome engineering. Elsevier, pp 125–141. https://doi.org/10.1016/B978-0-12-819025-8.00005-3

Scheller HV, Ulvskov P (2010) Hemicelluloses. Annu Rev Plant Biol 61:263–289. https://doi.org/10.1146/annurev-arplant-042809-112315
Schmid JC, Benedikt F, Fuchs J, Mauerhofer AM, Hofbauer H (2021) Syngas for biorefineries from thermochemical gasification of lignocellulosic fuels and residues — 5 years' experience with an advanced dual fluidized bed gasifier design. Biomass Convers Biorefinery 11:2405–2442
Shahid MK, Batool A, Kashif A, Nawaz MH, Aslam M, Iqbal N, Choi Y (2021) Biofuels and biorefineries: development, application and future perspectives emphasizing the environmental and economic aspects. J Environ Manag 297:113268
Sharara MA, Clausen EC, Carrier DJ (2012) An overview of biorefinery technology. In: Biorefinery co-products: phytochemicals. primary metabolites and value-added biomass processing. Wiley
Singh A, Rodríguez Jasso RM, Gonzalez-Gloria KD, Rosales M, Belmares Cerda R, Aguilar CN, Singhania RR, Ruiz HA (2019) The enzyme biorefinery platform for advanced biofuels production. Bioresour Technol Rep 7(April):100257. https://doi.org/10.1016/j.biteb.2019.100257
Singh PK, Mohanty P, Mishra S, Adhya TK (2022) Food waste valorisation for biogas-based bioenergy production in circular bioeconomy: opportunities, challenges, and future developments. Front Energy Res 10(August). https://doi.org/10.3389/fenrg.2022.903775
Singhal A, Prashant. (2020) Biofuels: perspective for sustainable development and climate change mitigation. In: Kumar N (ed) Biotechnology for biofuels: a sustainable green energy solution. Springer, pp 1–22. https://doi.org/10.1007/978-981-15-3761-5_1
Solarte-Toro JC, González-Aguirre JA, Poveda Giraldo JA, Cardona Alzate CA (2021) Thermochemical processing of woody biomass: a review focused on energy-driven applications and catalytic upgrading. Renew Sust Energ Rev 136:110376. https://doi.org/10.1016/j.rser.2020.110376
Stoilova I, Krastanov A, Stanchev V (2010) Properties of crude laccase from Trametes versicolor produced by solid-substrate fermentation. Adv Biosci Biotechnol 01(03):208–215. https://doi.org/10.4236/abb.2010.13029
Sulaiman C, Abdul-Rahim AS, Ofozor CA (2020) Does wood biomass energy use reduce CO2 emissions in European Union member countries? Evidence from 27 members. J Clean Prod 253:119996. https://doi.org/10.1016/j.jclepro.2020.119996
Supramono D, Edgar J, Nasikin M (2018) Hydrogenation of non-polar fraction of bio-oil from co-pyrolysis of corn cobs and polypropylene for bio-diesel production. E3S Web Conf 02030:1–7
Tamošiūnas A, Gimžauskaitė D, Aikas M, Uscila R, Snapkauskienė V (2023) Biomass gasification to syngas in thermal water vapor arc discharge plasma. Biomass Convers Biorefinery 13(18):16373–16384. https://doi.org/10.1007/s13399-023-03828-3
Tsegaye B, Balomajumder C, Roy P (2019) Microbial delignification and hydrolysis of lignocellulosic biomass to enhance biofuel production: an overview and future prospect. Bull Natl Res Cent 43
Uzoejinwa BB, Nwoke OA, Asoiro FU, Ohagwu CJ (2023) Overview of biorefinery technology. In: Value-added products from algae: phycochemical production and applications. Springer, pp 157–190
Venderbosch R, Prins W (2010) Fast pyrolysis technology. Biofuel Bioprod Biorefineing 1:178–208. https://doi.org/10.1002/bbb.205
Vorwerk S, Somerville S, Somerville C (2004) The role of plant cell wall polysaccharide composition in disease resistance. Trends Plant Sci 9(4):203–209. https://doi.org/10.1016/j.tplants.2004.02.005
Weselake RJ, Chen G, Singer SD (2018) An integrated approach to plant bioproduct production. In: Chen G, Weselake RJ, Singer SD (eds) Plant bioproducts. Springer, New York. https://doi.org/10.1007/978-1-4939-8616-3
Woiciechowski AL, Bianchi A, Medeiros P, Rodrigues C, Porto L, Vandenberghe DS (2016) Feedstocks for biofuels. In: Green fuels technology. https://doi.org/10.1007/978-3-319-30205-8
Yogalakshmi KN, Mohamed Usman TM, Kavitha S, Sachdeva S, Thakur S, Adish Kumar S, Rajesh Banu J (2023) Lignocellulosic biorefinery technologies: a perception into recent advances in biomass fractionation, biorefineries, economic hurdles and market outlook. Fermentation 9(3):238. https://doi.org/10.3390/fermentation9030238

# Strain Engineering for Circular Bioeconomy

4

Karan Murjani, Dharmisha Solanki, Suresh Ramakrishna, and Vijai Singh

**Abstract**

As modernisation and urbanisation rise rapidly, they generate significant environmental challenges, including changes in the climate, lack of resources, and a buildup of untreated waste. In addition to this, there is a pressing need for economic models that focuses on sustainability and waste management. The circular bioeconomy can be a better solution to this global challenges which is increasing every day. Circular bioeconomy is based on two concepts: One is utilisation of renewable biological resources, and second is to minimise the waste by increasing the recycling methods and to improve the resource efficiency. Researchers are engaged in design of microorganisms that utilises waste as a form of substrate for production of valuable compounds. The utilisation of agro-industrial waste can mitigate environmental impacts, which promotes a sustainable source of valuable products. There are various products that can be produced by using lignocellulosic waste such as biofuels, biodiesel, proteins, amino acids, lactic acid, bioplastics, biopolymers and many. Currently, strain engineering of microorganisms can be the good strategy to produce these products. The microorganisms are known to be the cell factories for production of valuable compounds by using lignocellulosic waste as a substrate such as *Escherichia coli*, *Saccharomyces cerevisiae*, *Bacillus subtilis*, and many more. Another perspective of using these

K. Murjani · D. Solanki · V. Singh (✉)
Department of Biosciences, School of Science, Indrashil University, Rajpur, Mehsana-382715, Gujarat, India
e-mail: vijai.singh@indrashiluniversity.edu.in

S. Ramakrishna
College of Medicine, Hanyang University, Seoul, South Korea

Graduate School of Biomedical Science and Engineering, Hanyang University, Seoul, South Korea



V. Singh (ed.), *Sustainable Waste Management Towards Circular Bioeconomy*,
https://doi.org/10.1007/978-981-95-2227-9_4

microbial cell factories for degradation of plastic and polymers by modifying the microorganisms using genetic engineering methods. Thus, by using the lignocellulosic waste, researchers can contribute towards the greener economy by reducing the waste that is otherwise left untreated.

**Keywords**

Microbial cell factories · Polymers · Lignocellulosic waste · Plastics · Strain engineering · Yield

## 4.1 Introduction

The United Nations 2030 Agenda for Sustainable Development Goals (SDG) 12, Responsible Consumption and Production, draws attention to the need for sustainable systems that find an acceptable balance between what people require and how it affects the environment. The present world economy is approximately 8.6% circular; thus, over 90% of resources are ultimately wasted (Goal 12: Responsible Consumption and Production 2025). In parallel, over 650 million people live in severe poverty without access to basic needs, including foods, water, and housing. This noticeable imbalance highlights the crucial requirement to reduce the difference between wasteful consumption and excess production, which is a basic SDG 12 tenet. A change to a circular economy (CE) is crucial to achieve this balance. An important aims of the CE model is to limit waste generation, recycle materials, and ensure the constant flow of resources across the system. First proposed by economist Kenneth Boulding in 1966, this concept sought to build a "closed-loop" economy following natural ecosystems.

The circular bioeconomy is a mixture of a circular economy and bio-based production to encourage the long-term use of renewable biological resources. The bioeconomy focuses on turning resources and their waste streams into products that are worth more such as foods, feeds, bio-based materials, and bioenergy. The process not only minimises the dependence on the fossil fuels but also transforms waste into economic and ecological opportunities (Fig. 4.1) (Venkatesh 2022).

A circular bioeconomy has been proposed to boost economic growth in developed countries by combining a desirable "what" (circular economy) and desirable "how" (bioeconomy). A circular bioeconomy is regarded as a greater effective way to handle bio-based renewable resources by integrating principles of the circular economy into the bioeconomy. Any one of the five relationships listed below can also be used to summarise the debate surrounding the connection between the bioeconomy and the circular economy: (1) The bioeconomy and the circular economy are distinct but complementary; (2) both ideas are fully integrated; (3) both strategies are somewhat opposed; (4) the bioeconomy is a baseline for a circular economy; or (5) the circular economy is a means to shift from a fossil fuel-based economy to a bioeconomy (Tan and Lamers 2021).

The Intergovernmental Panel on Climate Change (IPCC) acknowledges that products derived from biomass replace petroleum-based carbon with natural carbon, which is considered carbon-neutral at the end of their life cycle. Biodegradable bioplastics, for instance, can turn almost 90% of organic carbon into $CO_2$ in 180 days, subsequently finalising a natural carbon loop used for biopower and biofuels. Thus, biomass also provides renewable energy that can lower environmental effects and fossil fuel reliance. The circular bioeconomy supplies a strong approach for sustainable, low-carbon, and regenerative development by aligning the life cycles of living things according to the fundamentals of a circular economy. By matching biological life cycles with circular economy ideas, the circular bioeconomy offers a strong model for sustainable, low-carbon, and regenerative development (Tan and Lamers 2021).

Biomass is biodegradable and functionally varied that is used for energy, nutrition, and materials, which makes its circular management of both exciting and difficult. Its perishability inhibits reuse and storage; poor disposal (e.g. landfilling) increases environmental issues, including methane emissions. Notwithstanding these difficulties, biomass often spans sectors including foods, water, sanitation, and energy, enabling various circular activities such as paper recycling, biogas generation, and nutrient recovery. Circular bioeconomy plans depend on this cross-sector integration (Schipfer et al. 2024).

In the past few years, waste valorisation has become a substitute to conventional waste disposal and landfilling. It encompasses any industrial procedure where waste is transformed into energy sources or useful products, as well as its recycling, composting, or reuse. Common methods of valorisation include turning leftovers, by-products, or undesirable materials into raw materials or incorporating them into the life cycle of another product. This method is in close alignment with the circular economy (CE) principles, which prioritise the closure of material loops, waste minimisation, and efficiency of the resources (Kanani et al. 2020). Among the many waste streams, municipal solid waste (MSW), which consists of packaging, paper, yard waste, food waste, wood, and plastics, has drawn the most focus. Usually, sewage sludge is handled separately. Generally, present waste management techniques fit into four groups: landfilling, recycling, thermal treatment, and biological treatment. Even with these alternatives, landfilling is still the most common practice worldwide, particularly in developing nations (Rajendran et al. 2021).

Just 15% of the waste is being picked up around the world and is recycled. The remaining unused materials are either dumped in landfills or went away in open spaces, which is very harmful for the environment. One of the biggest concerns is methane ($CH_4$) emissions from municipal solid waste (MSW) landfills, which rank among the top global $CH_4$ sources and contribute around 4–5% to annual global warming. This highlights the urgent need for sustainable waste treatment methods and integrating carbon footprint assessments into waste management policies (Hamdan et al. 2025; Lino et al. 2023). In India, the world's most populous country (17.5% of the global population), MSW management is a significant issue, of the ~160,000 tonnes of solid waste are generated every day, and about 95% is collected, but only 20–50% undergoes treatment. A significant portion remains untreated or is dumped without scientific safeguards, exacerbating environmental and health risks (Hamdan et al. 2025).

Despite significant advancements in synthetic biology, it is ability to consistently modify organisms for desired phenotypes which is still restricted. This is largely due to the complexity of biological systems and the absence of a universal method for successful strain engineering. Rather, to maximise strains for actual manufacturing conditions requires a multifaceted strategy that combines rational design, empirical testing, and the creation of high-dimensional data. To make bio-based products commercially, it feasibly depends on lowering the cost and time of strain development. Leveraging large-scale datasets allows researchers to predict and select high-performing strains, indicating a shift toward a more comprehensive, systems-level approach to strain engineering. Ultimately, this comprehensive approach holds the key to unlocking more reliable and scalable strain engineering for industrial applications (Abbate et al. 2023).

Widely used, the design–build–test–learn (DBTL) cycle design is a method of developing strain that is repeated till the desired level of performance is achieved. The design and build phases involve gene modifying in cells. The design phase usually shows the blueprint, and the build phase shows the instruments and procedures used to physically insert sequence variation. Test refers to phenotyping techniques and processes, while learn describes the computational tools utilised to analyse the data collected during testing. To cut down on development time and costs, every part of the DBTL cycle needs to work well together, which helps speed up and improve the process of enhancing strains for bioeconomy applications (Abbate et al. 2023).

The microbial engineering method modifies the genetic composition of the microbes. Due to peculiar qualities and ease of manipulation through recombinant DNA technology, microbial engineering has become a key model in biotechnology. For microbes to be effectively utilised in manufacturing processes, their native strains often require enhancement to lower production costs and to utilise inexpensive or complex raw materials (Sanghavi et al. 2020; Ali et al. 2020). While optimising fermentation dynamics can enhance production, the sustained success of industrial processes ultimately hinges on the improvement of the microbial strain itself. Strain improvement through genetic, evolutionary, or synthetic biology methods ensures consistent and scalable production of desired compounds. Strain engineering refers to the development of strains that exhibit the following characteristics such as (a) effective absorption of complex and inexpensive raw materials, (b) changing the product ratios and eliminating by-products, (c) excessive production and excretion of native and foreign goods, (d) easy scale-up and short fermentation time, (e) tolerance of several produced metabolites, and (f) morphological changes in cells appropriate for product separation (Sanghavi et al. 2020).

## 4.2 An Introduction to Strain Engineering

Strain engineering is a synthetic biology method that involves modification of genes in microorganisms or cells that improve the cell to produce a desired metabolite or to perform specific targeted function. This aims to modify strain performance, such as high yield of desired metabolite, substrate range, stress tolerance, or metabolic efficiency. Strain engineering facilitates the ability of bacteria to (1) non-traditional substrates or waste feedstocks such as lignocellulose hydrolysates; (2) produce valuable compounds such as biofuel and bioplastic; and (3) stress tolerance in low pH, high salinity, and the availability of harmful toxins (Mukhopadhyay 2015; Yao et al. 2022). The engineered strains not only enhance the sustainable production but also contribute to reduced environmental impacts. By optimising the microbial strains for different conditions, biomanufacturers can achieve more efficient and cost-effective production methods.

## 4.3 Approaches for Strain Engineering

There are different methods to engineer the desired microbial strain for production of desired product or desired character; these are as follows:

A. *Conventional methods:* These are older traditional methods that made the base of modern biotechnology. However, these methods are less precise and are still used in combination with modern techniques.

   (i) *Random mutagenesis:* Random mutagenesis is a laboratory technique that creates random changes in DNA to produce different mutant for studying how genes work or for developing proteins with better traits. Unlike targeted methods, it does not need a detailed knowledge of the structure of genes. In contrast to targeted approaches, it does not require prior understanding of the gene's structure. It includes several approaches, such as error-prone PCR, rolling circle amplification, chemical mutagens, physical mutagens, and insertional mutagenesis (Fujii et al. 2004).

   (ii) *Adaptive laboratory evolution (ALE):* ALE is a biotechnological procedure that promotes microbial evolution in regulated laboratory conditions to cultivate strains with improved characteristics for industrial, medicinal, or environmental purposes. Researchers induce selection pressures, such as substrate restrictions, toxin exposure, or temperature extremes, to direct microbial populations in accumulating advantageous mutations across generations, therefore improving desired phenotypes such as productivity, substrate usage, or stress tolerance (Sandberg et al. 2019; Wang et al. 2023).

B. *Modern methods:* With the progress made in genomics, bioinformatics, and molecular biology techniques, strain engineering becomes rapid, precise, and targeted.

(i) *Metabolic engineering:* It refers to the modification or redesign of the metabolic biosynthetic pathways that redirects the carbon flux towards desired metabolite which further increases its production. Modification of metabolic pathways can be achieved through overexpression or knockout of genes, addition of expression of heterologous pathways, and co-factor balancing.

(ii) *CRISPR-Cas based genome editing:* It is a precise molecular approach that facilitates targeted changes, including gene knockout, promoter editing, and gene deletion. It requires a unique guide RNA (gRNA) that directs the Cas nuclease to a desired DNA sequence by complementary base pairing (Murjani et al. 2024).

Thus, by using advanced random mutagenesis, and modern methods such as metabolic engineering and CRISPR-Cas systems, researchers can selectively target the gene of interest to engineer the desired bacterial strain for enhanced production of valuable products.

## 4.4 Engineering of Microbes for Waste Valorisation

Agriculture significantly influences the socioeconomic development of business in India. Over 50% of the population remains entirely dependent on agriculture for their primary source of income. Post-harvest, a substantial quantity of agricultural leftovers is left in the fields, frequently discarded through incineration, resulting in significant environmental risks. Harvesting crops produces a lot of agricultural waste generated, which usually gets burned, posing significant environmental risks. This remaining organic substance contains a considerable amount of carbon that is an affordable raw substance for producing alternative foods, feeds, and energy, as well as value-added compounds (Narnoliya et al. 2017). Consequently, in view of environmental concerns, the development of bio-sustainable and renewable resource technology for the handling of agro-industrial waste is crucial. Lignocellulose and starch are the predominant agro-industrial waste and renewable organic biomass on earth, generated annually worldwide across extensive regions from forests to oceans (da Silva 2016).

The bioeconomy depends on a number of products which use microbes that are capable of breaking down trees and shrubs to make biofuels and bioproducts (Asemoloye et al. 2023). Lignocellulosic waste can be digested by bacteria and subsequently released into the environment, ultimately integrating into the carbon cycle within the ecosystem, in which it is transformed to $CO_2$ and $H_2O$. Despite this, the natural breakdown of lignocellulosic waste is a prolonged process, requiring several years for the degradation of the existing quantity of lignocellulosic waste

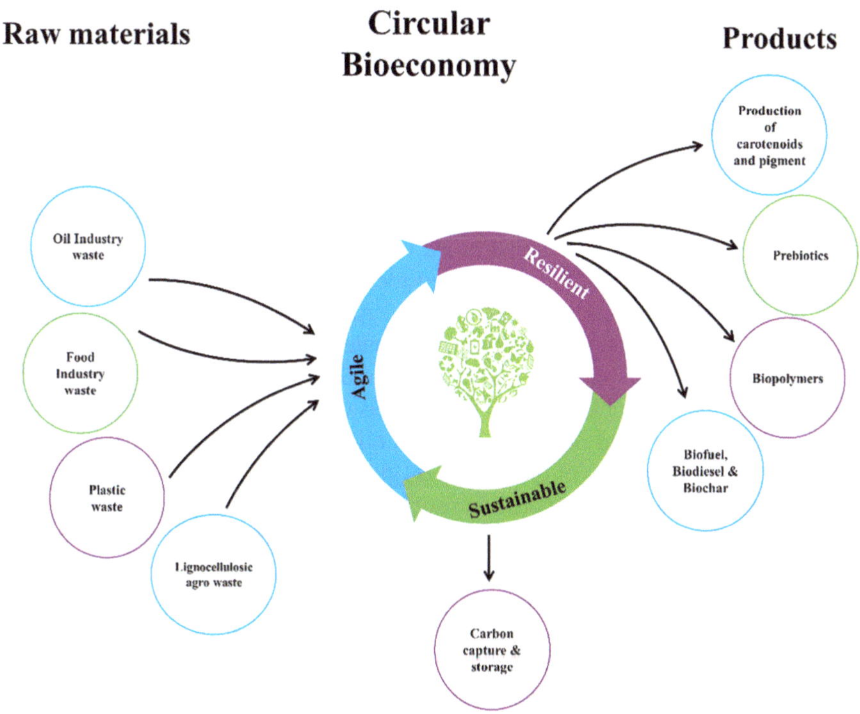


**Fig. 4.1** Schematic representation of circular bioeconomy approach: agro-industrial waste serves as an initial substrate for microbial growth, which leads to the production of important value-added products

(Narnoliya et al. 2017). Lignocellulosic biomass primarily comprises of three types of polymers, cellulose, hemicellulose, and lignin, which are complex linked and connected through covalent cross-linkages and non-covalent interactions (Singh et al. 2012). The two main enzymatic systems are used by microbial species to break down the lignocellulosic biomass that are hydrolytic and ligninolytic. While ligninolytic enzymes break down lignin and cleave phenolic rings, hydrolytic enzymes help to break down cellulose and hemicellulose (Dhillon and Kaur 2016) (Fig. 4.1).

In recent years, various technologies have been devised that deal with lignocellulosic biomass. Synthetic biology is used to make new metabolic pathways and species or to modify existing genetic circuits, natural biological systems, and operons. Microorganisms, particularly bacteria, are the principal organisms employed in engineered systems for the production of valuable chemicals from agro-industrial waste (Cameron et al. 2014; Chen 2014). There are many articles that have showed various potential uses of synthetic biology by using agro-industrial waste and converting it into end-products such as zero-calorie sugar, medicinal herbs, and prebiotics (Narnoliya et al. 2017).

### 4.4.1 Utilising Engineered Microbes for Sustainable Biofuel Production from Agro-Industrial Residues

Biofuels are significant products of lignocellulosic biomass (LCB) with substantial potential for tackling the worldwide energy generation crisis resulting from limited fossil fuel reserves (Reshmy et al. 2022). Both first-generation and second-generation food-based and non-food-based crops can serve as energy feedstocks for the synthesis of biofuels. The food-based feedstocks supply sugars, starches, and vegetable oils, while non-food-based feedstocks mostly consist of lignocellulosic biomass with elevated carbohydrate content. As the world progresses towards a carbon-free environment, biofuels may contribute to the reduction of carbon emissions in atmosphere. In addition to this, comparatively low concentrations of nitrogen, sulphur, and ash in LCB lead to the biofuels that is more safer and cleaner than the fossil based fuels and other conventional oils (Asemoloye et al. 2023). Due to the widespread availability of plant biomass globally, biofuels are derived from lignocellulosic biomass (LCBs) constitute sustainable energy sources.

A biofuel is good for the environment because it has a lot of hydrogen in it, which makes it less flammable (Feser and Gupta 2021). Nonetheless, there are many barriers in the process of conversion of lignocellulose into gasoline. A typical method for transforming lignocellulose into sugars, employed in biofuel production, by using chemical hydrolysis using concentrated acid. According to Zhou et al., this hydrolysis process offers various disadvantages, including elevated reaction temperatures and extensive energy consumption (Zhou et al. 2021).

*Saccharomyces cerevisiae* and *Zymomonas mobilis* are expected to serve as effective engineered hosts for biofuel production due to their higher ethanol yield and stability, in addition to this, the established genetic engineering techniques present for their manipulation. However, they are not advanced as a viable host due to the absence of the pentose metabolic pathway (Vinuselvi et al. 2011). The lack of capacity to employ pentose diminishes productivity and raises the costs associated with downstream processing required to extract unfermented sugars from the final product being produced (Nichols et al. 2001). The consumption of pentose may not possess an issue for *E. coli* and certain yeast species, including *Pichia stipitis*, as these organisms can naturally use both pentose and hexose carbohydrates (Vinuselvi et al. 2011).

Most types of yeast can only use a small range of sugars because they cannot break down pentose sugars. *S. cerevisiae* has been genetically modified to metabolise xylose by the expression of xylose reductase and xylose dehydrogenase genes derived from *P. stipitis*. Even though these mutants are better at using xylose, they generated a significant proportion of the by-product xylitol instead of ethanol, perhaps due to the lower activity of xylulokinase (XK). Overexpressing the endogenous XK made these mutants better at turning xylose into ethanol (Toivari et al. 2001). Nonetheless, the rate of growth of these mutants on xylose was insufficient due to various metabolic barriers, especially the reduced affinity of hexose transporters for xylose and reduced activity of pentose phosphate pathway enzymes (Karhumaa et al. 2005; Thanvanthri Gururajan et al. 2007).

Other disadvantage of *E. coli*, *S. cerevisiae*, and *Z. mobilis* are their significant preference for glucose metabolism as compared to other carbohydrates (Vinuselvi et al. 2011). Glucose inhibits the production of genes necessary for the catabolic system that facilitates the use of secondary metabolites (Görke and Stülke 2008). This occurrence is known as carbon catabolite repression (CCR). This results in an extended time of fermentation, which makes this approach noneconomic. It leads to other sugars such as xylose, arabinose, mannose, and galactose, which are generated from hemicelluloses, usually not more favourable substrates for biofuel production, hence reducing the productivity and yield (Jojima et al. 2010). Therefore, to enhance the output and yield of biofuels while reducing fermentation duration and costs, it is imperative to eradicate CCR. Numerous modifications of metabolism approaches have been investigated on behalf of eradication of carbon catabolite repression in *E. coli* and yeast but with minimal achievement (Vinuselvi et al. 2011).

Carbon catabolite repression extended beyond intrinsic metabolic pathways. The integrated heterologous metabolic pathway within the engineered organism exhibited catabolite inhibition (Van Zyl et al. 2007). The mechanisms of catabolite suppression in yeast have not been well investigated. Unlike bacteria, which utilises transcription factor interactions for catabolite suppression, yeast employs greater complexity in signal-transduction pathways to regulate the promoters of catabolite genes (Vinuselvi et al. 2011). For example, the usage of xylose necessitates a minimal amount of glucose. Excess glucose or the absence of glucose entirely suppresses the xylose consumption. A constitutive derepressed yeast strain lacking the MIG1 gene (multicopy inhibitor of GAL gene expression) metabolises xylose more efficiently in high-glucose conditions (Roca et al. 2004). The integration of the fungal xylose biosynthetic pathway with the bacterial arabinose metabolic pathway in *S. cerevisiae* facilitated concurrent utilisation of arabinose and xylose. In these mutant strains, all ingested xylose was fermented to ethanol; however, arabinose was transformed to arabitol instead of ethanol, due to the minimal selectivity of L-arabinose reductase (Karhumaa et al. 2006) (Table 4.1).

**Table 4.1** Illustrates the genes that are engineered for production of ethanol in *Saccharomyces cerevisiae* by using lignocellulosic waste

| Engineered gene | Gene product | Function of gene | References |
|---|---|---|---|
| *PsXR* | Xylose reductase | Catalyses the transformation of xylose into xylitol | Toivari et al. (2001) |
| *XDH* | Xylose dehydrogenase | Catalyses the conversion of xylitol to xylose | Toivari et al. (2001) |
| *XKS1* | Xylulokinase | Catalyses the transformation of xylulose to xylulose-5-phosphate | Toivari et al. (2001) |
| *MIG1* | C2H2 zinc finger protein | Regulates the gene expression involved in glucose metabolism | Roca et al. (2004) |
| *araA*, *araB*, *araC*, and *araD* | Arabinose metabolic pathway | *araA* transforms L-arabinose into L-ribulose, *araB* converts L-ribulose into L-ribulose-5-phosphate, *araD* converts L-ribulose-5-phosphate into D-xylulose-5-phosphate, and *araC* regulates the expression of *araBAD* | Karhumaa et al. (2006) |

Medium- and long-chain fatty acids, along with their ethyl or methyl derivatives, serve as essential intermediate metabolites in the biosynthesis of diesel ($C_9$–$C_{23}$ hydrocarbons) and jet fuel ($C_8$–$C_{16}$ hydrocarbons) (Peralta-Yahya and Keasling 2010). Numerous microorganisms, including bacteria, fungus, yeast, and algae, possess a capacity to accumulate lipids within growing circumstances. Agro-industrial waste serves as an excellent substrate for cultivating various microorganisms due to its diverse composition and high concentration of carbon. To enhance the efficiency of the microbes, strain engineering can be a good strategy. Certain yeast species possess increased oil content relative to algae or bacteria, and their rapid development rate facilitates their utilisation in biofuels production. Certain lipid-accumulating yeasts strains, such as *Rhodosporidium* sp., *Lipomyces* sp., *Cryptococcus* sp., and *Rhodotorula* sp., acquire substantial lipid levels, particularly when glucose serves as the carbon source (Hu et al. 2009).

Synthesis of biodiesel from microorganisms has mainly focused on the in vitro transesterification of triacylglycerols derived from different oleaginous microbes (Table 4.2). Earlier, Kalscheuer et al. described the synthesis of fatty acid ethyl esters via in vivo transesterification processes in terms of metabolically engineered *Escherichia coli* using the simultaneous expression of the ethanol production pathway derived from *Z. mobilis* and the wide range of substrate specificity acyltransferase from *Acinetobacter baylyi* ADP1 strain (Kalscheuer et al. 2006). Recently, the modifiing of *E. coli* was to synthesise fatty acid ethyl esters, fatty alcohols, and waxes using just basic sugars (Sorokina et al. 2009). The majority of fatty acids that are generated in microorganisms are employed in the

**Table 4.2** Illustrates the genes that are used to engineer the microbes for sustainable biofuels production from agro-industrial residues

| Engineered gene | Gene product | Function of gene | References |
|---|---|---|---|
| *pdc* | Pyruvate decarboxylase | Catalyses the decarboxylation of pyruvate to acetaldehyde | Kalscheuer et al. (2006) |
| *adhB* | Alcohol dehydrogenase | Catalyses the transformation of acetaldehyde to ethanol | Kalscheuer et al. (2006) |
| *fadD* | Acyl-CoA synthetase | Converts fatty acid into fatty-acyl CoA | Lennen et al. (2010) and Lu et al. (2008) |
| *fadE* | Acyl-CoA dehydrogenase | Converts fatty-acyl CoA into trans-enoyl-CoA | Lennen et al. (2010) and Lu et al. (2008) |
| *fadB* | Enoyl-CoA hydratase and 3-hydroxyacyl-CoA dehydrogenase | Enoyl-CoA hydratase converts trans-enoyl-CoA into L-3-hydroxyacyl-CoA, and 3-hydroxyacyl-CoA dehydrogenase converts L-3-hydroxyacyl-CoA into 3-ketoacyl-CoA | Lennen et al. (2010) and Lu et al. (2008) |
| *fadA* | 3-Ketoacyl-CoA thiolase | Transforms 3-ketoacyl-CoA into acetyl-CoA and shorter fatty acyl-CoA | Lennen et al. (2010) and Lu et al. (2008) |

production of phospholipids and triglycerides, while a certain amount undergoes degradation via the β-oxidation pathway. The generation of free fatty acids has been increased through the identification of FadD, the removal of β-oxidation, and the high level expression of host-derived acetyl-CoA carboxylase or by enhancing the functional activity of malonyl-CoA synthase in *E. coli* (Lennen et al. 2010; Lu et al. 2008) (Table 4.2).

## 4.5 Engineering of Microbes for Degradation of Polymers

Polymeric materials play an important role in our daily lives. Completely eliminating them from our livelihoods is thus impossible. However, strategies can be put into action to reduce the production of synthetic polymers. There are two main approaches to implement these strategies: (1) by changing the structure of the polymer and (2) by increasing the biodegradation of polymers. These goals facilitate the substitution of materials with extended lifespans, such as polyethylene (PE), polyethylene terephthalate (PET), polyurethane (PUR), and polystyrene (PS) with advanced biodegradable polymers (Mubayi et al. 2024). While designing, these polymers should be based on naturally occurring polymers with established degradation mechanisms or by changing the structure of polymer backbone to make it easier for microbes to break it down and use it for energy. Furthermore, there is an important need to enhance existing recycling methods through the advancement of chemical catalysis (not addressed in this research) or through the recognition and engineering of novel microbes and microbial consortia capable of polymer metabolism (Yu et al. 2019; Peng et al. 2021; Yang et al. 2014; Lankiewicz et al. 2023; Yoshida et al. 2016; Sulaiman et al. 2012). These engineered microbes and their product as enzymes that can degrade polymers more effectively (Mubayi et al. 2024).

Researchers were identified microorganisms that break down plastics, while advancements in genetic modification technologies facilitate their improvement. *Ideonella sakaiensis* was engineered through different methods to produce important enzymes such as PETase and MHETase that improve the polymer degradation efficiency (Mubayi et al. 2024). The crystal structure of PET has revealed some drawbacks which were rectified through different variants. One drawback was addressed in the central β-sheet that was broken up by an irregularity in one of the nearby β-strands in IsPETase. The disruption inhibited the creation of hydrogen bonds between proline and leucine (Son et al. 2019). The IsPETase$^{P181A}$ mutant solved this issue and elevated its melting temperature; however, there was insignificant advantage as relative to the wild-type PETase. Subsequent cycles of rational design, utilising the 3-D conformation of the wild-type enzyme and its variations, resulted in IsPETase$^{S121E/D186H/R280A}$, which demonstrated markedly enhanced thermal stability and degrading activity in contrast to the wild-type PETase (Son et al. 2019).

Most genetic engineering strategies discussed focus on increasing the biodegradation capabilities of microorganisms that are already efficient in plastic degradation. An alternative approach involves the expression of these enzymes in model microorganisms such as *E. coli* and *Bacillus subtilis*. *E. coli* produced PETases that used sec-dependent signal peptides that converts the PET into terephthalic acid and ethylene glycol (Seo et al. 2019; Shi et al. 2021). Random mutagenesis of PelB, a signal peptide that transports proteins to the periplasm, resulted in a 1.7 times more secretion efficiency (Shi et al. 2021). Extracellular PETase secretion was achieved using *B. subtilis*, *Phaeodactylum tricornutum*, and *Chlamydomonas reinhardtii*, while *Pichia pastoris* GS115 and *Vibrio natriegens* were transformed as biocatalysts that are whole cells by expressing PETases on their cell surfaces (Mubayi et al. 2024). *E. coli* was genetically modified to convert PET into vanillin (Sadler and Wallace 2021) and adipic acid (Valenzuela-Ortega et al. 2023). Among them, one is utilised in food and cosmetics, and the second one serves as a precursor for nylon. The host strain, *E. coli* MG1655 RARE (reduced aromatic aldehyde reduction), was introduced with plasmids containing enzymes terephthalate 1,2-dioxygenase (TPADO), dihydroxy-3,5-cyclohexadiene-1,4-dicarboxylic acid dehydrogenase (DCDDH), carboxylic acid reductase (CAR), and catechol O-methyltransferase (COMT) to convert terephthalic acid into vanillin with 79% yield (Sadler and Wallace 2021). Similar methodologies were employed to modify *E. coli* that possesses the genes needed to transform PET into adipic acid (Valenzuela-Ortega et al. 2023).

Engineering microbes for production of enzymes is a time-consuming process due to these researchers who are taking time to improve polymer degrading enzymes. Additionally, using artificial intelligence and machine learning approaches can significantly improve the product yield.

## 4.6 Economic Aspects with Techno-economic Analysis (TEA) and Life Cycle Assessment (LCA)

To address challenges during scale-up, systems biology is used to understand how microbes respond to fluctuating bioreactor conditions such as nutrient limitations and stress. Synthetic biology complements this approach by enabling precise control over cellular responses and improving strain robustness. Despite progress, maintaining strain performance at industrial scales remains difficult. Scale-down models help test strains under realistic conditions, improving selection and scalability. Integrating TEA, LCA, and systems-level tools supports the development of strains that are both high-performing and commercially feasible (Wehrs et al. 2019).

Sustainable process design depends on each of techno-economic analysis (TEA), life cycle assessment (LCA), and environmental regulatory systems (Mahmud et al. 2021). TEA assesses the technical viability and economic efficacy of technologies. On the other hand, LCA evaluates the environmental effects linked to every stage of a product's life cycle, from the extraction of raw materials (cradle) to the final disposal of the product (grave). Essential LCA measures are carbon footprint, water use, energy consumption, land use, and pollution emissions, while TEA examines important metrics including capital expenditures (CAPEX), operational expenditures (OPEX), market potential, return on investment (ROI), and cost-benefit ratios (Ferdous et al. 2023). A cradle-to-gate LCA, for instance, assesses effects from raw material extraction to the point where the product departs the manufacturing plant (Cao 2017). In contrast to TEA studies, which usually use a gate-to-gate boundary and emphasise only processing and extraction phases. Both LCA and TEA play different yet complementary functions: while LCA shows environmental trade-offs, TEA emphasises economic viability (Ferdous et al. 2023).

## 4.7 Regulation, Biosafety, and Global Perspectives on Genetically Engineered Organisms

Genetic engineering involves modifying DNA of an organism in the laboratory by adding, removing, or altering base pairs to introduce new traits. This can include inactivating virulence genes or inserting recombinant genes for desired functions (Mahdizade Ari et al. 2024). While GMOs include modified plants, animals, and microbes, genetically modified microorganisms (GEMs) refer specifically to bacteria, fungi, and yeasts altered using molecular biology techniques to gain new phenotypes or produce useful products (Hassan and Saleh 2023). Different countries have developed diverse regulatory frameworks for genetically engineered (GE) plants and foods, shaped by scientific, cultural, environmental, and political factors. Although the core elements of scientific risk assessment are largely similar worldwide, policy decisions often reflect differing societal values, levels of public acceptance, and risk tolerances. Some policy decisions differ, and some countries take a permissive stance, while others follow a precautionary approach. Many use a process-based regulatory model, requiring pre-market review for GE products but not for similar traits developed through traditional breeding. Some systems also address broader issues like consumer labelling, economic impacts, and protection of non-GE farming (Hassan and Saleh 2023).

The primary international forum addressing the maintenance of synthetic biology is the convention on biological diversity (CBD) and its related agreements, the Cartagena Protocol on Biosafety, which governs the safe handling of living engineered organisms (LMOs), and the Nagoya Protocol, which deals with access and benefit distribution of genetic resources. Discussions on synthetic biology have been ongoing under the CBD for nearly a decade, covering emerging issues including digital sequence information, risk assessment, and biosafety (Keiper and Atanassova 2020).

In response to the rise of GMOs since the late 1980s, many countries (e.g. Australia, Brazil, China, EU, India, and South Africa) created or changed biotech regulatory systems to manage risks to health and the environment (Nap et al. 2003). The Cartagena Protocol serves as a global framework for the regulation of LMOs/GMOs. While synthetic biology is presently regulated under existing GMO frameworks at the international, regional, and national levels, regulatory bodies agree that these systems are largely sufficient but it should be regularly reviewed as the technology evolves (Dederer and Hamburger 2019).

Trade, Labelling, and Societal Views: The Agreement on Technical Barriers to Trade (TBT Agreement) is also important, particularly with regard to the labelling and packaging of products that contain genetically modified organisms. It is permissible for technical regulations to be enacted to safeguard human health or the environment; however, these regulations must not be more restrictive on trade than is necessary and it must apply in the same manner to both domestic and imported goods (Komen 2012). Additionally, researchers are looking into the idea of using plants (like fruits and vegetables) to create oral vaccines that people can eat, which could help stop the spreading of disease in poorer countries and lower the costs of large vaccination efforts. Researchers are presently working on creating plant-based vaccines for hepatitis B virus (HBV), enterotoxigenic *Escherichia coli* (ETEC), and Norwalk virus in potatoes and lettuce. In addition to food and agricultural applications, GMOs are now being used to produce medically important substances such as edible vaccines and therapeutic proteins. Researchers have even employed genetically modified animals to cultivate transplant tissues and human organs, a practice known as xeno-transplantation. The diverse applications of GMOs offer numerous advantages to humanity; however, many individuals express concerns regarding potential risks (Phillips 2008).

Genetic engineering enables plants to synthesise pharmacologically active proteins, such as mammalian antibodies, blood product substitutes, vaccines, hormones, cytokines, and various other therapeutic agents (Goldstein and Thomas 2004). The community has articulated two distinct perspectives: On the one hand, recombinant DNA technology is regarded as a powerful instrument for improving crop yield (first-generation GMOs) and the quality of food (second-generation GMOs), as well as serving as "drug factories" for the manufacture of vaccines and therapeutic medicines (third-generation GMOs). Supporters of GMOs claim that they are vital for advancing sustainable agriculture, as they can diminish the environmental impact of farming by lowering pesticide usage, conserving fossil fuels, reducing CO2 emissions, and preserving soil and moisture (Buiatti et al. 2013).

## 4.8 Conclusion and Future Perspective

The alteration from circular economy to a circular bioeconomy signifies a crucial role on how societies produce, consume, and manage the biological resources which also represent the management of wastes that are generated from residential and commercial areas. Through the utilization of metabolic modification, synthetic biology, systems biology, and microbial strain modification, these methods have emerged as beneficial for valorising agro-industrial and lignocellulosic wastes into high-value bioproducts. These methods are not only improved the efficiency but also the cost-effectiveness which can reduce the dependency on fossil fuel.

Microbial cell factories such as *Escherichia coli*, *Saccharomyces cerevisiae*, and *Corynebacterium glutamicum* have been successfully engineered to convert heterogeneous expression systems into biofuels, bioplastics, platform chemicals, and nutraceuticals. By using the techniques such as CRISPR-Cas systems, biosensors have improved the precision and scalability of process. Furthermore, the effectiveness of utilisation of the biomass waste can be increased by using bioprocess and co-culture techniques which offers promising results for high yield production.

In spite of the fact that there have been numerous advancements, challenges persist in the feedstock heterogeneity, process economics, product toxicity, and regulatory bottlenecks. Forthcoming research goals must be focussed on robust microbial strains and improvement of these strains by increasing the stress tolerance efficiency and by using the genome-scale metabolic models (GSMM) to increase the metabolic flux towards targeted product.

In conclusion, the manufacture of valuable compounds such as biofuels, biodiesel, bioplastics, and enzyme degrading plastics by using lignocellulosic waste as a substrate by using various microbes is an economical and sustainable process to produce these products. Using genetic engineering techniques to improve the production of these compounds can be cost-effective and it can improve the efficiency of these products. Thus, following the regulatory guidelines of national or international forums for utilising engineered microbes can improve the efficacy and safety of the product.

**Acknowledgement** K.M. acknowledges financial assistance from Indrashil University, Rajpur, Mehsana. Financial assistance from the National Research Foundation (NRF) (2021M3A9H3015390, RS-2023-00279214 and RS-2024-00341469) to S.R. is acknowledged.

## References

Abbate E, Andrion J, Apel A, Biggs M, Chaves J, Cheung K, Ciesla A, Clark-ElSayed A, Clay M, Contridas R (2023) Optimizing the strain engineering process for industrial-scale production of bio-based molecules. J Ind Microbiol Biotechnol 50(1):kuad025

Ali R, Zulaykha KD, Sajjad N (2020) Genetically modified microbes as biofertilizers. In: Bioremediation and biotechnology, Techniques for noxious substances remediation, vol 4, pp 275–293

Asemoloye MD, Bello TS, Oladoye PO, Remilekun Gbadamosi M, Babarinde SO, Ebenezer Adebami G, Olowe OM, Temporiti MEE, Wanek W, Marchisio MA (2023) Engineered yeasts and lignocellulosic biomaterials: shaping a new dimension for biorefinery and global bioeconomy. Bioengineered 14(1):2269328

Buiatti M, Christou P, Pastore G (2013) The application of GMOs in agriculture and in food production for a better nutrition: two different scientific points of view. Genes Nutr 8:255–270

Cameron DE, Bashor CJ, Collins JJ (2014) A brief history of synthetic biology. Nat Rev Microbiol 12(5):381–390

Cao C (2017) Sustainability and life assessment of high strength natural fibre composites in construction. In: Advanced high strength natural fibre composites in construction. Elsevier, pp 529–544

Chen H (2014) Biotechnology of lignocellulose. In: Theory and practice. Chemical Industry Press and Springer, China

da Silva LL (2016) Adding value to agro-industrial wastes. Ind Chem 2(2)

Dederer H-G, Hamburger D (2019) Regulation of genome editing in plant biotechnology. Springer, Cham

Dhillon GS, Kaur S (2016) Agro-industrial wastes as feedstock for enzyme production: apply and exploit the emerging and valuable use options of waste biomass. Academic Press

Ferdous J, Bensebaa F, Pelletier N (2023) Integration of LCA, TEA, process simulation and optimization: a systematic review of current practices and scope to propose a framework for pulse processing pathways. J Clean Prod 402:136804
Feser J, Gupta A (2021) Performance and emissions of drop-in aviation biofuels in a lab-scale gas turbine combustor. J Energy Resour Technol 143(4):042103
Fujii R, Kitaoka M, Hayashi K (2004) One-step random mutagenesis by error-prone rolling circle amplification. Nucleic Acids Res 32(19):e145–e145
Goal 12: Responsible Consumption and Production (2025) UN environment programme. Retrieved 01/05 from https://www.unep.org/topics/sustainable-development-goals/why-do-sustainable-development-goals-matter/goal-12
Goldstein D, Thomas J (2004) Biopharmaceuticals derived from genetically modified plants. QJM 97(11):705–716
Görke B, Stülke J (2008) Carbon catabolite repression in bacteria: many ways to make the most out of nutrients. Nat Rev Microbiol 6(8):613–624
Hamdan A, Panda S, Jain MS, Raj V, Mathew S (2025) Assessing municipal solid waste in Indian smart cities: a path towards waste-to-energy. Heliyon 11(6):e42770
Hassan AI, Saleh HM (2023) Production of amino acids and nucleic acids from genetically engineered microbial cells and their relevance to biodegradation. Green Energy Environ Technol 2:1–47
Hu C, Zhao X, Zhao J, Wu S, Zhao ZK (2009) Effects of biomass hydrolysis by-products on oleaginous yeast Rhodosporidium toruloides. Bioresour Technol 100(20):4843–4847
Jojima T, Omumasaba CA, Inui M, Yukawa H (2010) Sugar transporters in efficient utilization of mixed sugar substrates: current knowledge and outlook. Appl Microbiol Biotechnol 85:471–480
Kalscheuer R, Stolting T, Steinbuchel A (2006) Microdiesel: Escherichia coli engineered for fuel production. Microbiology 152(9):2529–2536
Kanani F, Heidari MD, Gilroyed BH, Pelletier N (2020) Waste valorization technology options for the egg and broiler industries: a review and recommendations. J Clean Prod 262:121129
Karhumaa K, Hahn-Hägerdal B, Gorwa-Grauslund MF (2005) Investigation of limiting metabolic steps in the utilization of xylose by recombinant Saccharomyces cerevisiae using metabolic engineering. Yeast 22(5):359–368
Karhumaa K, Wiedemann B, Hahn-Hägerdal B, Boles E, Gorwa-Grauslund M-F (2006) Co-utilization of L-arabinose and D-xylose by laboratory and industrial Saccharomyces cerevisiae strains. Microb Cell Factories 5:1–11
Keiper F, Atanassova A (2020) Regulation of synthetic biology: developments under the convention on biological diversity and its protocols. Front Bioeng Biotechnol 8:310
Komen J (2012) The emerging international regulatory framework for biotechnology. GM Crops Food 3(1):78–84
Lankiewicz TS, Choudhary H, Gao Y, Amer B, Lillington SP, Leggieri PA, Brown JL, Swift CL, Lipzen A, Na H (2023) Lignin deconstruction by anaerobic fungi. Nat Microbiol 8(4):596–610
Lennen RM, Braden DJ, West RM, Dumesic JA, Pfleger BF (2010) A process for microbial hydrocarbon synthesis: overproduction of fatty acids in Escherichia coli and catalytic conversion to alkanes. Biotechnol Bioeng 106(2):193–202
Lino FA, Ismail KA, Castañeda-Ayarza JA (2023) Municipal solid waste treatment in Brazil: a comprehensive review. Energy Nexus 11:100232
Lu X, Vora H, Khosla C (2008) Overproduction of free fatty acids in E. coli: implications for biodiesel production. Metab Eng 10(6):333–339
Mahdizade Ari M, Dadgar L, Elahi Z, Ghanavati R, Taheri B (2024) Genetically engineered microorganisms and their impact on human health. Int J Clin Pract 2024(1):6638269
Mahmud R, Moni SM, High K, Carbajales-Dale M (2021) Integration of techno-economic analysis and life cycle assessment for sustainable process design–a review. J Clean Prod 317:128247
Mubayi V, Ahern CB, Calusinska M, O'Malley MA (2024) Toward a circular bioeconomy: designing microbes and polymers for biodegradation. ACS Synth Biol 13(7):1978–1993

Mukhopadhyay A (2015) Tolerance engineering in bacteria for the production of advanced biofuels and chemicals. Trends Microbiol 23(8):498–508
Murjani K, Tripathi R, Singh V (2024) An overview and potential of CRISPR-Cas systems for genome editing. Prog Mol Biol Transl Sci 208:1–17
Nap JP, Metz PL, Escaler M, Conner AJ (2003) The release of genetically modified crops into the environment: part I. Overview of current status and regulations. Plant J 33(1):1–18
Narnoliya LK, Jadaun JS, Singh SP (2017) Management of agro-industrial wastes with the aid of synthetic biology. In: Biosynthetic technology and environmental challenges. Springer, pp 11–28
Nichols N, Dien B, Bothast R (2001) Use of catabolite repression mutants for fermentation of sugar mixtures to ethanol. Appl Microbiol Biotechnol 56:120–125
Peng X, Wilken SE, Lankiewicz TS, Gilmore SP, Brown JL, Henske JK, Swift CL, Salamov A, Barry K, Grigoriev IV (2021) Genomic and functional analyses of fungal and bacterial consortia that enable lignocellulose breakdown in goat gut microbiomes. Nat Microbiol 6(4):499–511
Peralta-Yahya PP, Keasling JD (2010) Advanced biofuel production in microbes. Biotechnol J 5(2):147–162
Phillips T (2008) Genetically modified organisms (GMOs): transgenic crops and recombinant DNA technology. Nat Educ 1(1):213
Rajendran N, Gurunathan B, Han J, Krishna S, Ananth A, Venugopal K, Priyanka RS (2021) Recent advances in valorization of organic municipal waste into energy using biorefinery approach, environment and economic analysis. Bioresour Technol 337:125498
Reshmy R, Philip E, Madhavan A, Sirohi R, Pugazhendhi A, Binod P, Awasthi MK, Vivek N, Kumar V, Sindhu R (2022) Lignocellulose in future biorefineries: strategies for cost-effective production of biomaterials and bioenergy. Bioresour Technol 344:126241
Roca C, Haack MB, Olsson L (2004) Engineering of carbon catabolite repression in recombinant xylose fermenting Saccharomyces cerevisiae. Appl Microbiol Biotechnol 63:578–583
Sadler JC, Wallace S (2021) Microbial synthesis of vanillin from waste poly (ethylene terephthalate). Green Chem 23(13):4665–4672
Sandberg TE, Salazar MJ, Weng LL, Palsson BO, Feist AM (2019) The emergence of adaptive laboratory evolution as an efficient tool for biological discovery and industrial biotechnology. Metab Eng 56:1–16
Sanghavi G, Gupta P, Rajput M, Oza T, Trivedi U, Singh NK (2020) Microbial strain engineering. In: Engineering of microbial biosynthetic pathways. Springer, pp 11–32
Schipfer F, Burli P, Fritsche U, Hennig C, Stricker F, Wirth M, Proskurina S, Serna-Loaiza S (2024) The circular bioeconomy: a driver for system integration. Energy Sustain Soc 14(1):34
Seo H, Kim S, Son HF, Sagong H-Y, Joo S, Kim K-J (2019) Production of extracellular PETase from Ideonella sakaiensis using sec-dependent signal peptides in E. coli. Biochem Biophys Res Commun 508(1):250–255
Shi L, Liu H, Gao S, Weng Y, Zhu L (2021) Enhanced extracellular production of is PETase in Escherichia coli via engineering of the pelB signal peptide. J Agric Food Chem 69(7):2245–2252
Singh R, Kapoor V, Kumar V (2012) Utilization of agro-industrial wastes for the simultaneous production of amylase and xylanase by thermophilic actinomycetes. Braz J Microbiol 43:1545–1552
Son HF, Cho IJ, Joo S, Seo H, Sagong H-Y, Choi SY, Lee SY, Kim K-J (2019) Rational protein engineering of thermo-stable PETase from Ideonella sakaiensis for highly efficient PET degradation. ACS Catal 9(4):3519–3526
Sorokina O, Kapus A, Terecskei K, Dixon LE, Kozma-Bognar L, Nagy F, Millar AJ (2009) A switchable light-input, light-output system modelled and constructed in yeast. J Biol Eng 3:1–16
Sulaiman S, Yamato S, Kanaya E, Kim J-J, Koga Y, Takano K, Kanaya S (2012) Isolation of a novel cutinase homolog with polyethylene terephthalate-degrading activity from leaf-branch compost by using a metagenomic approach. Appl Environ Microbiol 78(5):1556–1562
Tan EC, Lamers P (2021) Circular bioeconomy concepts—a perspective. Front Sustain 2:701509

Thanvanthri Gururajan V, Gorwa-Grauslund M-F, Hahn-Hägerdal B, Pretorius IS, Cordero Otero RR (2007) A constitutive catabolite repression mutant of a recombinant Saccharomyces cerevisiae strain improves xylose consumption during fermentation. Ann Microbiol 57:85–92

Toivari MH, Aristidou A, Ruohonen L, Penttilä M (2001) Conversion of xylose to ethanol by recombinant Saccharomyces cerevisiae: importance of xylulokinase (XKS1) and oxygen availability. Metab Eng 3(3):236–249

Valenzuela-Ortega M, Suitor JT, White MF, Hinchcliffe T, Wallace S (2023) Microbial upcycling of waste PET to adipic acid. ACS Central Sci 9(11):2057–2063

Van Zyl WH, Lynd LR, den Haan R, McBride JE (2007) Consolidated bioprocessing for bioethanol production using Saccharomyces cerevisiae. Biofuels:205–235

Venkatesh G (2022) Circular bio-economy—paradigm for the future: systematic review of scientific journal publications from 2015 to 2021. Circ Econ Sustain 2(1):231–279

Vinuselvi P, Park JM, Lee JM, Oh K, Ghim C-M, Lee SK (2011) Engineering microorganisms for biofuel production. Biofuels 2(2):153–166

Wang G, Li Q, Zhang Z, Yin X, Wang B, Yang X (2023) Recent progress in adaptive laboratory evolution of industrial microorganisms. J Ind Microbiol Biotechnol 50(1):kuac023

Wehrs M, Tanjore D, Eng T, Lievense J, Pray TR, Mukhopadhyay A (2019) Engineering robust production microbes for large-scale cultivation. Trends Microbiol 27(6):524–537

Yang J, Yang Y, Wu W-M, Zhao J, Jiang L (2014) Evidence of polyethylene biodegradation by bacterial strains from the guts of plastic-eating waxworms. Environ Sci Technol 48(23):13776–13784

Yao X, Liu P, Chen B, Wang X, Tao F, Lin Z, Yang X (2022) Synthetic acid stress-tolerance modules improve growth robustness and lysine productivity of industrial Escherichia coli in fermentation at low pH. Microb Cell Factories 21(1):68

Yoshida S, Hiraga K, Takehana T, Taniguchi I, Yamaji H, Maeda Y, Toyohara K, Miyamoto K, Kimura Y, Oda K (2016) A bacterium that degrades and assimilates poly (ethylene terephthalate). Science 351(6278):1196–1199

Yu K, Yi S, Li B, Guo F, Peng X, Wang Z, Wu Y, Alvarez-Cohen L, Zhang T (2019) An integrated meta-omics approach reveals substrates involved in synergistic interactions in a bisphenol A (BPA)-degrading microbial community. Microbiome 7:1–13

Zhou Z, Liu D, Zhao X (2021) Conversion of lignocellulose to biofuels and chemicals via sugar platform: an updated review on chemistry and mechanisms of acid hydrolysis of lignocellulose. Renew Sust Energ Rev 146:111169

# Circular Design and Biomanufacturing

5

Kamini Pandey, Madhu Kumari, Alok Prapati, Sonali,
Tanushri Dixit, and Barkha Singhal

**Abstract**

Sustainability is a much-needed approach in the present scenario. The rising population leads to an exhaustive demand for resources and infrastructure. Therefore, integrating circular design principles with biomanufacturing processes provides a transformative approach to sustainable production within the circular bioeconomy framework. The technological advancements in biomanufacturing, such as fermentation, metabolic engineering, synthetic biology, and waste valorisation, enable the conversion of waste biomasses into valuable products of regular and commercial use. The other physicochemical approaches, including hydrothermal liquefaction (HTL), gasification, and anaerobic digestion, add dynamic synergy in this conversion process. This chapter delves into the integration of various strategies, such as design for disassembly (DfD), upcycling, and sustainable material selection, along with life cycle assessment (LCA), that can minimise the environmental impact in the life cycles of the product. The closed-loop production models encourage efficiency and sustainability by using waste products from one process as inputs for another. The case studies mentioned in the chapter highlight the practical applications and advantages of integrating circular design with biomanufacturing. The chapter also explores the technological and economic limitations, along with policy hurdles. To overcome these obstacles, cooperation, legislative backing, and ongoing innovation are needed. The chapter ends by highlighting how this integrated approach can promote a resilient bioeconomy and sustainable industrial practices.

K. Pandey · M. Kumari · A. Prapati · Sonali · T. Dixit · B. Singhal (✉)
School of Biotechnology, Gautam Buddha University, Greater Noida, UP, India
e-mail: barkha@gbu.ac.in



V. Singh (ed.), *Sustainable Waste Management Towards Circular Bioeconomy*,
https://doi.org/10.1007/978-981-95-2227-9_5

**Keywords**

Biomanufacturing · Circular design · Circular bioeconomy · Waste valorisation · Sustainable materials · Closed-loop systems

## 5.1 Introduction

The overwhelming anthropogenic activities have escalated environmental challenges due to the excessive accumulation of waste, depletion of fossil resources, and greenhouse gas emissions. There is a pressing need to transform the fossil-based economy into the circular economy. This transformation encompasses the sustainable use of resources, targeting the minimisation of waste and extending the life cycle of products through innovative design and manufacturing practices. These are circular design principles mimicking nature's state of sustainability, approaches focusing on strategies that facilitate product longevity, reparability, and recyclability, and advocating the inclusion of renewable resources (Dumée 2022). The plan for circular design includes disassembly for easier separation of components, modular design, replacing or upcycling parts, and selecting recyclable or biodegradable materials (Desing et al. 2021). This is mediated through the technological approach known as life cycle assessment (LCA), which covers a comprehensive assessment of the environmental impact of all stages of a product's life, from raw material extraction to end-of-life disposal (Osman et al. 2021).

Biomanufacturing is the key driver that assimilated circular design principles by leveraging biological systems to produce value-added products from renewable biomass (Whitford et al. 2022). The key processes, such as microbial fermentation, anaerobic digestion, enzymatic catalysis, gasification, and hydrothermal liquefaction, transform waste into energy and materials such as biofuels, bioplastics, and biochemicals (Sarker et al. 2025; Okolie et al. 2022). Integrating these biomanufacturing processes with circular design principles leads to the development of closed-loop production models. These models target maximum resource efficiency by optimising the biological process for sustainable products with minimal environmental impact, and they are circular. These synergies are essential to achieve environmental sustainability, foster innovation, and advance the circular bioeconomy (Salvador et al. 2021). Through circularity-based biomanufacturing design, industries can guarantee that products are sustainable throughout their whole life cycle, including end-of-life care. Sustainable practices become economically feasible by improving the efficiency and scalability of bioprocesses within these systems. This chapter explores the foundational ideas, technological developments, and practical applications of the intersection of biomanufacturing and circular design. It examines how these ideas are implemented through in-depth case studies, identifies obstacles and difficulties, and discusses potential future developments. The chapter hopes to add to the present discussion about realising a circular bioeconomy and sustainable industrial development (Fig. 5.1).

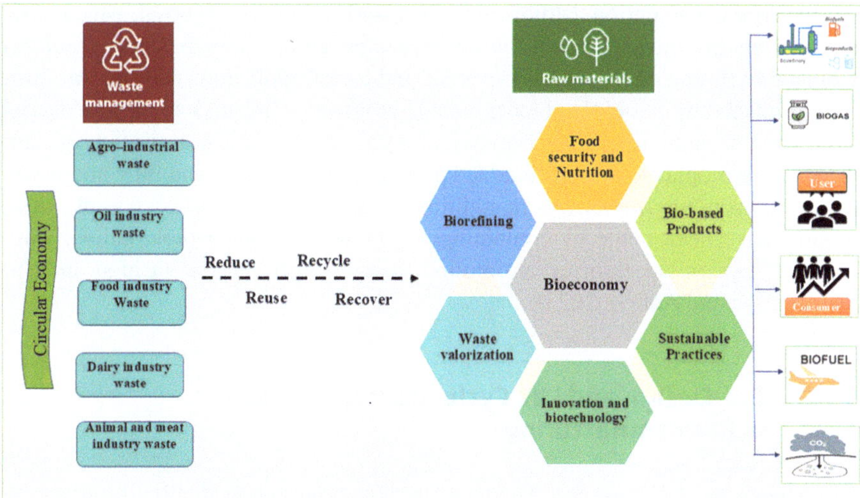


**Fig. 5.1** Sustainable development in circular bioeconomy

## 5.2 Circular Design Principles

### 5.2.1 Design for Disassembly and Upcycling

In biomanufacturing, design for disassembly (DfD) focuses on developing simple systems and products that disassemble at the end of their useful lives. This makes it easier to recover, reuse, or recycle parts, aligning with the core waste reduction objective in the circular economy. The upcycling converts waste materials into goods of greater quality or enhanced value. Presently, various studies have been conducted around the globe to adopt this concept. Recently, an extrudable paste has been developed by combining agricultural waste materials like bamboo microfibres and chitosan with mycelium from *Ganoderma lucidum*. This composite structure is used for packaging, insulation, and architectural applications (Velenturf and Purnell 2021). It is a classic example of upcycling in which waste materials are transformed into value-added products. Furthermore, fungal furniture has been developed using agricultural waste substrates by growing *Ganoderma lucidum* (reishi mushroom). The resulting furniture is composed of strong, lightweight, and biodegradable fibres, exemplifying the upcycling of waste through a biological process, aligning with the principles of the circular economy (Bofylatos 2022).

### 5.2.2 Material Selection for Sustainability

The demand for bio-based materials has seen significant growth around the globe. Biomanufacturing can be sustainable by selecting materials based on renewability, biodegradability, and eco-friendliness. These materials can be assessed through their energy consumption, carbon emissions, and end-of-life disposal, which posits minimal environmental impact. Various research groups have recently developed sustainable biomaterials with widespread industrial applications. Bioplastic films were produced by combining biodegradable polymers with amyloid fibrils extracted from food proteins like whey (Peydayesh et al. 2021). This valorisation of dairy waste into a bioplastic as a value-added product is eco-friendly and fully aligns with a circular economy framework. In addition, ornamental materials such as bio-based sequins have been made by combining flower waste with inedible animal waste used in fashion accessories, promoting sustainability in the fashion industry (Aishwariya and Adhiya 2023).

### 5.2.3 Eco-Design and Life Cycle Assessment (LCA) in Biomanufacturing

Eco-design and LCA are critical steps for designing and evaluating bioprocesses and bioproducts with environmental sustainability. The eco-design process starts with the very early stages of bioproducts. It integrates environmental considerations into the design process that will help to minimise the negative impact on the surrounding ecosystems throughout their life cycle. This concept has enhanced the recyclability and life span of the bioproducts. Life cycle assessment (LCA) is an analytical tool that comprehensively evaluates the process and products from raw material to end-of-life disposal. Various computational tools, such as SimaPro, GaBi, OpenLCA, and Umberto, and databases, such as Ecoinvent, Agri-footprint, and ELCD, have been developed to perform a comprehensive life cycle assessment of bioprocesses. Together, these processes lead to the correct decision-making aspects for sustainable innovation, empower industries for regulatory compliance, minimise their environmental impact, and align with the circular bioeconomy.

## 5.3 Biomanufacturing in the Circular Bioeconomy

### 5.3.1 Biotechnological Processes in Biomanufacturing

#### 5.3.1.1 Microbial Fermentation

Contemporary food systems are encountering a critical juncture, struggling to meet the escalating global demand for nutritious and protein-rich food, driven by rapid population growth. Conventional components of the agrifood sector, including crop agriculture, livestock farming, and aquaculture, are increasingly constrained by factors such as soil degradation, climate variability, and the intensifying frequency of environmental catastrophes. While transitioning toward plant-based diets presents a relatively sustainable alternative, it remains heavily reliant on seasonal agricultural cycles and demands extensive inputs of land, water, and agrochemicals. In contrast, microorganisms present a resilient and environmentally sustainable alternative to traditional protein sources like animal-derived products. Their rapid growth rates, minimal resource requirements, and adaptability to controlled environments make them promising candidates for revolutionising food production. This paradigm shift opens the door to what may be termed a microbial revolution in food systems, wherein microbial processes have the potential to supplement or even supplant conventional food production strategies.

At the core of this transformation lies microbial biotransformation, a term encompassing both classical fermentation and broader metabolic conversions facilitated by microorganisms. This process involves the biochemical modification of food substrates via microbial metabolism, resulting in the synthesis of novel bioactive compounds or the enhancement of existing nutritional and sensory qualities. Microbial biotransformation is particularly significant in fermented foods, where it contributes to both preservation and the development of desirable organoleptic and functional properties. Historically, fermentation emerged as a natural method of food preservation, with traditional dairy fermentations relying on spontaneous microbial populations present in raw milk. With advances in microbiology and biotechnology, modern food fermentation employs well-characterised, standardised starter cultures selected for their reproducible metabolic traits and commercial applicability.

Microbial fermentation utilises bacteria, yeasts, and filamentous fungi to bioconvert substrates, ranging from simple sugars to complex biomolecules, into a variety of value-added products. These include not only food items but also pharmaceuticals (e.g. antibiotics), industrial enzymes, biofuels, and fine chemicals, highlighting the versatility and industrial importance of microbial systems (Sawant et al. 2025) (Fig. 5.2).

### Process of Microbial Fermentation

One of the most important processes that takes place in a bioreactor, usually a closed cylinder where microorganisms are combined with a nutritional medium, is fermentation. To supply the required oxygen, air, occasionally oxygen-saturated, is pumped through. Steam or water moving via heat exchanger tubes regulates the temperature. The architecture of the bioreactor makes it possible to precisely regulate the growth parameters, such as temperature, dissolved oxygen concentration, and pH (acidity or alkalinity).

### The Steps Involved in Fermentation

1. *Preparation of Culture Medium*: The culture medium supplies organic carbon as well as vital nutrients (such as phosphorus, nitrogen, and other elements) for microbial growth. It needs to be sterilised to get rid of alien microbes, which is usually accomplished with chemicals, heat, radiation, or filtration.
2. *Obtaining Pure Microbial Strains*: Pure and extremely productive microbe cultures are acquired before fermentation and kept in a controlled environment to preserve viability and avoid contamination.
3. *Actual Fermentation Process*: Fermentation refers to the whole biotechnology process, from adding microorganisms to a heated and prepared medium to the conclusion of biosynthesis. This process occurs in the fermenter, where microor-

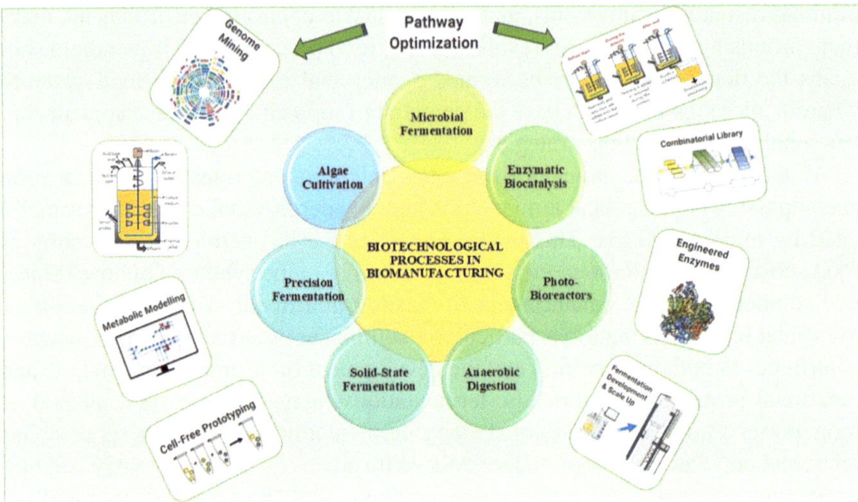


**Fig. 5.2** Role of biotechnological processes in biomanufacturing

ganisms grow or transform nutrients into desirable products. Culture liquid, often broth, is the resultant concoction of microorganisms, nutrients, and biosynthetic products.

4. *Isolation and Purification of Final Product*: After fermentation, the target product is purified from the remaining ingredients in the culture broth. The desired product is separated and purified using methods like filtration, centrifugation (the separation of particles by centrifugal force), and chemical precipitation.

Fermentation entails meticulous culture medium preparation, the addition of pure microbial strains, regulated biosynthesis in a bioreactor, and the subsequent purification of the end product for a range of industrial uses. In biotechnology, this procedure is crucial for creating useful compounds, including vitamins, enzymes, antibiotics, and biofuels (Dhande 2024).

#### 5.3.1.2 Enzymatic Catalysis

The application of biocatalysts, including isolated enzymes and metabolically engineered microbial cells, has gained considerable traction in the chemical and biotechnological sectors. This growing interest stems from the environmentally sustainable nature of biocatalytic processes, which typically operate under mild reaction conditions and reduce reliance on toxic reagents and non-renewable resources (Lee et al. 2019). As a cornerstone of sustainable biomanufacturing, enzymatic catalysis aligns closely with the principles of the circular bioeconomy. By leveraging enzymes' high specificity, catalytic efficiency, and biodegradability, industries are increasingly capable of converting biomass and waste streams into high-value products, thereby reducing their ecological footprint and dependence on fossil-derived inputs (Roland Wohlgemuth et al. 2024).

#### Process Overview

1. *Enzyme Production*: Enzyme biosynthesis is predominantly done through microbial fermentation using well-characterised hosts such as *Escherichia coli*, *Bacillus subtilis*, or *Saccharomyces cerevisiae*. Following cultivation, the target enzymes are harvested through cell disruption and subjected to downstream purification processes to yield functionally active biocatalysts.
2. *Enzyme Immobilisation*: To enhance operational stability, reusability, and ease of separation, enzymes are frequently immobilised on solid matrices (e.g. polymers, silica, or resins). Immobilisation techniques such as adsorption, covalent bonding, entrapment, or encapsulation allow enzymes to retain activity over multiple catalytic cycles, making the process economically viable for industrial applications.
3. *Catalysis*: Under optimised physicochemical conditions (e.g. pH, temperature, and substrate concentration), the immobilised enzymes facilitate specific biotransformations. These reactions produce target compounds, such as bio-based chemicals, pharmaceuticals, or food additives, with high efficiency and minimal by-product formation.

#### Enzymatic Valorisation of Organic Waste

The transformation of organic waste into useful bioproducts is a key use of enzymatic catalysis in the circular bioeconomy. Complex polymers found in food and agricultural waste can be hydrolysed by enzymes like lipases, cellulases, amylases, and proteases to produce simpler molecules like sugars, fatty acids, and amino acids. The manufacturing of biofuels, bioplastics, and other bio-based compounds uses these monomers as feedstocks. For example, cellulase converts cellulose to glucose, which can be fermented to produce bioethanol, a sustainable fuel. In addition to reducing trash disposal problems, this enzymatic method helps produce sustainable materials and energy (Chakraborty et al. 2023).

#### Bioconversion of Lignocellulosic Biomass

Cellulose, hemicellulose, and lignin make up lignocellulosic biomass, a plentiful and renewable resource. These complex carbohydrates are broken down into fermentable sugars with the help of enzymatic catalysis. The synthesis of bio-based products, including organic acids, solvents, and biopolymers, is made possible by enzymes like xylanases and ligninases. In addition to adding value to agricultural leftovers, this bioconversion supports the circular bioeconomy's waste reduction and resource efficiency goal (Pattarawan Intasian et al. 2021).

### 5.3.2 Feedstock Utilisation

Waste biomass, such as food scraps, industrial by-products, and agricultural residues, can be used as a valuable feedstock for biomanufacturing, providing an economical and environmentally friendly way to produce a range of goods, including chemicals, biomaterials, and biofuels. By turning these materials, which would otherwise be regarded as waste, into useful resources, waste disposal problems can be avoided and a circular economy can be encouraged (Adetunji et al. 2023).

#### 5.3.2.1 Types of Biomass Feedstocks

##### Agricultural Residues

Bioactive chemicals are abundant in agricultural leftovers. These wastes can serve as a substitute source for the manufacture of numerous goods, such as biogas, biofuel, mushrooms, and tempeh, which are employed as raw materials in various sectors and research projects. The production cost and the environmental pollution burden can be decreased using agro-industrial wastes as raw materials. Solid-state fermentation (SSF) produces biofuels, enzymes, vitamins, antioxidants, animal feed, antibiotics, and other compounds from agro-industrial wastes. Field residues and process residues are two additional categories for agricultural residues. Field residues are leftovers from crop harvesting that are found in the field. In contrast to process residues, which are leftovers after the crop has been converted into another helpful resource, these field residues comprise leaves, stalks, seed pods, and stems. Molasses, husks, bagasse, seeds, leaves, stems, straw, shells, pulp, stubble, peel, roots, and other wastes are utilised in manufacturing, animal feed, soil enhancement, fertilisers, and other operations (Sadh et al. 2018).

##### Food Waste

Food waste acts as a great fuel for creating bio-based products. Food waste is generated in many varieties, including liquid, solid, and semi-solid, such as wastewater, fats, spent oil, harmful home products, and others. It is well known that these wastes can potentially have detrimental effects on the environment and human health. Because so much water is used for many purposes, such as cooking, cleaning, sanitation, and transportation, liquid waste is produced. Additionally, lignin, cellulose, amylose, and monosaccharides compress solid wastes, expressing nutrients in a polluted form. By turning trash into useful commodities, its use not only lessens the environmental problems related to landfill disposal but also advances the creation of a circular bioeconomy (Sufficiency et al. 2022; Tsegaye et al. 2021).

Industrial by-products are essential to biomanufacturing due to their rich composition and abundance of useful molecules, particularly those from the food processing industry. These by-products, produced at various stages of food production, include items such as fruit peels, pulp, seeds, bran, and spent grains. They contain bioactive substances like proteins, dietary fibres, lipids, polyphenols, and other phytochemicals. These by-products play a significant role in biomanufacturing. For instance, fruit and vegetable leftovers are beneficial for creating functional foods and nutraceuticals because they are high in fibre and antioxidants. Cereal by-products, such as wheat bran, also offer essential nutrients that can be utilised in various industrial processes (Jaouhari et al. 2023).

## 5.3.3 Conversions Technologies

### 5.3.3.1 Anaerobic Digestion

Anaerobic digestion is a complex biological process wherein anaerobic microorganisms metabolise organic substrates in the absence of molecular oxygen, leading to the production of biogas a mixture primarily comprising methane ($CH_4$) and carbon dioxide ($CO_2$). This process occurs naturally in anoxic environments such as sediments of water bodies, saturated soils, and the gastrointestinal tracts of ruminants and termites. For successful anaerobic digestion, the microbial consortia involved require specific physiological and environmental conditions, including:

(a) Degradable organic carbon source serving as a metabolic substrate
(b) Adequate moisture content to facilitate microbial activity and solubilisation of organic matter
(c) An optimal environmental milieu characterised by oxygen exclusion, mesophilic or thermophilic temperature ranges, and near-neutral pH conditions

Anaerobic digestion is facilitated through a syntrophic collaboration of four distinct microbial functional groups, each responsible for a sequential biochemical transformation:

1. Hydrolysis is referred to as liquefaction; this initial phase involves hydrolytic bacteria enzymatically cleaving complex macromolecules such as polysaccharides, proteins, and lipids from substrates like agricultural residues, food waste, and manure. The resulting monomeric products include monosaccharides, amino acids, and long-chain fatty acids.
2. Acidogenesis: Acidogenic bacteria further metabolise the hydrolysis products into volatile fatty acids (VFAs), alcohols, ketones, hydrogen gas ($H_2$), and carbon dioxide ($CO_2$), contributing to the acidification of the system.
3. Acetogenesis: Acetogenic microorganisms oxidise VFAs and alcohols into acetic acid, $CO_2$, $H_2$, and ammonia. These intermediates are essential precursors for methanogenesis and serve to regulate hydrogen partial pressure within the system.
4. Methanogenesis: This terminal stage is carried out by methanogenic archaea, which are strict obligate anaerobes. They utilise acetic acid, $CO_2$, and $H_2$ to produce methane. These archaea play a crucial role in reducing system toxicity by consuming hydrogen and VFAs, thus maintaining the redox balance.

These stages occur concurrently, with microbial consortia functioning in a tightly interdependent and dynamic equilibrium. Disruption in any one stage due to inhibitory conditions or imbalance in microbial populations can negatively affect the overall process efficiency. While most microbial participants are facultative anaerobes capable of surviving under variable redox conditions, methanogens require strictly anoxic environments to thrive (Felton et al. 2024).

#### 5.3.3.2 Gasification

Along with an oxidising agent in a gasifier, the wet biomass is dehydrated by being exposed to a low temperature of 150 °C and subsequently heated to a high temperature of 800 to 900 degrees Celsius. Before the volatilisation and breakdown of solid complex hydrocarbons into combustible gases like syngas, hydrogen, tar, etc., the dry waste residue initially degrades with increased heat (Kshirsagar et al. 2021). The gasification of dry biomass is a thermochemical process that transforms organic material into a combustible gas mixture known as syngas, primarily composed of carbon monoxide (CO) and hydrogen ($H_2$). This process unfolds through several sequential stages: drying, pyrolysis, combustion, and reduction. Initially, in the drying phase, biomass with a moisture content typically between 10% and 20% is introduced into the gasifier. Here, the heat generated within the system evaporates the moisture, producing steam and leaving behind dry biomass particles. This step is crucial, as excessive moisture can reduce the efficiency and calorific value of the resulting syngas.

Following drying, the biomass enters the pyrolysis zone, which is heated to temperatures ranging from 200 °C to 500 °C without oxygen. This thermal decomposition, called devolatilisation, breaks down complex organic molecules into volatile gases, tar, and solid char. The volatile gases produced during this stage can be further processed in subsequent steps. In the combustion stage, a controlled amount of oxygen is introduced, allowing some volatile gases and char to combust. This exothermic reaction generates the necessary heat to sustain the endothermic processes of pyrolysis and reduction. The combustion primarily yields carbon dioxide ($CO_2$) and carbon monoxide (CO), contributing to the overall energy balance of the system. Finally, in the reduction phase, the remaining char reacts with the $CO_2$ and steam produced earlier. These reactions, facilitated by high temperatures, convert the char into additional CO and $H_2$, enriching the syngas composition. The resulting syngas can then be utilised for various applications, including electricity generation, heating, or as a feedstock for chemical synthesis. This multistage process efficiently converts biomass into a versatile, cleaner-burning fuel, offering a sustainable alternative to fossil fuels.

##### Hydrogen Production Through Gasification

Treating syngas may produce more hydrogen, and 99.9% pure hydrogen can be obtained by removing contaminants using membrane separation or adsorption shift. Two important variables that impact process performance are temperature and the ratio of biomass to oxidising agent. To make 1 kilogram of hydrogen, about 12–14 kilograms of bone-dry biomass is needed. This technology's efficiency ranges from 35% to 50%.

##### Hydrothermal Liquefaction

One popular thermochemical conversion technique used to turn biomass feedstocks with a high-water content into biofuels and numerous other beneficial industrial chemicals is hydrothermal liquefaction. The three main categories of hydrothermal processes are carbonisation, liquefaction, and gasification. Hydrothermal liquefaction occurs at intermediate temperatures between 250 °C and 374 °C and pressures between 4 MPa and 25 MPa. Producing a primary product, such as bio-oil, charcoal, gases, or industrial chemicals, like adhesives, benzene, toluene, and xylene, is the primary goal of the hydrothermal process (Mathanker et al. 2021). In the depolymerisation phase, macromolecular structures such as cellulose, hemicellulose, and lignin are broken down into smaller molecules. This process mimics natural geological transformations but occurs at elevated temperatures and pressures, typically between 200 °C and 350 °C, and in the presence of water. The breakdown of these polymers results in shorter-chain hydrocarbons, enhancing the thermal stability of the resultant biofuel. The subsequent decomposition phase involves several chemical reactions, including deamination (removal of amino groups), dehydration (elimination of water molecules), and decarboxylation (release of carbon dioxide). These reactions facilitate the removal of oxygen from the biomass, primarily as $CO_2$ and $H_2O$, thereby increasing the energy density of the bio-crude. Under the high-temperature and high-pressure conditions of HTL, water acts as a solvent and reactant, hydrolysing biomass components into monomers like glucose. These monomers can further degrade into compounds such as furfurals, phenols, glycolaldehydes, and organic acids, many of which are highly soluble in water. Finally, the repolymerisation phase involves the recombination of intermediate fragments into larger, more complex molecules. In scenarios where hydrogen is scarce or the concentration of reactive intermediates is high, these fragments can polymerise into high-molecular-weight compounds, leading to the formation of char or coke. This step is critical as it influences the yield and quality of the bio-crude, with excessive repolymerisation potentially reducing the efficiency of the HTL process (Table 5.1).

## 5.3.4 Integrated Biomanufacturing Systems

Integrated biomanufacturing systems (ICBs) are becoming the norm in the biopharmaceutical sector, and they are superior to the traditional batch operations. ICBs involve the combination of a perfusion bioreactor (upstream process) with a continuous downstream process (DSP). Through their combination, there is faster response to market demands, potential cost savings, and enhanced product quality (Hubert Schwarz et al. 2022).

### 5.3.4.1 Biorefineries Where Multiple Biotechnological Processes Are Combined

#### Lignocellulosic Biorefinery

A lignocellulosic biorefinery is a green process that converts plant biomass to a variety of products including biofuels, chemicals, and biomaterials. It utilises the renewable and vast material of lignocellulose, a lignin-hemicellulose-cellulose complex, to produce valuable products. It is a highly promising substitute for fossil

**Table 5.1** Diverse biomass feedstocks and their conversion process in the circular bioeconomy

| Process | Feedstock | Operating conditions | Primary outputs | References |
|---|---|---|---|---|
| Anaerobic digestion | Agricultural residues, food waste, and animal manure | Temperature: mesophilic (35–40 °C) or thermophilic (50–60 °C)<br>Anaerobic (oxygen-free) environment | Biogas (methane and $CO_2$)<br>Digestate (nutrient-rich fertiliser) | Eboibi et al. |
| Gasification | Dry biomass (e.g. wood chips, crop residues)<br>Municipal solid waste | Temperature: 700–1500 °C Partial oxidation with controlled oxygen or steam | Syngas (CO, $H_2$, $CH_4$) and biochar | David Ballenghien et al. |
| Hydrothermal liquefaction | Wet biomass (e.g. sewage sludge, algae, food waste) | Temperature: 250–374 °C Pressure: 4–22 MPa Subcritical or supercritical water conditions | Bio-crude oil<br>Aqueous phase containing organics<br>Gas phase ($CO_2$, $H_2$)<br>Solid residues (hydrochar) | Stian Hersvik Hegdahl et al.<br>Raskevi Karka et al. |
| Fast pyrolysis | Corn stover | Temperature: 450–600 °C<br>Heating rate: Rapid residence time: <2 seconds | Bio-oil<br>Biochar<br>Syngas | |
| Combustion | Switchgrass (*Panicum virgatum*) | Direct combustion in boilers<br>Moisture content: ~30% after drying | Heat-power | *Panicum virgatum* https://en.wikipedia.org/wiki/Panicum_virgatum |
| Torrefaction | Miscanthus × giganteus | Temperature: 200–320 °C<br>Atmosphere: inert (absence of oxygen)<br>Duration: 20–60 minutes | Torrefied biomass (bio-coal)<br>Volatile gases | Wikipedia (2023). Miscanthus × giganteus. https://en.wikipedia.org/wiki/Miscanthus_%C3%97_giganteus.<br>Wikipedia (2023). Torrefaction. https://en.wikipedia.org/wiki/Torrefaction |

fuel and conventional chemical industries and promotes a circular economy by minimising waste and optimising the use of resources.

Key aspects of lignocellulosic biorefineries:

- Feedstock: The primary feedstock is lignocellulosic biomass in the form of crop residues (wheat straw, maize stalk), forest residues, and other plant vegetation-based materials.
- Pretreatment: Pretreatment steps, like physical or chemical treatments, need to break down the lignocellulose complex structure and make it more accessible for further processing.

Various methods are used to convert lignocellulose into valuable products, including:

- Enzymatic hydrolysis: Using enzymes to break down cellulose and hemicellulose into fermentable sugars that can be used to produce biofuels like ethanol or to produce other chemicals.
- Thermochemical conversion: Pyrolysis, gasification, and hydrothermal liquefaction are the techniques used to yield biofuels, bio-oils, and other useful products.
- Lignin valorisation: Lignin is a recalcitrant polymer and a rich source of chemicals and materials.
- Product purification and recovery: Techniques such as distillation, filtration, and chromatography separate and recover target products (Wagh et al. 2024).

### Conventional Biorefineries

Conventional biorefineries are large-scale industrial facilities that integrate multiple biotechnological processes to convert biomass into various high-value products, such as biofuels, biochemicals, and bioenergy. By integrating multiple conversion technologies, these biorefineries optimise resource recovery and reduce waste, in line with the principles of a circular bioeconomy. In order to effectively transform biomass into useful products, traditional biorefineries incorporate several biotechnological processes:

- Pretreatment: For disintegrating the intricate biomass structure, pretreatment operations include steam explosion and hydrothermal treatments. These operations enhance the accessibility of cellulose and hemicellulose to further enzymatic hydrolysis.
- Enzymatic hydrolysis: Utilises specific enzymes to hydrolyse cellulose and hemicellulose into fermentable sugars, primarily glucose and xylose.
- Fermentation: Yeast and bacteria microbes ferment the sugars into biofuels (e.g. ethanol, butanol) and biochemicals (e.g. lactic acid, succinic acid).
- Anaerobic digestion: Degrades organic matter without oxygen to produce biogas (a mixture of methane and carbon dioxide), which can be used for power and heat.

- Gasification and pyrolysis: Thermochemical conversion of biomass to syngas (a blend of carbon monoxide and hydrogen) and bio-oil, which are intermediates in the production of fuels and chemicals.

The incorporation of various biotechnological operations into conventional biorefineries enables the overall use of biomass components:

- Carbohydrates: Hydrolysed to fermentable sugars and further to biofuels and biochemicals
- Lignin: Typically used as a renewable energy source by burning or otherwise treated as a value-added product
- Proteins and lipids: Used for food, animal feed, or as bioplastic and other material precursor
- Residual biomass: Transferred into biochar, which has the potential as a soil conditioner or for the sequestration of carbon (Leong et al. 2021; Alvira et al. 2010; Kumar et al. 2009; Annevelink et al. 2022).

## 5.4 Circular Design and Biomanufacturing Synergies

### 5.4.1 Closed-Loop Production Model

A closed-loop production model is a system where nothing is wasted. In a closed loop, the products are designed in such a way that they can be repaired, reused, recycled, or composted. Waste from one process becomes the input (resource) for another. The goal is to keep materials circulating instead of ending up in landfills or being burned.

A "closed-loop production model for circular design and biomanufacturing synergies" means: A system where products are designed from the start to be reused or recycled, using biological materials and processes that fit naturally into circular systems. Waste is minimised, and materials are constantly looped back into the system, often using natural or bio-based methods (Fig. 5.3).

Through reuse, recycling, and remanufacturing, loop production (CLPS) maximises resource efficiency and minimises waste in order to balance environmental and economic objectives. One of the most important challenges for sustainability is the move from a linear and wasteful flow economy to a circular economy (Stikane et al. 2022). Multi-company networks known as "sustainable supply chain networks" (SSCNs) collaborate to close the material loop, increase eco-efficiency, and incorporate waste as a useful input into other production processes. These networks broaden the scope of traditional supply chain duties from a single company to a systemwide viewpoint. With an emphasis on locating conceptual and methodological gap in the existing literature, it examines how complexity and uncertainty are handled in closed-loop supply chain (CLSC) management. It acknowledges that although combining forward and reverse logistic makes CLSCs essential for improving sustainability, their complex structure and unpredictability of supply

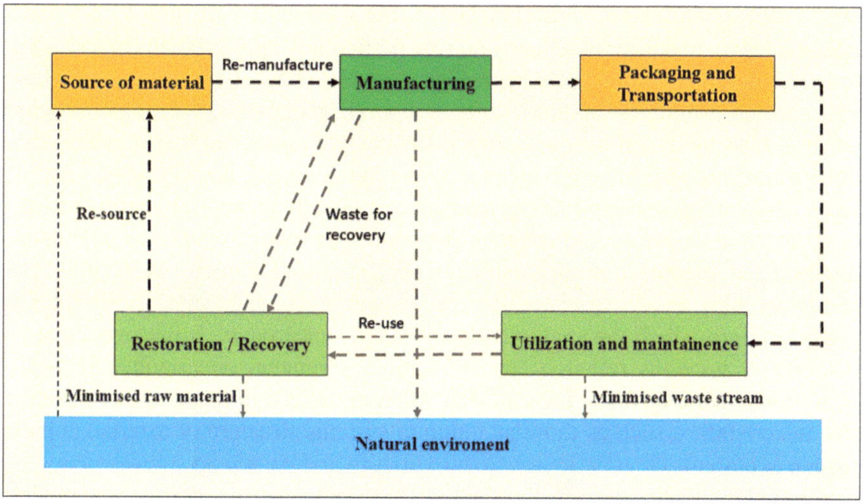


**Fig. 5.3** Schematic diagram of closed-loop production manufacturing

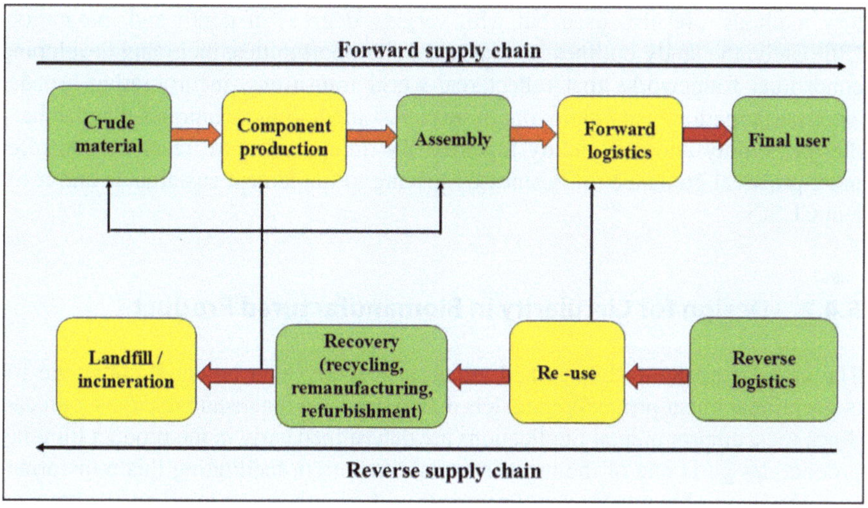


**Fig. 5.4** General framework of a closed-loop supply chain

chain variable provide serious obstacle. In order to facilitate product recovery (reuse recycling and remanufacturing), closed-loop supply chains (CLSCs) combined forward and reverse logistic promoting sustainability and value regeneration (Coenen et al. 2018) (Fig. 5.4).

In CLSC, there are two categories of complexity and uncertainty: deep uncertainty where stakeholders lack consensus or adequate knowledge about outcome parameter or models and dynamic complexity, which describes erratic interactions across time, space, and organisational level. These principles are drawn from transition management and system theory and are particularly applicable in context where future conditions are uncertain or unclear (Coenen et al. 2018). Although it takes into account a number of dimensions (temporal, geographic, and organisational) when discussing dynamic complexity, it frequently ignores the way in which these dimensions interact. Although they are important concepts, non-linearity, feedback loops, and emergent behaviours are not always used. It also draws attention to lack of focus on anticipatory and adaptive steering mechanisms, which are instruments that could aid organisations in better navigating change. Parameter uncertainty, context uncertainty, model structure uncertainty, and outcome uncertainty are the four categories into which the study divides source of uncertainty. It discovers that the majority of previous research concentrates on parameter uncertainty, frequently with probabilistic or fuzzy logic model, whereas more significant and comprehensive uncertainties, such as those attaining to systems structure or external context, remain explored.

Methodologically, the review identifies a dominance of optimisation models, often relying on mixed-integer linear programming, but notes a lack of tools that simultaneously address deep uncertainty and dynamic complexity. Other approaches, such as system dynamics modelling, multi-criteria decision analysis, and participatory methods, are also used but with varying degrees of depth and integration. Ultimately, the study outlines several research opportunities, including developing conceptual frameworks that reflect real-world transitions, incorporating broader actor participation, and employing more robust modelling techniques that can handle uncertainty and complexity together. By doing so, future research can offer more practical guidance for businesses striving to implement sustainable and resilient CLSCs.

### 5.4.2 Design for Circularity in Biomanufactured Product

The circular economy has gained attention recently as a potential substitute for conventional linear production models due to the drive for sustainable development. Since most environmental implications are determined early in the product lifetime, product design is one of the most important factors in facilitating this transformation. Designing biomanufactured products with circularity in mind entails creating system and product that fit into closed-loop or natural cycle, minimise waste, and maximise reuse, recycling, or regeneration. When planning for circularity in biomanufacturing which uses biological systems like bacteria, enzyme, or grown cells to create goods, the primary goal is to create sustainable product from beginning to end (Fig. 5.5).

The circualr design highlights sustainibility in biomanufacturing circular design highlights sustainability by incorporating fundamental ideas at every stage of the product life cycle one of the main recommendations is to replace fossil fuel with renewable biological feedstock such as algae, waste biomass, or plant-based

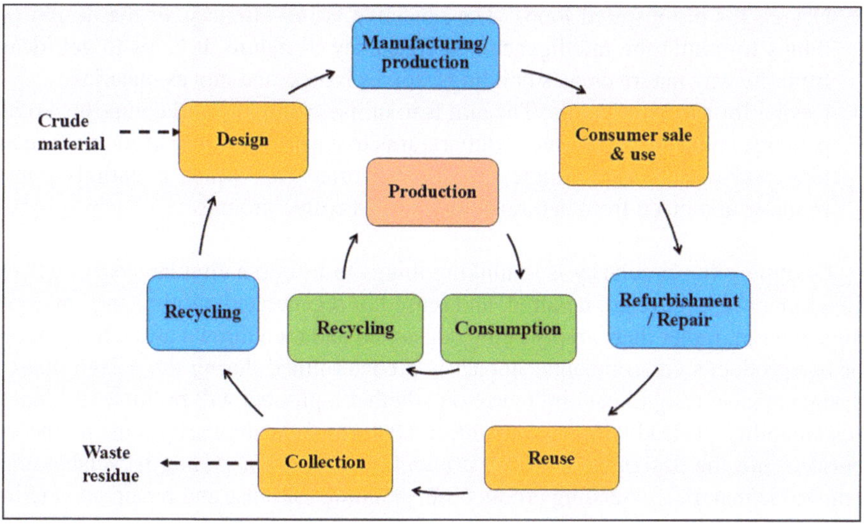


**Fig. 5.5** General framework for circular economy

resources. Product that may be readily recycled or biodegraded, for instance, bio-plastic that decomposes in compost, should be made such that they can be taken apart and reused or decomposed organically without endangering ecosystem. In order to improve processes for low energy consumption and minimal waste, micro-organisms that efficiently create beneficial chemicals while reducing hazardous by-products must be employed. Planning for a product's end of life from the start is also crucial in order to make sure that it can be safely broken down, reused, or converted into something else. Lastly, in order to reduce waste and increase the value of resources, materials should be recycled, repaired, reused, or remanufactured in a closed loop.

Waste is a worthy input particularly crucial as a biotech expand, circular design helps fight climate change, lesser dependency or limited resources and build more robust, regenerative system. it suggests harnessing materials that fit into "multiple loop system" rather than ones that are used and thrown away. This way materials continue to flow and cycle through the economy and nature rather than ending up as a waste (Mestre and Cooper 2017).

Included are the four major circular design strategies:

1. Design to slow the loops: The objective is to extend the product's life span so that fewer replacements are required. It provides robust, repairable, and upgradable items, enhancing the sentimental connection between the consumer and goods.
2. Design to close the loops: When a product is no longer usable, the objective is to recover and reuse the material. For material recovery, consider recycling, remanufacturing, and disassembly.

3. Design for bio-inspired loops: The objective is bio-mimicry, or the design of things to mimic the intelligence and efficiency of nature. It helps to get ideas from the way nature creates buildings, solves issues, and moves materials.
4. Design for bio-based loops: The aim is to utilise biological and composite components, ensuring the product safely returns to nature at the end of its life (much like food waste). An example is the C2C coffee cup, which is entirely compostable and made from natural materials and coffee grounds.

Designing for circularity is rethinking things to fit into a circular system where components are recycled, repaired, and reused or regenerated, minimising environmental harm, rather than simply being consumed and then thrown away. Since 80% of our product's environmental impact is predetermined during the design phase, designers have a significant influence on whether a product will promote or hinder sustainability. Instead of being introduced after the fact, circularity must be incorporated into the design process (Mestre and Cooper 2017). Using recyclable and renewable material, extending product life, planning for reuse and repair, and facilitating product remanufacturing or compositing are some of the many aspects of circularity. Circularity is difficult to design for. Complex study supply chains, unclear product life cycles, trade-off during durability and recyclability, and a lack of precise design tools or indicators are some of the difficulties. It discusses the difficulties in applying circular design concepts to the standard process of product development. However, many of the present circular design techniques are exclusively complicated or too narrowly focused, making it challenging for designers to apply them successfully inside conventional design processes. As a result, they frequently lack practical applicability. A thorough and understandable paradigm that operationalises circular design using the four primary techniques of reduce, reuse, recycle, and rethink is put forth in order to address this difficulty. The purpose of this 4R model is to assist designer in accessing and enhancing the circular potential of their goods. A case study featuring the redesign of a product that combines a name card and a USB flash drive is provided to demonstrate the usefulness of this framework. By comparing the original and updated design using life cycle analysis (LCA), the study shows how environmental gains can be made without sacrificing functionality (Shekar et al. 2024). This framework supports susutanibility-focused decision making for uplifiting economic growth. The framework seeks to close the gap between theory and application by providing a balance between conceptual clarity and useful advice. It encourages the creation of sustainable goods that continue to be useful and relevant throughout the duration of their original life cycle by considering on early design choices.

### 5.4.3 Bioprocess Optimisation for Circular System

Bioprocess optimisation for a circular model involves enhancing biological production processes to demand less input, yield more output, generate less waste, and ideally reuse or regenerate materials within a closed-loop framework, thus creating a sustainable approach. This optimisation method explores the synergistic integration of the bioeconomy (utilising biological resources for energy, materials, etc.) and the circular economy (reducing waste and reusing materials) to tackle global environmental and economic challenges, particularly concerning sustainability, energy, and waste management. Technology alone is insufficient; rather it should be optimised, integrated, and scaled intelligently, highlighting the area that needs to be emphasised:

1. Smart energy storage and recovery: This area focuses on how to save energy for future use, especially exploring things like salt-hydrate-based systems, which can efficiently store industrial waste or a lot of heat for a long time. Additionally, this area improves the performance of batteries and cooling systems, especially for electric vehicles and solar panels, so that they perform better and last longer (Srivastav et al. 2025).
2. Optimising energy systems: Conducting studies to maximise the efficiency of renewable energy systems, including improving photovoltaic cooling to boost efficiency by 10% or more, as well as developing advanced power plant design and heat pump setups for enhanced waste heat utilisation (Srivastav et al. 2025).
3. Bioenergy and circular supply chains: This area concentrates on the use of biomass as a sustainable alternative instead of fossil fuel and the development of supply chain model for the conversion of a biomass to fuel accounting for real-world uncertainties (Srivastav et al. 2025).
4. Toward EU's carbon neutrality (by 2050): Some studies have explored the strategies on how to manage renewable energy source like solar power and wind across different countries especially in Europe in context to the reduction of pollution up to 2050 (Srivastav et al. 2025).
5. Energy equality and sustainability in developing regions: case studies spanning five Asian nations demonstrates how education and income have a significant impact on energy stability in developing regions; rural communities are starting to transition to sustainable cooking energy, but more assistance is required (Srivastav et al. 2025).
6. After COVID-19, the waste produced: The effect of COVID-19, particularly the waste produced by gloves and mask and other personal protective equipment bracket (PPE) instead of disposing this garbage in landfill smart system offer solution for converting into energy and usable items (Srivastav et al. 2025).

In order to convert biological resources into high-value product with least possible negative impact on the environment, bioprocess optimisation is essential. In the reduced waste generation and closing resource loop optimization strategies concentrate on enhancing the stability, scalability, and efficiency of biotechnological processes. The methodological improvement of biological and chemical processes to increase yield, lower energy consumption, and limit by-product is known as a bioprocess optimisation. This entails extending product life cycle encouraging bioconversion including waste as a feedstock and optimising product matrix in a circular system. Bioprocess in circular system is optimised using a variety of methods including process integration, waste valorisation, adaptive control system, and biocatalyst engineering. Process the integration of process has been considered to save energy and resources; upstream (like fermentation) and downstream (like separation purification) steps are combined. Waste valorisation using food scraps, industrial by-product, or agricultural waste as a source material for enzymatic or microbiological activities. Adaptive control systems use machine learning algorithm and real-time monitoring to dynamically modify parameters like temperature, pH, and substrate level for the best possible microbial growth and product creation. Biocatalyst engineering includes enhancing the microbial strain or enzymes for increased production, substrate specificity, and stress tolerance (Table 5.2).

**Table 5.2** Recent studies performed for bioprocess optimisation for circular system by evaluating various parameters

| S. No. | Parameter | Description | References |
|---|---|---|---|
| 1. | Type of feedstock | Selection of renewable or waste-derived materials (e.g. lignocellulosic biomass, agro-industrial residues) as substrates for bioprocesses | De Medeiros et al. (2022) |
| 2. | Methods of pretreatment | Techniques applied to feedstocks to enhance their digestibility and fermentation efficiency, such as mechanical, chemical, or enzymatic treatments | Iram et al. (2020) |
| 3. | Microbial strain selection | Choosing robust and efficient microbial strains (wild-type or genetically engineered) capable of converting substrates into desired products under specific conditions | Shepelin et al. (2018) |
| 4. | Energy recovery systems | Implementing systems like heat exchangers or anaerobic digesters to recover and reuse energy within the process, thereby improving overall energy efficiency | Beegle and Borole (2018) |
| 5. | Water recycling | Strategies to treat and reuse process water, minimising freshwater consumption and wastewater generation | Dhanwani et al. (2025) |
| 8. | Nutrient recovery | Techniques to recover and recycle nutrients (e.g. nitrogen, phosphorus) from waste streams back into the process or for other applications | Mehta et al. (2015) |
| 7. | Product recovery efficiency | Optimisation of downstream processing steps to maximise the yield and purity of the desired product while minimising losses | Javani et al. (2011) |
| 8. | Waste valorisation | Converting process waste into valuable by-products, such as using residual biomass for bioenergy or as soil amendments | Lee et al. (2019) |
| 9. | Economic analysis | Assessing the cost-effectiveness of the process, including capital and operational expenditures, to ensure economic viability | Dormady et al. (2022) |
| 10. | Life cycle assessment (LCA) | Evaluating the environmental impacts of the bioprocess throughout its life cycle, from raw material acquisition to end-of-life disposal | Zibek et al. (2024) |

## 5.5 Case Studies of Circular Design and Biomanufacturing

### 5.5.1 Sustainable Bioplastic Production from Agricultural Residues: A Thai Case Study

The growing global concerns about plastic waste and intense fossil resource depletion have intensified the interest in the development of sustainable material alternatives such as bioplastics. These are derived from renewable biomass sources and offer the potential for biodegradability while significantly reducing the environmental footprint. Thailand, known for its agricultural abundance and increasing sustainability initiatives, has emerged as a global leader in bioplastic production. With its robust feedstock availability and strategic government policies, the country presents a compelling model for bioplastic development within the framework of the circular economy. Thailand produces a large quantity of agricultural biomass annually, particularly from sugarcane and cassava cultivation. These serve as primary feedstock for the synthesis of polylactic acid (PLA) and polyhydroxyalkanoates (PHA). In the evaluated system, the sugar and starch-rich residues are hydrolysed into fermentable sugars and converted via microbial fermentation. Lactic acid is polymerised into PLA, while mixed organic substrates, like food processing waste, are used to produce PHA using strains such as *Cupriavidus necator*. Residual biomass and sludges from fermentation processes are valorised through anaerobic digestion, supplying biogas for energy requirements (Iram et al. 2020). The integrated process chain demonstrates favourable material and energy yields. From PLA production, lactic acid yields up to 0.85 grams per gram of sugar with mechanical performance compared to conventional plastics like polystyrene. PHA production reaches yields of 0.4 grams per gram of volatile solids under optimised conditions. Additionally, biogas production (Iram et al. 2020) from waste biomass covers around 60% of the facility's thermal demand, further enhancing energy efficiency and circularity. Overall, over 70% of the original biomass carbon is retained in high-value products or energy. Environmentally, substituting petrochemical plastics with PLA and PHA leads to a significant reduction in greenhouse gas emissions, estimated between 55% and 75%. These savings stem from the use of renewable inputs and avoid landfill emissions while having the potential for industrial composting. Furthermore, residues of lignin from pretreatment processes are processed into biochar, contributing to carbon sequestration and improved soil health. Economically, Thailand's bioplastic industries show moderate but growing viability. The production costs of PLA range from €2.00 to €2.80 per kilogram, which is above those of fossil-based plastics but competitive within the premium biodegradable product market. With approximately 90% of bioplastic production presently exported, the industry benefits from strong international demand. Financial support under the Bio-Circular-Green (BCG) economic model and foreign investment has helped reduce payback periods to an estimated 7–10 years for medium-scale facilities.

Despite these positive trends, several barriers remain. Infrastructure limitations, particularly the scarcity of industrial composting facilities, restrict the realisation of full environmental benefits. Market competition with cheaper conventional plastics also hinders domestic expansion. Additionally, public misunderstanding regarding biodegradability often leads to improper disposal, undermining its ecological advantages. Developing labelling standards and public education programmes is critical in addressing these issues. Social acceptance and regulatory coherence are essential for the sector's sustained growth. Although public sentiment towards biodegradable plastics is generally favourable, concerns about economic feasibility and technological readiness still persist. Government strategies, including tax incentives, public procurement mandates, and R&D grants, could enhance investor confidence and sector resilience. Public demonstrations and stakeholder engagement will also play a pivotal role in legitimising the value of bioplastics. In summary, this case study highlights Thailand's significant potential in the bioplastics sector. With its integrated production approach, favourable environmental outcomes, and supportive policy environment, the country is well positioned to lead bioplastic innovation in Southeast Asia. Continued investment in infrastructure, consumer awareness, and technological innovation will be crucial to unlocking the full sustainability benefits of this emerging industry.

### 5.5.2 Circular Biofuel Production from Waste Biomass: A Norwegian Case Study

Climate change is the present issue, which has heightened interest in green energy, such as biofuels obtained from organic waste. In regions such as heavy transport and industrial activities where electrification is difficult, these renewable fuels can replace fossil fuels. Norway, famous for its hydropower and oil industry, is attempting to incorporate other sources of energy to diversify its hydropower-dominated power system. Although bioenergy presently comprises only a small share of Norway's energy consumption, the country's existing biomass resources can be better utilised. Every year, Norway generates a noteworthy figure of organic waste, around 450,000 tonnes of food waste and similarly sized agricultural by-products, along with some spent grain from breweries. Efficient utilisation of waste is limited. In the system examined, each type of waste goes through the optimal treatment. For food waste and brewery waste, which are moist and readily consumable, anaerobic digestion is performed, which results in the production of biogas and nutrient-rich digestate. In the case of dry agricultural residue, gasification turns it into biochar and syngas. From semi-solid residues like digestate, crude bio-oil is obtained through hydrothermal liquefaction (HTL). When these processes are combined, the mechanisms form a system aiming to bio-mimic nutrient and energy extraction. This system had a high rate of conversion and energy efficiency. For instance, anaerobic digestion generated approximately 0.45 cubic metres of methane-rich biogas for every kilogram of volatile solids. Gasification of crop residues produced close to 2 cubic metres of syngas per kilogram, and HTL of wet digestate was about 0.25 litres of crude bio-oil per kilogram of output. Approximately 78% of the feedstock's original energy content was retained. The estimated energy return on investment

(EROI) was approximately 3.8, signalling a favourable net energy output compared to conventional fuels (DeMedeiros et al. 2022). From an environmental standpoint, the system offers numerous advantages. It was estimated that using biogas instead of fossil fuels would cut emission levels by approximately 55–70%. In addition, syngas and HTL-derived fuels achieved reductions of about 65% and 80%, respectively. These reductions stem from the renewable inputs and not landfilling organic remains, which avoids methane emissions. Moreover, using digestate and biochar substitutes helps reduce synthetic fertilisers and enhances soil while capturing carbon, which assists in combating climate change.

From an economic perspective, it could become feasible or possibly propitious depending on certain parameters. It was determined generation costs would range from €0.95 to €1.25 (generally most like €1.00 to €1.25) per litre of fuel produced, which is near the market price of retail fuel prices but higher than the retail price of fossil fuel costs. Along with revenue generated from waste collection fees and sales of biofertiliser and other by-products, a small- to medium-sized facility could return the upfront purchase in an estimated 6–9 years. However, given the large capital requirements and operational complexity, which private investors find to be delicate value proposition to indicate merit without supportive legislation or public funding opportunities, it is less likely private funders will take the plunge without legislative support or funding source. There are many barriers preventing exit to some. Even though the technical research presented technical feasibility and the results were promising again, it is still a daunting place for mass uptake—especially given that historically we have been holding policy support with caution—which is again something to which investors and funders do and will be required to consider. Establishing reliable long-term carbon pricing, biofuel blending guidelines, or production subsidies could provide the stability needed for sustained investment. Compounding the situation is that complex plant operations and capital intensive equipment such as gasifiers and HTL reactors render costs and risks higher, implying an even larger financial challenge. Revealing all infrastructure needs to a functioning waste collection system and waste output distribution will always need to take place. The third area is the social acceptance. The vast majority of people would want to convert waste into fuel; the drawback comes from locating the facility and such factors as odour or truck traffic displacement to another area. Therefore, work in providing public awareness of all aspects of the system, transparent operations in all areas of the industrial process, and community benefits to include pristine air quality and new jobs will be needed to gain public acceptance. Proposals to address some public questions regarding waste-based fuel or the safety or quality of the fuel will require demonstration projects or to provide clear discussion about the sustainability and reliability of the system and quality of the product. In the final analysis, this case study demonstrates how circular biofuel systems in Norway possess the potential to be successful. By combining anaerobic digestion, gasification, and hydrothermal liquefaction, a range of organic waste can be transformed into useful energy products. Significant energy recovery, reduced greenhouse gas emissions, and beneficial by-products are all achieved by this process.

### 5.5.3 Sustainable Biosurfactant Production from Agro-Industrial Residues: A Brazilian Perspective

The need for renewable and ecological alternatives has been ignited by rising concerns about the environmental impacts of surfactants derived from petrochemicals. Microbes utilise the organic substrates to produce biosurfactants, which are amphiphilic molecules that offer a practical solution with plenty of uses in bioremediation, agriculture, and medicine. Brazil has a suitable environment for developing sustainable biosurfactants based on the concepts of the "circular economy", due to its plentiful agricultural resources and developing bioeconomy policies. The national agricultural sector produces significant amounts of by-products such as molasses, soybean cake, and waste oils that are ideal substrates for microbial biosurfactant synthesis. In this case, *Pseudomonas aeruginosa* and *Bacillus subtilis* are being widely used to synthesise rhamnolipids and lipopeptides, respectively. These residues undergo a controlled aerobic fermentation in industrial fermenters to produce biosurfactants. The anaerobic digestion process treats the leftover biomass and fermentation broth, which produces biogas to satisfy on-site energy needs. Process integration has the potential to enhance resource efficiency. Rhamnolipid obtained from glycerol substrates was of 3.2 g/L under optimised fed-batch conditions, while the surfactin production derived from soybean oil derivatives approached to 1.8 g/L. The concurrent biogas generation offset over 50% of thermal demand that supports the sustainability of operations. Moreover, residual biomass has been valorised into soil amendments that supports the regenerative agricultural practices (Ciurko et al. 2022). Environmental assessments suggest that biosurfactant substitution can reduce the greenhouse gas emissions by 45–70% as compared to synthetic counterparts. These benefits are primarily due to the utilisation of the renewable feedstock while avoiding the petrochemical processing with the reduced toxicity. Additionally, the use of biochar from digest residues proves the enhancement of soil carbon retention and also promotes the long-term ecological balance.

From an economic perspective, biosurfactant production in Brazil is gaining popularity. However, the present cost, which ranges between $3.50 and $6.00 per litre, remains an issue compared to synthetic surfactants. Nevertheless, value-added properties such as biodegradability, low toxicity, and a high emulsification index justify the premium in niche markets. Export-oriented strategies and public-private R&D funding under Brazil's National Bioeconomy Policy have improved market penetration and investor confidence, reducing the average payback periods to under 8 years for medium-scale units. Nonetheless, challenges persist due to limited downstream processing infrastructure and inconsistent product quality, which hinder scalability. Regulatory clarity on biosurfactant classification and its use in food or cosmetics continues to be a concern. Consumer misconceptions about "biological" ingredients further complicate acceptance. Therefore, the development of quality standards, market awareness campaigns, and stakeholder collaboration is vital for long-term growth. Public perception of biosurfactants is broadly optimistic, but sustained development depends on policy stability and infrastructure development. Incentives such as green procurement mandates, tax credits, and targeted education initiatives can help bridge the gap between innovation and commercialisation. Community-based pilot programmes and academic-industry partnerships will also be of considerable importance in demonstrating the real-world benefits of biosurfactants. In conclusion, Brazil's unique convergence of biomass abundance, microbial diversity, and policy support creates fertile ground for biosurfactant innovation. By leveraging its agro-industrial base within a circular economy framework, the country can stand out as a global leader in environmentally benign surfactant technologies. Continued investment in scale-up technologies, regulatory frameworks, and market awareness will be key to realising the full potential of this green chemistry frontier.

## 5.6 Challenges and Barriers

The concept has gained widespread traction in recent times. It holds tremendous potential for achieving sustainability but is hindered by several technological, economic, and legal challenges. The conversion of value-added products from waste-derived feedstocks suffers from low efficiency due to the recalcitrant and heterogeneous nature of biomass. There are mass transfer limitations when scaling up various commercial bioprocesses. The expensive downstream processing needed to obtain purified products from complex substrates presents significant engineering hurdles. The higher costs and intensive energy requirements associated with bioprocessing hinder the improvement of the cost-effectiveness of technological development. The high capital and process development costs often discourage investors' interest in financial support. The underdeveloped, localised, and fragmented status of supply chains for bio-feedstocks leads to unpredictability in process design and execution. Furthermore, concerns about establishing clear policies and standardisation for circular bio-based products may delay market entry. A lack of skilled manpower for sustainability and environmental impact assessment of bioprocesses and bioproducts is also evident. Therefore, for sustainable circular design and biomanufacturing, bridging these gaps is essential for enabling a systems-level transformation.

## 5.7 Opportunities for Advancing Circular Design and Biomanufacturing

Despite the challenging road to leading the circulatory and biomanufacturing fields, several drivers propel this concept toward advancements and effective implementation. Biotechnological developments, such as metabolic engineering, synthetic biology, and systems biology, have created more efficient microbial cell factories tailored to specific waste as feedstocks. Advancements in downstream processing technologies and the development of high-throughput purification platforms allow the commercial sector to adopt biomanufacturing technologies more easily. The adoption of a greener policy framework, along with the implementation of carbon taxation, emissions trading, and greener procurement, stimulates demand for biomanufacturing and sustainable bio-productions. Public-private partnerships and funding from various government sectors also boost long-term investment and reduce risks in financial assets, helping create circular biomanufacturing infrastructure. Simultaneously, capacity-building efforts and educational initiatives that integrate industrial biotechnology, eco-design principles, and circular economy must be implemented to apprise the present generations and innovators of taking this legacy into their future framework.

## 5.8 Conclusion

Biomanufacturing and circular design are two convergent approaches to a sustainable and regenerative industrial future. They offer a framework for converting wasteful, linear systems into closed-loop operations prioritising socioeconomic equity, environmental resilience, and resource efficiency. By incorporating concepts like life cycle thinking, sustainable material selection, and design for disassembly into creating new products and processes, sustainability is a primary goal rather than an afterthought. Biomanufacturing presents exceptional opportunities to turn wasteful or low-value inputs into high-value products when integrated into integrated biorefinery models. These systems combine enzymatic catalysis, microbial fermentation, and creative feedstock utilisation methods to optimise material and energy recovery. However, enduring systemic and technical obstacles must be removed to fully achieve this synergy's potential. The emergence of enabling technologies, progressive legislation, and an increasing focus on sustainability in both business and academia have laid the groundwork for significant advancement. Circular design and biomanufacturing can propel the shift to a genuinely sustainable bioeconomy through innovation, education, and policy alignment.

## References

Adetunji AI, Oberholster PJ, Erasmus M (2023) From garbage to treasure: a review on biorefinery of organic solid wastes into valuable biobased products. Bioresour Technol Rep 24:101610

Aishwariya S, Adhiya R (2023) Eco-couture: redefining fashion with renewable textiles and sustainable printing

Alvira P, Tomás-Pejó E, Ballesteros M, Negro MJ (2010) Pretreatment technologies for an efficient bioethanol production process based on enzymatic hydrolysis: a review. Bioresour Technol 101(13):4851–4861

Annevelink B, Garcia Chavez L, van Ree R, Vural Gursel I, Bell G, Mandl M, Lindorfer J, Hesser F, Hilz X, Stern T, Mussatto S, Stichnothe H, Leahy JJ, De Bari I, Motola V, Giuliano A, de Jong E, Mossberg J, Shmorhun M (2022) Global biorefinery status report 2022. IEA Bioenergy Task 42

Beegle JR, Borole AP (2018) Energy production from waste: evaluation of anaerobic digestion and bioelectrochemical systems based on energy efficiency and economic factors. Renew Sustain Energy 96:343–351

Bofylatos S (2022) Upcycling systems design: Developing a methodology through design. Sustainability 14(2):600

Chakraborty D, Chatterjee S, Althuri A, Palani SG, Venkata Mohan S (2023) Sustainable enzymatic treatment of organic waste in a framework of circular economy. Bioresour Technol 370:128487

Ciurko D, Czyżnikowska Ż, Kancelista A, Łaba W, Janek T (2022) Sustainable production of biosurfactant from agro-industrial oil wastes by Bacillus subtilis and its potential application as antioxidant and ACE inhibitor. IJMS 23(18):10824

Coenen J, Van Der Heijden RECM, Van Riel ACR (2018) Understanding approaches to complexity and uncertainty in closed-loop supply chain management: past findings and future directions. J Clean Prod 201:1–13

Dumée LF (2022) Circular materials and circular design—review on challenges towards sustainable manufacturing and recycling. Circ Econ Sustain 2(1):9–23

Desing H, Braun G, Hischier R (2021) Resource pressure -a circular design method. Resour Conserv Recycl 164:105179

De Medeiros TDM, Dufossé L, Bicas JL (2022) Lignocellulosic substrates as starting materials for the production of bioactive biopigments. Food Chem: X 13:100223

Dhande S (2024) What is microbial fermentation? Biology Notes

Dormady NC, Rose A, Roa-Henriquez A, Morin CB (2022) The cost-effectiveness of economic resilience. Int J Prod Econ 244:108371

Felton G, Lansing S, Moss A, Klavon K (2024) Anaerobic digestion: basic processes for biogas (FS-994). University of Maryland Extension

Hasheminejad M, Tabatabaei M, Mansourpanah Y, Far MK, Javani A (2011) Upstream and downstream strategies to economize biodiesel production. Bioresour Technol 102(2):461–468

Iram A, Cekmecelioglu D, Demirci A (2020) Ideal feedstock and fermentation process improvements for the production of Lignocellulolytic enzymes. Processes 9(1):38

Jaouhari Y, Travaglia F, Giovannelli L, Picco A, Oz E, Oz F, Bordiga M (2023) From industrial food waste to bioactive ingredients: a review on the sustainable management and transformation of plant-derived food waste. Foods 12(11):2183

Jivani J, Srinivasarao M, Dhanwani A (2025) Modified sustainable water management strategy in batch process plants: a case study on reducing freshwater consumption. Water Supply 25(2):212–227

Kshirsagar S, Nallapaneni A, Malladi K (2021) Biomass gasification: a circular economy enabler for hydrogen production. WRI India

Kumar P, Barrett DM, Delwiche MJ, Stroeve P (2009) Methods for pretreatment of lignocellulosic biomass for efficient hydrolysis and biofuel production. Ind Eng Chem Res 48(8):3713–3729

Lee EY (2019) Recent advances on biocatalysis and metabolic engineering for biomanufacturing. Catalysts 9(9):707

Lee SY, Sankaran R, Chew KW, Tan CH, Krishnamoorthy R, Chu D-T, Show P-L (2019) Waste to bioenergy: a review on the recent conversion technologies. BMC Energy 1(1):4

Leong HY, Chang CK, Khoo KS, Chew KW, Chia SR, Lim JW, Show PL, Lam MK (2021) Waste biorefinery towards a sustainable circular bioeconomy: a solution to global issues. Biotechnol Biofuels 14(1):87

Mathanker A, Das S, Pudasainee D, Khan M, Kumar A, Gupta R (2021) A review of hydrothermal liquefaction of biomass for biofuels production with a special focus on the effect of process parameters, co-solvents, and extraction solvents. Energies 14(16):4916

Mehta CM, Khunjar WO, Nguyen V, Tait S, Batstone DJ (2015) Technologies to recover nutrients from waste streams: a critical review. Crit Rev Environ Sci Technol 45(4):385–427

Mestre A, Cooper T (2017) Circular product design. A multiple loops life cycle design approach for the circular economy. Des J 20(sup1):S1620–S1635

Ntasian, P., Prakinee, K., Phintha, A., Trisrivirat, D., Weeranoppanant, N., Wongnate, T., & Chaiyen, P. (2021) Enzymes, in vivo biocatalysis, and metabolic engineering for enabling a circular economy and sustainability. ACS Catal 11(14):8842–8875

Osman AI, Mehta N, Elgarahy AM, Al-Hinai A, Al-Muhtaseb AH, Rooney DW (2021) Conversion of biomass to biofuels and life cycle assessment: A review. Environ Chem Lett 19(6):4075–4118.

Okolie JA, Epelle EI, Tabat ME, Orivri U, Amenaghawon AN, Okoye PU, Gunes B (2022) Waste biomass valorization for the production of biofuels and value-added products: a comprehensive review of thermochemical, biological and integrated processes. Renew Sust Energ Rev 159:323–344

Oraby A, Briem A-K, Bippus L, Rupp S, Zibek S (2024) Life cycle assessment as a driver for process optimisation of cellobiose lipids fermentation and purification. Int J Life Cycle Assess 29(9):1652–1671

Peydayesh M, Bagnani M, Mezzenga R (2021) Sustainable bioplastics from amyloid fibril-biodegradable polymer blends. ACS Sustain Chem Eng 9(35):11916

Sadh PK, Duhan S, Duhan JS (2018) Agro-industrial wastes and their utilization using solid state fermentation: a review. Bioresour Bioprocess 5(1):1–15

Sarker TR, Ethen DZ, Asha HH et al (2025) Transformation of municipal solid waste to biofuel and bio-chemicals—a review. Int J Environ Sci Technol 22:3811–3832

Salvador R, Puglieri FN, Halog A, Andrade FG, Piekarski CM, De Francisco AC (2021) Key aspects for designing business models for a circular bioeconomy. J Clean Prod 278:124341

Sawant SS, Park H-Y, Sim E-Y, Kim H-S, Choi H-S (2025) Microbial fermentation in food: impact on functional properties and nutritional enhancement—a review of recent developments. Fermentation 11(1):15

Schwarz H, Gomis-Fons J, Isaksson M, Scheffel J, Andersson N, Andersson A, Chotteau V (2022) Integrated continuous biomanufacturing on pilot scale for acid-sensitive monoclonal antibodies. Biotechnol Bioeng 119(8):2152–2166

Shepelin D, Hansen A, Lennen R, Luo H, Herrgård M (2018) Selecting the best: evolutionary engineering of chemical production in microbes. Genes 9(5):249

Srivastav AL, Bagherian A, Ghosh D (2025) The circular bioeconomy: pathways to sustainability and resource optimization. Clean Techn Environ Policy 27(2):815–832

Stikane A, Dace E, Stalidzans E (2022) Closing the loop in bioproduction: spent microbial biomass as a resource within circular bioeconomy. New Biotechnol 70:109–115

Sufficiency E, Qamar SA, Ferreira LFR, Franco M, Iqbal HMN, Bilal M (2022) Emerging biotechnological strategies for food waste management: a green leap towards achieving high-value products and environmental abatement. Energy Nexus 6:100077

Tsegaye B, Jaiswal S, Jaiswal AK (2021) Food waste biorefinery: pathway towards circular bioeconomy. Foods 10(6):1174

Whitford B, Jones D, Kinnane S (2022) Biomanufacturing design: Reducing the environmental burden. Curr Opin Biotechnol 76:102717

Wagh MS, Sa S, Nath PC, Chakraborty A, Amrit R, Mishra B, Mishra AK, Mohanta YK (2024) Valorisation of agro-industrial wastes: circular bioeconomy and biorefinery process—a sustainable symphony. Process Saf Environ Prot 183:708–725

Wohlgemuth R (2024) Enzyme catalysis for sustainable value creation using renewable biobased resources. Molecules 29(23):5772

# Resource Recovery from Waste for Sustainable Biorefineries

# 6

Prakash Kumar and Ajay Kumar

**Abstract**

Waste has attracted immense interest as a co-feedstock for sustainable biorefinery technologies, as highlighted in Chapter 6, which examines efficient conversion of waste streams into value-added bio-based products and energy carriers. The chapter gets readersstarted with an overview of conversion pathways, detailing thermochemical and biochemical pathways. Comprehensive case studies highlight success across the agricultural, municipal, and industrial sectors, including waste-to-energy projects, industrialsymbiosis collaborations, and advanced waste valorization processes. Techno-economic assessments and life cycle evaluations are underscored as crucial instruments for analyzing the practicality and sustainability of waste-derived biorefinery systems. Case studies demonstrate successful applications, reinforcing the significance of interdisciplinary cooperation among engineers, policymakers, and ecologists. By compiling present research, technological innovations, and future pathways, it offers a thorough guide for moving from traditional waste management to groundbreaking resource recovery ecosystems. It supports strengthened public–private collaborations, uniform regulations, and educational initiatives to hasten the implementation of sustainable biorefineries, ultimately aiding global decarbonization and resource resilience objectives. This document acts as an essential resource for researchers, industry participants, and policymakers dedicated to reimagining waste as a fundamental aspect of sustainable industrial metabolism.

P. Kumar
Department of Biotechnology, Sharda University Agra, Agra, Uttar Pradesh, India

A. Kumar (✉)
Department of Biotechnology, Faculty of Engineering and Technology, Rama University, Kanpur, Uttar Pradesh, India



V. Singh (ed.), *Sustainable Waste Management Towards Circular Bioeconomy*,
https://doi.org/10.1007/978-981-95-2227-9_6

**Keywords**

Resource recovery · Sustainable biorefineries · Renewable energy · Biofuels · Waste valorization

## 6.1 Introduction

Worldwide waste generation has recently risen due to fast industrialization, urbanization, and population expansion, which has provided multidimensional environmental, economic, and societal problems. Traditional waste management methods, such as landfills and incineration, are not only unsustainable but also contribute to resource depletion and the release of greenhouse gases. As a result, there has been a rising trend toward creative methods that emphasize resource recovery, valorization, and waste minimization; therefore, it gives a path for sustainable biorefineries to become a foundation of present waste management. A resource recovery from garbage is an interdisciplinary sector paying attention to turning waste products—biomass, chemicals, and bio-based materials—into useful resources. In support of environmentally friendly policies seeking to lessen damage on our environment, this groundbreaking technology fits well with the circular bioeconomy as it enables sustainable resource consumption and waste minimization. Biorefineries maximize the use of several waste products by means of present technologies such anaerobic digestion, pyrolysis, enzymatic hydrolysis, and fermentation, given their integrated nature. By this technique can use of organic waste and agricultural residues for biofuel production (Leong et al. 2021).

Advances in artificial intelligence (AI) and process optimization further enhance the scalability and financial viability of biorefineries. Marzban et al. (2023) argue that through the incorporation of artificial intelligence systems, biorefineries are becoming more able to achieve self-sufficient energy production, lower carbon footprints, and improved efficiency (Shree et al. 2024). In Chap. 6, an investigation of the most recent developments, difficulties, and possibilities in this sector is conducted. It emphasizes how policy frameworks, interdisciplinary research, and cooperation among stakeholders drive the growth and acceptance of sustainable biorefinery technologies (Shree et al. 2024) for meeting a worldwide environmental and economic goals (Keharika et al. 2024).

### 6.1.1 Concept of Sustainable Biorefineries

Advanced processing systems are specifically created to change several kinds of biomass and waste products into fuels, chemicals, and materials in line with important guidelines for sustainable biorefineries. By means of this change, energy vectors and value-added goods can be created in a way that reduces environmental impact and supports resource efficiency. The base of the sustainable biorefinery idea is the incorporation of many transformation pathways in one plant like a combination of thermochemical, biochemical, and catalytic techniques; these systems can handle heterogeneous biomass—including industrial by-products, municipal waste, and agricultural remnants—and valorize each component. This full-scale strategy improves economic sustainability by generating a wide range of bioproducts as well as by maximizing energy recovery (SolarteToro et al. 2023). Sustainable biorefineries' close matching with circular economy ideas is one important element. Solarte Toro and Cardona Alzate (2023) claim that this design philosophy lessens dependency on virgin resources, cuts greenhouse gas emissions, and promotes long-term environmental stewardship. Techno-economic analysis, life cycle assessments, and environmental impact evaluations guarantee that every point meets demanding sustainability criteria—from feedstock sourcing and transformation processing to product distribution. Integrating these assessment methods enables biorefineries to offer openness and responsibility in resource recovery and bio-based product production in addition to process optimization (Jiao et al. 2024). They are sophisticated, interconnected systems that guide the shift toward a resilient, bio-based, and circular economy, not just processing units. Under a philosophy of environmental accountability and economic viability, these biorefineries are paving the path for a more sustainable future by converting garbage into riches.

### 6.1.2 Importance of Waste Utilization in Biorefineries

The application of waste in biorefineries is a basis of sustainable development that addresses immediate environmental issues while supporting circular bioeconomy and resource recovery. They are as follows:

(a) *Environmental benefits*: Biorefineries reduce environmental pollution by diverting waste from landfills and reducing greenhouse gas emissions. For instance, the conversion of agricultural waste into biofuels and chemicals reduces dependency on fossil fuels (Leong et al. 2021). This approach reinforces worldwide efforts to address climate change.

(b) *Resource recovery*: By drawing useful products from waste streams, biorefineries enable the conversion of these resources into bioenergy, bioplastics, and organic fertilizers (Lee et al. 2024b) by combining advanced technology like pyrolysis and anaerobic digestion.

(c) *Financial opportunity*: By transforming trash into biorefineries, the value of trash creates revenue from goods produced from it and lowers waste management expenses. One option could be to process municipal solid waste so as to help energy security by generating biogas and bioethanol (Kim et al. 2023).

(d) *Contribution to circular bioeconomy*: Waste use in biorefineries shows the concept of a circular bioeconomy, whereby waste is considered a resource rather than an issue. By promoting sustainable production and use patterns, this paradigm shift supports long-term financial and environmental sustainability (Leong et al. 2021).

(e) *Achieve sustainability goals*: The waste recovery in biorefineries could help meet more general sustainability objectives, including lowering greenhouse gas emissions, supporting the circular economy, and reducing waste sent to landfills. Biorefineries can assist industries in lowering their environmental footprints and support initiatives to combat climate change by extracting value from waste streams (Zhang et al. 2023).

(f) *The technological innovations enabling waste recovery*: Find the technological for resource recovery processes as well as enable waste recovery technologies that involves the creation of effective conversion tools by anaerobic digesting, enzymatic hydrolysis, biological fermenting, and thermochemical conversion. All of these approaches can be used on various kinds of waste to recover valuable substances suitable for conversion into biofuels or chemicals (Bajpai 2022).

### 6.1.3 Integrative Approach

We can adopt an integrative approach by combining theoretical models with practical case studies and techno-economic analyses. It emphasizes:

- *Interdisciplinary research*: Linking process engineering, environmental science, and economics to deliver a holistic view
- *Sustainability metrics*: Incorporating life cycle assessments and environmental impact studies to validate the sustainability credentials of biorefinery operations
- *Circular economy principles*: Shows ways of transforming waste streams back into resources, therefore decreasing dependence on virgin materials and lowering environmental footprints.

## 6.2 Types of Waste and Their Potential in Biorefineries

By means of biorefineries, waste integration is a revolutionary strategy for attaining sustainability and promoting a circular bioeconomy. Here, several kinds of waste and their possible uses in biorefineries are examined.

### 6.2.1 Agriculture Waste

Abundant and renewable feedstocks for biorefineries include agricultural residues—crop stalks, husks, and animal waste. Processes such anaerobic digestion and pyrolysis can turn these residuals into biofertilizers, biochemicals, and biofuels (Kavitha Shree et al. 2025). Valorizing of agricultural waste not only helps to reduce environmental damage but also improves resource recovery.

### 6.2.2 Industrial Waste

Lignocellulosic biomass and chemical residues are among industrial by-products with great biorefinery potential. Sophisticated biotechnological techniques allow for the transformation of these wastes into valuable commodities including bio-based polymers and organic acids (Leong et al. 2021). Including industrial waste into biorefineries helps to lower reliance on fossil fuels (Fig. 6.1).

### 6.2.3 Municipal Solid Waste (MSW)

Composed of organic and inorganic parts, municipal solid waste poses a major obstacle for urban waste management. Processing the organic fraction of MSW in biorefineries yields biogas and ethanol, therefore supporting energy recovery and waste minimization (Dawar et al. 2025). To maximize MSW management plans, machine learning methods are more and more employed.

### 6.2.4 Food Waste

A serious worldwide problem with important environmental and financial consequences is food waste. By turning food waste into biofuels, bioplastics, and other value-added products, biorefineries provide a green solution (Tsegaye et al. 2021). This strategy reduces landfill dependency and greenhouse emissions and is in keeping with the ideas of a circular economy.

**Fig. 6.1** Illustrating types of waste use in biorefinery

### 6.2.5 Organic Sludge and Wastewater

Rich in nutrients and organic content, wastewater and organic sludge are ideal for biorefinery uses. Pyrolysis and microbial fuel cells help the recycling of bioenergy and chemicals from this rubbish (Pelagalli et al. 2024). These developments improve the sustainability of processes of treating wastewater.

## 6.3 Technologies for Resource Recovery

The increasing international focus on sustainability has spurred the creation of cutting-edge resource recovery technologies turning garbage into goods of value. Reducing dependence on limited resources, promoting a circular bioeconomy, and solving environmental problems all depend on these technologies. As integrated systems, biorefineries use advanced techniques to recover bioenergy, chemicals, and several other high-value goods from various waste streams (Lee et al. 2024b). Waste valorization feasibility and efficiency has been much improved by latest developments in resource recovery technologies like anaerobic digestion, pyrolysis, and bioelectrochemical systems (Jiao et al. 2024). These systems help not only reduce environmental degradation but also promote energy security and economic development. Machine learning and artificial intelligence's integration further improves resource recovery methods by providing accurate waste classification and process control (Kavitha Shree et al. 2025).

### 6.3.1 Waste Conversion to Bioenergy

A key building block of sustainable biorefineries is the transformation of trash into bioenergy; this simultaneously meets energy needs lowers environmental degradation. Three main technologies for waste-to-bioenergy (WtB) conversion—anaerobic digestion, gasification and pyrolysis, and fermentation processes—are discussed in this part.

#### 6.3.1.1 Anaerobic Digestion

Decomposing organic garbage in the absence of oxygen, anaerobic digestion (AD) generates biogas—a combination of methane and carbon dioxide—as well as digestate by means of a chemical reaction. Agricultural residues, municipal solid waste, and wastewater sludge are all commonly treated with this technology. Recent AD developments have been centered on maximizing biogas production via co-digestion and pretreatment techniques (Alengebawy et al. 2024). Besides improving energy recovery, the incorporation of AD into biorefineries helps with nutrient recycling and reduction of greenhouse gases.

#### 6.3.1.2 Gasification and Pyrolysis

Under controlled circumstances, pyrolysis and gasification are thermochronological techniques transforming organic waste into syngas, bio-oil, and biochar. Whereas pyrolysis takes place in the lack of oxygen, gasification is partial oxidation at very high temperatures. For plastic waste and lignocellulosic biomass, these are especially potent solutions. Developing gasification methods have shown better efficiency and scalability, therefore making them suitable for large-scale bioenergy generation (Jafri et al. 2020).

#### 6.3.1.3 Fermentation Technologies

Microbes in fermentation technologies turn organic substrates into biofuels such bioethanol, biobutanol, and biogas. For waste high in carbohydrates—that is, food waste and agricultural waste—this method is especially well suited. Fermentation has lately been advanced by means of genetically modified microorganisms and enzymatic hydrolysis to improve yield and efficiency (Shahzad et al. 2024) (Fig. 6.2).

### 6.3.2 Recovery of Valuable Chemicals and Bioproducts

Modern biorefineries do not only produce power; they also concentrate on the integrated restoration of valuable compounds potentially substitution fossil-based materials. Recovering bioproducts from waste enables the transformation of underutilized biomass into molecules used in agriculture, food, pharmaceuticals, and packaging. Three key products are as follows:

#### 6.3.2.1 Bio-Based Polymers

Promising substitutes to artificial plastics are bio-based polymers derived from waste streams. Fruit processing remnants, sugarcane bagasse, rice husks, and other lignocellulosic materials as food and agricultural waste offer a great source of

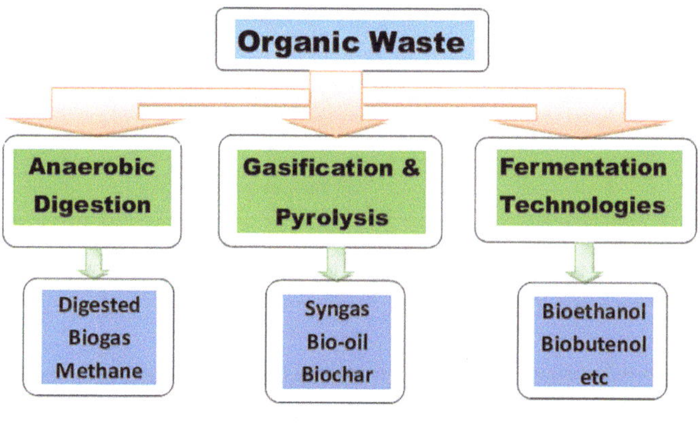


**Fig. 6.2** Illustrating bio-conversion of waste into bioenergy

carbohydrates and lignin to be used in polymers. Controlled catalytic, enzymatic, or fermentative reactions can transform these waste feedstocks into biodegradable polymers with customized physical characteristics, recent research has shown (Maraveas 2020; Kaur et al. 2024).

#### 6.3.2.2 Organic Acids and Alcohols

Bioplastics, solvents, and food ingredients are synthesized from organic acids and alcohols; biorefineries can use advanced fermentation and catalytic hydrolysis, and novel hydrothermal and microbial fermentation techniques have been used to turn carbohydrate-rich waste from food and agriculture into citric, acetic, and lactic acids as well as alcohols including bioethanol (Fang et al. 2025). Moreover, integrated process methods have shown how sequential fermentation and reactive extraction can effectively recover these components (Sharma et al. 2021) (Fig. 6.3).

#### 6.3.2.3 Proteins, Lipids, and Enzymes

From waste streams comes great possibility as industrial catalysts and dietary supplements in extracted proteins, lipids, and enzymes. Advanced methods of separation and protein recovery from agricultural waste (Fatinee et al. 2021). Recent studies on fishery and aquaculture residues have shown that high-value enzymes with distinctive cold-adapted and catalytic properties can be effectively synthesized and purified, hence improving industrial biocatalysis (Khiari 2024).

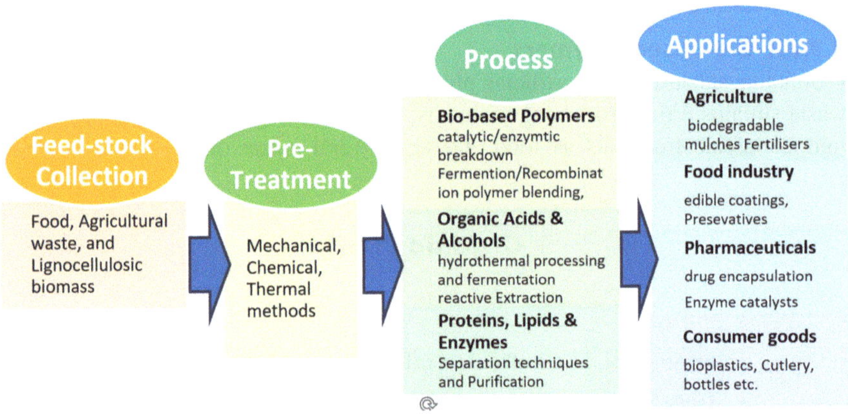


**Fig. 6.3** Illustrating recovery of valuable chemicals and bioproducts from waste

### 6.3.3 Waste-to-Biofuel Technologies

By transforming waste natural product into renewable energy carriers, waste-to-biofuel technologies assist in addressing waste management challenges and reducing reliance on fossil fuels (Lee et al. 2019). Two promising conversion pathways within this expansive sector are the production of bioethanol and the creation of biodiesel from waste oils. Both pathways demonstrate how innovative processes and integrated biorefinery concepts can contribute to a circular bioeconomy (Quevedo Amador et al. 2024).

#### 6.3.3.1 Bioethanol Production

Due to its ability to utilize low-value residues for generating high-value renewable fuels, bioethanol from waste biomass is receiving considerable attention. Generally, this process entails several stages:

(a) *Feedstock selection and collection*: Different waste streams—including agricultural residues, municipal solid waste, and industrial by-products—are chosen based on their carbohydrate content. Utilizing these waste streams minimizes waste and recovers resources (Lee et al. 2019).
(b) *Pretreatment*: Waste biomass is often resistant to decomposition due to its lignocellulosic structure. Pretreatment techniques (acidic, alkaline, or enzymatic) are employed to dismantle complex cell wall matrices, thus freeing fermentable sugars. This phase is vital for improving enzyme accessibility and overall process yield (Ferrusca et al. 2023).
(c) *Enzymatic hydrolysis*: After treating, a mix of enzymes like cellulase and hemicellulase breaks down sugars into simple ones. Optimizing saccharification requires tuning enzyme assemblages and process parameters, occasionally employing altered enzyme versions for enhanced productivity (Lee et al. 2019).
(d) *Fermentation*: Yeast turns sugar into alcohol during fermentation. Cutting-edge tactics like synchronous saccharification and fermentation decrease expenses and enhance the output by coupling hydrolysis and fermentation into a unified procedure (Leong et al. 2021).
(e) *Distillation and purification*: Finally, clean ethanol is made by heating the fermentation mixture and removing water. Purified bioethanol may intermix with gasoline to lessen greenhouse gas outputs or function as an autonomous sustainable energy source (Quevedo-Amador et al. 2024) (Fig. 6.4).

Integrate this methodology to elevate biofuel output and simultaneously diminish the ecological impact, promoting bioethanol as a viable substitute for conventional petroleum-based energy sources.

#### 6.3.3.2 Biodiesel from Waste Oils

Production of biodiesel from oils generates from leftover fats and lipids, transforming them into renewable fatty acid methyl esters (FAME) that serve as diesel

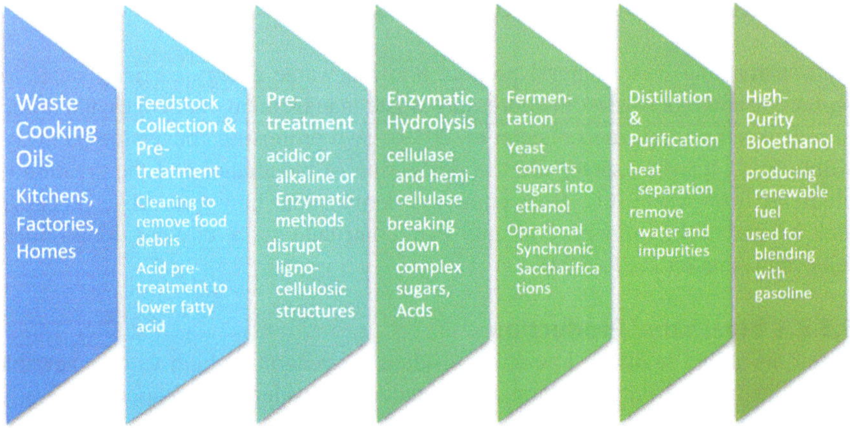


**Fig. 6.4** Illustrating production of bioethanol from waste

replacements. The mechanism principally revolves around the transesterification procession and typically involves the following stages:

(a) *Feedstock collection and pretreatment*: Waste cooking oils come from kitchens, factories, and homes. Before processing, these oils are cleansed to discard food debris and could receive acid pretreatment to decrease free fatty acid quantities, thus avoiding catalyst incapacitation in the ensuing reaction (Leong et al. 2021).
(b) *Transesterification reaction*: In this important part, oils with fat can mix with water (typically alcohol) by a helper. Homogeneous catalysts (e.g., sodium hydroxide) and innovative heterogeneous catalysts are utilized to enhance the production of fatty acid methyl esters (FAME), which represent biodiesel. Increasing studies target bifunctional catalysts to manage both free fatty acids and triglycerides together, thereby improving conversion rates and minimizing ecological effects (Lee et al. 2019).
(c) *Separation and purification*: After making biodiesel, the liquid mixture usually has leftover glycerol, some unused methanol, and bits from the chemical mix used. Biodiesel is subsequently refined via sedimentation, centrifugation, and desiccation to conform to the standards for fuel purity. This cleansing procedure confirms that resulting biofuel meets national and global purity standards, endorsing its use in transit and industry (Quevedo-Amador et al. 2024).

Innovations in processes, like in-place filtration and catalytic progress, have markedly increased biodiesel output from waste cooking oil (WCO) and boosted its viability costs. Increased utilization of these innovations not only eases waste oil disposal challenges but also provides a sustainable substitute for regular diesel, subsequently reducing ecological emissions (Lee et al. 2024a).

## 6.4 Sustainability Assessment of Waste Utilization in Biorefineries

This comprehensive evaluation is crucial to guarantee that the adoption of waste-to-product methods simultaneously reduces landfill buildup and emissions of greenhouse gases, as well as conforms to norms of a circular bioeconomic system. Lee et al. (2019) highlighted the key role of renewable fuels and chemicals, presenting a sustainability evaluation of waste-to-energy conversion at biorefineries considering ecological, economical, and societal effects.

This technique guarantees that advantages like diminished carbon impact, minimal water consumption, and decreased environmental disruptions are meticulously measured (Quevedo-Amador et al. 2024). Alongside life cycle analysis (LCA), techno-economic evaluation (TEA) assesses the cost-efficiency and financial viability of biorefinery processes, incorporating social impact analyses that highlight community advantages, employment generation, and stakeholder involvement, guaranteeing waste valorization measures support comprehensive sustainable progression (Leong et al. 2021). By skillfully utilizing refuse substances, these bio-facilities act as inventive centers that convert potential contaminants into concrete economic and ecological benefits—a vital move for achieving lasting sustainability and energy independence (Lee et al. 2024a).

### 6.4.1 Environmental Effects

Life cycle assessment (LCA) examines the full life cycle of biorefinery operations as well as their environmental impact. Starting with waste accumulation and processing, the whole supply chain—from application to ultimate disposal of the product—is assessed in the thorough review. Among their main environmental worries are the effects of global warming (risk), the rise of nutrient pollution, acidifying water pollution, and energy consumption (Gupta et al. 2023). According to Singh et al. (2022), feedstock selection, power use, and waste conversion all have great impact on environmental sustainability (2022). Projects like consequential LCA and intermittent LCA offer a more sophisticated understanding of long-lasting ecological interactions connected to bioprocesses (Kumar et al. 2024a, b). Among the most important impact categories are eutrophication, acidification, energy consumption, and global warming potential (Gupta et al. 2023). LCA analysis of biorefineries shows that chemical source selection, process energy use, and waste valorization all very affect environmental sustainability (Singh et al. 2022). More sophisticated LCA techniques like consequential and hybrid LCA provide more thorough information on the long-term environmental trade-offs linked to biorefinery activities (Kumar et al. 2024a, b).

### 6.4.2 Economic Viability

Cost–Benefit Review. One very important factor determining whether or not biorefineries can recover resources is economic feasibility. Using a systematic method to evaluate financial returns against investment costs—including capital expenses (CAPEX), operating costs (OPEX), and potential income from bio-based products—cost-benefit analysis (CBA) helps. Including subsidies, carbon credits, and incentives from the circular economy can help to make these initiatives more economically viable, according to Patel et al. (2022). Recent research shows that the financial feasibility of biorefineries improves when several income sources—including platform chemicals, biofertilizer, and bioenergy—are combined (Zhang et al. 2024a, b, c, d).

### 6.4.3 Social and Policy Constraints

Sustainable biorefineries will be successful based not just on economic and environmental circumstances but also on social acceptance and legal structure. The adoption of waste-derived goods is greatly influenced by public opinion, job creation, and ethical issues (Mishra et al. 2023). Governments all over the world are designing laws to encourage waste valorization via incentives, requirements, and waste management rules (Fernandez et al. 2023). Harmonizing global sustainability standards and certifications—including ISCC (International Sustainability and Carbon Certification)—can increase public confidence and commercial demand for bio-based products (Choudhury et al. 2024).

### 6.4.4 Carbon Footprint Reduction

Resource recovery in biorefineries aims primarily to help lower the carbon footprint. Reducing greenhouse gas (GHG) emissions strongly depends on methods like process improvement, energy-efficient technologies, and carbon sequestration (Das et al. 2023). According to Lee et al. (2024b), replacing fossil-based feedstocks with bio-based choices helps much toward carbon neutrality. Often combined with life cycle assessment, carbon footprint analysis measures direct and indirect emissions; this helps leaders to create waste valorization policies friendly with the climate.

## 6.5 Case Studies of Resource Recovery for Biorefineries

Dynamic centers are rising up at biorefineries for turning several waste streams into clean energy, bio-based chemicals, and high-valued products. These plants support a sustainable circular economy by including activities that extract materials from waste.

### 6.5.1 Biorefineries Using Agricultural Wastes as Source

Readily available feedstock for biorefineries includes agricultural wastes such as corn stover, rice husks, crop peels, and animal excrement. Thermochemical, biochemical, or hybrid processes will produce bioenergy, biofuel, and specialty chemicals from these residues. Anaerobic decomposition of animal waste and crop remains, for example, generates biogas that can be used locally for power or heat; the digestate serves as a nutrient-rich soil amendment (Leong et al. 2021). Furthermore, pretreatment methods such as acid hydrolysis and steam explosion increase the release of fermentable sugars from lignocellulosic biomass, thereby allowing for efficient fermentation processes generating bioethanol and bio-based solvents (Liu et al. 2016). Many times, integrated process designs are used to optimize resource recovery and reduce waste, thereby bringing agricultural waste conversion into line with circular economy principles (Leong et al. 2021).

### 6.5.2 Municipal Waste to Energy Schemes

A complicated mix of organic and inorganic compounds constitutes municipal solid waste. Technology such as bio-methanation and burning helps to transform the organic part and therefore greatly assists in renewable energy generation and waste reduction. Municipal waste to energy initiatives usually requires organized gathering, sorting, and preprocessing of garbage so as to extract the biodegradable part. After that, anaerobic digestion turns this fraction into biogas, or in some cases gasification and pyrolysis are used to create syngas for power generation (Lee et al. 2020). For example, several urban facilities have successfully integrated combined heat and power (CHP) systems that not only produce energy but also reduce the burden on landfills and lower greenhouse gas emissions (Marzban et al. 2025).

### 6.5.3 Industrial Symbiosis and Used Value Discovery

A cooperative strategy in which the raw materials for one industry become the by-products or waste for another is known as industrial symbiosis. This interindustry exchange would go a long way in lowering environmental effects and raising combined resource efficiency. Marzban et al. (2025) give a noteworthy case in the industrial ecosystem in Kalundborg, Denmark, where waste heat, extra water, and chemical by-products are exchanged among neighboring facilities, therefore reducing energy use and demand for raw materials. Advanced processes in the waste valorization sector convert industrial waste including spent catalysts, process residues, and off-spec products into high-value intermediates and chemicals (Shree et al. 2024). Catalytic upgrading, gasification, and hydrothermal refinement among other methods turn these waste products into useful products including syngas, biochar, and platform chemicals, which provide a basis for green manufacture. Companies can save much money and help the environment by including waste valorization plans inside industrial symbiosis networks.

### 6.5.4 Creative Ideas and New Technologies

Recent advances using the most recent biotechnology, process integration, and digital optimization are stretching the limits of resource recovery in biorefineries to reach almost zero waste. Rising technology covers:

(a) *Advanced pretreatment methods*: According to Lee et al. (2020), new approaches using ionic liquids, supercritical $CO_2$ extraction, and hydrothermal liquefaction are raising feedstock digestibility as well as efficiency in biomass useful chemical extraction.
(b) *Developing custom microbial strains*: The ability to turn complex waste streams into particular bio-based chemicals and fuels will improve yield and conversion rates, therefore enhancing yield and conversion rates.
(c) *Process intensification and digitalization*: The use of sensor networks and real-time data analytics enables exact process control, adaptive operations, and optimization of production rates in biorefineries, hence offering the agility needed for variable waste inputs (Marzban et al. 2025).
(d) *Hybrid process integration*: Combining thermochemical and biochemical conversion techniques at a single site can maximize the recovery of both energy and high-value chemicals, leading to increased general performance and sustainability (Shree et al. 2024).

Not only do these developing technologies help biorefineries to be more cost effective, but they also advance the dream of a completely synergized and sustainable circular bioeconomy in which waste is seen as a scarce commodity.

## 6.6 Challenges and Opportunities

Although it presents a range of technical, financial, legislative, and market difficulties, the road toward sustainable biorefineries that retrieve resources from waste is full of promise. To create a dynamic circular bioeconomy, one must tackle these obstacles while taking advantage of new possibilities. Transitioning toward sustainable biorefineries involves addressing a range of challenges.

### 6.6.1 Technical Roadblocks in Waste Management

Biorefineries should handle very variable waste streams usually containing pollutants and changing moisture levels. This variation might lower product quality and hinder conversion efficiency. Incomplete breakdown of lignocellulosic feedstock results from suboptimal pretreatment processes, lowering the output of biofuels and biochemistry (Smith et al. 2022). Furthermore, needed to handle feedstock variation and maximize reactor output are sophisticated process integration and strong control systems (Zhao et al. 2021).

### 6.6.2 Economic Difficulties with Scale-Up

High capital costs, unknown feedstock supply, and market volatility make it difficult to scale up biorefinery operations from pilot or laboratory size to commercial production. The competitive pricing of traditional fossil-based goods sometimes erodes the economic viability, so biorefineries have to seek niche markets or take advantage of governmental assistance (Kumar et al. 2023). Creative financial models and public–private partnerships can help to reduce these financial dangers and encourage the move to full-scale operations.

### 6.6.3 Policy and Regulatory Frameworks

The growth of sustainable biorefineries depends much on a supportive regulatory environment. Still, investment and innovation could be slowed by varied rules and legal ambiguity on waste classification, incentives for renewable energy, and bio-based product standards. Clear, harmonized standards help to create an environment of commerce where technology adoption and cross-sector partnerships are encouraged and where conditions are somewhat predictable. Support for the changing scenario of resource recovery calls for flexible policies and international collaboration.

### 6.6.4 Market Opportunities for Resource Recovery Products

Notwithstanding the difficulties mentioned above, there is great market opportunity for waste-sourced high-value goods. The growing emphasis on sustainability and circular economy ideas on a world scale has driven call for specialty items, renewable fuels, and bio-based chemicals. By transforming low-quality waste into marketable goods, companies that effectively innovate in waste valorization are creating fresh sources of income (Lee et al. 2023). Strict environmental regulations and consumers' growing demand for environmentally friendly items also help to push market possibilities (Patel et al. 2022).

## 6.7 Future Perspectives

Looking ahead to the future of sustainable biorefineries, the value of technological advancements, circular economy principles, synergy with renewable energy systems, and artificial intelligence and automation possibilities cannot be overstated. Taken together, these elements will very much improve waste to energy production and support more sustainable, affordable, and efficient biorefinery systems. Here we examine how progress in every one of these respects will influence the development of biorefinery science.

### 6.7.1 Developments in Waste to Bioenergy Processes

Now days have seen outstanding advances in waste-to-bioenergy (WtB) technologies, by using enzymatic and microbial methods for transforming lignocellulosic biomass into biofuels is among the major fields of development. According to Li et al. (2023), the creation of genetically engineered microbes and enzymes able to more effectively break down complex biomass is transforming the efficiency of bioenergy generation. Furthermore, being improved to handle a larger selection of feedstocks—such as agricultural residues, municipal solid waste (MSW), and even plastic waste (Zhang et al. 2024a, b, c, d) are advancements in pyrolysis and gasification technologies, that maximize methane yield from different waste streams including food waste and agricultural by-products have improved also biogas output by means of anaerobic digestion (Gao et al. 2023a, b). An emerging emphasis is on incorporating carbon capture and storage (CCS) into WtB systems, which enables biorefineries to be carbon negative and therefore supports the lowering of greenhouse gas emissions (Ghosh et al. 2024).

### 6.7.2 Role of Circular Economy in Biorefinery

For biorefineries, one of the most hopeful approaches for attaining sustainability is the shift to a circular economy. In designing biorefineries, the implementation of circular economy ideas promotes closed-loop systems of waste streams from food production that might be converted into biofuels, while the leftover biomass might be used as a soil fertilizer (Korol et al. 2023). Further improvement of resource use and decrease of environmental impact come from the integration of industrial symbiosis, whereby several industries cooperate to share waste streams (Kumar et al. 2024a, b). Biorefineries can also change their asset allocation to turn waste streams into high-value chemicals and bioplastics. This not only raises their economic sustainability but also lowers reliance on fossil energy (Zhou et al. 2024).

### 6.7.3 Renewable Energy Systems Integration in Combination

Energy self-sufficient and sustainability would be impossible lacking the synergy of biorefineries and renewable energy systems. Using renewable energy sources like solar, wind, and hydropower, biorefineries can cut down their dependence on fossil fuels and therefore be more environmentally friendly in the long term. Biogas generation, for example, can serve as an alternative power source when solar or wind energy output varies (Hosseini et al. 2023). Including biomass energy systems into biorefineries can also offer a constant power supply for ongoing activities, therefore improving both energy independence and biorefinery robustness (Fang et al. 2024). Hybrid renewable energy systems, which combine biomass, wind, and solar power, will serve to improve the overall efficiency and resilience of biorefineries (Hosseini et al. 2023). These developments will enable biorefineries to become major players in attempts to slow down climate change by achieving net zero emissions.

### 6.7.4 Potential of AI and Automation in Waste Management

Waste management, like many other sectors, is fast changing, thanks to automation technologies and artificial intelligence (AI). Automation and artificial intelligence could be key to predicting ideal circumstances for turning waste into biofuels or biogas using machine learning algorithms that examine waste composition. By improving process parameters and feedstock selection, this strategy can raise bioenergy yields and lower operational costs (Kumar et al. 2024a, b). Using sophisticated robotics and AI-driven sorting technologies, automation in waste sorting and collection would greatly increase the precision and speed of waste management in biorefineries. Automated systems can identify organic garbage from recycling, cutting human labor and raising the rate at which waste that could be transformed into bioenergy is processed (Zhou et al. 2024). The development of sustainable biorefineries depends much on progress in waste-to-bioenergy technologies, the acceptance of circular economy principles, integration with renewable energy systems, and the use of AI and automation in waste management. These advancements will result in more sustainable, cost-effective, and efficient biorefineries.

### 6.7.5 Recommendations for Future Research

While notable advancements have been achieved, numerous obstacles persist in enhancing and expanding biorefinery technologies. Upcoming research should concentrate on the following areas to realize the complete capabilities of biorefineries:

(a) *Feedstock optimization*: A primary issue in waste-to-bioenergy systems is the variance of feedstocks. Future investigations should aim at creating more effective approaches for preprocessing various waste materials and improving feedstock quality, particularly from municipal solid waste (MSW) and agricultural by-products (Zhang et al. 2024a, b, c, d). Studies into innovative feedstocks, such as food waste, marine algae, and industrial waste, could further broaden the spectrum of materials utilized in bioenergy generation (Ghosh et al. 2024).
(b) *Improved Enzymatic and Microbial Systems*: Research should persist in examining the genetic modification of microorganisms and enzymes to enhance the efficiency of converting complex organic waste materials. Creating more resilient microbial consortia that can decompose lignocellulosic biomass and plastic waste will be essential for scaling up biorefinery operations (Li et al. 2023). Furthermore, progressing in the application of synthetic biology may result in even higher efficiencies in waste conversion.
(c) *Integration with Energy Storage Systems*: As biorefineries progressively blend renewable energy, investigation into energy storage technologies becomes crucial. By enhancing biogas storage and biofuel storage, biorefineries could more effectively manage the unpredictable nature of renewable energy sources and deliver more stable and dependable energy output (Fang et al. 2024).

(d) *AI and Automation in Process Optimization*: Future research should emphasize the creation of AI algorithms for ongoing monitoring and process optimization in biorefineries. This encompasses predictive maintenance, resource optimization, and automated control systems to boost the overall efficiency and sustainability of operations (Kumar et al. 2024a, b). In addition, robotic waste sorting and AI-based analytics could improve the sorting and processing of waste materials, rendering biorefineries more cost-efficient.
(e) *Life Cycle Assessment and Sustainability Metrics*: There is a necessity for thorough life cycle assessments (LCAs) to assess the environmental, economic, and social effects of biorefineries. Research should strive to establish standardized sustainability metrics that can direct the design and functioning of biorefineries and assist stakeholders in making informed decisions (Korol et al. 2023). By tackling these research deficiencies, the complete potential of biorefineries can be attained, resulting in more efficient, sustainable, and economically feasible systems for resource recovery from waste.

## 6.8 Conclusion

For the future of sustainable biorefineries by the waste-to-bioenergy (WtB) technologies, innovations include the enhancement of microbial and enzymatic systems for the decomposition of complex waste materials such as lignocellulose (Li et al. 2023). Innovative techniques in gasification, pyrolysis, and anaerobic digestion are enhancing conversion effectiveness, resulting in increased production of biofuels and biogas (Zhang et al. 2024a, b, c, d). Furthermore, incorporating carbon capture into biorefineries can lead to carbon negativity, aligning them with objectives for climate change mitigation (Gao et al. 2023a, b). Applying circular economy principles in biorefinery design fosters a zero-waste approach by transforming waste into a valuable resource. By employing ideas such as industrial symbiosis, biorefineries can significantly reduce their environmental impact while expanding the range of products they offer, including biochemicals and bioplastics (Ghosh et al. 2024). Combining biorefineries with renewable energy sources like solar, wind, and biomass has the potential to make biorefineries energy independent and more robust. Hybrid renewable energy systems can stabilize energy availability by supporting variable renewable sources like wind and solar (Fang et al. 2024). The recent use of AI and automation in waste sorting, predictive analytics, and process optimization is revolutionizing biorefinery operations. AI technologies are enhancing the precision and effectiveness of waste conversion processes, while automation reduces labor costs and improves the accuracy of waste management (Kumar et al. 2024a, b). These advancements underscore that biorefineries, backed by state-of-the-art technologies, possess significant potential in converting waste into a vital resource, fostering both energy production and economic development while minimizing environmental impact.

## References

Alvarado A, Pérez M, Rodríguez M (2024) Production of bioplastics from agricultural waste: economic and environmental considerations. J Environ Manag 256:113557. https://doi.org/10.1016/j.jenvman.2024.113557

Bajpai P (2022) Biomass to biofuels: strategies for global industries. Elsevier. https://doi.org/10.1016/B978-0-12-821574-3.00010-2

Chandra R, Singh S, Kumar P (2024) Biogas production from industrial organic waste via anaerobic digestion: optimization of operational parameters. Bioresour Technol 184:125452. https://doi.org/10.1016/j.biortech.2024.125452

Choudhury M, Lee H, Das R (2024) Certification and standardization of bioproducts. J Ind Ecol 28(1):124–140. https://doi.org/10.1111/jiec.13241

Das R, Zhang Y, Patel A (2023) Strategies for carbon footprint reduction in biorefineries. Renew Energy 210:1284–1298. https://doi.org/10.1016/j.renene.2023.08.065

Dawar I, Srivastava A, Singal M, Dhyani N, Rastogi S (2025) A systematic literature review on municipal solid waste management using machine learning and deep learning. Artif Intell Rev. https://doi.org/10.1007/s10462-025-11196-9

Fang W, Zhang L, Wu H (2024) Hybrid renewable energy systems for biorefineries: enhancing resilience and sustainability. Renew Energy 173:556–565. https://doi.org/10.1016/j.renene.2024.02.032

Fang Z, Smith RL Jr, Guo H (2025) Production of organic acids and alcohols from agricultural residues and food wastes. Springer. https://doi.org/10.1007/s40974-023-00312-6

Fatinee Mohd Yusree FI, Peter AP, Mohd Nor MZ, Show PL, Mokhtar MN (2021) Latest advances in protein-recovery technologies from agricultural waste. Foods 10(11):2748. https://doi.org/10.3390/foods10112748

Fernandez G, Choudhury R, Kumar P (2023) Global policies on waste valorization. Waste Manag Res 41(9):1012–1030. https://doi.org/10.1177/0734242X23111235

Ferrusca MC, Romero R et al (2023) Biodiesel production from waste cooking oil: a perspective on catalytic processes. Processes. https://doi.org/10.3390/pr11071952

Gao F, Li J, Zhang S (2023a) Enzymatic hydrolysis and fermentation of food waste for bioethanol production. Renew Energy 162:1104–1113. https://doi.org/10.1016/j.renene.2020.09.105

Gao F, Zhang T, Li S (2023b) The integration of carbon capture and storage with waste-to-bioenergy systems. Renew Energy 171:1243–1250. https://doi.org/10.1016/j.renene.2023.02.041

Ghosh S, Sharma M, Roy K (2024) Circular economy applications in biorefinery design and operation. J Clean Prod 292:125040. https://doi.org/10.1016/j.jclepro.2024.125040

Gupta P, Kumar A, Singh R (2023) Life cycle assessment of biorefineries: a comprehensive review. Renew Sust Energ Rev 185:113805. https://doi.org/10.1016/j.rser.2023.113805

Hosseini SE, Ghazi M, Yeganeh B (2023) Integration of renewable energy with biorefinery systems for sustainable production. Energy 237:121567. https://doi.org/10.1016/j.energy.2023.121567

Jiao H, He X, Sun J, Elsamahy T, Al-Tohamy R, Kornaros M, Ali SS (2024) A critical review on sustainable biorefinery approaches and strategies for wastewater treatment and production of value-added products. Energy Ecol Environ. https://doi.org/10.1007/s40974-023-00312-6

Kaur R, Pathak L, Vyas P (2024) Biobased polymers of plant and microbial origin and their applications—a review. Biotechnol Sustain Mater. https://doi.org/10.1186/s44316-024-00014-x

Kavitha Shree GG, Arokiamary S, Kamaraj M, Aravind J (2025) Biorefinery approaches for converting fruit and vegetable waste into sustainable products. Int J Environ Sci Technol. https://doi.org/10.1007/s13762-024-06202-6

Keharika K, Pothu R, Boddula R (2024) Advances in waste valorization for sustainable biorefineries. Environ Res. https://doi.org/10.1016/j.envres.2024.117095

Khiari Z (2024) Enzymes from fishery and aquaculture waste: research trends in the era of artificial intelligence and circular bio-economy. Mar Drugs 22(9):411. https://doi.org/10.3390/md22090411

Korol L, Kumar S, Yang J (2023) Industrial symbiosis for sustainable biorefinery development. Renew Sust Energ Rev 169:112971. https://doi.org/10.1016/j.rser.2023.112971

Kumar A, Gupta R, Singh P (2023) Economic feasibility and scale-up challenges in sustainable biorefineries. Energy Econ 105:105456. https://doi.org/10.1016/j.eneco.2023.105456

Kumar V, Sharma B, Das R (2024a) Advanced LCA approaches in waste valorization. Resour Conserv Recycl 198:106424. https://doi.org/10.1016/j.resconrec.2024.106424

Kumar P, Singh S, Patel R (2024b) Predictive analytics for optimizing biorefinery operations using AI. Biofuels Bioprod Biorefin 18(2):1153–1166. https://doi.org/10.1002/bbb.2215

Lee SY, Sankaran R, Chew KW et al (2019) Waste to bioenergy: a review on the recent conversion technologies. BMC Energy. https://doi.org/10.1186/s42500-019-0004-7

Lee H, Choi Y, Park J (2023) Market opportunities for bio-based products from waste valorization. J Sustain Chem 12(4):275–290. https://doi.org/10.1016/j.suschem.2023.112345

Lee YY, Srinivaas M, Li IC, Keharika K, Pothu R, Boddula R, Al-Qahtani N, Huang BW, Chang-Chien GP (2024a) Enhancing sustainable energy through cutting-edge waste biorefinery technologies. Reactions.https://doi.org/10.3390/reactions5040059

Lee P, Rahman M, Fernandez G (2024b) Bio-based feedstocks for carbon neutrality. Biofuels Bioprod Biorefin 18(2):345–362. https://doi.org/10.1002/bbb.2412

Leong H, Y, Chang C-K et al (2021) Waste biorefinery towards a sustainable circular bioeconomy: a solution to global issues. Biotechnol Biofuels Bioprod. https://doi.org/10.1186/s13068-021-01939-5

Li X, Zhang S, Wang Y (2023) Genetically engineered microorganisms for efficient lignocellulosic biomass degradation. Biotechnol Biofuels 16(1):45–58. https://doi.org/10.1186/s13068-023-02120-w

Liu Z, Liao W, Liu Y (2016) A sustainable biorefinery to convert agricultural residues into value-added chemicals. https://doi.org/10.1186/s13068-016-0609-8

Maraveas C (2020) Production of sustainable and biodegradable polymers from agricultural waste. Polymers 12(5):1127. https://doi.org/10.3390/polym12051127

Marzban N, Hoffmann T, Sturm B (2023) Smart integrated biorefineries in bioeconomy: a concept toward zero-waste, emission reduction, and self-sufficient energy production. Biofuel J. https://doi.org/10.18331/BRJ2025.12.1.4

Marzban N, Psarianos M, Herrmann C (2025) Smart integrated biorefineries in bioeconomy: a concept toward zero-waste, emission reduction, and self-sufficient energy production. https://doi.org/10.18331/BRJ2025.12.1.4

Mishra A, Rahman M, Das S (2023) Social perception of bio-based waste valorization. Sustainability 15(7):5124. https://doi.org/10.3390/su15075124

Patel R, Fernandez L, O'Connor D (2022) Emerging trends in resource recovery from waste for biorefineries. Bioresour Technol 345:127–139. https://doi.org/10.1016/j.biortech.2022.134567

Pelagalli V, Langone M, Matassa S, Race M, Tuffi R, Papirio S, Lens PNL, Lazzazzara M, Frugis A, Petta L, Esposito G (2024) Pyrolysis of municipal sewage sludge: challenges, opportunities and new valorization routes for biochar, bio-oil, and pyrolysis gas. Environ Sci: Water Res Technol. https://doi.org/10.1039/D4EW00278D

Quevedo-Amador RA, Escalera-Velasco BP et al (2024) Application of waste biomass for the production of biofuels and catalysts: a review. Clean Techn Environ Policy. https://doi.org/10.1007/s10098-023-02728-4

Shahzad HMA, Asim Z, Khan SJ, Almomani F, Mahmoud KA, Mustafa MRU, Rasool K (2024) Thermochemical and biochemical conversion of agricultural waste for bioenergy production: an updated review. Discov Environ. https://doi.org/10.1007/s44274-024-00171-w

Sharma N, Sahota PP, Singh MP (2021) Organic acid production from agricultural waste. In: Waste to energy: prospects and applications. https://doi.org/10.1007/978-981-33-4347-4_17

Shree GK, Arokiamary S, Aravind J (2024) Biorefinery approaches for converting fruit and vegetable waste into sustainable products. Int J Environ Sci Technol. https://doi.org/10.1007/s13762-024-06202-6

Singh S, Verma D, Mehta K (2022) Environmental impact assessment of biomass-based biorefineries. J Clean Prod 347:131261. https://doi.org/10.1016/j.jclepro.2022.131261

Smith J, Doe A, Brown C (2022) Overcoming technical barriers in waste processing for biorefineries. J Clean Prod 320:128637. https://doi.org/10.1016/j.jclepro.2022.128637

Solarte-Toro JC, Alzate CAC (2023) Sustainability of biorefineries: challenges and perspectives. Energies 16(9):3786. https://doi.org/10.3390/en16093786

Tsegaye B, Jaiswal S, Jaiswal AK (2021) Food waste biorefinery: pathway towards circular bioeconomy. Foods. https://doi.org/10.3390/foods10061174

Zhang L, Fernandez M, Lee J (2024a) Multi-product strategy for biorefineries. Energy Convers Manag 297:117633. https://doi.org/10.1016/j.enconman.2024.117633

Zhang X, Li J, Liu Y (2024b) Optimization of feedstock quality in municipal solid waste-based biorefineries. Waste Manag 124:345–354. https://doi.org/10.1016/j.wasman.2024.01.017

Zhang Y, Li X, Zhao L (2024c) Pyrolysis of municipal solid waste: bio-oil production and energy recovery. Bioresour Technol 358:127672. https://doi.org/10.1016/j.biortech.2024.127672

Zhang X, Wang J, Zhao J (2024d) Algal biorefinery for $CO_2$ capture and biofuel production: pathways and perspectives. Algal Res 53:102543. https://doi.org/10.1016/j.algal.2024.102543

Zhao L, Nguyen T, Kim S (2021) Advances in pretreatment technologies for lignocellulosic waste valorization. Renew Energy 180:1143–1155. https://doi.org/10.1016/j.renene.2021.01.089

Zhou Y, Yang Y, Zhang L (2024) Robotic waste sorting and automation in biorefinery processes. Autom Constr 147:103550. https://doi.org/10.1016/j.autcon.2023.103550

# Progress in Conversion of Biomass into Biofuels

7

Nitin Chauhan, Sujata Adhana, Deepak Gola, Shweta Kalia, Anil Kumar Mavi, Priyadarshini Dey, Abhishek Choudhary, and Indra Mani

**Abstract**

A sustainable biorefinery for biofuel production is a facility designed to convert biomass into various types of biofuels while minimizing environmental impact and maximizing resource efficiency. The approach of sustainable biorefinery is based on utilizing the potential of renewable resources instead of fossil fuels as used by traditional petroleum refineries. The main objective is to develop closed-loop system that minimizes waste and environmental impact but maximizes the

N. Chauhan
Department of Zoology, Sri Aurobindo College, University of Delhi, New Delhi, India

S. Adhana
Department of Biomedical Sciences, Bhaskaracharya College of Applied Sciences, University of Delhi,
New Delhi, India

D. Gola
Trinity International, New Delhi, India

S. Kalia
Department of Biotechnology, IMS Engineering College Ghaziabad, Ghaziabad, UP, India

A. K. Mavi
Department of Botany, Sri Aurobindo College, University of Delhi, New Delhi, India

P. Dey
Department of Biotechnology, MS Ramaiah Institute of Technology, Bengaluru, Karnataka, India

A. Choudhary
Galgotias University, Greater Noida, UP, India

I. Mani (✉)
Department of Microbiology, Gargi College, University of Delhi, New Delhi, India
e-mail: indra.mani@gargi.du.ac.in



V. Singh (ed.), *Sustainable Waste Management Towards Circular Bioeconomy*,
https://doi.org/10.1007/978-981-95-2227-9_7

potential of renewable resources. For sustainable biorefinery, the key ingredients would be agricultural residues, forestry residues, energy crops, and organic waste. The ideal approach is the use of nonfood biomass to prevent competition with food production and also ensure sustainable land-use practices. Moreover, a more advanced strategy called integrated biorefinery is now focusing on utilizing the biomass to its maximum efficiency and converting it into value-added bio-based products (biocomposites, fibers, biopolymers) and biochemicals (organic acids, alcohols, enzymes, bioplastics) along with biofuels (ethanol, biodiesel, biogas, biojet fuel). The process of biorefinery is based on pretreatment which helps in initial processing to break down biomass into its fundamental components, followed by primary conversion and secondary conversion to form intermediate products and high-value products, respectively. Further, waste valorization, which is the process of converting by-products and wastes into products with added value, such as producing biogas from residues, is performed. More research is needed to standardize the processing steps, create efficient waste management, monitor the quality and sustainability, and evaluate economic viability.

**Keywords**

Sustainable biorefinery · Biofuels · Feedstock · Biochemical · Valorization

## 7.1 Introduction

The high energy demand creates huge burden on nonrenewable resources. Biorefinery has now become a crucial participant in sustainable strategy. A biorefinery for biofuel production is a facility that utilizes biomass feedstocks to produce biofuels through various conversion processes. Biomass feedstocks typically need to undergo preprocessing to remove impurities, reduce particle size, and enhance accessibility for subsequent conversion processes. This may involve processes like milling, grinding, or chipping. In order to replace large quantities of petroleum fuel, there should be a high availability of feedstock source. The conversion of biomass into biofuels and biochemicals generally requires upgrading biorefinery technologies and biomass. Processes such as fractionation, liquefaction, pyrolysis, hydrolysis, fermentation, and gasification are considered under biomass upgrading (Rodionova et al. 2022). The bio-oil received after pyrolysis of biomass can act as a fuel for vehicles. Biomass feedstocks can include lignocellulosic biomass, agricultural residues (like corn and wheat straw), food waste, onion waste, potato peel waste, forestry waste (sawdust, wood chips), and algae. Literature suggested that due to the structural complexity of the feedstock, a suitable pretreatment method has to be employed for efficient enzymatic and microbial conversion. The first-generation biofuel (bioethanol) does not need rigorous pretreatment processes, but the major disadvantage is the release of greenhouse gases. The second-generation biofuel (bioethanol, biohydrogen, and biomethane) requires extensive pretreatment methods, but the main advantage is that it utilizes waste as feedstock. The third-generation biofuel (bioethanol) utilizes the capacity of microalgae to provide good yield. Further, genetically modified feedstock with carbon capture and sequestration technology is associated with fourth-generation biofuel. Finally, in order to produce several generations of biofuels in one process, with added advantage of valorization of feedstock and bioprocess, a more sustainable method called integrated biorefinery is developed and utilized (Moodley 2021). There are several advantages associated with integrated biorefinery which also depends on different feedstocks and products (Demirbas 2009). In this chapter, we have provided state-of-the-art overview about the diverse feedstocks with their appropriate process mechanisms needed for biofuel and associated product development. To provide efficient biorefinery products to the end users, a large investment with better infrastructure facilities is required.

## 7.2 Algal Biorefinery for Biofuel Production

Algal biorefinery stands at the forefront of sustainable energy solutions, offering a revolutionary approach to biofuel production through the utilization of algae, a versatile and prolific source of biomass (Quereshi et al. 2017; Chye et al. 2018). This innovative concept integrates various components and processes to transform algal biomass into a range of valuable products, with biofuels taking the spotlight (Mathimani and Pugazhendhi 2019). At its core, algal biorefinery embodies the principles of circular economy and resource efficiency, capitalizing on the inherent abilities of algae to convert sunlight and carbon dioxide into organic compounds through photosynthesis (Sharma et al. 2021a; Zabochnicka et al. 2022). The journey of algal biomass through the biorefinery begins with cultivation systems, which encompass diverse setups ranging from open ponds to closed-loop photobioreactors, each tailored to optimize productivity, scalability, and economic viability (Yen et al. 2019; Angeles et al. 2020; Paladino and Neviani 2021). Once cultivated, algae undergo harvesting and dewatering processes to concentrate biomass and remove excess water, employing methods such as filtration, centrifugation, and flocculation to achieve efficient separation (Zhu et al. 2018). Biomass pretreatment aims to enhance the accessibility of intracellular components through drying, milling, and cell disruption, thus facilitating downstream processing (Bharte and Desai 2021).

Fractionation and extraction emerge as pivotal steps in the algal biorefinery workflow, where various biomass fractions including lipids, proteins, and carbohydrates are separated and recovered for further utilization (Zhou et al. 2022). Extraction techniques like solvent extraction, mechanical pressing, and enzymatic hydrolysis are deployed to isolate these valuable components, laying the foundation for subsequent conversion processes (Show et al. 2015). In the realm of biofuel production, algal biorefinery offers a multitude of pathways, with lipids serving as a precursor for biodiesel production through transesterification, carbohydrates undergoing fermentation to yield bioethanol, and biomass itself being subjected to hydrothermal liquefaction to generate bio-oil, which can be further refined into drop-in fuels (Gu et al. 2020).

However, the significance of algal biorefinery transcends mere biofuel production, extending into the realm of coproduct recovery and value addition (Fig. 7.1). Nutrient-rich residues generated during biomass processing can be recycled back into cultivation systems, fostering a closed-loop approach and enhancing the sustainability of the entire process. Moreover, algal biorefinery yields a spectrum of coproducts ranging from high-protein animal feed to specialty chemicals and pharmaceuticals, thus diversifying revenue streams and enhancing the economic viability of the biorefinery model. Furthermore, the carbon sequestration potential of algae emerges as a critical aspect, with algae sequestering carbon dioxide from industrial sources during cultivation, thereby offering a promising pathway toward carbon-negative biofuel production (Chandra et al. 2019).

Despite its immense potential, algal biorefinery encounters several challenges on the road to commercialization. Scalability, cost-effectiveness, and technological limitations pose significant hurdles, necessitating continuous research and development efforts to overcome these barriers (Lam et al. 2019). Strain selection, cultivation optimization, and process integration emerge as key areas for innovation, with advancements in bioreactor design, genetic engineering, and conversion technologies poised to unlock the full potential of algal biorefinery (Larkum et al. 2012; Coleman et al. 2014). Moreover, regulatory frameworks and market incentives play a pivotal role in fostering the widespread adoption of algal biorefinery, facilitating its integration into the existing energy infrastructure, and accelerating the transition toward a sustainable, low-carbon future.

Algal biorefinery represents a paradigm shift in the production of biofuel, harnessing the inherent capabilities of algae to address the pressing challenges of energy security, environmental sustainability, and climate change mitigation. Through its holistic approach to biomass utilization, algal biorefinery offers a pathway toward a circular bioeconomy, where waste is minimized, resources are

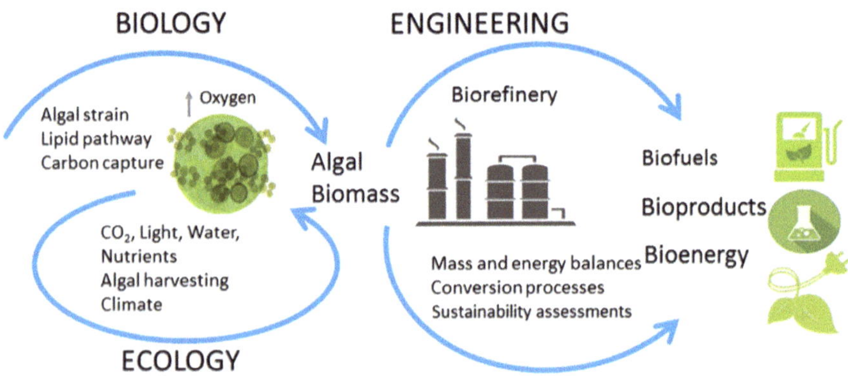


**Fig. 7.1** Sustainable integrated approach highlighting biology, ecology and its integration with algal biorefinery. (Adapted from Allen et al. 2018)

maximized, and value is created at every stage of the process. As research advances and technology matures, algal biorefinery promises to transform the landscape of renewable energy, ushering in a new era of sustainable development and prosperity.

## 7.3 Potato Peel Waste Biorefinery

Potato peel waste has been shown to be a good feedstock utilizing biological and thermochemical techniques and produces various important by-products. The refinery of potato peel waste can efficiently convert agricultural waste into biofuels, thereby helping to accomplish the goal of sustainable energy production. For the production of biofuel, processes such as detoxification, pretreatment, hydrolysis, and fermentation are performed (Ebrahimian et al. 2022). Potato peel waste generally contains starch, lignin, cellulose, and hemicellulose, and because of these constituents, it is considered as an efficient feedstock for biorefinery industry. A biorefinery of potato peel produces bioethanol and biogas, with additional solid residue which can be used as a fertilizer, contributing to its maximum utilization (Barampouti et al. 2023). For the bioethanol production, potato peel waste undergoes stages of liquefaction and saccharification. In order to attain these, several pretreatment stages need to be performed, i.e., thermal, acid, basic, or enzymatic ones (Rodríguez-Martínez et al. 2023; Ben Atitallah et al. 2019; Zhang et al. 2010). Pretreatment steps play a significant role as they disrupt the stearic hindrances of lignin and hemicellulose which further helps to increase the accessibility of enzyme and digestion of cellulose (Soni et al. 2023).

As mentioned earlier, potato peel waste is rich in starch, pectin, cellulose, and hemicelluloses; therefore, the combination of various enzymes like carbohydrases, including endo-1, 2-beta-D-glucanase, exo-1,4-beta-D-glucanase, beta-1,4-glucosidase, endo-beta-1,4-xylanase, pectinase, amylase, etc., can be utilized to convert biomass into free sugars. This kind of enzyme combination is easily not available; therefore, it is advisable to use appropriate microorganisms for the production of mentioned enzymes. Further, it is important to regulate the processes of pretreatments and enzymatic hydrolysis with the production of necessary cocktail of enzymes for the cost-effective and sustainable manufacturing of ethanol from potato peel waste (Soni et al. 2023).

## 7.4 Onion Waste-Based Biorefinery

In order to obtain biofuel and other useful biochemical, onion waste-based biorefinery has emerged as an innovative approach for sustainable energy production. Onion has been the second largest crop produced globally for food, oil, pickles, etc. However, during its processing, large quantity of waste material in the form of skin, stalk, and root is generated; moreover, onion waste is also produced as postharvest losses. The analysis of onion constitutes suggested the presence of dietary fibers, phenolics, flavonoids, cellulose, hemicellulose, and simple sugars. The waste obtained from onions can be utilized for the development of biofuels, organic acids, polysaccharides, and polyphenols. For onion waste biorefinery, the processes, i.e., enzymatic hydrolysis, fermentation, and hydrothermal carbonization, have been highlighted (Sagar et al. 2022). In a research, Shahid et al. (2023) studied the potential of bioethanol production from onion waste by *Saccharomyces cerevisiae* and *Pichia pastoris.* The study highlighted that the monoculture of *S. cerevisiae* has the potential to produce maximum bioethanol concentration of 30.56 g/L, with a productivity and energy at 0.355 g/L/h and 916.8 kJ/L, respectively, after 86 h of fermentation at 30 °C and pH 5. On the other hand, a maximum of 21.06 g/L bioethanol concentration with productivity and energy of 0.26 g/L and 631.8 kJ/L, respectively, after 72 h of fermentation at 30 °C and pH 5 was obtained from *P. pastoris*. In another study, Kim et al. (2019) performed a 5-day experiment for the acetic acid production from onion waste. It was observed that onion waste is efficient in acetic acid production. The simultaneous saccharification and two-step fermentation method highlighted better performance than simultaneous saccharification and cofermentation process. The simultaneous saccharification and two-step fermentation produced 19.3 g/L of acetic acid with highest conversion yield of 90.5% from onion waste when 6% substrate loading (w/v) was used.

## 7.5 Lignocellulosic Biorefineries for the Sustainable Production of Biofuels

Agro-industrial practices globally generate significant amounts of lignocellulosic biomass waste (Sharma et al. 2021b). According to the Department of Energy, the United States annually generates 1.3 billion tons of dry lignocellulosic biomass. Of this total, agricultural residues account for 933 million tons per year, while forest wastes contribute 368 million tons per year (De Bhowmick et al. 2018). It is the largest source of energy production; however, conventional methods for energy production using waste biomass have adverse environmental effects (Sharma et al. 2023). To meet rising energy demands sustainably, a systematic approach utilizing lignocellulosic biomass for "green" fuels and biochemicals through lignocellulosic biorefineries is recommended (Kumar and Verma 2021). This approach aligns with the economic policy of zero waste, distinguishing it from traditional petrochemical refineries (Velidandi et al. 2023).

Lignocellulose, produced by plant cells, consists of cellulose, hemicellulose, and lignin, along with pectin, proteins, and inorganic compounds in smaller fraction (Brandt et al. 2013). Cellulose accounts for 30–50% of the dry weight, with a high degree of polymerization making it rigid and insoluble. Hemicellulose (20–40% dry matter), with a lower degree of polymerization, is branched and contains various functional groups, while lignin (15–25% dry matter) provides structural stability to plant tissue (Cherubin et al. 2018; Menon and Rao 2012). Hemicellulose is more easily hydrolyzed than cellulose to produce ethanol, while lignin can be extracted for other chemicals in a biorefinery context. The complex structures of lignocellulosic constituents make them resistant to degradation, requiring pretreatment processes to release polysaccharides for conversion, which can be costly and involve particle size reduction, enzymatic hydrolysis, and fermentation.

The pretreatment process is required to increase the surface area of cellulose and hemicellulose, which improves their accessibility and permeability. Various physical, chemical, and biological treatment methods are reported in the literature for lignocellulosic pretreatment (Table 7.1). Physical treatment method disrupts the refractory structure of biomass to increase its specific surface area, thereby rendering it more accessible to enzymes for hydrolysis (Sharma et al. 2022). It includes milling (reduces cellulose crystallinity and particle size), extrusion (shearing, heating, shearing, and mixing of biomass), microwave, and ultrasound (removal of lignin and yields high sugar). Chemical pretreatment techniques greatly reduce the degree of polymerization and crystallinity of the lignocellulosic biomass and successfully extract lignin. It includes acid and alkali treatment (increase the disruption of biomass structure before hydrolysis), green pretreatment solvents, ionic liquid treatment (dissolution-precipitation process to dissolve cellulose and also eliminate lignin), and deep eutectic solvent treatment (DES pretreatment which increases lignocellulose's solubility and digestibility) (Yadav et al. 2023). However, physical and chemical pretreatment methods involve some drawbacks like energy intensive, toxic substance formation, longer residence time, etc. (De Bhowmick et al. 2018). Biological method (lignocellulolytic enzymes, fungus) is an eco-friendly option from the pretreatment of lignocellulosic biomass. The combining of two or three pretreatment methods to maximize biomass conversion efficiency while minimizing costs and environmental impacts.

After pretreatment, the biomass is hydrolyzed into fermentable sugars for ethanol production using acid at optimum pH and temperature. Concentrated acid hydrolysis necessitates operating at temperatures below 50 °C, atmospheric pressures, and acid concentrations ranging from 10% to 30% (Machineni 2020). Enzymatic hydrolysis is commonly utilized for ethanol production from lignocellulosic residues. It avoids by-product formation that could inhibit enzyme activity (Álvarez et al. 2016; Waghmare et al. 2018).

The lignocellulosic biomass conversion led to the formation of integral components of biofuels such as bioethanol, hydrogen, and methane through two possible methods. First is thermochemical conversion, where biomass is gasified or pyrolyzed to produce syngas which is further converted to biofuel using biological process. Second is biochemical conversion, where complex carbohydrates are converted to biofuels using microorganisms. Third is hybrid conversion where products formed after biochemical conversion are further converted into high-yield products using thermochemical conversion (Balan 2014). Various lignocellulosic wastes, such as wheat straw, rice husk, rice straw, corn cob, sunflower hulls, potato peels, etc., are used for ethanol production. Raj and Krishnan (2019) reported 91% of ethanol yield from sugarcane bagasse, and Zhu et al. (2019) reported 68% bioethanol yield from poplar wood.

Although lignocellulosic biorefineries have many benefits and play a significant role in the circular bioeconomy, significant barriers to profitability and commercialization exist. Life cycle assessment (LCA) of the biofuel production process

**Table 7.1** Various pretreatment methods used for conversion of lignocellulosic biomass

| Substrate | Pretreatment method | Conditions | Hydrolysis | Biofuel production | Remarks | References |
|---|---|---|---|---|---|---|
| Sugarcane straw | Microwave-assisted acid pretreatment | 220 °C and 0.3% (*w/v*) $CH_3COOH$ | Enzymatic hydrolysis | Butyric acid: 19.5 $gL^{-1}$ yield (0.46 g butyric acid/g of sugar) | – | Fonseca et al. (2020) |
| Rice straw | Ultrasound-assisted acid pretreatment | 0.55 g $MnO_2$, 56.6 mM HCl, 192.5 °C, 34 min ultrasonication | Catalyst ($MnO_2$) | 52.1% 2,5-furandicarboxylic acid | Original high content of Fe and Mn in the solution causes moderate to heavy pollution; high Zn concentration causes serious pollution threat | Chai et al. (2021) |
| Rice straw | Microwave-assisted alkali pretreatment | Microwave irradiation at 120 °C for 30 min | Enzymatic hydrolysis | Sugar yield: 350 $mgg^{-1}$ | Pretreatment method consume 60.05 Wh/kg energy | Jasmine et al. (2023) |
| *Ipomoea carnea* and *Arundo donax* | Ultrasound-assisted alkali pretreatment | Ultrasonication at 400 W for 15 min | Enzymatic hydrolysis | Reducing sugars: 294.08 $mgg^{-1}$ and bioethanol 85.04% | – | Muthuvelu et al. (2019) |
| Wheat straw | Ultrasound-assisted ionic liquid pretreatment | 130 °C for 15 and 30 min at 100 W | Enzymatic hydrolysis | Ethanol yield 38.9 $gL^{-1}$ and 42.0 $gL^{-1}$ | 148.1 and 172.5 g ethanol: 1 kg Initial wheat straw pretreated for 15 and 30 min | Ziaei-Rad et al. (2023) |
| Hemp fibers | Microwave-assisted eutectic solvent pretreatment | Radiation at 100 °C for 5 min | Enzymatic hydrolysis | Biochar yield: 26% | – | Ahmed et al. (2023a, b) |
| Wheat straw | Microwave-assisted hydrothermal treatment | 120 °C for 10 min | Anaerobic digestion | Methane production: 21.53% | The energy profit value is 11.8 confirming co-digestion improved energy efficiency of the process | Zhao et al. (2023) |

estimates the environmental impacts associated with it (Ubando et al. 2022). It estimates the total emissions produced throughout the whole process concluding how the process affects the environment and decision-makers can take action on it (Saravanan et al. 2023). Zoppi et al. (2023) performed LCA on a biorefinery process that combined aqueous phase reforming (APR) and hydrothermal liquefaction (HTL) to produce biofuels from corn stover, a lignin-rich lignocellulosic feedstock. The LCA findings indicated that the bioprocess exhibited a global warming potential of 56.1 and 58.4 g $CO_2$ eq/MJ biofuel, respectively. Furthermore, in comparison to fossil diesel, the integration of APR-HTL resulted in a 37% reduction in environmental impact, which further decreased to 80% when green energy was utilized, particularly with high lignin content. Also, Song et al. (2023) performed LCA to study the impact of eutectic solvent on biofuel production from microalgae. The results showed NER (net energy ratio) value increases from 0.65 to 0.83 increased energy output. These results offer policymakers a framework for evaluating the energy benefits of enhanced bioprocessing methods for converting lignocellulosic biomass into biofuels and other bio-products.

## 7.6 Food Waste-Based Sustainable Biorefineries for Generation of Biofuels

Food waste (FW) is a serious worldwide problem that exacerbates resource depletion, greenhouse gas emissions, and environmental pollution. It is one of the biowastes that is produced the most globally (Perez et al. 2023). The term "food losses" refers to "the decrease in edible food mass throughout the part of the supply chain that specifically leads to edible food for human consumption" by the Food and Agriculture Organization (FAO) (Gustavsson et al. 2011). Food waste presents ethical concerns as well as environmental and financial ones since it could be utilized to address global hunger (Melikoglu et al. 2013). The materials that are deteriorated, contaminated, thrown away, or lost and cannot be used in any way for human consumption are usually referred to as food waste (Tonini et al. 2018). Food waste generation is expected to worsen in the coming years as a result of global population growth and economic development in many Asian countries (Aleluia and Ferrão 2016). Through the production of biofuels, which can replace fossil fuels and reduce greenhouse gas emissions, food waste can be managed sustainably in biorefineries (Singh et al. 2016; Tsegaye et al. 2021). Each year, 13 billion and one-half tons of food (Paritosh et al. 2017) and a third of the food produced worldwide is wasted, harming the world economy to the tune of about \$750 billion, or INR 47 trillion (Gustavsson et al. 2011). It is projected that between 2005 and 2025, the production of urban food waste (FW) in Asian countries will rise annually by 278 to 416 million tons (Paritosh et al. 2017). India ranks seventh globally in terms of total food waste, while the Russian Federation tops the list. The environment is suffering from the massive production of FW, mainly because of greenhouse gas (GHG) emissions (Gustavsson et al. 2011). This chapter examines the possibilities for producing biofuel from food waste-based biorefineries, emphasizing their contribution to renewable energy production and sustainable waste management.

## 7.7 Concepts and Technologies of Food Waste Biorefineries

Biorefineries that process food waste use different conversion technologies to get value out of waste streams. Anaerobic digestion, fermentation, and thermochemical procedures like pyrolysis and gasification are some of these technologies. One popular method for turning organic waste into biogas, a sustainable energy source, is anaerobic digestion. Bioethanol, which can be mixed with gasoline or utilized as vehicle fuel, is created when food waste is fermented. Food waste can be converted by thermochemical processes into syngas or biodiesel, which can then be processed further to produce liquid fuels or chemicals (Ahmed et al. 2023b).

### 7.7.1 Biofuel Production from Food Waste

**Biodiesel** Transesterification is the process by which food waste oils and fats can be converted into biodiesel. Grease trap and leftover cooking oil are common feedstocks used in the manufacturing of biodiesel. Biodiesel can be used in existing diesel engines without modification and has properties similar to conventional diesel (Ahmed et al. 2023b).

**Bioethanol** Food waste with high sugar or starch content can be converted to bioethanol. Sugars are converted to ethanol through the processes of saccharification and fermentation. Fruit and vegetable wastes, as well as leftovers from food processing, are sources of food waste that can be used to produce bioethanol (Ahmed et al. 2023b).

**Biogas** The process of organic matter is digested anaerobically in food waste yields biogas. Biogas with high methane content can be converted to biomethane for use as fuel for vehicles, or heat and power can be produced from this renewable energy source (Ahmed et al. 2023b).

## 7.8 Sustainable Valorization of Sugarcane Residues for Production of Biofuels

Sugarcane is a source of sucrose, jaggery, and syrup. After successful extraction of main components, residues in the form of bagasse (the fibrous material left after extracting juice) and straw (leaves and tops) are abundant by-products of the sugarcane industry. The presence of good amount of lignocellulosic, mineral, and other organic nutrients and biochemicals in leftover residues makes it an efficient biofuel producer and also promotes economic growth. By using the leftover of sugarcane industry for the development of biofuel, it is considered as an affordable sustainable solution. Yadav et al. (2022) highlighted that sugarcane bagasse has low ash content with the presence of cellulose 45%, hemicellulose 32%, and lignin 17%. The main problem in using bagasse is the need of large quantity of enzymes for the conversion of lignocellulosic biomass into fermentable sugars. As per Reena et al. (2022), in order to produce bioethanol for bagasse carries processes, i.e., pretreatment, cellulose hydrolysis, fermentation, separation, and effluent treatment needs to be carried out. It has been identified that lignocellulosic biomass of sugarcane bagasse is made of complex intrinsic structure and also lacks monosaccharaides; therefore, it is difficult for microorganisms to convert into fermentable sugars. It has been suggested that before the introduction of hydrolytic enzymes, a pretreatment process to break the material through physiological process is highly required.

Wani et al. (2023) highlighted that the use of metagenomics and meta-omics techniques is helpful in the identification and characterization of appropriate microorganisms which can play a key role in the breakdown of sugarcane bagasse. With recent advances in sugarcane biorefinery, it is now identified as an important eco-friendly and sustainable process.

## 7.9 Conclusion and Future Perspectives

The green, safe, sustainable, and efficient technologies are the main requirements to achieve the sustainable development goals. In this direction, biorefineries are playing important role in the circular bioeconomy as they are harnessing the potential of feedstock such as lignocellulosic biomasses, food waste, microalgae, etc. These materials are used for the production of biofuels through various intermediate processing steps. The integrated biorefinery is further utilizing the residual waste and using the biomass to its maximum level by utilizing latest advanced technologies and strategies. By these methods, along with biofuels, the production of various high-value commercial products takes place which is contributing greatly to the circular bioeconomy. With establishment of biorefineries globally, it is equally important to keep upgrading and identifying the most feasible design, feedstock, and processes as part of research and development exercise. The sustainable design needs proper monitoring and evaluation. Moreover, it is also important to visualize the type of products that will be generated and how can it be useful for future generations. There is a tremendous scope of research in the field of biorefineries and its upgradation.

**Conflict of Interest** Authors declare that they do not have any conflict of interest.

## References

Ahmed B, Gwon J, Thapaliya M, Adhikari A, Ren S, Wu Q (2023a) Combined effects of deep eutectic solvent and microwave energy treatments on cellulose fiber extraction from hemp bast. Cellulose. https://doi.org/10.1007/s10570-023-05081-3

Ahmed SF, Kabir M, Mehjabin A, Oishi FTZ, Ahmed S, Mannan S, Mofijur M, Almomani F, Badruddin IA, Kamangar S (2023b) Waste biorefinery to produce renewable energy: bioconversion process and circular bioeconomy. Energy Rep 10:3073–3091

Aleluia J, Ferrão P (2016) Characterization of urban waste management practices in developing Asian countries: a new analytical framework based on waste characteristics and urban dimension. Waste Manag 58:415–429

Allen J, Unlu S, Demirel Y, Black P, Riekhof W (2018) Integration of biology, ecology and engineering for sustainable algal-based biofuel and bioproduct biorefinery. Bioresour Bioprocess 5(1):1–28

Álvarez C, Reyes-Sosa FM, Díez B (2016) Enzymatic hydrolysis of biomass from wood. Microb Biotechnol. https://doi.org/10.1111/1751-7915.12346

Angeles R, Arnaiz E, Gutiérrez J, Sepúlveda-Muñoz CA, Fernandez-Ramos O, Muñoz R, Lebrero R (2020) Optimization of photosynthetic biogas upgrading in closed photobioreactors combined with algal biomass production. J Water Process Eng 38:101554

Balan V (2014) Current challenges in commercially producing biofuels from Lignocellulosic biomass. ISRN Biotechnol. https://doi.org/10.1155/2014/463074

Barampouti EM, Christofi A, Malamis D, Mai S (2023) A sustainable approach to valorize potato peel waste towards biofuel production. Biomass Convers Biorefinery 13(9):8197–8208

Ben Atitallah I, Antonopoulou G, Ntaikou I, Alexandropoulou M, Nasri M, Mechichi T, Lyberatos G (2019) On the evaluation of different saccharification schemes for enhanced bioethanol production from potato peels waste via a newly isolated yeast strain of Wickerhamomyces anomalus. Bioresour Technol 289:121614

Bharte S, Desai K (2021) Techniques for harvesting, cell disruption and lipid extraction of microalgae for biofuel production. Biofuels 12(3):285–305

Brandt A, Gräsvik J, Hallett JP, Welton T (2013) Deconstruction of lignocellulosic biomass with ionic liquids. Green Chem. https://doi.org/10.1039/c2gc36364j

Chai Y, Yang H, Bai M, Chen A, Peng L, Yan B, Zhao D, Qin P, Peng C, Wang X (2021) Direct production of 2, 5-Furandicarboxylicacid from raw biomass by manganese dioxide catalysis cooperated with ultrasonic-assisted diluted acid pretreatment. Bioresour Technol. https://doi.org/10.1016/j.biortech.2021.125421

Chandra R, Iqbal HM, Vishal G, Lee HS, Nagra S (2019) Algal biorefinery: a sustainable approach to valorize algal-based biomass towards multiple product recovery. Bioresour Technol 278:346–359

Cherubin MR, Oliveira DMDS, Feigl BJ, Pimentel LG, Lisboa IP, Gmach MR, Varanda LL, Morais MC, Satiro LS, Popin GV, De Paiva SR, Dos Santos AKB, De Vasconcelos ALS, De Melo PLA, Cerri CEP, Cerri CC (2018) Crop residue harvest for bioenergy production and its implications on soil functioning and plant growth: a review. Sci Agric. https://doi.org/10.1590/1678-992x-2016-0459

Chye JTT, Jun LY, Yon LS, Pan S, Danquah MK (2018) Biofuel production from algal biomass. In: Bioenergy and biofuels. CRC Press, pp 87–118

Coleman AM, Abodeely JM, Skaggs RL, Moeglein WA, Newby DT, Venteris ER, Wigmosta MS (2014) An integrated assessment of location-dependent scaling for microalgae biofuel production facilities. Algal Res 5:79–94

De Bhowmick G, Sarmah AK, Sen R (2018) Lignocellulosic biorefinery as a model for sustainable development of biofuels and value added products. Bioresour Technol. https://doi.org/10.1016/j.biortech.2017.09.163

Demirbas MF (2009) Biorefineries for biofuel upgrading: a critical review. Appl Energy 86:S151–S161

Ebrahimian F, Denayer JFM, Karimi K (2022) Potato peel waste biorefinery for the sustainable production of biofuels, bioplastics, and biosorbents. Bioresour Technol 360:127609. https://doi.org/10.1016/j.biortech.2022.127609

Fonseca BC, Reginatto V, López-Linares JC, Lucas S, García-Cubero MT, Coca M (2020) Acetic acid as catalyst for microwave-assisted pretreatment of sugarcane straw aids highly specific butyric acid bioproduction. Ind Crop Prod. https://doi.org/10.1016/j.indcrop.2020.112936

Gu X, Martinez-Fernandez JS, Pang N, Fu X, Chen S (2020) Recent development of hydrothermal liquefaction for algal biorefinery. Renew Sust Energ Rev 121:109707

Gustavsson J, Cederberg C, Sonesson U (2011) Global food losses and food waste. Save Food at Interpack Düsseldorf, Germany

Jasmine A, Rajendran M, Thirunavukkarasu K, Abinandan S, Vaidyanathan VK, Krishnamurthi T (2023) Microwave-assisted alkali pre-treatment medium for fractionation of rice straw and catalytic conversion to value-added 5-hydroxymethyl furfural and lignin production. Int J Biol Macromol. https://doi.org/10.1016/j.ijbiomac.2023.123999

Kim HM, Choi IS, Lee S, Yang JE, Jeong SG, Park JH et al (2019) Biorefining process of carbohydrate feedstock (agricultural onion waste) to acetic acid. ACS Omega 4(27):22438–22444

Kumar B, Verma P (2021) Biomass-based biorefineries: an important architype towards a circular economy. Fuel. https://doi.org/10.1016/j.fuel.2020.119622

Lam MK, Khoo CG, Lee KT (2019) Scale-up and commercialization of algal cultivation and biofuels production. In: Biofuels from algae. Elsevier, pp 475–506

Larkum AW, Ross IL, Kruse O, Hankamer B (2012) Selection, breeding and engineering of microalgae for bioenergy and biofuel production. Trends Biotechnol 30(4):198–205

Mathimani T, Pugazhendhi A (2019) Utilization of algae for biofuel, bio-products and bio-remediation. Biocatal Agric Biotechnol 17:326–330

Machineni L (2020) Lignocellulosic biofuel production: review of alternatives. Biomass Conversion and Biorefinery, 10(3):779–791

Melikoglu M, Lin CSK, Webb C (2013) Analysing global food waste problem: pinpointing the facts and estimating the energy content. Cent Eur J Eng 3:157–164

Menon V, Rao M (2012) Trends in bioconversion of lignocellulose: biofuels, platform chemicals & biorefinery concept. Prog Energy Combust Sci. https://doi.org/10.1016/j.pecs.2012.02.002

Moodley P (2021) Sustainable biofuels: opportunities and challenges. Sustain Biofuels:1–20

Muthuvelu KS, Rajarathinam R, Kanagaraj LP, Ranganathan RV, Dhanasekaran K, Manickam NK (2019) Evaluation and characterization of novel sources of sustainable lignocellulosic residues for bioethanol production using ultrasound-assisted alkaline pre-treatment. Waste Manag. https://doi.org/10.1016/j.wasman.2019.02.015

Paladino O, Neviani M (2021) Airlift photo-bioreactors for Chlorella vulgaris cultivation in closed-loop zero waste biorefineries. Biomass Bioenergy 144:105926

Paritosh K, Kushwaha SK, Yadav M, Pareek N, Chawade A, Vivekanand V (2017) FW to energy: an overview of sustainable approaches for FW management and nutrient recycling. Biomed Res Int 2017:1–19

Pérez T, Vergara SE, Silver WL (2023) Assessing the climate change mitigation potential from food waste composting. Sci Rep 13(1):1–14

Quereshi S, Ahmad E, Pant KK, Dutta S (2017) Recent advances in production of biofuel and commodity chemicals from algal biomass. In: Algal biofuels: recent advances and future prospects. Springer, Cham, pp 393–419

Raj K, Krishnan C (2019) Improved high solid loading enzymatic hydrolysis of low-temperature aqueous ammonia soaked sugarcane bagasse using laccase-mediator system and high concentration ethanol production. Ind Crop Prod. https://doi.org/10.1016/j.indcrop.2019.01.032

Reena R, Alphy MP, Reshmy R, Thomas D, Madhavan A, Chaturvedi P, Pugazhendhi A, Awasthi MK, Ruiz H, Kumar V, Sindhu R, Binod P (2022) Sustainable valorization of sugarcane residues: efficient deconstruction strategies for fuels and chemicals production. Bioresour Technol 361:127759. https://doi.org/10.1016/j.biortech.2022.127759

Rodionova MV, Bozieva AM, Zharmukhamedov SK, Leong YK, Lan JCW, Veziroglu A et al (2022) A comprehensive review on lignocellulosic biomass biorefinery for sustainable biofuel production. Int J Hydrog Energy 47(3):1481–1498

Rodríguez-Martínez B, Coelho E, Gullón B, Yáñez R, Domingues L (2023) Potato peels waste as a sustainable source for biotechnological production of biofuels: process optimization. Waste Manag 155:320–328

Sagar NA, Kumar Y, Singh R, Nickhil C, Kumar D, Sharma P, Om Pandey H, Bhoj S, Tarafdar A (2022) Onion waste based-biorefinery for sustainable generation of value-added products. Bioresour Technol 362:127870. https://doi.org/10.1016/j.biortech.2022.127870

Saravanan A, Karishma S, Senthil Kumar P, Rangasamy G (2023) A review on regeneration of biowaste into bio-products and bioenergy: life cycle assessment and circular economy. Fuel. https://doi.org/10.1016/j.fuel.2022.127221

Shahid I, Hussain G, Anis M, Farooq MU, Usman M, Fouad Y, Krzywanski J (2023) Enzymatic co-fermentation of onion waste for bioethanol production using Saccharomyces cerevisiae and Pichia pastoris. Energies 16(5):2181

Sharma GK, Khan SA, Shrivastava M, Bhattacharyya R, Sharma A et al (2021a) Circular economy fertilization: Phycoremediated algal biomass as biofertilizers for sustainable crop production. J Environ Manag 287:112295

Sharma S, Nargotra P, Sharma V, Bangotra R, Kaur M, Kapoor N, Paul S, Bajaj BK (2021b) Nanobiocatalysts for efficacious bioconversion of ionic liquid pretreated sugarcane tops biomass to biofuel. Bioresour Technol. https://doi.org/10.1016/j.biortech.2021.125191

Sharma V, Tsai ML, Chen CW, Sun PP, Patel AK, Singhania RR, Nargotra P, Di Dong C (2022) Deep eutectic solvents as promising pretreatment agents for sustainable lignocellulosic biorefineries: a review. Bioresour Technol. https://doi.org/10.1016/j.biortech.2022.127631

Sharma S, Tsai ML, Sharma V, Sun PP, Nargotra P, Bajaj BK, Chen CW, Di Dong C (2023) Environment friendly pretreatment approaches for the bioconversion of Lignocellulosic biomass into biofuels and value-added products. Environ - MDPI. https://doi.org/10.3390/environments10010006

Show KY, Lee DJ, Tay JH, Lee TM, Chang JS (2015) Microalgal drying and cell disruption–recent advances. Bioresour Technol 184:258–266

Singh V, Chaudhary DK, Mani I, Dhar PK (2016) Recent advances and challenges of the use of cyanobacteria towards the production of biofuels. Renew Sust Energ Rev 60:1–10. https://doi.org/10.1016/j.rser.2016.01.099

Song W, He Y, Huang R, Li J, Yu Y, Xia P (2023) Life cycle assessment of deep-eutectic-solvent-assisted hydrothermal disintegration of microalgae for biodiesel and biogas co-production. Appl Energy. https://doi.org/10.1016/j.apenergy.2023.120758

Soni SK, Sharma B, Sharma A, Thakur B, Soni R (2023) Exploring the potential of potato peels for bioethanol production through various pretreatment strategies and an in-house-produced multi-enzyme system. Sustainability 15(11):9137

Tonini D, Albizzati PF, Astrup TF (2018) Environmental impacts of food waste: learnings and challenges from a case study on UK. Waste Manag 76:744–766

Tsegaye B, Jaiswal S, Jaiswal AK (2021) Food waste biorefinery: pathway towards circular bioeconomy. Foods 10(6):1174

Ubando AT, Anderson S, Ng E, Chen WH, Culaba AB, Kwon EE (2022) Life cycle assessment of microalgal biorefinery: a state-of-the-art review. Bioresour Technol. https://doi.org/10.1016/j.biortech.2022.127615

Velidandi A, Kumar Gandam P, Latha Chinta M, Konakanchi S, Reddy Bhavanam A, Raju Baadhe R, Sharma M, Gaffey J, Nguyen QD, Gupta VK (2023) State-of-the-art and future directions of machine learning for biomass characterization and for sustainable biorefinery. J Energy Chem 81:42–63. https://doi.org/10.1016/j.jechem.2023.02.020

Waghmare PR, Khandare RV, Jeon BH, Govindwar SP (2018) Enzymatic hydrolysis of biologically pretreated sorghum husk for bioethanol production. Biofuel Res J. https://doi.org/10.18331/BRJ2018.5.3.4

Wani AK, Rahayu F, Fauziah L, Suhara C (2023) Advances in safe processing of sugarcane and bagasse for the generation of biofuels and bioactive compounds. J Agric Food Res 12:100549

Yadav P, Anu, Kumar Tiwari S, Kumar V, Singh D et al (2022) Sugarcane bagasse: an important lignocellulosic substrate for production of enzymes and biofuels. Biomass Convers Biorefinery 14(5):6111–6142

Yadav A, Sharma V, Tsai ML, Chen CW, Sun PP, Nargotra P, Wang JX, Di Dong C (2023) Development of lignocellulosic biorefineries for the sustainable production of biofuels: towards circular bioeconomy. Bioresour Technol. https://doi.org/10.1016/j.biortech.2023.129145

Yen HW, Hu IC, Chen CY, Nagarajan D, Chang JS (2019) Design of photobioreactors for algal cultivation. In: Biofuels from algae. Elsevier, pp 225–256

Zabochnicka M, Krzywonos M, Romanowska-Duda Z, Szufa S, Darkalt A, Mubashar M (2022) Algal biomass utilization toward circular economy. Life 12(10):1480

Zhang L, Chen Q, Jin Y, Xue H, Guan J, Wang Z, Zhao H (2010) Energy-saving direct ethanol production from viscosity reduction mash of sweet potato at very high gravity (VHG). Fuel Process Technol 91:1845–1850

Zhao Z, Shao Z, Qu Q, Ji M, Cheng D, Guo X (2023) Promoting the overall energy profit through using the liquid hydrolysate during microwave hydrothermal pretreatment of wheat straw as co-substrate for anaerobic digestion. Sci Total Environ. https://doi.org/10.1016/j.scitotenv.2022.159463

Zhou Y, Liu L, Li M, Hu C (2022) Algal biomass valorisation to high-value chemicals and bioproducts: recent advances, opportunities and challenges. Bioresour Technol 344:126371

Zhu L, Li Z, Hiltunen E (2018) Microalgae Chlorella vulgaris biomass harvesting by natural flocculant: effects on biomass sedimentation, spent medium recycling and lipid extraction. Biotechnol Biofuels 11:1–10

Zhu J, Chen L, Gleisner R, Zhu JY (2019) Co-production of bioethanol and furfural from poplar wood via low temperature (≤90 °C) acid hydrotropic fractionation (AHF). Fuel. https://doi.org/10.1016/j.fuel.2019.05.155

Ziaei-Rad Z, Pazouki M, Fooladi J, Azin M, Gummadi SN, Allahverdi A (2023) Investigation of a robust pretreatment technique based on ultrasound-assisted, cost-effective ionic liquid for enhancing saccharification and bioethanol production from wheat straw. Sci Rep. https://doi.org/10.1038/s41598-022-27258-9

Zoppi G, Tito E, Bianco I, Pipitone G, Pirone R, Bensaid S (2023) Life cycle assessment of the biofuel production from lignocellulosic biomass in a hydrothermal liquefaction—aqueous phase reforming integrated biorefinery. Renew Energy. https://doi.org/10.1016/j.renene.2023.02.011

# Integration and Optimization of Biorefinery

# 8

Ajay Kumar and Anuj Yadav

**Abstract**

The integration and optimization of biorefinery processes are crucial for developing sustainable alternatives to conventional energy and chemical production, reducing reliance on finite fossil resources. Optimization strategies, informed by techno-economic analyses, are crucial for enhancing the economic viability and environmental friendliness of biorefineries. These strategies are designed to enhance the efficiency of biorefinery processes, reduce waste generation, and optimize resource utilization. One of the primary advantages of integrating and optimizing biorefinery processes is the mitigation of environmental impact. Life cycle assessments are essential for evaluating the environmental impact of biorefinery operations, covering the entire process from raw material extraction to product disposal. This comprehensive analysis helps biorefineries make informed decisions to minimize their environmental footprint and promote sustainable practices. Additionally, integration and optimization efforts drive technological innovation in biorefineries. Researchers are constantly exploring new technologies, processes, and materials to improve efficiency and sustainability. This continuous innovation pushes the boundaries of what is achievable in terms of conversion efficiency, product yields, and process scalability. Overall, the integration and optimization of biorefinery processes are essential for advancing sustainable and renewable solutions for meeting the world's energy and chemical needs. As technology advances and markets increasingly embrace sustainability, biorefineries are poised to play a pivotal role in the shift towards a more sustainable and circular bioeconomy.

A. Kumar (✉) · A. Yadav
Department of Biotechnology, Faculty of Engineering and Technology, Rama University, Kanpur, Uttar Pradesh, India



V. Singh (ed.), *Sustainable Waste Management Towards Circular Bioeconomy*,
https://doi.org/10.1007/978-981-95-2227-9_8

**Keywords**

Biorefinery · Integration · Optimization · Sustainability · Renewable resources

## 8.1 Introduction

The integration and optimization of biorefinery processes represent a revolutionary approach in sustainable and renewable energy and chemical production. Biorefineries are advanced facilities that utilize biological, chemical, and physical processes to transform biomass into a diverse range of valuable products, such as biofuels, biochemicals, and biomaterials (Dwi Prasetyo et al. 2020). This integrated approach is crucial for enhancing overall efficiency, reducing environmental impact, and ensuring the economic viability of biorefinery operations. It signifies a significant shift in resource utilization, moving away from traditional fossil-based processes towards the sustainable and eco-friendly conversion of biomass resources (Meramo et al. 2022).

Biorefineries distinguish themselves from conventional oil refineries by employing biological processes to convert renewable resources like agricultural residues, wood, and algae into bio-based products. This differentiation is crucial for reducing dependence on finite fossil fuels and moving towards more sustainable production methods (Takkellapati et al. 2018). By integrating various processes within biorefineries, such as biomass pretreatment, fermentation, and product separation, it becomes possible to achieve higher overall yields and lower environmental impact (Bušić et al. 2018).

The optimization of biorefinery processes is a complex and multifaceted endeavor. It involves addressing key challenges such as feedstock variability, economic competitiveness, and adherence to regulatory standards. Through continuous improvement and fine-tuning of processes, researchers and engineers strive to achieve cost-effectiveness, reduce environmental impact, and meet the evolving demands of the market (Flores et al. 2022). Optimization efforts encompass diverse areas, including process modeling, waste valorization, and the development of advanced control systems. One of the key advantages of biorefineries is their ability to capitalize on a diverse range of feedstocks (Yadav et al. 2022). Unlike conventional refineries that rely on crude oil, biorefineries can utilize agricultural residues, forestry waste, and energy crops. This diversity not only reduces waste but also promotes the sustainable management of biomass resources (Saleem 2022). Additionally, the use of biomass feedstocks can help revitalize rural economies by creating new markets for agricultural and forestry products (Popp et al. 2016).

Integrating various processes within biorefineries is essential for maximizing efficiency and reducing waste. By combining biomass pretreatment, fermentation, and product separation, biorefineries can achieve higher overall yields and lower environmental impact (Bertacchi et al. 2022). Optimization strategies, including process modeling, simulation, and control, play a crucial role in improving product quality, reducing energy consumption, and increasing the overall profitability of

**Table 8.1** The various aspects of biorefinery processes

| Aspect | Description |
|---|---|
| Feedstock | A range of biomass sources, including agricultural residues, forestry waste, and energy crops, serve as raw materials. |
| Pretreatment | Biomass undergoes pretreatment processes such as milling, drying, and size reduction to enhance its suitability for conversion processes. |
| Conversion processes | Biomass undergoes enzymatic or microbiological conversion into sugars, which can then be fermented to produce biofuels and other products. |
| Product separation | Products are separated and purified from the reaction mixture using techniques like distillation, filtration, and chromatography. |
| Environmental impact | Life cycle assessments are performed to assess the environmental impact of biorefinery operations and promote sustainable practices. |

biorefinery operations. Ongoing research and development in the field of biorefineries are crucial for unlocking their full potential in the transition to a more sustainable future (Godvin Sharmila et al. 2021). Through the utilization of biological processes and renewable resources, biorefineries present a promising path toward a greener and more sustainable future. They signify a notable stride toward adopting a more sustainable and environmentally friendly approach to energy and chemical production (Bhatia et al. 2022). The integration and optimization of biorefinery processes offer a revolutionary approach to sustainable and renewable energy and chemical production. Through the strategic design and coordination of biological, chemical, and physical processes, biorefineries can maximize efficiency, reduce waste, and minimize environmental impact, as illustrated in Table 8.1. Continuous research and development efforts are essential for maximizing the potential of biorefineries and driving forward sustainable energy and chemical production.

## 8.2 Biorefinery Process Integration

The integration of biorefinery processes involves the meticulous design and harmonization of various biological, chemical, and physical processes within a biorefinery to optimize performance, enhance efficiency, and maximize the production of valuable products from biomass feedstocks (Takkellapati et al. 2018). This integration is essential for minimizing waste, reducing energy consumption, and lessening environmental impact, while also maximizing resource utilization and product yields (Purchase et al. 2021). A critical aspect of integration is the selection and preparation of feedstocks, which can include agricultural residues, energy crops, forestry waste, and algae (Caicedo et al. 2016).

These feedstocks undergo pretreatment to make them suitable for further processing, such as size reduction, drying, and chemical treatment. Following pretreatment, biomass undergoes conversion processes like fermentation, pyrolysis, and hydrolysis, depending on the desired end products (Kumar et al. 2017). Fermentation converts sugars into biofuels or biochemicals, while pyrolysis and hydrolysis convert biomass into bio-oil and sugars, respectively. Product recovery and purification

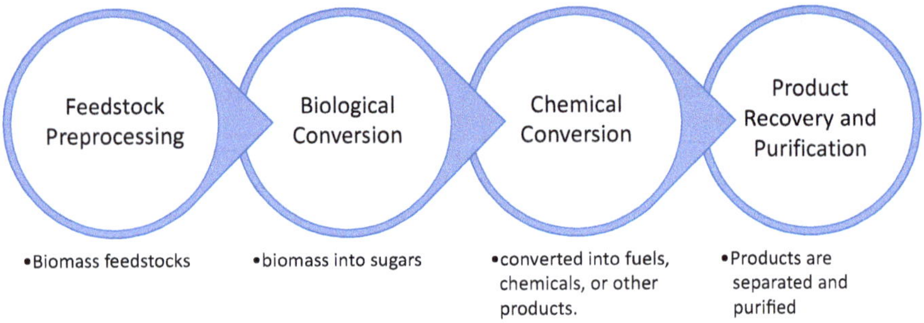


**Fig. 8.1** Integration of biorefinery processes

are then carried out to isolate the desired products from the conversion processes (Robak and Balcerek 2018). This may involve separation techniques such as distillation, filtration, and chromatography. Biorefineries also aim to minimize waste by utilizing byproducts and residues from the conversion processes. For instance, lignin, a byproduct of biofuel production, can be repurposed as a fuel or chemical feedstock, minimizing waste and enhancing overall efficiency (Elmekawy et al. 2013).

Energy integration is crucial, as biorefineries often integrate energy production and utilization processes to improve efficiency. The surplus heat generated during conversion processes can be utilized to dry feedstocks or produce electricity, decreasing dependence on external energy sources. Continuous process optimization and control are essential for maximizing efficiency and productivity (Cherubini 2010). This involves monitoring key process parameters and adjusting operations to ensure optimal performance. Environmental impact assessments are conducted to ensure compliance with regulatory standards and minimize the environmental footprint of biorefineries. This includes monitoring emissions, waste generation, and resource consumption (Sharma et al. 2021). Integration with existing infrastructure is another important aspect, as biorefineries can be integrated with existing industrial infrastructure to leverage resources and infrastructure (Takkellapati et al. 2018). Finally, economic viability is crucial, and biorefineries must assess the costs and benefits of integration to ensure competitiveness with traditional fossil-based processes. The integration of biorefinery processes is a complex and multifaceted endeavor that requires careful design, coordination, and optimization (Zetterholm et al. 2020). Biorefineries play a crucial role in sustainable energy and chemical production by maximizing efficiency, minimizing waste, and utilizing renewable resources (Narisetty et al. 2023). Here are key aspects of the integration of biorefinery processes, illustrated in Fig. 8.1:

**Feedstock Preprocessing** Biomass feedstocks require careful preparation and preprocessing to eliminate impurities and optimize their suitability for conversion processes in biorefineries. These feedstocks comprise a wide array of materials, including agricultural residues like corn stover and wheat straw, forestry waste such as sawdust and wood chips, algae, and dedicated energy crops like switchgrass and miscanthus (Nguyen et al. 2020). The preparation of biomass involves several key steps. Milling, or grinding, reduces the size of the biomass particles, increasing their surface area and improving their accessibility to enzymes and other conversion agents. Drying removes moisture from the biomass, which is essential for preventing microbial growth and ensuring efficient conversion. Size reduction further breaks down the biomass into smaller particles, facilitating the subsequent conversion processes (Broda et al. 2022). Biorefineries often employ advanced monitoring systems to collect real-time data on process variables such as temperature, pressure, and composition (Balan 2014). This data allows operators to adjust process parameters in real time to maintain optimal conditions and maximize efficiency. For example, if the moisture content of the biomass is too high, operators can adjust drying parameters to reduce moisture levels and improve conversion efficiency (Gao et al. 2023).

The integration of preprocessing and monitoring technologies is crucial for the efficient operation of biorefineries. Through the optimization of biomass feedstock preparation and continuous monitoring and adjustment of process conditions, biorefineries can maximize the production of valuable products while minimizing waste and environmental impact. This approach is essential for the advancement of sustainable and renewable energy and chemical production (Khan et al. 2018).

**Biological Conversion** Enzymes or microorganisms are key players in breaking down biomass into sugars and other components, primarily through fermentation processes. Biomass, comprising various organic materials like agricultural residues, forestry waste, and dedicated energy crops, undergoes enzymatic or microbiological conversion into sugars (Vasić et al. 2021). Enzymatic conversion utilizes enzymes to catalyze the breakdown of complex biomass molecules into simpler sugars. These enzymes are typically derived from microorganisms or produced through recombinant DNA technology (Adrio and Demain 2014). Microbial conversion, on the other hand, employs microorganisms such as bacteria or fungi to break down biomass into sugars through fermentation processes (Voidarou et al. 2020).

After sugars are produced through enzymatic or microbiological conversion, they can undergo further processing through fermentation to yield biofuels like ethanol or other high-value chemicals. Fermentation is a metabolic process wherein microorganisms convert sugars into alcohol or acids under anaerobic conditions (Maicas 2020). The enzymatic or microbiological conversion of biomass into sugars, followed by fermentation into biofuels and chemicals, constitutes a crucial aspect of biorefinery operations. These processes provide a sustainable and environmentally friendly alternative to conventional fossil fuel-based production methods, thereby advancing renewable energy and chemical production (Jeswani et al. 2020).

**Chemical Conversion** Sugars and intermediates derived from biomass are crucial feedstocks for producing biofuels, biochemicals, and various other products. These sugars and intermediates undergo chemical conversion through catalytic processes, which are crucial for transforming biomass into valuable end products. Two common catalytic processes used in this conversion are hydrothermal liquefaction and pyrolysis (Díaz-Pérez and Serrano-Ruiz 2020). Hydrothermal liquefaction is a thermochemical conversion process that involves the degradation of biomass in a hot, pressurized water environment. This process converts the complex organic compounds in biomass into simpler, more valuable products like bio-oil, which can be refined into biofuels or used as a feedstock for biochemical production. Hydrothermal liquefaction offers several advantages, including high efficiency, reduced greenhouse gas emissions, and the ability to process a wide range of biomass feedstocks (Grande et al. 2021).

Pyrolysis is a thermochemical conversion process that entails heating biomass in the absence of oxygen to break it down into biochar, bio-oil, and gases like hydrogen and methane. Biochar can be used as a soil amendment, while bio-oil can be further refined into biofuels or used as a precursor for biochemical production (Egbosiuba 2022). Pyrolysis is a versatile process capable of handling various types of biomass feedstocks and producing a range of valuable products. The catalytic processes such as hydrothermal liquefaction and pyrolysis are essential for converting biomass-derived sugars and intermediates into biofuels, biochemicals, and other valuable products. These processes provide a sustainable and environmentally friendly alternative to traditional fossil fuel-based production methods, thereby advancing renewable energy and chemical production (Jeswani et al. 2020).

**Product Recovery and Purification** Products derived from biomass conversion processes are separated and purified from the reaction mixture using a variety of techniques, including distillation, filtration, and chromatography (Bušić et al. 2018). These separation and purification processes are essential for obtaining high-purity products that meet the quality standards required for commercial applications. Distillation is a widely employed technique that utilizes differences in boiling points to separate components within a mixture. It is especially effective for separating volatile compounds, like biofuels, from non-volatile components (Zhang et al. 2018).

Filtration is employed to separate solids from liquids or gases by passing the mixture through a porous material that retains the solid particles. This technique is frequently used to eliminate impurities from biofuels or biochemicals. Chromatography is a versatile technique that separates components based on their affinity for a stationary phase (Mauerhofer et al. 2019). It is used to separate complex mixtures into individual components, making it useful for purifying biofuels, biochemicals, and other products derived from biomass.

Purification processes are crucial for ensuring that the final products meet the quality standards required for commercial applications. These processes may involve multiple steps, including further distillation, filtration, or chromatography, to remove impurities and contaminants (Gharby 2022). In addition to ensuring product quality, biorefineries must also align their product portfolio with market demands to maintain a competitive edge and ensure stable revenue generation. This involves understanding market trends and consumer preferences to produce products that meet market needs (Moncadam et al. 2016).

Optimizing separation processes is key to enhancing the recovery of valuable products while minimizing energy consumption. This can be accomplished by employing advanced process control strategies and integrating energy-efficient technologies. Similarly, the development of efficient purification methods is crucial for obtaining high-purity products. This may involve the use of novel separation techniques or the optimization of existing processes to improve product purity and yield (Mitra and Murthy 2022). Overall, the separation and purification of products derived from biomass conversion processes are critical steps in the biorefinery value chain. Through the use of effective separation and purification techniques, biorefineries can manufacture high-quality products that meet market demands and advance sustainable energy and chemical production.

## 8.3 Continuous Improvement

Ongoing research in the field of biorefineries is dedicated to developing and implementing novel technologies, catalysts, and processes. These advancements aim to enhance efficiency, reduce costs, and broaden the range of products that can be derived from biomass feedstocks (Kundas et al. 2018). One area of focus is the development of new conversion technologies that can more effectively break down biomass into its constituent components. These technologies may include novel pretreatment methods, more efficient enzymatic processes, or innovative reactor designs. By improving the efficiency of biomass conversion, researchers hope to increase the yield of valuable products such as biofuels and biochemicals (Vasić et al. 2018).

Catalysts play a crucial role in many biorefinery processes, and researchers are actively working to develop new catalysts that are more efficient and cost-effective. These catalysts may be used to facilitate reactions such as hydrolysis, fermentation, or catalytic cracking, allowing for faster and more selective conversion of biomass into desired products (Maroa and Inambao 2021). In addition to developing new technologies and catalysts, researchers are also exploring new process configurations that can improve overall biorefinery efficiency. This may include integrating different conversion processes, optimizing process conditions, or designing more efficient separation and purification methods (Solarte-Toro et al. 2023).

Advanced monitoring and control systems are essential for optimizing biorefinery operations in real time. These systems allow operators to closely monitor key process parameters and adjust operations as needed to maintain optimal conditions and ensure consistent product quality. By implementing these systems, biorefineries can improve efficiency, reduce waste, and minimize downtime. Feedback from ongoing operations is crucial for identifying areas where improvements can be made (Molina-Peñate et al. 2024). Researchers use data collected from biorefinery operations to identify bottlenecks, optimize process parameters, and implement continuous improvements. This iterative process of research, development, and implementation is essential for advancing the integration of biorefinery processes and promoting the sustainable production of bio-based products.

## 8.4 Optimization of Biorefinery Processes

Optimization of biorefinery processes involves the systematic improvement and fine-tuning of various stages within a biorefinery to enhance efficiency, reduce costs, and maximize the production of valuable products from biomass (Liu et al. 2021). One key aspect of optimization is feedstock selection and preparation. This involves choosing biomass types that yield high-quality products and are sustainable. The preparation process, including size reduction and drying, is optimized to improve the efficiency of subsequent conversion processes (Liu and Yu 2021).

Conversion processes are also optimized to maximize product yields and reduce energy consumption. Parameters such as temperature, pressure, and enzyme concentrations are adjusted to achieve these goals. Product recovery and purification techniques, such as distillation, filtration, and chromatography, are optimized to recover high-purity products while minimizing energy and resource consumption (Guajardo and Schrebler 2023).

Waste utilization is another important aspect of optimization, where byproducts and residues are efficiently converted into valuable products or used for energy generation. Energy integration is also optimized to minimize energy consumption and reduce reliance on external sources (Bhatia et al. 2023). Advanced monitoring and control systems are employed to optimize process parameters in real time, ensuring consistent product quality and efficiency. Environmental impact is minimized through optimization, including reducing greenhouse gas emissions and waste generation (Brundage et al. 2018). Economic viability is a key consideration in optimization efforts, aiming to reduce production costs and increase profitability. By systematically optimizing these key aspects, biorefineries can improve their overall efficiency, reduce their environmental impact, and enhance their economic viability, thus contributing to the advancement of sustainable energy and chemical production (Romero-Perez et al. 2021). Key aspects of this optimization include, as shown in Fig. 8.2:

**Process Modeling** Mathematical models are essential for understanding and optimizing biorefinery processes. These models represent the complex interactions within biorefineries, enabling researchers to simulate their behavior and identify areas for improvement (Fischer 2008). One of the key aspects of using mathematical models in biorefineries is the development of accurate models that can simulate various processes. These models allow for virtual experimentation, helping researchers identify optimization opportunities without the need for costly and time-consuming physical experiments. Another crucial aspect is the identification of key

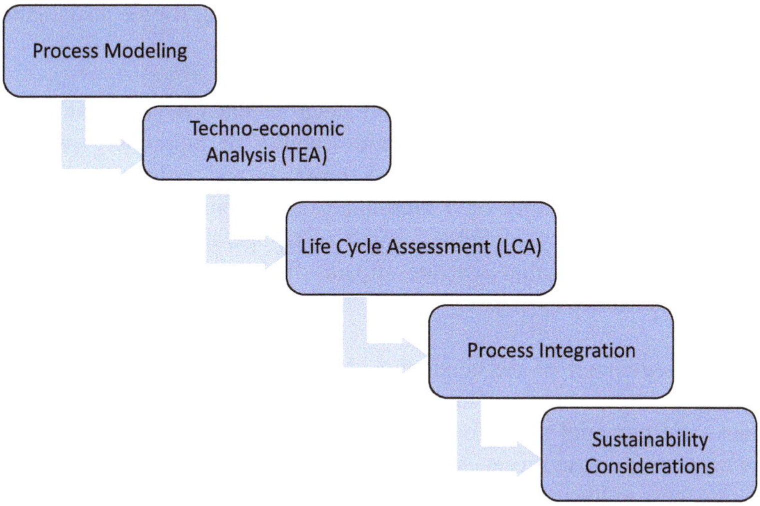


**Fig. 8.2** Optimization of biorefinery processes

parameters affecting process performance (Vora et al. 2023). Mathematical models help researchers pinpoint these parameters, allowing them to focus on optimizing them to improve efficiency and reduce costs. Additionally, mathematical models assist in the selection of biomass feedstock by considering factors such as availability, cost, and suitability for desired end products. This optimization helps biorefineries maximize the use of renewable resources and minimize environmental impact (Dwi Prasetyo et al. 2020).

Efficient methods for preparing and handling feedstock are also developed using mathematical models. These methods enhance the accessibility of feedstock for subsequent conversion processes, reducing energy consumption and improving overall process efficiency (Palkovits and Delidovich 2018). Furthermore, mathematical models are used for scale-up, allowing biorefineries to plan for expansion and improve process reliability. Mathematical modeling also plays a role in catalyst development and reaction engineering. By exploring new catalysts and improving reaction engineering in chemical conversion processes, researchers can enhance process efficiency and reduce environmental impact (Shen et al. 2022). Additionally, mathematical models help identify new uses for by-products and residues generated during biorefinery processes, minimizing waste and improving overall resource utilization.

In biological conversion processes, bioinformatics and systems biology approaches are used to optimize processes. These approaches provide a deeper understanding of metabolic pathways and microorganism behavior, leading to improved efficiency and product yields (Helmy et al. 2020). Overall, mathematical modeling is crucial for advancing biorefinery processes, enabling researchers to optimize operations, reduce costs, and minimize environmental impact. Continued development and application of these models are essential for the sustainable production of bio-based products.

**Techno-economic Analysis (TEA)** Total Environmental Assessment (TEA) is a critical tool for evaluating the economic viability of biorefinery processes. It involves a comprehensive analysis of costs, revenues, and profitability to assess the feasibility and sustainability of a biorefinery operation. TEA considers various aspects of the biorefinery, including capital costs, operating costs, revenues from product sales, and potential revenue streams from by-products or co-products (Rajendran and Murthy 2017).

Regular assessment of the economic viability of the biorefinery is crucial to ensure its long-term success. This includes conducting periodic reviews of costs, revenues, and potential areas for cost reduction. By identifying opportunities to reduce costs or increase revenues, biorefineries can improve their profitability and competitiveness in the market (Bartling et al. 2021). TEA also includes identifying and mitigating potential risks that could affect the economic performance of the biorefinery. These risks may include fluctuations in feedstock prices, shifts in market demand for biorefinery products, or regulatory changes affecting production costs. By identifying these risks early, biorefineries can develop strategies to mitigate them and protect their economic viability (Wiatrowski et al. 2022). Overall, TEA is essential for ensuring the economic viability of biorefinery processes. By providing a comprehensive assessment of costs, revenues, and profitability, TEA helps biorefineries make informed decisions to improve their financial health and sustainability.

**Life Cycle Assessment (LCA)** Life Cycle Assessment (LCA) is a crucial methodology for evaluating the environmental impact of biorefinery processes. It considers various factors such as greenhouse gas emissions, energy consumption, water usage, and resource depletion to assess the overall environmental sustainability of a biorefinery operation (Sajid 2021). The primary objective of LCA in the context of biorefineries is to ensure sustainability by evaluating the environmental impact and identifying opportunities for improvement. This assessment helps ensure that biorefineries operate in an environmentally sustainable manner, minimizing their impact on the environment (Sevigné-Itoiz et al. 2021).

An essential aspect of LCA is optimizing biorefinery processes to reduce resource consumption and waste generation. Optimization strategies can include improving energy efficiency, reducing water consumption, or finding ways to reuse or recycle waste materials. These efforts help biorefineries minimize their environmental footprint and align with sustainable practices (Gonçalves and Maximo 2022). Overall, LCA plays a crucial role in ensuring the environmental sustainability of biorefinery processes. By evaluating the environmental impact and identifying opportunities for improvement, LCA helps biorefineries operate in a more sustainable and environmentally friendly manner.

**Process Integration** Optimizing the overall energy and material flows within the biorefinery is crucial for reducing waste, improving efficiency, and minimizing environmental impact. This optimization involves identifying opportunities to streamline processes, reduce energy consumption, and maximize the use of resources (Bocker and Silva 2024). For example, by analyzing the energy requirements of different processes, biorefineries can identify ways to use waste heat from one process as an energy source for another, improving overall energy efficiency (Mukamwi et al. 2023). In biological conversion processes, optimizing enzymes or microorganisms can improve conversion rates and yields. This optimization may involve selecting or engineering enzymes or microorganisms that are more efficient or resistant to inhibitors, thereby improving overall process efficiency (Adrio and Demain 2014). In chemical conversion processes, fine-tuning temperature, pressure, and other parameters can optimize product formation. By carefully controlling these parameters, biorefineries can improve product yields and reduce the formation of unwanted by-products, leading to more efficient and sustainable processes (Almeida et al. 2012). Efficient use and recycling of process streams are also essential for minimizing waste and resource consumption. By recycling water, nutrients, and other process streams, biorefineries can reduce their environmental footprint and improve overall sustainability. Overall, optimization of biorefinery processes is essential for reducing waste, improving efficiency, and minimizing environmental impact. By carefully analyzing and fine-tuning processes, biorefineries can achieve more sustainable operations and contribute to a greener future (Bhatia et al. 2023).

**Sustainability Considerations** Assessing the social, economic, and environmental sustainability of biorefinery processes is essential to ensure their alignment with sustainable development goals. This evaluation considers the impact of biorefinery operations on local communities, the economy, and the environment (Abu-Bakar et al. 2023). Social sustainability involves assessing how biorefineries contribute to the well-being of local communities, such as through job creation, community engagement, and support for local businesses. Economic sustainability focuses on the financial viability of biorefineries, considering factors such as profitability, cost-effectiveness, and contribution to economic growth (Bansal et al. 2023). Environmental sustainability assesses the impact of biorefinery operations on the environment, including factors such as greenhouse gas emissions, resource depletion, and waste generation. By evaluating these factors, biorefineries can identify areas for improvement and implement strategies to enhance their overall sustainability (Falano et al. 2014). Ensuring compliance with environmental and safety standards is crucial for biorefineries to meet regulatory requirements. Biorefineries must adhere to strict regulations regarding emissions, waste management, and worker safety to minimize their impact on the environment and protect the health and well-being of their employees and surrounding communities (Gupta et al. 2023). Compliance with these standards not only helps biorefineries avoid legal penalties but also demonstrates their commitment to responsible and sustainable operations.

Optimizing biorefinery processes is a dynamic and ongoing effort that involves a combination of technological innovation, process control, economic analysis, and sustainable practices. Technological innovation plays a key role in optimizing biorefinery processes, as new technologies can improve efficiency, reduce waste, and lower costs. Process control ensures that biorefineries operate at peak efficiency, minimizing energy and resource consumption (Babcock-Jackson et al. 2023). Economic analysis helps biorefineries identify cost-saving opportunities and assess the financial viability of new technologies and practices. Sustainable practices, such as recycling and waste reduction, are also essential for optimizing biorefinery processes and minimizing their environmental impact (Aidonojie et al. 2023).

Continuous improvement and adaptation to evolving technologies and market demands are crucial for the long-term success of biorefinery operations. By continuously improving their processes and practices, biorefineries can enhance their efficiency, reduce costs, and improve their overall sustainability. This continuous improvement is essential for biorefineries to remain competitive in the market and contribute to a more sustainable future (Barragán-Ocaña et al. 2023).

## 8.5 Challenges and Research Areas

The integration and optimization of biorefinery processes present several challenges that researchers and engineers are actively addressing. These challenges span technical, economic, and environmental aspects, requiring innovative solutions and continuous improvement efforts, as shown in Fig. 8.3 (Calvo Flores and Martin-Martinez 2022). A major challenge is the variability of biomass feedstocks, which can affect process efficiency and product quality. Researchers are investigating ways to manage feedstock variability effectively, aiming to enhance process stability and overall performance (Balan 2014). Additionally, integrating multiple processes within a biorefinery complex requires meticulous planning and optimization. Researchers are developing advanced techniques for process integration to improve efficiency and reduce waste (Varbanov et al. 2021).

Another challenge is achieving product diversification, as biorefineries strive to produce a range of valuable products. This challenge involves overcoming technical and economic constraints to expand the portfolio of sustainable products. Cost competitiveness is a critical consideration for biorefineries, as they must compete with traditional fossil-based processes (Lange et al. 2021). Researchers are exploring cost-effective technologies and processes to enhance the economic viability of biorefinery operations. Sustainability metrics play a crucial role in assessing the environmental, economic, and social impacts of biorefinery processes. Researchers are working to refine and standardize these metrics to facilitate more effective decision-making (Venkata Mohan and Katakojwala 2020). Waste valorization is also a key challenge, as biorefineries produce various waste streams that can be challenging to manage. Researchers are exploring innovative methods to valorize these waste streams, turning them into valuable products or energy sources.

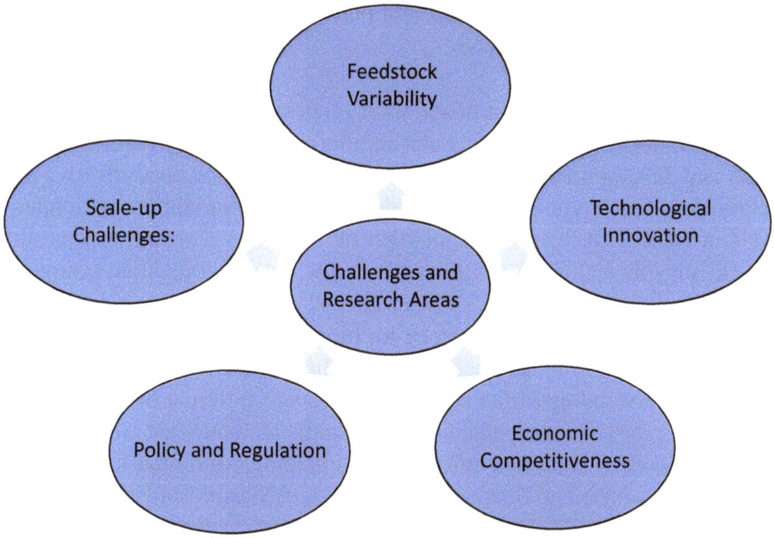


**Fig. 8.3** Challenges and research areas in the integration and optimization of biorefinery processes

Process scale-up from laboratory to commercial scale is a significant challenge for biorefineries. Researchers are developing strategies to scale up processes efficiently and cost-effectively, ensuring smooth transition and optimal performance at larger scales. Technological innovation is fundamental to addressing these challenges. Researchers are exploring advanced catalysts, novel process designs, and other innovative technologies to improve efficiency and sustainability in biorefinery operations (Fuchs et al. 2023). The integration and optimization of biorefinery processes require addressing a range of challenges through innovative research and collaboration. By overcoming these challenges, biorefineries can enhance their efficiency, reduce environmental impact, and contribute to a more sustainable future. Here are some key challenges and research areas in the integration and optimization of biorefinery processes:

**Feedstock Variability** The varying compositions of different biomass feedstocks present a significant challenge for biorefineries. This variability can affect the efficiency and effectiveness of biorefinery processes, as different feedstocks may require different processing methods or conditions (Smit et al. 2022). For example, some feedstocks may have higher moisture content or contain more difficult-to-break-down components, which can impact the overall efficiency of biomass conversion processes. One specific challenge is the need for flexible and adaptive processing technologies that can accommodate these variations in feedstock composition. Biorefineries must be able to adjust their processes to effectively handle different types of biomasses without compromising efficiency or product quality (Freitas et al. 2021). Developing versatile biorefinery platforms requires a thorough understanding of the chemical and physical properties of various biomass feedstocks. It also involves the development of innovative processing technologies capable of adapting to these differences.

Another challenge is the optimization of processing methods for different feedstocks to maximize product yields and minimize waste. This requires continuous research and development to identify the most efficient and cost-effective processing methods for each type of biomass. Additionally, the variability of biomass feedstocks can impact the overall economics of biorefinery operations, as different feedstocks may have different costs associated with their collection, transportation, and processing (Kim et al. 2013). Overall, the varying compositions of biomass feedstocks present a complex challenge for biorefineries. Addressing this challenge requires a combination of advanced research, innovative technology development, and a deep understanding of the unique properties of different biomass feedstocks. By developing flexible and adaptive processing technologies, biorefineries can overcome these challenges and unlock the full potential of biomass as a sustainable feedstock for bio-based products (Calvo Flores and Martin-Martinez 2022).

**Technological Innovation** The development and implementation of advanced technologies, such as genetically modified organisms (GMOs) and novel catalysts, pose several challenges for the biorefinery industry. One challenge is the complexity and novelty of these technologies, which may require significant research and development efforts to optimize and scale up for industrial use. Additionally, the regulatory environment surrounding GMOs and other advanced technologies can be stringent, requiring biorefineries to navigate complex regulatory processes to ensure compliance. Another challenge is the requirement for interdisciplinary collaboration and expertise to effectively develop and deploy these technologies (Karalis et al. 2020). GMOs, for example, require expertise in molecular biology, genetics, and bioprocessing, highlighting the need for collaboration between biologists, chemists, and engineers. Additionally, the development of novel catalysts and reactor designs requires expertise in chemistry, materials science, and chemical engineering, highlighting the interdisciplinary nature of biorefinery research and development (Ribeiro et al. 2019).

Furthermore, the cost and scalability of these technologies can be significant challenges. Developing and implementing GMOs and novel catalysts can require substantial investment in research, development, and infrastructure. Scaling up these technologies from the laboratory to industrial scale can also be challenging, requiring careful optimization and engineering to ensure efficiency and cost-effectiveness. Overall, while advanced technologies offer exciting opportunities to enhance the efficiency and sustainability of biorefineries, they also present significant challenges that must be overcome through collaboration, innovation, and careful planning. Successfully addressing these challenges will be crucial for implementing advanced technologies in the biorefinery industry (Shukla et al. 2018).

**Economic Competitiveness** To achieve commercial viability, biorefinery processes must be cost-competitive with traditional fossil-based processes. This requires careful consideration of the entire value chain, from sourcing feedstock to distributing products. Techno-economic analysis (TEA) is essential for identifying cost-effective strategies and optimizing process parameters to maximize profitability. Reducing production costs is critical for achieving cost competitiveness (Gadkari et al. 2023). This can be achieved through process optimization, scale-up, and integrating co-products and by-products. Additionally, exploring new revenue streams and market opportunities can enhance the overall economics of biorefinery operations. This may involve diversifying product lines, targeting niche markets, or forming strategic partnerships.

Government policies and incentives also significantly influence the economic viability of biorefineries. Supportive policies for renewable energy and bio-based products can help level the playing field with traditional fossil-based processes. However, regulatory uncertainty and fluctuating market conditions can pose challenges that need to be carefully managed (Calvo Flores and Martin-Martinez 2022). The attaining cost-competitiveness with traditional fossil-based processes is a complex but vital objective for the biorefinery industry. Through TEA, exploration of new revenue streams, and optimization of production processes, biorefineries can enhance their economic viability and contribute to a more sustainable future.

**Policy and Regulation** The regulatory environment and government policies present significant challenges for the development and deployment of biorefinery technologies. One major challenge is the complexity and variability of regulations across different jurisdictions. Biorefineries must navigate a complex web of regulations related to environmental protection, waste management, health and safety, and land use, among others (Kumar et al. 2017). This complexity is further compounded for biorefineries operating in multiple regions or countries, as they must comply with different sets of regulations. Another challenge is the lack of consistent or harmonized regulations for biorefinery technologies. The regulatory landscape for biorefineries is often fragmented, with regulations varying widely between regions and even within the same country. This creates uncertainty for biorefinery developers and investors, as they must navigate a patchwork of regulations that may be inconsistent or conflicting.

Additionally, regulatory requirements for biorefineries are constantly evolving as new technologies emerge and our understanding of environmental and health impacts evolves. This dynamic regulatory environment makes it challenging for biorefineries to plan and invest in long-term projects, as regulations may change unexpectedly, requiring costly modifications to comply.

Furthermore, the lack of specific regulations tailored to biorefinery technologies can pose challenges (Usmani et al. 2021). Many existing regulations were developed with traditional industries in mind and may not fully address the unique characteristics of biorefineries. This can create regulatory gaps or ambiguities that biorefineries must navigate, potentially leading to delays or increased costs. Overall, the regulatory environment and government policies present significant challenges for the development and deployment of biorefinery technologies. Addressing these challenges will require collaboration between government, industry, and other stakeholders to develop clear, consistent, and supportive regulations that encourage innovation and investment in biorefineries.

**Scale-up Challenges** Scaling up biorefinery operations from laboratory to industrial scale poses several challenges that must be addressed for successful commercialization. One of the primary challenges is maintaining efficiency and cost-effectiveness when transitioning from small-scale to large-scale operations (Balan 2014). This requires careful optimization of process parameters, equipment design, and operating conditions to ensure that the process can handle larger volumes of feedstock while maintaining product quality and yield. Another challenge is developing scalable and robust technologies that can reliably operate at industrial scale. This involves selecting and optimizing the right combination of technologies for each step of the biorefining process, from biomass pretreatment to product recovery. It also requires designing equipment and infrastructure that can withstand the harsh operating conditions of industrial-scale operations while minimizing energy and resource consumption (Meegoda et al. 2018).

Securing funding and investment for scale-up activities can also be a significant challenge, especially for small or medium-sized companies. The capital investment required for scaling up biorefinery processes can be substantial, and securing funding from investors or financial institutions can be difficult, particularly for innovative technologies that may be perceived as high-risk. Additionally, navigating the complex regulatory environment for industrial-scale operations can be challenging. Biorefineries must comply with environmental regulations, health and safety standards, and other regulatory requirements, which can vary depending on the location and scale of operations (Tsagkari et al. 2016). In conclusion, scaling up biorefinery operations requires careful planning, innovative technology development, and collaboration between industry, government, and research institutions. Addressing these challenges is essential for realizing the full potential of biorefineries in contributing to a sustainable bioeconomy.

## 8.6 Advanced Process Control

Advanced Process Control (APC) is a critical element in optimizing biorefinery operations. APC systems utilize real-time data for monitoring and adjusting process parameters, ensuring efficient and cost-effective operations (Zainurin et al. 2022). By implementing APC, biorefineries can maximize product yields, minimize waste generation, and reduce energy consumption. Research in biorefineries also focuses on developing and refining sustainability metrics and standards. These metrics assist in evaluating the environmental and social impacts of biorefinery processes, ensuring alignment with sustainable development objectives. Establishing clear metrics and standards enables biorefineries to track their progress toward sustainability and make informed decisions to reduce their environmental footprint (Sacramento Rivero et al. 2022).

Genetic engineering and selective breeding are key research areas in biorefineries. By enhancing the characteristics of energy crops, such as yield, resilience, and suitability for biorefinery processes, researchers can develop more efficient and sustainable feedstocks. This research ensures that biorefineries have access to high-quality feedstocks that can be efficiently converted into valuable products (Kumar and Sharma 2017). In-depth market analysis is another crucial aspect of advancing biorefinery technologies. By understanding market trends and demands, biorefineries can tailor their product portfolios to meet consumer needs. This analysis helps biorefineries identify emerging opportunities and diversify their product offerings, ensuring long-term viability and competitiveness in the market. Addressing these challenges and advancing research in these areas is crucial for the continued growth and success of biorefinery technologies. By optimizing processes, developing sustainable feedstocks, and aligning with market demands, biorefineries can play a significant role in transitioning to a more sustainable and environmentally friendly economy (Silva et al. 2022).

## 8.7 Significance of Biorefinery Integration and Optimization

Integrated and optimized biorefinery processes are vital for tackling global challenges of sustainability and resource depletion. By utilizing renewable biomass resources, biorefineries offer a sustainable alternative to conventional fossil-based energy and chemical production. This shift reduces the reliance on finite fossil resources, which are environmentally damaging and contribute to climate change (Saleem 2022). Optimization strategies, guided by techno-economic analyses, are essential for ensuring the viability and competitiveness of biorefineries (Dwi Prasetyo et al. 2020). These strategies aim to maximize the efficiency of biorefinery processes, minimize waste generation, and optimize resource utilization. By optimizing their operations, biorefineries can reduce costs, improve product quality, and enhance overall process efficiency (Bušić et al. 2018).

Furthermore, integration and optimization efforts in biorefineries are aimed at minimizing their environmental footprint. Life cycle assessments are utilized to evaluate the environmental impact of biorefinery processes, encompassing the entire process from raw material extraction to product disposal. This comprehensive approach assists biorefineries in making informed decisions to minimize their environmental impact and operate in an environmentally responsible manner. Ongoing research in biorefinery integration and optimization is essential for driving technological innovation (Meramo et al. 2022). Researchers are constantly exploring new technologies, processes, and materials to improve the efficiency and sustainability of biorefineries. These efforts push the boundaries of what is achievable in terms of conversion efficiency, product yields, and process scalability, driving the biorefinery industry towards greater sustainability and competitiveness (Tong et al. 2023). In conclusion, the integration and optimization of biorefinery processes are crucial for advancing sustainable and renewable solutions for meeting the world's energy and chemical needs. As technology evolves and markets shift towards more sustainable practices, biorefineries are well-placed to play a central role in the transition to a more sustainable and circular bioeconomy.

## References

Abu-Bakar NA, Roslan AM, Hassan MA, Rahman MHA, Ibrahim KN, Abd Rahman MD, Mohamad R (2023) Environmental impact assessment of rice mill waste valorisation to glucose through biorefinery platform. Sci Rep 13(1):14767

Adrio JL, Demain AL (2014) Microbial enzymes: tools for biotechnological processes. Biomolecules 4(1):117–139

Aidonojie PA, Ukhurebor KE, Oaihimire IE, Ngonso BF, Egielewa PE, Akinsehinde BO, Kusuma HS, Darmokoesoemo H (2023) Bioenergy revamping and complimenting the global environmental legal framework on the reduction of waste materials: a facile review. Heliyon 9(1):e12860

Almeida JR, Fávaro LC, Quirino BF (2012) Biodiesel biorefinery: opportunities and challenges for microbial production of fuels and chemicals from glycerol waste. Biotechnol Biofuels 5(1):48

Babcock-Jackson L, Konovalova T, Krogman JP, Bird R, Díaz LL (2023) Sustainable fertilizers: publication landscape on wastes as nutrient sources, wastewater treatment processes for nutrient recovery, biorefineries, and green ammonia synthesis. J Agric Food Chem 71(22):8265–8296

Balan V (2014) Current challenges in commercially producing biofuels from lignocellulosic biomass. ISRN Biotechnol 2014:463074

Bansal S, Garg I, Vasa L (2023) Can social enterprises aid sustainable development? Evidence from multi-stage investigations. PloS One 18(2):e0281273

Barragán-Ocaña A, Merritt H, Sánchez-Estrada OE, Méndez-Becerril JL, Del Pilar Longar-Blanco M (2023) Biorefinery and sustainability for the production of biofuels and value-added products: A trends analysis based on network and patent analysis. PLoS One 18(1):e0279659

Bartling AW, Stone ML, Hanes RJ, Bhatt A, Zhang Y, Biddy MJ, Davis R, Kruger JS, Thornburg NE, Luterbacher JS, Rinaldi R, Samec JSM, Sels BF, Román-Leshkov Y, Beckham GT (2021) Techno-economic analysis and life cycle assessment of a biorefinery utilizing reductive catalytic fractionation. Energy Environ Sci 14(8):4147–4168

Bertacchi S, Jayaprakash P, Morrissey JP, Branduardi P (2022) Interdependence between lignocellulosic biomasses, enzymatic hydrolysis and yeast cell factories in biorefineries. Microb Biotechnol 15(3):985–995

Bhatia L, Jha H, Sarkar T, Sarangi PK (2023) Food waste utilization for reducing carbon footprints towards sustainable and cleaner environment: a review. Int J Environ Res Public Health 20(3):2318

Bocker R, Silva EK (2024) Pulsed electric field technology as a promising pre-treatment for enhancing orange agro-industrial waste biorefinery. RSC Adv 14(3):2116–2133

Broda M, Yelle DJ, Serwańska K (2022) Bioethanol production from Lignocellulosic biomass-challenges and solutions. Molecules (Basel, Switzerland) 27(24):8717

Brundage MP, Bernstein WZ, Hoffenson S, Chang Q, Nishi H, Kliks T, Morris KC (2018) Analyzing environmental sustainability methods for use earlier in the product lifecycle. J Clean Prod 187:10

Bušić A, Marđetko N, Kundas S, Morzak G, Belskaya H, Ivančić Šantek M, Komes D, Novak S, Šantek B (2018) Bioethanol production from renewable raw materials and its separation and purification: a review. Food Technol Biotechnol 56(3):289–311
Caicedo M, Barros J, Ordás B (2016) Redefining agricultural residues as bioenergy feedstocks. Materials (Basel, Switzerland) 9(8):635
Calvo Flores FG, Martin-Martinez FJ (2022) Biorefineries: achievements and challenges for a bio-based economy. Front Chem 10:973417
Cherubini F (2010) The biorefinery concept: using biomass instead of oil for producing energy and chemicals. Energy Convers Manag 51:1412–1421
Díaz-Pérez MA, Serrano-Ruiz JC (2020) Catalytic production of jet fuels from biomass. Molecules (Basel, Switzerland) 25(4):802
Dwi-Prasetyo W, Putra ZA, Bilad MR, Mahlia TMI, Wibisono Y, Nordin NAH, Wirzal MDH (2020) Insight into the sustainable integration of bio- and petroleum refineries for the production of fuels and chemicals. Polymers 12(5):1091
Egbosiuba TC (2022) Biochar and bio-oil fuel properties from nickel nanoparticles assisted pyrolysis of cassava peel. Heliyon 8(8):e10114
Elmekawy A, Diels L, De Wever H, Pant D (2013) Valorization of cereal-based biorefinery byproducts: reality and expectations. Environ Sci Technol 47(16):9014–9027
Falano T, Jeswani HK, Azapagic A (2014) Assessing the environmental sustainability of ethanol from integrated biorefineries. Biotechnol J 9(6):753–765
Fischer HP (2008) Mathematical modeling of complex biological systems: from parts lists to understanding systems behavior. Alcohol Res Health 31(1):49–59
Freitas EN, Salgado JCS, Alnoch RC, Contato AG, Habermann E, Michelin M, Martínez CA, Polizeli MLTM (2021) Challenges of biomass utilization for bioenergy in a climate change scenario. Biology 10(12):1277
Fuchs W, Rachbauer L, Rittmann SKR, Bochmann G, Ribitsch D, Steger F (2023) Eight up-coming biotech tools to combat climate crisis. Microorganisms 11(6):1514
Gadkari S, Narisetty V, Maity SK, Manyar H, Mohanty K, Jeyakumar RB, Pant KK, Kumar V (2023) Techno-economic analysis of 2,3-butanediol production from sugarcane bagasse. ACS Sustain Chem Eng 11(22):8337–8349
Gao Y, Wang M, Raheem A, Wang F, Wei J, Xu D, Song X, Bao W, Huang A, Zhang S, Zhang H (2023) Syngas production from biomass gasification: influences of feedstock properties, reactor type, and reaction parameters. ACS Omega 8(35):31620–31631
Gharby S (2022) Refining vegetable oils: chemical and physical refining. Sci World J 2022:6627013
Godvin Sharmila V, Dinesh Kumar M, Pugazhendi A, Bajhaiya AK, Gugulothu P, Rajesh Banu J (2021) Biofuel production from Macroalgae: present scenario and future scope. Bioengineered 12(2):9216–9238
Gonçalves MLMBB, Maximo GJ (2022) Circular economy in the food chain: production, processing and waste management. Circ Econ Sustain:1–19
Grande L, Pedroarena I, Korili SA, Gil A (2021) Hydrothermal liquefaction of biomass as one of the most promising alternatives for the synthesis of advanced liquid biofuels: a review. Materials (Basel, Switzerland) 14(18):5286
Guajardo N, Schrebler RA (2023) Upstream and downstream bioprocessing in enzyme technology. Pharmaceutics 16(1):38
Gupta PP, Bankar NJ, Mishra VH, Sanghavi S, Badge AK (2023) The efficient disposal of biomedical waste is critical to public health: insights from the central pollution control board guidelines in India. Cureus 15(10):e47303
Helmy M, Smith D, Selvarajoo K (2020) Systems biology approaches integrated with artificial intelligence for optimized metabolic engineering. Metab Eng Commun 11:e00149
Jeswani HK, Chilvers A, Azapagic A (2020) Environmental sustainability of biofuels: a review. Proc Math Phys Eng Sci 476(2243):20200351
Karalis DT, Karalis T, Karalis S, Kleisiari AS (2020) Genetically modified products, perspectives and challenges. Cureus 12(3):e7306

Khan MI, Shin JH, Kim JD (2018) The promising future of microalgae: current status, challenges, and optimization of a sustainable and renewable industry for biofuels, feed, and other products. Microb Cell Factories 17:36

Kim J, Sen SM, Maravelias C (2013) An optimization-based assessment framework for biomass-to-fuel conversion strategies. Energy Environ Sci 6:1093–1104

Kumar AK, Sharma S (2017) Recent updates on different methods of pretreatment of lignocellulosic feedstocks: a review. Bioresour Bioprocess 4(1):7

Kumar S, Smith SR, Fowler G, Velis C, Kumar SJ, Arya S, Rena, Kumar R, Cheeseman C (2017) Challenges and opportunities associated with waste management in India. R Soc Open Sci 4(3):160764

Lange L, Connor KO, Arason S, Bundgård-Jørgensen U, Canalis A, Carrez D, Gallagher J, Gøtke N, Huyghe C, Jarry B, Llorente P, Marinova M, Martins LO, Mengal P, Paiano P, Panoutsou C, Rodrigues L, Stengel DB, van der Meer Y, Vieira H (2021) Developing a sustainable and circular bio-based economy in EU: by partnering across sectors, upscaling and using new knowledge faster, and for the benefit of climate, environment & biodiversity, and people & business. Front Bioeng Biotechnol 8:619066

Liu WJ, Yu HQ (2021) Thermochemical conversion of lignocellulosic biomass into mass-producible fuels: emerging technology progress and environmental sustainability evaluation. ACS Environ Au 2(2):98–114

Liu ZH, Hao N, Wang YY, Dou C, Lin F, Shen R, Bura R, Hodge DB, Dale BE, Ragauskas AJ, Yang B, Yuan JS (2021) Transforming biorefinery designs with 'Plug-In Processes of Lignin' to enable economic waste valorization. Nat Commun 12(1):3912

Makepa DC, Chihobo CH (2024) Barriers to commercial deployment of biorefineries: A multi-faceted review of obstacles across the innovation chain. Heliyon, 10(12):e32649

Maicas S (2020) The role of yeasts in fermentation processes. Microorganisms 8(8):1142

Maroa S, Inambao F (2021) A review of sustainable biodiesel production using biomass derived heterogeneous catalysts. Eng Life Sci 21(12):790–824

Mauerhofer LM, Pappenreiter P, Paulik C, Seifert AH, Bernacchi S, Rittmann SKR (2019) Methods for quantification of growth and productivity in anaerobic microbiology and biotechnology. Folia Microbiol 64(3):321–360

Meegoda JN, Li B, Patel K, Wang LB (2018) A review of the processes, parameters, and optimization of anaerobic digestion. Int J Environ Res Public Health 15(10):2224

Meramo S, Fantke P, Sukumara S (2022) Advances and opportunities in integrating economic and environmental performance of renewable products. Biotechnol Biofuels Bioprod 15(1):144

Mitra S, Murthy GS (2022) Bioreactor control systems in the biopharmaceutical industry: a critical perspective. Syst Microbiol Biom 2(1):91–112

Molina-Peñate E, Artola A, Sánchez A (2024) Exploring biorefinery alternatives for biowaste valorization: a techno-economic assessment of enzymatic hydrolysis coupled with anaerobic digestion or solid-state fermentation for high-value bioproducts. Bioengineered 15(1):2307668

Moncadam J, Aristizábal Marulanda V, Cardona CA (2016) Design strategies for sustainable biorefineries. Biochem Eng J 116:10

Mukamwi M, Somorin T, Soloha R, Dace E (2023) Databases for biomass and waste biorefinery—a mini-review and SWOT analysis. Bioengineered 14(1):2286722

Narisetty V, Reshmy R, Maitra S, Tarafdar A, Alphy MP, Kumar AN, Madhavan A, Sirohi R, Awasthi MK, Sindhu R, Varjani S, Binod P (2023) Waste-derived fuels and renewable chemicals for bioeconomy promotion: a sustainable approach. Bioenergy Res 16(1):16–32

Nguyen QA, Smith WA, Wahlen BD, Wendt LM (2020) Total and sustainable utilization of biomass resources: a perspective. Front Bioeng Biotechnol 8:546

Palkovits R, Delidovich I (2018) Efficient utilization of renewable feedstocks: the role of catalysis and process design. Philos Transact A Math Phys Eng Sci 376(2110):20170064

Popp J, Harangi-Rákos M, Gabnai Z, Balogh P, Antal G, Bai A (2016) Biofuels and their co-products as livestock feed: global economic and environmental implications. Molecules (Basel, Switzerland) 21(3):285

Purchase CK, Al Zulayq DM, O'Brien BT, Kowalewski MJ, Berenjian A, Tarighaleslami AH, Seifan M (2021) Circular economy of construction and demolition waste: a literature review on lessons, challenges, and benefits. Materials (Basel, Switzerland) 15(1):76

Rajendran K, Murthy GS (2017) How does technology pathway choice influence economic viability and environmental impacts of lignocellulosic biorefineries? Biotechnol Biofuels 10:268

Ribeiro LF, Amarelle V, Alves LF, Viana de Siqueira GM, Lovate GL, Borelli TC, Guazzaroni ME (2019) Genetically engineered proteins to improve biomass conversion: new advances and challenges for tailoring biocatalysts. Molecules (Basel, Switzerland) 24(16):2879

Robak K, Balcerek M (2018) Review of second-generation bioethanol production from residual biomass. Food Technol Biotechnol 56(2):174–187

Romero-Perez JC, Vergara L, González-Delgado ÁD (2021) Development of a methodology for the synthesis of biorefineries based on incremental economic and exergetic return on investment. ACS Omega 6(9):6112–6123

Sacramento-Rivero JC, Mwampamba TH, Navarro-Pineda FS, Musule R, García CA, Martínez-Bravo RD, Morales-García AL, Equihua-Sánchez M, Fuentes-Gutiérrez AF, Gallardo-Álvarez RM, Ruiz Camou CR, Grande-Acosta GK, Manzini F, Islas-Samperio JM, Camarillo R (2022) A methodological framework for assessing the sustainability of solid biofuels systems. Bioenergy Res 15(4):1797–1819

Sajid Z (2021) A dynamic risk assessment model to assess the impact of the coronavirus (COVID-19) on the sustainability of the biomass supply chain: a case study of a U.S. biofuel industry. Renew Sustain Energy Rev 151:111574

Saleem M (2022) Possibility of utilizing agriculture biomass as a renewable and sustainable future energy source. Heliyon 8(2):e08905

Sevigné-Itoiz E, Mwabonje O, Panoutsou C, Woods J (2021) Life cycle assessment (LCA): informing the development of a sustainable circular bioeconomy? Philos Trans A Math Phys Eng Sci 379(2206):20200352

Sharma HB, Vanapalli KR, Samal B, Cheela VRS, Dubey BK, Bhattacharya J (2021) Circular economy approach in solid waste management system to achieve UN-SDGs: solutions for post-COVID recovery. Sci Total Environ 800:149605

Shen Y, Wang X, Lei J, Wang S, Hou Y, Hou X (2022) Catalytic confinement effects in nanochannels: from biological synthesis to chemical engineering. Nanoscale Adv 4(6):1517–1526

Shukla M, Al-Busaidi KT, Trivedi M, Tiwari RK (2018) Status of research, regulations and challenges for genetically modified crops in India. GM Crops Food 9(4):173–188

Silva TN, Thomas JB, Dahlberg J, Rhee SY, Mortimer JC (2022) Progress and challenges in sorghum biotechnology, a multipurpose feedstock for the bioeconomy. J Exp Bot 73(3):646–664

Smit AT, van Zomeren A, Dussan K, Riddell LA, Huijgen WJJ, Dijkstra JW, Bruijnincx PCA (2022) Biomass pre-extraction as a versatile strategy to improve biorefinery feedstock flexibility, sugar yields, and lignin purity. ACS Sustain Chem Eng 10(18):6012–6022

Solarte-Toro JC, Ortiz-Sanchez M, Inocencio-García P-J, Cardona Alzate CA (2023) Sustainable biorefineries based on catalytic biomass conversion: a review. Catalysts 13(5):902

Takkellapati S, Li T, Gonzalez MA (2018) An overview of biorefinery derived platform chemicals from a cellulose and hemicellulose biorefinery. Clean Techn Environ Policy 20(7):1615–1630

Tong KTX, Tan IS, Foo HCY, Show PL, Lam MK, Wong MK (2023) Sustainable circular biorefinery approach for novel building blocks and bioenergy production from algae using microbial fuel cell. Bioengineered 14(1):246–289

Tsagkari M, Couturier JL, Kokossis A, Dubois JL (2016) Early-stage capital cost estimation of biorefinery processes: a comparative study of heuristic techniques. ChemSusChem 9(17):2284–2297

Usmani Z, Sharma M, Awasthi A, Lukk T, Tuohy M, Gong L, Nguyen-Tri P, Goddard A, Bill R, Nayak S, Gupta VK (2021) Lignocellulosic biorefineries: the current state of challenges and strategies for efficient commercialization. Renew Sust Energ Rev 148:10

Varbanov PS, Jia X, Lim JS (2021) Process assessment, integration and optimisation: the path towards cleaner production. J Clean Prod 281:124602

Vasić K, Knez Ž, Leitgeb M (2021) Bioethanol production by enzymatic hydrolysis from different lignocellulosic sources. Molecules (Basel, Switzerland) 26(3):753

Venkata Mohan S, Katakojwala R (2020) A critical view on the environmental sustainability of biorefinery systems. Curr Opin Green Sustain Chem 27:10

Voidarou C, Antoniadou M, Rozos G, Tzora A, Skoufos I, Varzakas T, Lagiou A, Bezirtzoglou E (2020) Fermentative foods: microbiology, biochemistry, potential human health benefits and public health issues. Foods (Basel, Switzerland) 10(1):69

Vora LK, Gholap AD, Jetha K, Thakur RRS, Solanki HK, Chavda VP (2023) Artificial intelligence in pharmaceutical technology and drug delivery design. Pharmaceutics 15(7):1916

Wiatrowski M, Klein BC, Davis RW, Quiroz-Arita C, Tan ECD, Hunt RW, Davis RE (2022) Techno-economic assessment for the production of algal fuels and value-added products: opportunities for high-protein microalgae conversion. Biotechnol Biofuels Bioprod 15(1):8

Yadav V, Sarker A, Yadav A, Miftah AO, Bilal M, Iqbal HMN (2022) Integrated biorefinery approach to valorize citrus waste: a sustainable solution for resource recovery and environmental management. Chemosphere 293:133459

Zainurin SN, Wan Ismail WZ, Mahamud SNI, Ismail I, Jamaludin J, Ariffin KNZ, Kamil WA, W. M. (2022) Advancements in monitoring water quality based on various sensing methods: a systematic review. Int J Environ Res Public Health 19(21):14080

Zetterholm J, Bryngemark E, Ahlström J, Söderholm P, Harvey S, Wetterlund E (2020) Economic evaluation of large-scale biorefinery deployment: a framework integrating dynamic biomass market and techno-economic models. Sustainability 12:7126

Zhang QW, Lin LG, Ye WC (2018) Techniques for extraction and isolation of natural products: a comprehensive review. Chin Med 13:20

# Prospects on Bio-fuels from Agri-waste and Its Techno-economy

9

Mani Jayakumar, Arega Begna,
Sundramurthy Venkatesa Prabhu, and Magesh Rangaraju

**Abstract**

Agricultural residue (AR) is a major cradle of biomass that can be converted into valuable biofuels. Bio-fuels are renewable and sustainable and can help to lessen our dependence on fossil fuels. There are a number of different ways to convert agricultural waste into bio-fuels of bio-ethanol, biodiesel, biogas, and bio-hydrogen. The transformation of AR into bio-fuels is an encouraging technology that has the prospective to diminish our reliance on fossil fuels and help to protect the environment. Conversely, around are a sum of encounters that essential to be addressed earlier this approaches can be extensively accepted. These

M. Jayakumar (✉)
Department of Chemical Engineering, Haramaya Institute of Technology, Haramaya University, Haramaya, Dire Dawa, Ethiopia

Department of Biotechnology, Faculty of Engineering, Karpagam Academy of Higher Education, Coimbatore, Tamil Nadu, India

Centre for Natural Products and Functional Foods, Karpagam Academy of Higher Education, Coimbatore, Tamil Nadu, India
e-mail: drjayakumarmani@haramaya.edu.et

A. Begna
Department of Chemical Engineering, Haramaya Institute of Technology, Haramaya University, Haramaya, Dire Dawa, Ethiopia

S. Venkatesa Prabhu
Department of Biotechnology, Faculty of Engineering, Karpagam Academy of Higher Education, Coimbatore, Tamil Nadu, India

Centre for Natural Products and Functional Foods, Karpagam Academy of Higher Education, Coimbatore, Tamil Nadu, India

M. Rangaraju
Department of Chemical Engineering, Adhiparasakthi Engineering College, Melmaruvathur, Tamil Nadu, India



V. Singh (ed.), *Sustainable Waste Management Towards Circular Bioeconomy*,
https://doi.org/10.1007/978-981-95-2227-9_9

encounters comprise, the budget of production, environmental impact, and the availability of feedstock. Despite these challenges, the conversion of agricultural waste into bio-fuels is an encouraging strategy with the prospective to make a substantial influence to our energy future. As the technology continues to develop and the cost of production decreases, bio-fuels are likely to become more widely available and affordable. This will make them a more attractive option for consumers and businesses. In this chapter we will discuss on feasibility of raw materials, and technology development on bio-fuels and detailed discussion on ethanol to methane and its techno economy prospects.

**Keywords**

Agricultural waste · Bio-fuels · Biomass · Bio-resources · Prospect · Feedstock · Techno-economy

## 9.1 Introduction

Biofuel arose as a sparkling beacon of hope in a world in critical need of a sustainable solution for the environment. Biofuel, which is generated from organic-matter such as plants, agricultural leftovers, or even algae, is a more sustainable alternative and environmentally friendly to typical fossil-fuels (Periyasamy et al. 2024; Chozhavendhan et al. 2023). Its history began hundreds of years ago when people discovered that certain materials should be converted into energy. However, it wasn't until the nineteenth century that the concept of biofuels really gained roots. Initially, biofuel pioneers experimented with vegetable oils and animal fats as energy sources. Forward to the twentieth century, when environmental concerns intensified and the demand for renewable energy became more pressing. On a bigger scale, researchers and scientists began to investigate biofuel alternatives. They used plants such by way of soybeans, sugarcane, and corn to make ethanol and biodiesel. These biofuels showed promise, but there were still constraints in terms of supply and cost-effectiveness. Technology improvements and new thinking have brought biofuel into the forefront in recent years. Microorganisms that can directly manufacture biofuels have been identified by scientists, eliminating the requirement for plant-based sources totally.

Algae, for example, have emerged as biofuel powerhouses owing to their capability to proficiently convert sunlight and carbon dioxide into useable energy. Biofuel is now making substantial advances in transportation, aircraft, and even power generation. It contributes to the decrease in greenhouse gas emissions (GHGs) reduces our reliance on finite fossil fuel supplies, and mitigates climate change (Kumar et al. 2022). Consider a world in which cars zip down the street, planes soar through the sky, and homes run on clean, renewable biofuels. Mother Nature smiles on us in our world where sustainability and progress go hand in hand. In this chapter, we will examine the feasibility of feedstocks, mainly agricultural wastes, as well as biofuel technology development and a full review of ethanol to methane and its techno-economic prospects.

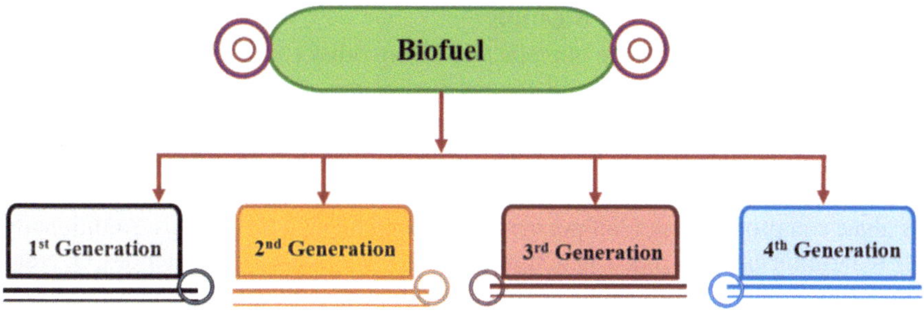


**Fig. 9.1** Biofuels and its various generations

### 9.1.1 Biofuel Classification

Biofuels are categorized into generations centered on the feedstock's utilized and the manufacturing processes involved. The concept of biofuels is still relatively unfamiliar to many. An agricultural fuel derived from biological sources (bioethanol, biodiesel) can either take the form of a liquid such as ethanol, biodiesel or biogas, or it can take the form of a gaseous such as biogas or hydrogen (Jayakumar et al. 2023c; Raju et al. 2009). Listed below are brief summaries of each generation (Chozhavendhan et al. 2022) and illustrated in Fig. 9.1.

#### 9.1.1.1 First-Generation Biofuels

Food crops or feedstocks that are typically used for food or animal feed are used to make first generation biofuels. The feedstocks include crops like corn, sugarcane, soybean, rapeseed, and vegetable oils. These feedstock's are converted into biofuels through processes like fermentation (for ethanol) or transesterification (for biodiesel) (Agarwal and Kumar 2018). First-generation biofuels have been commercially produced for many years and are widely used as alternatives to fossil fuels (Aro 2016). However, concerns are raised about their potential impact on food prices, competition with food production, and land use change.

#### 9.1.1.2 Second-Generation Biofuels

Generally biofuels of the second generation are made most widely from the non-food feedstocks, such by way of agricultural residues (like corn-stover, wheat straw), energy crops (like switch grass, miscanthus), and woody biomass (Aro 2016). These feedstock's contain cellulose and lignin, which can be transformed into bio-fuels through advanced process techniques like biochemical conversion (enzymatic hydrolysis) or thermochemical conversion (gasification and pyrolysis) (Menon and Rao 2012). Second-generation biofuels have the advantage of utilizing feedstock's that are not in direct competition with food production. They offer higher potential for GHGs reduction and can subsidize to ecological land use practices.

#### 9.1.1.3 Third-Generation Biofuels

Third-generation biofuels are manufactured from other types of microorganisms or algae. Algae have a higher oil content and can remain cultivated in various environments, including ponds, bioreactors, or even wastewater (Behera et al. 2015). The oil mined from algae can be transformed into biofuels through processes like transesterification. Third-generation biofuels have the advantage of high-yield potential, as algae can produce more oil per acre associated to customary crops. Additionally, algae farming can take place on non-arable land and utilize wastewater or $CO_2$ emissions, providing a sustainable approach to biofuel production (Chowdhury and Loganathan 2019).

#### 9.1.1.4 Fourth-Generation Biofuels

Fourth-generation biofuels (FGBs), often denoted to as cutting-edge biofuels, are manufactured from feedstock's that are not derived from plants or algae (Abdullah et al. 2019). These include synthetic biology approaches, such as genetically engineered microorganisms or artificial photosynthesis, which enable the manufacture of biofuels from non-biological bases like carbon dioxide or hydrogen (Cavelius et al. 2023). FGBs are still in the initial phases of development and hold great potential for achieving higher energy efficiencies and reducing overall environmental impacts. Each generation of biofuels represents advancements in technology and feedstock utilization, aiming for improved sustainability, reduced emissions, and minimized competition with food production. Continued research and development in biofuel production are crucial to further enhance efficiency, cost-effectiveness, and sustainability to come across the developing global mandate for renewable energy.

## 9.2 Benefits of Biofuels

Biofuels could have a big impact on our energy supply in the future. They have a lot of advantages, including lower greenhouse gas emissions, higher energy security, better air quality, and a favorable economic impact (Sethumadhavan et al. 2024; Gizaw et al. 2024). These advantages are anticipated to become even more obvious as biofuel production technology improves. The following diagram discusses several benefits of biofuel compared to traditional fossil fuels illustrated in Fig. 9.2.

Conversely, it is critical to emphasize that bio-fuels must be manufactured in an ecological way in order to minimize harmful consequences on food production, land use, and water resources (Jayakumar et al. 2022b). To ensure that biofuels provide overall environmental and social advantages, responsible and well-regulated manufacturing techniques are essential. By using biofuels, we can promote good change, contribute to a greener, and reduce our environmental impact, added ecological prospect for future generations.

Agricultural waste refers to residual materials produced by agricultural activities, such as residues from crops, manure from animals, food processing wastes, and other organic substances (Hamda et al. 2024). Agro-industry wastes, fruit wastes,

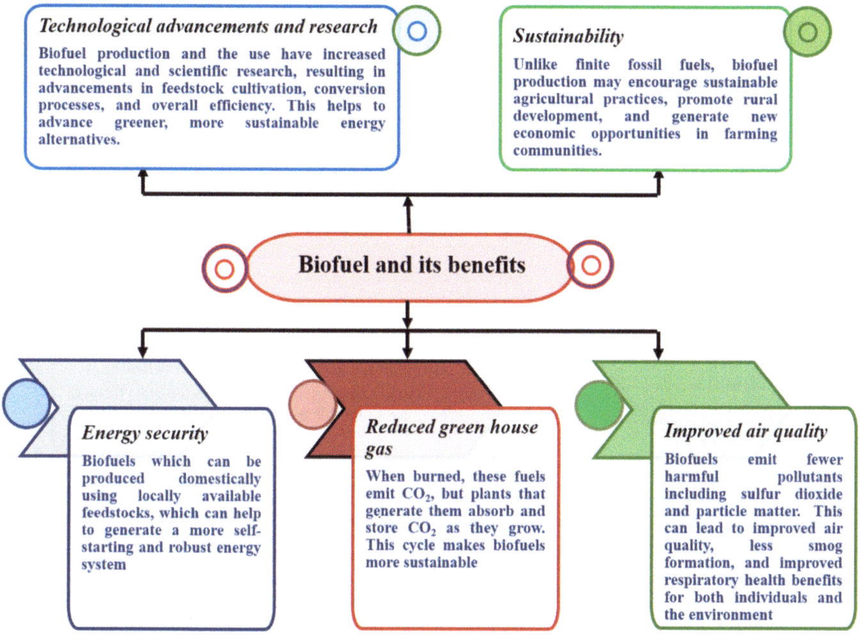


**Fig. 9.2** Biofuel and its benefits

harvest leftovers, livestock wastes, and vegetable-wastes are the four major types of agricultural waste (Jayakumar et al. 2023a, f; Pattanaik et al. 2019). While agricultural waste has many applications, it can also have negative environmental consequences if not managed appropriately. Here are some significant considerations about agricultural waste and its environmental consequences depicted in Fig. 9.3.

Given the enormous amount of waste produced by agriculture and its influence on animals, environment, and the individuals, it is preferable that it be converted into biofuel. Thus, transforming agro farm yard residues to biofuel is solitary of the long-term waste controlling approaches to carbon balance in the environment, assure sanitation, and resource recovery (Ibarra-Gonzalez and Rong 2019). In addition, 21% of all greenhouse gas emissions come from agriculture. A sustainable pathway for agriculture has yet to be discovered, with the expected increase in global population to almost 10 billion by 2050. Due to the limited availability of natural resources, this must be accomplished by utilizing less fossil fuel, reducing emissions, and reducing solid waste (Karić et al. 2022). There are several advantages to using agricultural waste to make biofuel. First, it reduces the amount of waste transferred to landfills or incinerators. Second, it has the potential to help create new agricultural jobs. Third, it has the probable to diminish dependence on fossil-fuel. However, there are significant drawbacks to using agricultural waste to generate biofuel. One issue is that the expense of treating agricultural waste might be prohibitively expensive. Another issue is that the grade of biofuel manufactured from agricultural waste varies dependent on the category of waste utilized. Despite

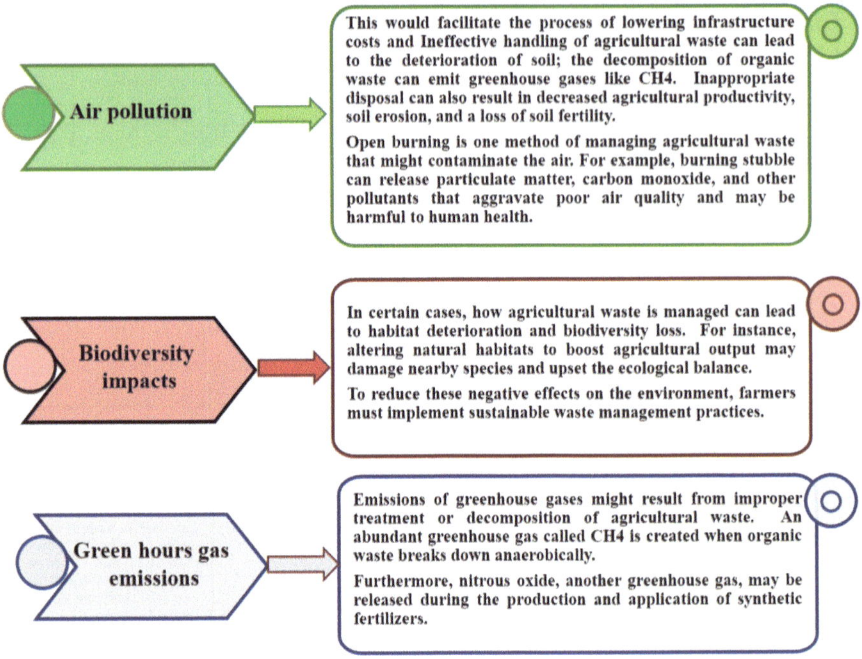


**Fig. 9.3** Agricultural waste and its environmental consequences

these obstacles, using agricultural waste to make biofuel is a potential area of study. As technology advances, the cost of processing agricultural waste is projected to fall, and the quality of the biofuel produced is likely to improve. This will make agricultural waste a more appealing possibility to produce biofuel.

## 9.3 Biofuels and Other Useful Items Might Potentially Be Made from Agricultural Waste

Agricultural waste has the possible to remain transformed into biofuels and other value added products through various processes (Gari et al. 2024). This approach provides an opportunity to not only manage agricultural waste effectively but also create additional economic and environmental benefits (Karimi Alavijeh et al. 2020). Agricultural waste be able to utilize as feedstock for the production of biofuels such as biodiesel and ethanol. Among other lignocellulosic biomass sources, the exploitation of AR for bio-fuel generation is the present focus of worldwide research (Zabed et al. 2017). Through processes like fermentation or thermochemical conversion, the organic matter in AR can be converted into fuel alternatives. This reduces dependence on fossil-fuels and helps mitigate GHGs. Another valuable product that can be derived from agricultural waste is biogas. By using anaerobic digestion, organic-waste-materials like manure, food processing byproducts, and crop residues can be decomposed to produce biogas, which is predominantly composed of methane. Biogas can be utilized as a fuel for vehicles or as a green energy source for heating and power generating (Achinas et al. 2017). The usage of agri-waste for biogas manufacture necessitates the integration of agri-waste management, biogas generation, and consumption, as well as the enactment of a circular-economy (Kapoor et al. 2020). Agro farm yard waste can also serve as a basis of biomaterials and chemicals (Duque-Acevedo et al. 2020). By employing advanced technologies, waste materials can be processed to extract valuable compounds that can be used in various industries. For example, lignocellulosic biomass from crop residues can be converted into platform chemicals like sugars, organic acids, or bio-based polymers. Fertilizers and Soil Amendments: Agricultural waste can be transformed into soil amendments or nutrient-rich organic fertilizers. Composting and biochar production are common methods to convert waste materials into beneficial produces that enrich soil fertility, retain moisture, and improve overall soil health. Some agricultural waste, such as crop byproducts or leftover grains, can be processed into animal feed. By extracting and refining nutrients from waste materials, these products can supplement animal diets, reducing the need for additional feed resources. By converting agro farm yard waste into other valuable products and biofuels, we be able to minimize environmental impacts associated with waste disposal while simultaneously promoting the circular economy. It's an exciting field with opportunities for sustainable waste management, energy production, and resource utilization.

## 9.4 Types of Biofuels

Biofuels are categorized depending on their physical state and content. Liquid biofuels, gaseous biofuels, and solid biofuels are the three major forms of biofuels (Nigam and Singh 2011). Figure 9.4 summarizes the various types of biofuels.

### 9.4.1 Liquid Biofuels

From biological or biodegradable biomass, liquid biofuels can be manufactured in a liquefied state. The most common category of fuel is liquid biofuel. There are several notable products in energy statistics associated with liquid biofuels (Jayakumar et al. 2023f; Ashine et al. 2023). They are further divided into two major subcategories: ethanol and biodiesel. Ethanol is often utilized as a fuel additive or can be combination with gasoline to minimize GHGs in transportation, whereas biodiesel be able to utilize as a standby for diesel fuel and is compatible with most diesel engines without any changes (Demirbas 2011).

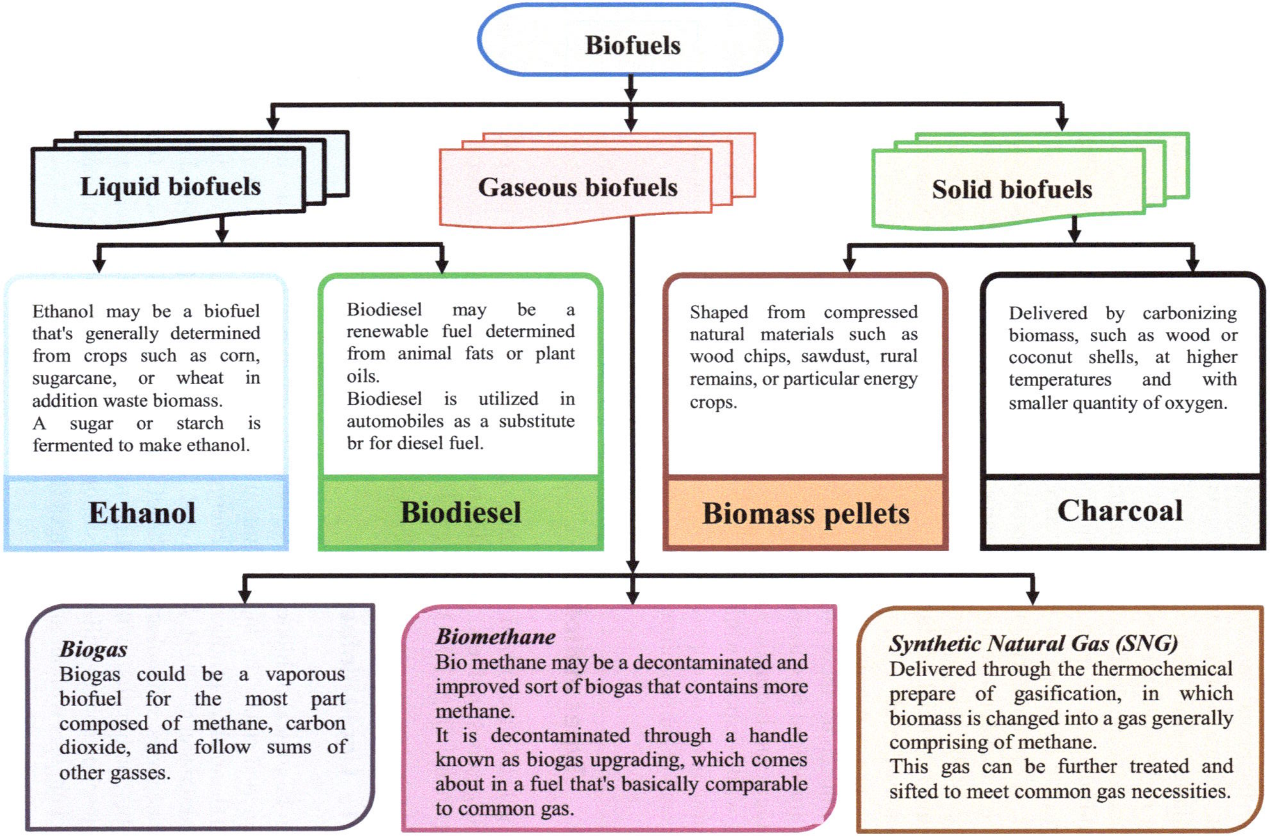


**Fig. 9.4** Biofuels and its types

### 9.4.2 Gaseous Biofuels

In addition to their low density, gaseous biofuels are also low-volume fuels. These biofuels are generally made of methane ($CH_4$) or hydrogen gas ($H_2$). Biogas and biomethane are examples of gaseous biofuels that are created by the anaerobic-digestion of organic-waste materials. These fuels be able to utilize to generate energy, heat homes, and even transport people.

Several advantages have been demonstrated for gaseous biofuels, such as greater reactivity, minimizing waste formation, requiring fewer oxidants, being easier to control, and requiring only a small reactor (Jayakumar et al. 2023c).

### 9.4.3 Solid Biofuels

Solid bio-fuels are resulting from sources such by way of wood, agro farm leftovers, and energy based crops such as switch grass. Solid biofuels can be produced from both processed and unprocessed biomass. Natural biofuels and synthetic biofuels are solid biofuels (Jayakumar et al. 2023c). These biofuels are commonly employed in residential and industrial settings for heating and power generation.

It is crucial to remember that each form of bio-fuel has its specific rewards and considerations in standings of availability, manufacturing procedures, and impact on the environment. Every form of biofuel has its own manufacturing procedures, applications, and environmental benefits. They all help to diversify the energy mix and reduce the discharge of GHGs.

## 9.5 Potential Feedstock's for Biofuel Making

There are various feedstocks that can be utilized to produce biofuels, and many of them come from agricultural fields. Agro farm yard waste used as a major feedstock for bio-fuel manufacture be able to classified as fruit waste, livestock waste, industrial processing waste, and crop residues (Awogbemi and Von Kallon 2022). Agro farm yard wastes are categorized as lignocellulosic biomass because they contain multifaceted molecular assemblies like ash, cellulose, certain extractives, hemicellulose, and lignin, primarily protein. Through digestion, the wastes' structures are foremost dissolved into simpler monomer units before being converted into other configurations (Guo et al. 2021). The following Table 9.1 shows the various potential feedstocks from agro farm yard waste illustrated in Fig. 9.5 and its chemical composition as assets for bio-fuel manufacture that reported by different researchers.

**Table 9.1** Various potential feedstock's from agricultural waste and its chemical composition

| S. No | Feedstock | Components | | | | Reference |
|---|---|---|---|---|---|---|
| | | Cellulose % | Hemicellulose % | Lignin % | Others % | |
| 1. | Almond shells | 16.7 | 31.6 | 29.7 | 6.8 | Morales et al. (2020) |
| 2. | Apple pomace | 32.62 | 23.6 | 25.38 | 0.14 | Leonel et al. (2020) |
| 3. | Banana peel | 7.5 | 74.9 | 7.9 | 0.01 | Aishah (2018) |
| 4. | Barley straw | 32.9 | 27.2 | 16.8 | N/a | Álvarez et al. (2021) |
| 5. | Cocoa pods | 26.1 | 4.82 | 21.29 | 0.064 | Antwi et al. (2019) |
| 6. | Coconut husks | 26.72 | 17.73 | 41.19 | N/a | Muharja et al. (2018) |
| 7. | Canola residues | 21.4 | 15.9 | 23.0 | N/a | Ranjan et al. (2023) |
| 8. | Corn stover | 34 | 19.53 | 18.6 | N/a | Yu et al. (2019) |
| 9. | Cotton stalks | 34 | 18 | 31.6 | N/a | Keshav et al. (2023) |
| 10. | Finola hemp stalks | 62 | 17 | 19 | N/a | Brar et al. (2022) |
| 11. | Grape pomace | 9.15 | 22.13 | 23.19 | 7 | Bender et al. (2020) |
| 12. | Jatropha fruit shell | 40 | 34 | 12.7 | 3 | Odetoye et al. (2018) |
| 13. | Oat hulls | 37 | 33 | 18 | 4.8 | Skiba et al. 2017) |
| 14. | Peanut shells | 35.5 | 20.6 | 34.1 | 4.4 | Ge et al. (2020) |
| 15. | Rice straw | 35.63 | 18.06 | 31.97 | 10.24 | Kaur and Kuhad (2019) |
| 16. | Rapeseed straw | 35.5 | 20.9 | 21.7 | 7.6 | Kuglarz et al. (2018) |
| 17. | Sorghum stalks | 27 | 25 | 11 | N/a | Shahzadi et al. (2014) |
| 18. | Sugarcane bagasse | 37.72 | 22.34 | 22.95 | 2.5 | Liu et al. (2021) |
| 19. | Sunflower stalks | 33.99 | 24.96 | 25.9 | N/a | Manmai et al. (2021) |
| 20. | Soybean straw | 34.09 | 16.05 | 21.6 | N/a | Wan et al. (2011) |
| 21. | Switch grass | 39 | 25 | 15 | N/a | Pasangulapati et al. (2012) |
| 22. | Wheat straw | 17.00 | 12.5 | 18.00 | 3 | Shrivastava and Sharma (2023) |

## 9.6 Potential Biofuel Production Practices

Biomass energy comprises liquid biofuels (butanol, biodiesel, biooil, ethanol, and methanol), biogas, and solid biofuels (however, any solid might be burned to provide heat energy, and usually wood). While liquid biofuels and biogas need to be converted before they be able to use as fuel, solid biofuels be able to charred straight to harvest energy. There are numerous potential practices that transform biomass to

**Fig. 9.5** Various biomass feedstocks from agro wastes for biofuel production

fuels that power manufacturing sectors, produce fuel for automobiles, and achieve other energy necessities. How biomass is handled based on the category of biomass (e.g., oilseed crops or manure) and how it will be utilized (e.g., power generators or to fuel cars). The potential biofuel production practices are illustrated in Fig. 9.6.

### 9.6.1 Transesterification Practice (TrEP)

Generally biodiesel is produced, through TrEP is most commonly used (Jayakumar et al. 2023b; Kasirajan et al. 2022). The perseverance of this practice is to reduce the viscosity of feedstock oil when alcohol is present. A variety of catalysts are used to catalyze this reaction, including acidic, basic, and enzymatic catalysts. The catalytic properties of these enzymes are superior to those of other enzymatic catalysts.

Before, biodiesel was made by combining free fatty acids from feedstock oil with homogenous acid or a basic catalyst such as HCl, $H_3PO_4$, $H_2SO_4$, KOH, and NaOH respectively (Jayakumar et al. 2021). Biodiesel production through trans-esterification has some advantages, such as its effectiveness and eco-friendliness (Ashine et al. 2023).

### 9.6.2 Hydrothermal Liquefaction (HtL)

By heating high moisture biomass at 200–400 degrees Celsius and applying 50–250 bar pressure to a high-pressure reactor, HTL liquefies biomass. $H_2O$ or a combination of solvent and a $H_2O$ (phenol, propanol, methanol, and ethanol) is present during the reaction. High moisture biomass has been found to be more

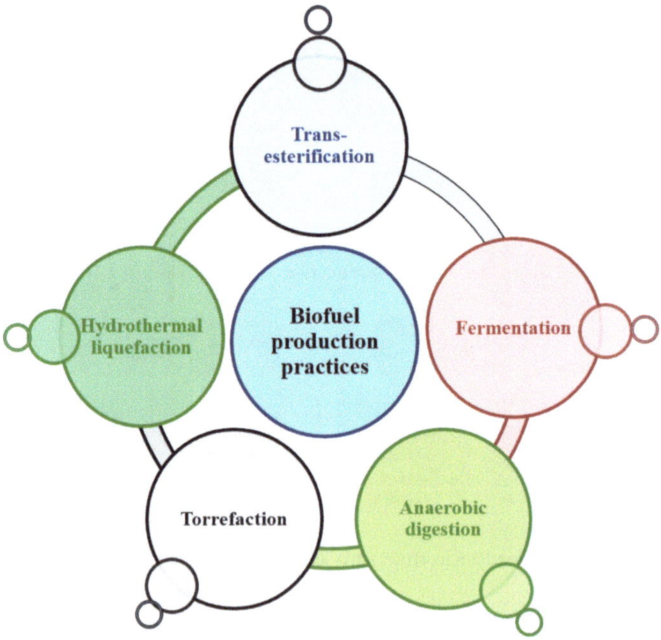


**Fig. 9.6** Potential biofuel production practice

successfully processed using the HtL approach (Gollakota et al. 2018). Here are a few benefits of the HtL procedure: (1) recycle of aqueous products, (2) it be able to comprise misty bio-mass from last practices, (3) it is independent of biomass type because traditional processes depend on biomass type to boost byproduct yields, and finally (4) it does not require drying of lignocellulosic biomass (Gundupalli et al. 2022).

### 9.6.3 Fermentation

Using microorganisms to convert carbohydrates like sucrose, glucose, and fructose into cellular-energy, fermentation is a biological process that yields carbon dioxide as a byproduct and the intended result bioethanol. This conversion is carried out by the yeast *Saccharomyces cerevisiae* in an anaerobic-environment; the fermentation of alcohol is considered an anaerobic practices (Jayakumar et al. 2024). Among the special and widely recognized techniques for producing bioethanol from ALCM biomass is fermentation process technology. Agricultural residues' potential higher glucan content biomass would theoretically produce taller ethanol yields. There are four distinct processes involved in the typical biomass conversion process: ALCM pretreatment, hydrolysis, fermentation, and flash distillation (Jayakumar et al. 2022c; Selvakumar et al. 2022). The following are some benefits of the fermentation process for ethanol. At atmospheric pressure and around 30 °C, the reaction occurs.

Because it uses so little energy, this is inexpensive. Since the glucose is obtained from plants, the procedure is carbon neutral (Jayakumar et al. 2022c).

### 9.6.4 Torrefaction

The recently developed thermochemical practice recognized as "torrefaction" comprises heating biomass feedstock in an inert atmospheric condition for a relatively long time (15–60 minutes) at low temperatures (200–300 °C) at a low heating rate. Here, the biomass precursors have been converted into liquid (bio-oil) and solid (biochar). Furthermore, fewer gaseous mixes ($CH_4$, CO, $CO_2$, and etc.) were also created. More constant in properties and with a higher heating value than raw biomass, torrefied biomass (biochar) is less expensive and more environmentally friendly (Jayakumar et al. 2023d). The following are some of the process' special benefits: over the years, torrefaction of biomass materials has drawn a lot of consideration because it can yield high-grade solid biofuel that is more durable, grinds more easily, has a higher calorific value and, and bulk density has a higher energy-density than the original, untreated biomass material (Acharya et al. 2012).

### 9.6.5 Anaerobic Digestion (AD)

AD has the potential to be a sustainable solution for the energy conversion of waste sources, going beyond the conventional waste management methods of land filling and burning. Various anaerobic microbe kinds separate the complex organic feedstocks in AD processes. The AD process uses a wide range of feedstocks as a substrate, including agricultural (energy crops, harvest residues), industrial (sludge), and municipal garbage. The amounts of protein and carbohydrates in each substrate vary, resulting in the production of biogas with varying quality. Numerous factors, including temperature, pH, organic loading rate, sludge retention time, sludge retention distribution, up flow velocity, total solids, volatile solids, and more, affect the composition of biogas. Here are a few advantages of the AD process: (1) efficient management of manure, (2) process benefits soil health, (3) reduction in methane emissions, (4) generation of energy from renewable resources, (5) diversion of organics from landfills.

## 9.7 Techno Economy Analysis (TeA) for Biofuel Production

To perform a thorough technology comparison, one must first determine the crucial elements that affect the product's total profitability. To achieve this, a Techno economy analysis (TeA) intends to be accompanied on numerous manufacture approaches established precisely for the larger scale biofuel manufacture (Srinadh and Neelancherry 2023). For a given system or procedure, there could be multiple options. For example, biofuels, such as biogas, bioethanol, bio oil, biodiesel or biobutanol, be able to formed from numerous biomass feedstocks, comprising agricultural residues, algae based biomass, cereal grains, dairy farm yard wastes, energy crops, fruit processing industry wastes, food wastes, forest residues, vegetable processing wastes, and valorization of sugar processing wastes (Jayakumar et al. 2023f; Gari et al. 2024; Gundupalli et al. 2022; Hamda et al. 2024; Gizaw et al. 2024; Temesgen et al. 2024).

However, any product development process should prioritize economic factors like maximizing profit and minimizing capital and operating expenditures. Determining manufacturing costs is the main goal of TeA. TeA is frequently used to set performance objectives for the development process and exclude less desired solutions during the early stages of product development (Shah et al. 2016). Following the procedures shown in Fig. 9.7, a TeA is carried out in detail for the construction and implementation of the best design alternative that has been selected.

It offers performance and cost boundaries which promote assistance in the decrease of excessive investment and costs associated risks. Usually, a TeA encompasses of a market investigation (analysis and data collection) comprising the study of energy balance and mass balance. In addition, a comprehensive budget evaluation is directed earlier to an economic investigation. Finally, a sensitivity and risk analysis is carried out. The TeA majorly comprises operation expenditure (OE) and capital expenditure (CE) alongside through material fabrication and processing cost. Moreover, OE encompasses expenses associated to energy, labour, and raw materials. CE comprises expenses related to procurements, construction, engineering, financing, and installation. Principal budget investments are intended by means of an anticipated market value of the functioning equipment. Moreover, the operating and capital cost reserves are appraised constructed on mutually of the energy

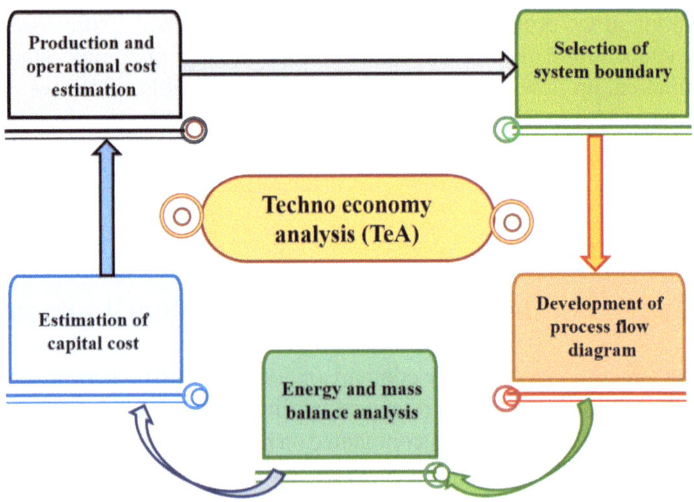


**Fig. 9.7** Techno economy analysis and its steps

balance and mass balance outputs from the handling models (Tschulkow et al. 2020; Van Dael et al. 2013). TeA benefits investigate both economic details and technical of a process technology or specific project (Mahmud et al. 2021). The TeA for a process or product comprises the valuation of its maintenance requirement, market availability, service life, skill obligation for maintenance and operation, variety of volumes offered, ease of installation and transportation, adaptableness in precise project conditions, aesthetic, the inherent risks, and material.

## 9.8 Prospects and Challenges

An increase in rapid industrialization and population growth be able to cause a rise in claim for fuel and food in the future centuries. Traditionally, fuels are manufactured from fossil-crude-oils over a conventional chemical practice. Conversely, fuels are also manufactured via an environmental friendly biochemical practice from numerous lignocellulosic biomass feedstocks. Nevertheless, present days feedstocks such by way of sugarcane and corn are most widely utilized for the manufacture of biofuel and other biomaterials. Though, a rise in global populace rises the demand for land and food. There will also be a total depletion of fossil fuels in the coming decades unless an alternative source can be found. Hence, in contemporary ages, alternate waste lignocellulosic bio-mass materials have been studied and utilized for the manufacture of various biofuels. Taking this into account, numerous kinds of lignocellulosic biomass materials are deliberated renewable and redundant as it is measured an ample carbon cradle available on earth (Gundupalli et al. 2022). As argued, there are numerous approaches (transesterification, hydrothermal liquefaction, fermentation, torrefaction, and anaerobic digestion) to transform waste lignocellulosic biomass feedstocks to value added products (biodiesel, biooil, bioethanol, solid fuel, and biogas).

In this chapter, comprehensive sides of various biofuel production approaches were debated. Biofuel production practices deal a benefit of using biomass with higher moisture content associated with supplementary practicing techniques. Consequently, the biofuel production practices eradicate the necessity to dry biomass earlier ensuing with the thermo-chemical practice. Despite several studies undertaken to date, future biofuel production studies must still address many challenges, including the following. Literature reported by numerous researchers related to the production various biofuel was conceded out in laboratory level. Consequently, it is essential to upgrade the processing practice to industrial scale and pilot scale to solve the present need. From side to side scale-up, the outcome of the biofuel practices on biofuel yield and biomass conversion be able to deliberate in high capacity. It has not yet been possible to analyze the cost-benefit of biofuel production. The yield of biofuel from diverse bio-mass has remained discoursed in several reported works, but not the cost of producing biofuel from them fuel properties of crude obtained from fossil fuels have hardly been addressed and compared in literature.

Numerous categories of lignocellulosic bio-mass constituents should be investigated for these properties. This will enable a larger scale of biofuel production using suitable biomass and production practices. In the process of extracting and recovering biofuel, solvents and catalysts are used. However, few studies have explored the possibility of recycling these materials. It is essential to discourse this in the future investigations to decrease the manufacturing cost of biofuel from waste lignocellulosic biomass feedstocks. There will be a few identified evidences in pathways and mechanism associated to the transformation of lignin and polysaccharides to biofuels such as biodiesel, bioethanol, biooil, biogas, biochar, and in addition aqueous crude are well known. Conversely, a comprehensive investigation on the pathway in accepting the biofuel production system and the process is quite a challenge for several investigators (Jayakumar et al. 2022c, 2023f).

A thorough investigation on the interface effects of cellulose, hemicellulose, lignin, and other extractives on growing the biofuel from waste lignocellulosic biomass materials through the production practice (transesterification, hydrothermal liquefaction, fermentation, torrefaction, and anaerobic digestion) is yet a great challenge to realize. Numerous investigations determined with higher heating value and its end product biofuel yield. Conversely, it is essential that a comprehensive measurable assay of the aqueous crude and biofuel is supported by utilizing LC-MS (liquid chromatography mass spectroscopy) and GC-MS (gas chromatography mass spectroscopy) profile, respectively (Jayakumar et al. 2022a, 2023d, e; Chozhavendhan et al. 2023). Additionally, the similar is associated to the biofuel production practices working parameter circumstances. Limited research investigations have described the effect and of significance co-solvents throughout production process. It has remained established in less investigations that amalgamation of organic solvents with water be able to lessen the reaction temperature and reaction time. As a result, this lowers the cost of producing biofuel from different lignocellulosic biomass feedstocks.

## 9.9 Summary

In this chapter, a comprehensive argument on the biofuel classification, benefits of biofuels, types of biofuels, numerous types of lignocellulosic biomass feedstocks with cellulose, hemicellulose, lignin and other biomaterials composition, in addition the pieces of literatures, enlighten us with the various biofuel production process (transesterification, hydrothermal liquefaction, fermentation, torrefaction, and anaerobic digestion) with its advantages, and techno economy analysis. Furthermore, the prospects of biofuel production process were also discussed briefly. The valorization of the numerous lignocellulosic biomass feedstocks improves the industries' ecologically pleasant environment, creating it economically achievable and socially tolerable. Likewise, it tried to encapsulate and show the claim of these lignocellulosic biomass feedstocks as a potential reserve of various biofuel production. This would lower infrastructure costs and makes it easier to solve potential industrial issues like product refinement and catalyst regeneration. Furthermore, the petrochemical industry's flexibility in responding to market demands would help with final commodity commercialization. This is strongly assumed to devise abundant significance for the upgraded echo-industrial nature of biorefineries.

## References

Abdullah B et al (2019) Fourth generation biofuel: a review on risks and mitigation strategies. Renew Sust Energ Rev 107:37–50
Acharya B, Sule I, Dutta A (2012) A review on advances of torrefaction technologies for biomass processing. Biomass Convers Biorefinery 2:349–369
Achinas S, Achinas V, Euverink GJW (2017) A technological overview of biogas production from biowaste. Engineering 3(3):299–307. https://doi.org/10.1016/J.ENG.2017.03.002
Agarwal S, Kumar A (2018) Historical development of biofuels. In: Biofuels: Greenhouse gas mitigation and global warming: Next generation biofuels and role of biotechnology. https://doi.org/10.1007/978-81-322-3763-1_2
Aishah F (2018) Banana Peel. Elseviar Ltd, pp 1–8
Álvarez C et al (2021) Production of xylooligosaccharides, bioethanol, and lignin from structural components of barley straw pretreated with a steam explosion. Bioresour Technol 342:125953. https://doi.org/10.1016/j.biortech.2021.125953
Antwi E et al (2019) Anaerobic digestion and the effect of hydrothermal pretreatment on the biogas yield of cocoa pods residues. Waste Manag 88:131–140. https://doi.org/10.1016/j.wasman.2019.03.034
Aro EM (2016) From first generation biofuels to advanced solar biofuels. Ambio 45(1):24–31. https://doi.org/10.1007/s13280-015-0730-0
Ashine F et al (2023) Biodiesel production from Argemone mexicana oil using chicken eggshell derived CaO catalyst. Fuel 332(P2):126166. https://doi.org/10.1016/j.fuel.2022.126166
Awogbemi O, Von Kallon DV (2022) Valorization of agricultural wastes for biofuel applications. Heliyon 8:e11117
Behera S et al (2015) Scope of algae as third generation biofuels. Front Bioeng Biotechnol 2(February):1–13. https://doi.org/10.3389/fbioe.2014.00090
Bender ABB et al (2020) Effects of micronization on dietary fiber composition, physicochemical properties, phenolic compounds, and antioxidant capacity of grape pomace and its dietary fiber concentrate'. Lwt 117(August 2019). https://doi.org/10.1016/j.lwt.2019.108652
Brar KK et al (2022) A paradigm shift towards production of sustainable bioenergy and advanced products from cannabis/hemp biomass in Canada. Biomass Convers Biorefinery. https://doi.org/10.1007/s13399-022-02570-6
Cavelius P et al (2023) The potential of biofuels from first to fourth generation. PLoS Biol 21(3):e3002063
Chowdhury H, Loganathan B (2019) Third-generation biofuels from microalgae: a review. Curr Opin Green Sustain Chem 20:39–44. https://doi.org/10.1016/j.cogsc.2019.09.003
Chozhavendhan S et al (2022) Potentials and challenges in biodiesel production from algae—technological outlook. In: Biofuels and bioenergy. Elsevier, pp 183–203
Chozhavendhan S et al (2023) Current and prognostic overview on the strategic exploitation of anaerobic digestion and digestate: a review. Environ Res 216:114526
Demirbas, M.F., 2011. Biofuels from algae for sustainable development. Applied energy, 88(10), pp.3473-3480. https://doi.org/10.1016/j.apenergy.2011.01.059
Duque-Acevedo M et al (2020) Agricultural waste: review of the evolution, approaches and perspectives on alternative uses. Glob Ecol Conserv 22:e00902
Gari MT et al (2024) Effective utilization of agricultural cereal grains in value-added products: a global perspective. In: Value added products from food waste. Springer, pp 41–58
Ge S et al (2020) High-pressure CO2 hydrothermal pretreatment of peanut shells for enzymatic hydrolysis conversion into glucose'. Chem Eng J 385(October 2019):123949. https://doi.org/10.1016/j.cej.2019.123949
Gizaw DG et al (2024) Commercial products derived from vegetable processing industrial wastes and their recent conversion studies. In: Value added products from food waste. Springer, pp 147–167

Gollakota ARK, Kishore N, Gu S (2018) A review on hydrothermal liquefaction of biomass. Renew Sust Energ Rev 81:1378–1392

Gundupalli MP et al (2022) Hydrothermal liquefaction of lignocellulosic biomass for production of biooil and by-products: current state of the art and challenges. In: Biofuels and bioenergy, pp 61–84

Guo H et al (2021) Application of machine learning methods for the prediction of organic solid waste treatment and recycling processes: a review. Bioresour Technol 319:124114. https://doi.org/10.1016/j.biortech.2020.124114

Hamda AS et al (2024) Valorization of fruit processing industry waste into value-added chemicals. In: Value added products from food waste. Springer, pp 107–126

Ibarra-Gonzalez P, Rong B-G (2019) A review of the current state of biofuels production from lignocellulosic biomass using thermochemical conversion routes. Chin J Chem Eng 27(7):1523–1535. https://doi.org/10.1016/j.cjche.2018.09.018

Jayakumar M et al (2021) Heterogeneous base catalysts: synthesis and application for biodiesel production—a review. Bioresour Technol 331(March):125054. https://doi.org/10.1016/j.biortech.2021.125054

Jayakumar M, Gebeyehu KB, Selvakumar KV, Parvathy S et al (2022a) Waste Ox bone based heterogeneous catalyst synthesis, characterization, utilization and reaction kinetics of biodiesel generation from Jatropha curcas oil. Chemosphere 288:132534

Jayakumar M, Kuppusamy Vaithilingam S, Karmegam N, Jabesa A (2022b) Bioethanol production from green biomass resources: emerging technologies. In: Encyclopedia of Green materials, pp 1–12

Jayakumar M, Kuppusamy Vaithilingam S, Karmegam N, Gebeyehu KB et al (2022c) Fermentation technology for ethanol production: current trends and challenges. In: Gurunathan B, Sahadevan RBT-B (eds) Biofuels and bioenergy. Elsevier, pp 105–131. https://doi.org/10.1016/b978-0-323-90040-9.00015-1

Jayakumar M, Asrat L et al (2023a) Agricultural waste as a source of fine chemicals. In: Agricultural waste to value-added products: technical, economic and sustainable aspects. Springer, pp 355–375

Jayakumar M, Bizuneh Gebeyehu K, Kuppusamy Vaithilingam S (2023b) Heterogeneous catalysts from biomass resources for biodiesel production. In: Encyclopedia of green materials. Springer, pp 1–9

Jayakumar M, Gebeyehu KB et al (2023c) A comprehensive outlook on topical processing methods for biofuel production and its thermal applications: current advances, sustainability and challenges. Fuel 349:128690

Jayakumar M, Hamda AS et al (2023d) Comprehensive review on lignocellulosic biomass derived biochar production, characterization, utilization and applications. Chemosphere 345:140515

Jayakumar M et al (2023e) A comprehensive outlook on topical processing methods for biofuel production and its thermal applications: current advances, sustainability and challenges. Fuel 349(May):128690. https://doi.org/10.1016/j.fuel.2023.128690

Jayakumar M, Gindaba GT et al (2023f) Bioethanol production from agricultural residues as lignocellulosic biomass feedstock's waste valorization approach: a comprehensive review. Sci Total Environ 879:163158

Jayakumar M et al (2024) Paddy straw as a biomass feedstock for the manufacturing of bioethanol using acid hydrolysis and parametric optimization through response surface methodology and an artificial neural network. Biomass Convers Biorefinery:1–23

Kapoor R et al (2020) Valorization of agricultural waste for biogas based circular economy in India: a research outlook. Bioresour Technol 304:123036. https://doi.org/10.1016/j.biortech.2020.123036

Karić N et al (2022) Bio-waste valorisation: agricultural wastes as biosorbents for removal of (in)organic pollutants in wastewater treatment. Chem Eng J Adv 9:100239. https://doi.org/10.1016/j.ceja.2021.100239

Karimi Alavijeh M, Yaghmaei S, Mardanpour MM (2020) Assessment of global potential of biohydrogen production from agricultural residues and its application in nitrogen fertilizer production. Bioenergy Res 13(2):463–476. https://doi.org/10.1007/s12155-019-10046-1

Kasirajan R et al (2022) Synthesis of Bio-based oxides nano-composites catalyst from croton macrostachyus leaves for biodiesel production from croton macrostachyus seed oil. Fuel 325(May):124900. https://doi.org/10.1016/j.fuel.2022.124900

Kaur A, Kuhad RC (2019) Valorization of rice straw for ethanol production and lignin recovery using combined acid-alkali pre-treatment. Bioenergy Res 12(3):570–582. https://doi.org/10.1007/s12155-019-09988-3

Keshav PK et al (2023) Lignocellulosic ethanol production from cotton stalk: an overview on pretreatment, saccharification and fermentation methods for improved bioconversion process. Biomass Convers Biorefinery 13(6):4477–4493. https://doi.org/10.1007/s13399-021-01468-z

Kuglarz M et al (2018) Integrated production of cellulosic bioethanol and succinic acid from rapeseed straw after dilute-acid pretreatment. Bioresour Technol 265:191–199. https://doi.org/10.1016/j.biortech.2018.05.099

Kumar MD et al (2022) Macroalgae-derived biohydrogen production: biorefinery and circular bioeconomy. Biomass Convers Biorefinery 12(3):769–791. https://doi.org/10.1007/s13399-020-01187-x

Leonel LV et al (2020) Valorization of apple pomace using bio-based technology for the production of xylitol and 2G ethanol. Bioprocess Biosyst Eng 43(12):2153–2163. https://doi.org/10.1007/s00449-020-02401-w

Liu Y et al (2021) Process optimization for deep eutectic solvent pretreatment and enzymatic hydrolysis of sugar cane bagasse for cellulosic ethanol fermentation. Renew Energy 177:259–267. https://doi.org/10.1016/j.renene.2021.05.131

Mahmud R et al (2021) Integration of techno-economic analysis and life cycle assessment for sustainable process design–a review. J Clean Prod 317:128247

Manmai N, Unpaprom Y, Ramaraj R (2021) Bioethanol production from sunflower stalk: application of chemical and biological pretreatments by response surface methodology (RSM). Biomass Convers Biorefinery 11(5):1759–1773. https://doi.org/10.1007/s13399-020-00602-7

Menon V, Rao M (2012) Trends in bioconversion of lignocellulose: Biofuels, platform chemicals & biorefinery concept. Prog Energy Combust Sci 38(4):522–550. https://doi.org/10.1016/j.pecs.2012.02.002

Morales A et al (2020) Multiproduct biorefinery based on almond shells: impact of the delignification stage on the manufacture of valuable products. Bioresour Technol 315:123896. https://doi.org/10.1016/j.biortech.2020.123896

Muharja M et al (2018) An integrated green process: subcritical water, enzymatic hydrolysis, and fermentation, for biohydrogen production from coconut husk. Bioresour Technol 249:268–275. https://doi.org/10.1016/j.biortech.2017.10.024

Nigam PS, Singh A (2011) Production of liquid biofuels from renewable resources. Prog Energy Combust Sci 37(1):52–68. https://doi.org/10.1016/j.pecs.2010.01.003

Odetoye TE et al (2018) Thermochemical characterization of Nigerian Jatropha curcas fruit and seed residues for biofuel production. Energy Ecol Environ 3(6):330–337. https://doi.org/10.1007/s40974-018-0104-0

Pasangulapati V et al (2012) Effects of cellulose, hemicellulose and lignin on thermochemical conversion characteristics of the selected biomass. Bioresour Technol 114:663–669. https://doi.org/10.1016/j.biortech.2012.03.036

Pattanaik L et al (2019) Biofuels from agricultural wastes. In: Second and third generation of feedstocks: The evolution of biofuels. Elsevier Inc. https://doi.org/10.1016/B978-0-12-815162-4.00005-7

Periyasamy S et al (2024) Influencing factors and environmental feasibility analysis of agricultural waste preprocessing routes towards biofuel production–a review. Biomass Bioenergy 180:107001

Raju SS, Shinoj P, Joshi PK (2009) Sustainable development of biofuels: prospects and challenges. Econ Polit Wkly:65–72

Ranjan A, Welz PJ, Mthethwa T (2023) Investigation of an effective acid pre-treatment method for the valorisation of canola fines. Biomass Convers Biorefinery. https://doi.org/10.1007/s13399-023-03946-y

Selvakumar P et al (2022) Optimization of binary acids pretreatment of corncob biomass for enhanced recovery of cellulose to produce bioethanol. Fuel 321(February):124060. https://doi.org/10.1016/j.fuel.2022.124060

Sethumadhavan A et al (2024) Food process industry waste biomass as a promising alternative for green energy production. In: Value added products from food waste. Springer, pp 275–290

Shah A, Baral NR, Manandhar A (2016) Technoeconomic analysis and life cycle assessment of bioenergy systems. In: Advances in bioenergy. Elsevier, pp 189–247

Shahzadi T et al (2014) Advances in lignocellulosic biotechnology: a brief review on lignocellulosic biomass and cellulases. Adv Biosci Biotechnol 05(03):246–251. https://doi.org/10.4236/abb.2014.53031

Shrivastava A, Sharma RK (2023) Conversion of lignocellulosic biomass: production of bioethanol and bioelectricity using wheat straw hydrolysate in electrochemical bioreactor. Heliyon 9(1):e12951. https://doi.org/10.1016/j.heliyon.2023.e12951

Skiba EA et al (2017) Dilute nitric-acid pretreatment of oat hulls for ethanol production. Biochem Eng J 126:118–125. https://doi.org/10.1016/j.bej.2016.09.003

Srinadh RV, Neelancherry R (2023) Technoeconomic and sustainability analysis of agricultural waste conversion technologies. In: Agricultural waste to value-added products: technical, economic and sustainable aspects. Springer, pp 47–68

Temesgen T et al (2024) Valorization of wastes and by-products of cane-based sugar industry. In: Value added products from food waste. Springer, pp 185–204

Tschulkow M et al (2020) Integrated techno-economic assessment of a biorefinery process: the high-end valorization of the lignocellulosic fraction in wood streams. J Clean Prod 266:122022

Van Dael M et al (2013) A techno-economic evaluation of a biomass energy conversion park. Appl Energy 104:611–622

Wan C, Zhou Y, Li Y (2011) Liquid hot water and alkaline pretreatment of soybean straw for improving cellulose digestibility. Bioresour Technol 102(10):6254–6259. https://doi.org/10.1016/j.biortech.2011.02.075

Yu J et al (2019) Process integration for ethanol production from corn and corn stover as mixed substrates. Bioresour Technol 279:10–16. https://doi.org/10.1016/j.biortech.2019.01.112

Zabed H et al (2017) Bioethanol production from renewable sources: current perspectives and technological progress. Renew Sust Energ Rev 71:475–501

# Recycling of Synthetic and Biobased Plastic Waste for Value Added Applications

# 10

Devesh Sane, Aravind Babu S., Abhik Chattopadhyay, Harish Raj Arumugam, Shamlan Reshamwala, and Anagha Sabnis

**Abstract**

Plastic waste has emerged as a critical environmental and sustainability challenge due to its persistent accumulation, complex degradation pathways, and inadequate waste management systems. This chapter provides an in-depth review of the generation, environmental impacts, and recycling strategies for both synthetic and biobased plastics. It covers the role of key industries, marine and microplastic pollution, and greenhouse gas emissions, and introduces the circular bioeconomy as a sustainable framework. Various recycling technologies are discussed in detail including mechanical, chemical (e.g., glycolysis, acidolysis, solvolysis, hydrolysis, and aminolysis), enzymatic, microbial, and energy recovery techniques. Special focus is placed on recent advances in enzyme engineering (PETase, MHETase) and microbial metabolic pathways to enhance plastic degradation. Challenges such as mixed plastic streams, chemical stability, and inadequate infrastructure are critically analyzed. Furthermore, value-added applications of recycled intermediates, sustainability assessments through life cycle and economic analyses, and evolving global policies (e.g., EU Circular Economy Plan, India's EPR) are discussed to frame the regulatory and economic context. The chapter concludes with a future outlook on integrating advanced

D. Sane · A. Babu S. · A. Sabnis (✉)
Department of Polymer & Surface Engineering, Institute of Chemical Technology, Mumbai, India
e-mail: as.sabnis@ictmumbai.edu.in

A. Chattopadhyay · H. R. Arumugam
Department of Pharmaceutical Sciences & Technology, Institute of Chemical Technology, Mumbai, India

S. Reshamwala
Department of Biological Sciences & Biotechnology, Institute of Chemical Technology, Mumbai, India



V. Singh (ed.), *Sustainable Waste Management Towards Circular Bioeconomy*,
https://doi.org/10.1007/978-981-95-2227-9_10

recycling technologies with policy frameworks to drive scalable, sustainable plastic waste valorization.

**Keywords**

Plastic waste · Chemical recycling · Bioplastics · Circular bioeconomy · Enzymatic degradation · Sustainability assessment · Policy frameworks

## 10.1 Introduction

### 10.1.1 Global Waste Production and Generation

Numerous plastic product varieties are now an essential component of daily life. The manufacture of plastic has grown significantly during the previous 50 years. Every year, almost 12 million tonnes of plastic products are used in India, with 70% of them being carelessly thrown away as waste (Singh et al. 2017). In 1860, the first plastic was created, and in 1907, the worldwide plastics industry got underway, but in the 1920s, industry began to grow, and in the 1940s, plastic output increased. The amount of plastic produced increased from about 2 million tons in 1950 to 368 million tons in 2019. Thus, between 1950 and 2018, the usage of plastic has increased by around 180 times. It is estimated that 400.3 million tons of plastic would be produced globally in 2022. Therefore, it is anticipated that the manufacturing of plastic will continue to grow tremendously. The packaging industry uses 42% of all plastics manufactured worldwide, while the construction industry uses 17%. Out of the 300 million tonnes of conventional plastics manufactured each year, only 1% are bioplastics (Kumar et al. 2018; Rafey and Siddiqui 2023).

#### 10.1.1.1 Trend in Synthetic and Biobased Plastics

Bioplastics made from natural components generated from oil or renewable resources are being developed to have a low carbon footprint, a high recycling value, and full biodegradability and compostability (Babu et al. 2013). Sustainable bioplastics, production, and recycling methods are thought to be very compatible with the European circular economy plan and the United Nations Sustainable Development Goals (SDGs). Recent developments are being made to create next-generation bioplastics made from biomass, microbial/microalgal cells, and renewable waste streams in order to ensure that there is no competition with food and agricultural resources. This will ultimately create carbon-neutral infrastructure for the production and management of bioplastics (Spierling et al. 2018).

Thermoplastic starch (TPS), cellulose derivatives, and their mixes are examples of naturally occurring polymers that have the highest manufacturing capacity and are displacing plastics, especially in the flexible film packaging industry (Khan et al. 2017). It is noticed that Poly Lactic Acid is getting wide attention due to its technological advances in productivity and functionality. Polyhydroxyalkanoates (PHA) ranks second in terms of global manufacturing capacity, after polybutylene succinate (PBS) and poly (butylene adipate-co-terephthalate) (PBAT). The current global production of PHA is approximately 25,320 tonnes, according to a recent market estimate. This represents 1.2% of the total, while PBAT and PBS hold 13.4% and 4.3%, respectively (*Global production capacities of bioplastics 2018–2024 Global production capacities of bioplastics 2019 (by material type) Global production capacities of bioplastics 2024 (by material type)*, n.d.).

Despite only being mass-produced for roughly 70 years, synthetic plastic has surpassed the majority of man-made materials. Even though there are many different types of polymers, at the end of 2017, eight of them accounted for 95% of the total amount of primary plastics produced, which had surpassed nine billion metric tonnes. Approximately 10% of the seven billion tonnes of plastic garbage that have been produced thus far have been recycled, 14% have been incinerated, and 76% have ended up in landfills, dumps, or the environment. Global primary plastic production would reach 1.1 billion tonnes in 2050 if it were to continue its historic development trajectory (Geyer 2020).

Due to their versatility and relative low cost, plastics are used in all industrial sectors. There is probably no product type that does not exist in a version that contains significant amounts of plastic. Plastic resins are used extensively in packaging, building and construction, transportation, electrical and electronic equipment, and agriculture. It is also widely used in furniture and household, leisure and sports goods, and medical supplies and equipment. Waste produced during the manufacturing process is referred to as *production* waste or *pre-consumer waste*. For instance, plastic scraps or rejects generated while manufacturing plastic components fall into this category. In contrast, waste generated during or after the usage phase of a product is known as *post-consumer waste* or *end-of-life waste* (Geyer 2020).

#### 10.1.1.2 Key Industries Contributing

According to estimates, the packaging industry produced more than 142 million metric tons of plastic garbage in 2019, the most of any industrial sector. In the same year, the building and construction industry produced almost 77 million metric tons of plastic garbage. Industrial plastic wastes are plastic wastes that come from the large manufacturing, processing, and packaging industries. Industrial plastics are produced by small and medium-sized businesses, the electrical and electronics sector, the automotive sector, the packaging sector, demolition and construction firms, and others. Plastics are also unavoidable in the agricultural industry, where they are utilized for post-harvest activities, crop protection, mulching, irrigation pipes, silage covering, seed and fertilizer packaging, and more. Nets and coverings made of a lot of plastic are used to shield crops from animals, insects, and unfavourable weather. Modern medical sophisticated technologies and health sector service coverage have resulted in a significant amount of medical plastic waste (MPW). MPW, which comprises plastics in the form of gloves, syringes, canula pipes, pharmaceutical coverings, personal protective equipment (PPE), and other items, urgently needs to be disposed of in a safer and cleaner manner (Kibria et al. 2023).

## 10.1.2 Environmental Impact of Plastic Waste

The widespread use of disposable plastic products, combined with low recycling rates, has led to severe pollution of both soil and marine ecosystems. Plastic waste has become a major environmental challenge worldwide. In response, the global community has increasingly recognized the urgency of this issue, prompting numerous countries and regions to implement policies and initiatives aimed at reducing plastic usage. Plastics are inexpensive, lightweight, and long-lasting materials that are easily moulded into a wide range of items. Due to this, plastics are used in so many different ways that their manufacturing has significantly expanded in recent decades. However, since large amounts of discarded end-of-life plastics are building up as garbage in landfills and natural habitats around the world, plastic's endurance may have negative environmental repercussions. The sustainable recycling of post-consumer plastic solid waste is an essential environmental necessity, as it reduces oil consumption, minimizes waste, lowers carbon dioxide emissions, and ultimately supports sustainable development (Nasrollahi et al. 2020).

### 10.1.2.1 Marine Pollution

Plastic waste-induced marine pollution has emerged as one of the most pressing environmental concerns of the twenty-first century. Across the globe, it is widely recognized as a significant anthropogenic threat to the health and sustainability of marine and coastal ecosystems. Anthropogenic sources of the unprecedented and ongoing buildup of plastic pollutants in aquatic environments disrupt ecosystem structure, functions, and, in turn, services and values, either directly or indirectly. Megaplastics, macroplastics, mesoplastics, and microplastics are the four general categories into which the accumulated plastics in ocean basins can be divided. Commonly produced, commercial goods like cosmetics and personal care items contain microplastics, or microplastic particles are created even when bigger plastics break down in the environment and are subsequently broken up by physical, chemical, and biological processes (Guo and Wang 2019). Multiple factors drive the accumulation of plastic waste in aquatic environments, both directly and indirectly. Pollution from land-based and marine sources enters water bodies through various pathways—whether originating locally (in-situ) or from distant regions (ex-situ)—posing a serious threat to coastal and marine ecosystems. Plastic input from fresh water that ends in marine ecosystem, domestic and residential activities, tourism, and other economic activities, such as harbour operations, are the main contributors of land-based plastic pollution. More than 75% of the plastic debris found in the ocean comes from land-based sources (Andrady 2011).

There are several issues associated with the continued presence of plastic waste in maritime habitats. Due to the possibility of entanglement, ingestion, and exposure to hazardous chemical additions, marine life is especially at risk. Beyond the effects on the environment, plastic pollution has a major influence on fisheries, tourism, and coastal industries. Tourists avoid beaches and coastal waters that are polluted with plastic garbage, which affects the local economies that depend on marine-based industries negatively. Additionally, plastic waste contaminates seafood sources, destroys fishing gear, and marine ecosystems that sustain international fisheries (Worm et al. 2017).

#### 10.1.2.2 Microplastics

Microplastics, defined as plastic fragments measuring 5 mm or less in size, pose a serious threat to the environment. These particles are categorized into primary microplastics, which are intentionally manufactured at microscopic sizes such as microbeads in cosmetics, microfibers from textiles, and industrial pellets called nurdles and secondary microplastics, which originate from the gradual breakdown of larger plastic debris like bottles, bags, and fishing gear over time (Collignon et al. 2014; Tang et al. 2021; Chassignet et al. 2021). The increasing accumulation of microplastics has led to ecological disruptions, including harm to marine life, habitat degradation, and overwhelmed waste disposal systems. Alarmingly, a study by Melanie Bergmann and colleagues reported the presence of microplastics even in Arctic snow previously believed to be unspoiled with concentrations exceeding 10,000 particles per liter (Bergmann et al. 2019). Microplastics are an environmental contaminant of concern. It has been documented that they are present in air samples and food intended for human consumption. Consequently, exposure to microplastics through food or inhalation may occur, with unknown health consequences for humans. Potential particle, chemical, and microbiological dangers are indicated by complementary current domains, despite the fact that the relationship between microplastics and human health is still developing. By triggering or strengthening an immune response, microplastics can collect and cause localized particle toxicity if breathed or consumed. The localized leaching of endogenous additives, component monomers, and absorbed ambient contaminants may result in chemical toxicity. Chronic exposure is expected to be more concerning due to the potential accumulation of effects (Wright and Kelly 2017).

#### 10.1.2.3 Greenhouse Gas Emissions

In addition to the growing worry over the effects of plastic on ecosystems, human health, and seas, there is another aspect of the plastic disaster that is mainly unknown: plastic's role in climate change and global greenhouse gas emissions. The problem of managing plastic waste is the focus of the entire world today. Mismanaged plastics make a significant contribution to the increase in carbon emissions as a result of the release of plastic chemicals that are exposed to sunlight or are burned. In addition to their environmental persistence, plastics especially microplastics are increasingly recognized as a significant contributor to greenhouse gas emissions and climate change. As global plastic production continues to accelerate, the plastic industry has emerged as one of the fastest-growing sources of industrial carbon emissions. These emissions originate not only during the manufacturing phase but span the entire plastic life cycle, including raw material extraction, processing, transportation, waste handling, and eventual environmental leakage (Shen et al. 2020). A life cycle analysis of municipal solid waste (MSW) management reveals that in many developing nations, over half of the collected waste is either openly burned or disposed of in landfills without proper treatment. Only about 10–40%

undergoes processes like recycling or composting. Waste management activities are responsible for nearly 5% of global greenhouse gas (GHG) emissions, including methane ($CH_4$), carbon dioxide ($CO_2$), and nitrous oxide ($N_2O$), with methane alone contributing around 1–2%. These emissions significantly impact climate systems, contributing to global warming and harming ecosystems. Hence, it is vital to adopt sustainable waste management practices—covering everything from collection to final disposal—with a strong focus on reducing GHG emissions to protect natural resources and the environment (Gautam and Agrawal 2021). A study led by Chart Chiemchaisri investigated the dominant solid waste disposal practices in Thailand, focusing on open dumping and landfilling. Site surveys identified 95 landfills and 330 open dump locations. Methane emissions from these disposal sites were estimated using three different approaches: the Intergovernmental Panel on Climate Change (IPCC) method, the Landfill Gas Emission Model (LandGEM), and the closed flux chamber technique. Among the three, the IPCC method yielded the highest methane emission estimate at 366 gigagrams per year (Gg/yr), while LandGEM and the closed flux chamber method produced lower estimates of 115 Gg/yr and 103 Gg/yr, respectively (Chiemchaisri and Visvanathan 2008).

Reducing consumption and, consequently, waste creation is the most effective strategy to lower greenhouse gas emissions in waste management. Recycling materials can also reduce greenhouse gas emissions, both directly and indirectly. Refusing to dispose of waste at landfills or in any other manner (such as burning) reduces direct emissions. Reducing the amount of energy used to produce and acquire raw materials as well as to manufacture the final product can help reduce indirect emissions. The emissions from producing a product from virgin material can be avoided if recovered material is used in place of virgin material. When recycled materials are used instead of virgin ones, net energy consumption and, consequently, greenhouse gas emissions from energy production often decline (Korhonen and Dahlbo n.d.).

### 10.1.3 Concept of Circular Bioeconomy

The circular bioeconomy represents a transformative approach to plastic production and waste management by integrating renewable biological resources with circular economy principles. This model seeks to decouple plastic systems from fossil fuel dependence, reduce greenhouse gas (GHG) emissions, and mitigate environmental pollution through closed-loop material flows (Rajvanshi et al. 2023).

The shift from fossil-derived polymers to bio-based alternatives is central to the circular bioeconomy. Bioplastics, derived from biomass such as plants (e.g., corn, sugarcane), microalgae, and cyanobacteria, leverage photosynthesis to convert $CO_2$ into polymers like polylactic acid (PLA) and polyhydroxyalkanoates (PHA). These materials can reduce carbon footprints by up to 80% compared to conventional plastics, depending on feedstock sourcing and production methods. Biowaste and agricultural residues are increasingly prioritized as feedstocks to avoid competition with food crops, aligning with the "biorefinery" concept that mirrors petroleum refineries but uses sustainable inputs (Karan et al. 2019; Ali et al. 2023; Rajvanshi et al. 2023).

Industrial composting offers an end-of-life pathway for compostable bioplastics. For example, Berlin's waste management system separates compostable bioplastics, diverting them from landfills and enhancing recycling efficiency. However, biodegradation requires controlled conditions, as no bioplastic degrades effectively in marine environments. The circular bioeconomy redefines plastic's lifecycle through elimination, renewable feedstocks, and advanced recycling. While bio-based materials offer scalable solutions, success hinges on infrastructure investment, policy coherence, and public engagement. By prioritizing systemic innovation, this model can transform plastics from pollutants to pillars of sustainability (Ali et al. 2023).

### 10.1.4 Objectives and Scope Recycling

Recycling plays a crucial role in mitigating environmental impacts and has become a rapidly evolving focus within the plastics sector. It helps conserve petroleum resources, lowers carbon dioxide emissions, and significantly decreases the volume of plastic waste that ends up in landfills or incineration (Hopewell et al. 2009).

Recycling can be broadly categorized into four main types: primary, secondary, tertiary, and quaternary (also referred to as energy recovery or enzymatic methods in some cases). Primary recycling, often called closed-loop recycling, involves reprocessing clean, uncontaminated plastic waste into products with properties comparable to the original material, allowing reuse without significant degradation in quality (Al-Salem et al. 2009). Secondary recycling involves mechanical processes in which the polymer's chemical structure remains intact. However, the material undergoes physical reprocessing such as melting and remolding and is typically repurposed for applications different from its original use (Ignatyev et al. 2014). Tertiary recycling, sometimes referred to as chemical recycling, uses chemical processes to break down the polymer into value-added commodities. Typical processes include hydrolysis, glycolysis, aminolysis, acidolysis, and solvolysis (Ignatyev et al. 2014). Quaternary or energy recovery includes pyrolysis, gasification, etc. Certain polymers can be degraded using fungi, bacteria, and enzymes as well.

Polymer recycling and its usage for value addition is important for reducing the amount of plastic waste, reducing greenhouse gas emissions, and reduce the use of fossil fuels for the production of plastics. The recycled polymer is further used in polymer processing in case primary and secondary recycling whereas in case of tertiary recycling can be used to reduce the polymer to is monomeric or oligomeric level and can be utilized in new product development.

## 10.2 Types of Plastics and Challenges in Their Recycling

Plastics can be classified into two main categories based on the raw materials used in their production synthetic and biobased polymers. This classification focuses on the origin of the raw materials, which can range from fossil fuels to renewable resources.

### 10.2.1 Synthetic Plastics

These are the most common plastics, derived from non-renewable resources such as crude oil, natural gas, and coal. They are produced through polymerization or polycondensation processes. This category of polymer includes Polyethylene (PE), Polypropylene (PP), Polyvinyl Chloride (PVC), Polystyrene (PS), and Polyethylene Terephthalate (PET). Their main applications are packaging, construction materials, automotive parts, and consumer goods (Desidery and Lanotte 2022).

### 10.2.2 Biobased Plastics

These plastics are derived from natural, renewable resources such as plants, algae, or microorganisms. Bio-based materials, derived from renewable plant or other biogenic sources, can be chemically transformed to yield compounds with distinctive functionalities suitable for developing sustainable polymers. These polymers offer an alternative to those made from fossil-based resources. Depending on their molecular structure, bio-based polymers may be either biodegradable such as polylactic acid (PLA) and polyhydroxyalkanoates (PHAs) or non-biodegradable, like bio-derived polyethylene (Bio-PE) and polyethylene terephthalate (Bio-PET). These materials find use in various sectors, including food packaging, healthcare, textiles, and disposable consumer products (Cywar et al. 2021).

### 10.2.3 Challenges

Plastics have become an integral component of modern life due to their versatile mechanical and chemical characteristics. Their adaptability makes them valuable across multiple sectors, including automotive, electronics, and packaging. However, despite their widespread utility, plastics present a significant environmental challenge because of their resistance to degradation (Chen and Selvinsimpson 2022). A major obstacle to the broader implementation of plastic recycling is the lack of adequate infrastructure for collection and sorting, which continues to hinder progress in this area (Sambyal et al. 2024).

Collection plays a crucial yet often overlooked role in the plastic recycling process. Its importance is highlighted by the fact that globally, only around 15% of the 353 million metric tons of plastic waste generated is gathered for recycling approximately 55 million metric tons. Of this amount, just 33 million metric tons can actually proceed to further recycling steps, largely due to inadequate or inefficient collection systems (Sambyal et al. 2024).

Sorting represents the second critical phase in plastic recycling, where significant material losses often occur. These losses are largely due to the intricate design of plastic products and the lack of advanced sorting infrastructure. The efficiency of this step greatly influences the quality of the final recycled material. Factors such as the type of plastic, its size and shape, density, and the presence of additives, adhesives, or pigments all play a vital role in determining the most suitable sorting technique (Pan et al. 2020; Picuno et al. 2021).

Mechanical recycling, the most widely used recycling method, degrades the quality of plastics over multiple cycles. When plastics are reprocessed, their polymer chains break down due to thermal and mechanical stress, leading to inferior mechanical properties such as reduced tensile strength, impact resistance, and flexibility. This degradation limits the number of times the plastics can be recycled before they become unsuitable for use in high-value applications. Moreover, contaminants such as additives, dyes, and residual food waste affect the properties of recycled plastics. Post-consumer recycled plastics often exhibit discoloration, unpleasant odors, and inconsistent quality, making them less attractive for manufacturers who prefer the reliability of virgin plastics. To mitigate degradation, stabilizers and compatibilizers can be added, but these increase processing costs (Li et al. 2022).

To preserve the quality of recycled plastics, removing ink from printed plastic materials before extrusion is essential. A thorough understanding of the ink's chemical composition is key to achieving efficient de-inking (Ügdüler et al. 2023). Hatice Gecol and her team studied the removal of water-based ink from commercial polyethylene films using surfactants, examining the influence of calcium ions (hardness ions) across different pH conditions (Gecol et al. 2004). The presence of contaminants restricts the application of the waste plastic it can either be substances that are added intentionally or substance that comes in contact during its life. Prior to recycling, sorted plastic materials require comprehensive cleaning, as contaminants can compromise the quality of the recycled output and reduce its commercial viability. Samuel S. Núñez and team investigated decontaminating the post-consumer low-density polyethylene, extraction with polyethylene glycol at different temperatures with and without agitation, extraction using subcritical and hot water at different temperatures in a basic medium (Núñez et al. 2023). One of the most significant challenges in plastic recycling is dealing with mixed waste streams. The presence of incompatible plastics in a single waste stream can lead to contamination and reduced recyclability. For instance, even trace amounts of PVC contamination in a PET recycling stream can deteriorate the PET resin due to the release of hydrochloric acid gas when processed at PET's higher melting temperature. Conversely, the presence of PET in a PVC stream leads to the formation of crystalline PET lumps, which disrupt the uniformity and significantly lower the quality and value of the recycled PVC (Hopewell et al. 2009).

One of the major challenges in monomer-to-monomer chemical recycling lies in handling mixed-polymer and mixed-material waste, which demands depolymerization and separation processes that can tolerate various additives and impurities capabilities that are not yet universally established or commercially widespread. Another significant challenge in plastic recycling is the inherent chemical stability of many polymers. Plastics are designed for durability, which makes them resistant to degradation under normal environmental conditions. This same stability that makes plastics useful in various applications also presents difficulties in recycling. Certain plastics, such as polyethylene (PE) and polypropylene (PP), exhibit strong resistance to chemical breakdown, requiring high-energy inputs for effective depolymerization. The presence of additives like flame retardants, UV stabilizers, and plasticizers alters the chemical composition of plastics, making their degradation unpredictable and complicating the recycling efforts. The breakdown of such additives can lead to the formation of hazardous compounds, affecting both the quality of recycled plastics and environmental safety (Alassali et al. 2021; Zou et al. 2023).

Recycling biopolymers presents several technical hurdles due to their inherent characteristics. A key challenge is the inconsistency in their composition. Unlike conventional plastics, biopolymers are derived from diverse biological sources, leading to variations in their chemical and physical behaviour. This heterogeneity complicates the recycling process, as different biopolymers may require tailored recycling approaches. In mechanical recycling, a significant concern is the degradation of polymer quality with each cycle. For instance, recycled polyhydroxyalkanoates (PHAs) typically experience a reduction in molecular weight and mechanical performance, such as tensile strength. Repeated mechanical processing can further deteriorate the polymer chains, diminishing the material's properties and limiting its potential applications (Narancic et al. 2018).

The lack of recycling infrastructure is a significant barrier to effective plastic recycling. The lack of proper collecting channels is one of the major reason in lack of recycling along with sorting of the collected plastics. A large investment is required to install new sorting techniques so that proper sorting can be done, which will eventually result in mono material availability and decrease the problem of mixed waste stream (Ragaert et al. 2017). While recycling is intended to be an environmentally friendly solution, it can still pose environmental risks. Chemical recycling processes and quaternary recycling process consume energy, generate emissions, and produce secondary pollutants. Plastic recycling faces numerous challenges that must be addressed to improve its effectiveness and sustainability. Inefficiencies in collection and sorting, material degradation, economic constraints, technical difficulties in chemical recycling, environmental concerns, and limited market demand all contribute to the low recycling rates observed globally. The lack of recycling infrastructure is a major obstacle to achieving sustainable plastic waste management. Addressing this issue requires a combination of technological advancements, policy measures, and global cooperation (Sambyal et al. 2024). Inconsistent or absent policies on plastic waste management and recycling exacerbate the problem. Effective regulatory frameworks are essential to incentivize recycling and hold producers accountable (Guerrero et al. 2021).

## 10.3 Recycling Technologies

In polymer industry today, recycling is one of the most active sectors and one of the most crucial measures that can be taken to lessen its effects on the environment. Opportunities for recycling include lowering carbon dioxide emissions and decreasing the amount of post-industrial and post-consumer waste that must be disposed [1], [2].

There are four basic methods for recycling polymeric waste, usually referred to as primary (mechanical reprocessing into a product with similar attributes), secondary (mechanical reprocessing into goods that need lower properties), tertiary (chemical constituent recovery), and quaternary (energy recovery) (Hopewell et al. 2009).

### 10.3.1 Mechanical Recycling (Fig. 10.1)

Mechanical and physical recycling, often categorized as primary and secondary methods, are among the simplest and most fundamental approaches for polymer recovery, aside from disposal in landfills. (Rossignolo et al. 2024). Mechanical recycling of plastics involves several techniques that transform plastic waste into secondary raw materials or products without significantly altering their chemical structure. The materials must first go through a number of processing steps in these recycling operations in order to reduce the initial particle size to a level suitable for reprocessing in subsequent manufacturing events. Mechanical recycling is the most widely adopted approach for processing plastic waste. It generally involves several key stages such as collection, sorting, cleaning, and size reduction, although the

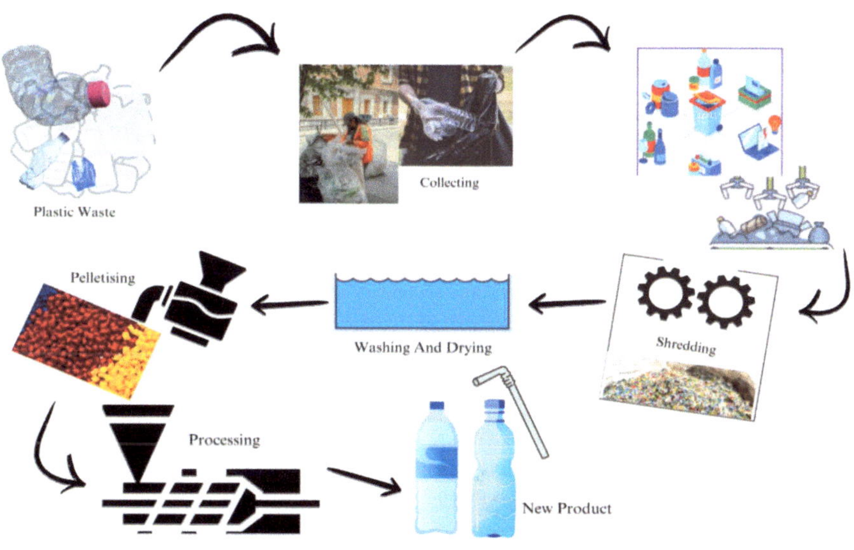


**Fig. 10.1** Schematic representation mechanical recycling of polymer

exact sequence and frequency of these steps may vary depending on the type and source of the plastic waste. The typical stages include the following: (Ragaert et al. 2017).

- *Sorting and separation*: Carried out based on attributes like shape, density, color, size, or chemical makeup.
- *Washing:* Aimed at eliminating impurities, especially organic residues.
- *Grinding:* Reduces the plastic waste into smaller flakes or particles for easier handling.
- *Compounding and pelletizing* (optional): In this step, ground flakes are reprocessed into uniform pellets, which are more suitable for further use by manufacturers than irregular flakes (Ragaert et al. 2017).

Extensive studies have been conducted in this area, particularly focusing on advanced collection and sorting technologies such as flotation, optical sorting, density-based separation, triboelectric (electrostatic) separation, magnetic density separation, electrical sorting, and X-ray detection (Yin et al. 2015). Once sorted, the recycled plastics whether in the form of pellets, powders, or flakes are melted and reprocessed into final products using various resin molding techniques. These include extrusion molding, injection molding, blow molding, vacuum molding, inflation molding, and melt spinning (Yin et al. 2015) (Table 10.1).

Recycling of both single-type (mono) and mixed plastic materials presents unique challenges. A key concern is the susceptibility of polymers to degradation under various conditions such as exposure to heat, oxygen, ultraviolet light, ionizing radiation, moisture (hydrolysis), and mechanical stress. In the context of mechanical recycling, polymers primarily undergo two major forms of degradation: one occurring during the material's service life and the other resulting from the thermal and mechanical stresses introduced during the reprocessing stages (Ravve 2000).

The most critical issue in plastic recycling is the thermal-mechanical degradation that occurs during reprocessing. This degradation, caused by the combined effects of heat and mechanical shear, affects both post-industrial (PI) and post-consumer (PC) plastic waste. In addition, plastics particularly post-consumer also undergo degradation over their service life due to prolonged exposure to environmental factors such as heat, oxygen, light, and moisture. While both forms of degradation

**Table 10.1** Difference between primary and secondary recycling

| | | |
|---|---|---|
| Primary Recycling | Mechanical recycling (closed-loop) by extrusion, compression and injection moulding. | Products with the same performance as virgin polymer |
| Secondary recycling | Size reduction (shredding, tearing, cutting, and grinding)<br>Mechanical recycling (open-loop) by extrusion, compression and injection moulding<br>Physical recycling (Rebonding, adhesive pressing) | Products with less demanding performance than virgin ones |

impact the quality of recycled materials, lifetime degradation is more relevant in the context of post-consumer plastics (Ragaert et al. 2017).

Thermal-mechanical degradation occurs when polymers are exposed to elevated temperatures and mechanical shear during melt processing. This simultaneous application of heat and shear stress triggers various degradation mechanisms within the polymer matrix, potentially altering its molecular structure and properties ('Beyler_Hirschler_SFPE_Handbook_3', n.d.). In commercial polymers, the primary degradation pathways during processing are chain scission and chain branching. The dominance of either mechanism depends on several factors, including the polymer type, initial molecular weight, and processing temperature. Thermal-mechanical degradation typically initiates with the homolytic cleavage of carbon-carbon bonds within the polymer backbone, producing free radicals. These radicals can then follow different reaction pathways leading either to disproportionation and subsequent chain scission or to intermolecular recombination, resulting in cross-linking or branching ('Beyler_Hirschler_SFPE_Handbook_3', n.d.).

The negative impacts of thermal-mechanical degradation during recycling can be mitigated through the use of various additives. A wide array of additives is available to enhance the recyclability and performance of plastics. Among these, heat stabilizers are commonly reintroduced during recycling processes, as they are often depleted during the material's initial use and processing (Ulutan 2003).

Throughout their lifecycle, plastic products are exposed to various environmental factors such as heat, oxygen, ultraviolet light, radiation, humidity, and mechanical forces. These combined effects lead primarily to degradation through photooxidative mechanisms (La Mantia 2000). The structural alterations occurring in polymers due to environmental exposure closely resemble those caused by thermo-mechanical degradation. However, the key distinction lies in the presence of atmospheric oxygen, which promotes the formation of oxygen-containing functional groups along the polymer backbone, ultimately impacting the material's performance. Both environmental aging and thermal-mechanical processes generate low-molecular-weight volatile compounds that often remain trapped in the solid polymer. When the material is reprocessed, these volatiles can migrate through the polymer melt, potentially interfering with the recycling process. Not only can these contaminants deteriorate the final product quality, but they may also affect the efficiency of processing by causing corrosion of equipment. Therefore, efficient degassing in recycling machinery is essential to minimize such issues (Xiang et al. 2002).

A significant hurdle in recycling mixed plastic waste lies in the varying melting points and processing requirements of the constituent polymers. During reprocessing, recyclers must typically use the processing temperature of the polymer with the highest melting point. This often results in overheating and thermal degradation of the lower-melting polymers, compromising the overall material properties. Additionally, such mixtures can form polymer blends during mechanical recycling. The compatibility of these blends is influenced by the Gibbs free energy of mixing, with many polymers exhibiting inherent immiscibility. To address this, compatibilization techniques are employed, the most common being the addition of compatibilizers to enhance interfacial adhesion and blend uniformity (Ragaert et al. 2017) (Table 10.2).

**Table 10.2** Some examples of mechanical recycling processes performed on different polymers

| Polymer | Mechanical/physical recycling methods | Next application of recycled polymer | References |
|---|---|---|---|
| PET (polyethylene Terephthalate) | Sorting, cleaning, grinding, extrusion, pelletizing | Bottles, fabric (polyester), packaging | Goring and Priestley (2023) |
| HDPE (high-density polyethylene) | Sorting, cleaning, grinding, extrusion, pelletizing | Bottles, plastic lumber, pipes | Zúñiga et al. (2023) |
| PVC (polyvinyl chloride) | Limited mechanical recycling; often downcycled | Flooring, panelling, gutters | Janajreh et al. (2015) |
| LDPE (low-density polyethylene) | Sorting, cleaning, grinding, extrusion, pelletizing (less common) | Trash can liners, floor tiles | Goring and Priestley (2023), van den Tempel and Picchioni (2024) |
| PP (polypropylene) | Sorting, cleaning, grinding, extrusion, pelletizing | Pallets, ice scrapers, rakes | van den Tempel and Picchioni (2024) |
| PS (polystyrene) | Mechanical recycling (less common); often downcycled | Insulation, packaging materials | Maharana et al. (2007), Thakur et al. (2018) |
| Polyurethane (Thermoset) | Mechanochemical method (crushing, grinding), cryomilling | Fillers in new products | Zia et al. (2007) |
| Epoxy (Thermoset) | Mechanical shredding for fillers; cryomilling | Fillers in new products | Kumar and Krishnan (2020) |
| PLA (polylactic acid) - biopolymer | Sorting, cleaning, grinding, extrusion, pelletizing | Biodegradable packaging, disposable cutlery | Beltrán et al. (2021) |
| PHA (polyhydroxyalkanoates) - biopolymer | Sorting, cleaning, grinding, extrusion, pelletizing | Biodegradable packaging, disposable cutlery | Vu et al. (2020) |

### 10.3.2 Chemical Recycling

Tertiary recycling, which involves converting plastic waste into chemicals and fuels, is drawing increasing global interest. The most favorable form of this recycling approach is to depolymerize plastics into their original monomers or other polymerizable intermediates that can be reused in polymer synthesis or transformed into valuable products. Plastic waste in recycling streams is generally categorized into thermoplastics and thermosets. Thermosetting polymers, due to their crosslinked structure, are not readily depolymerized and typically require pyrolysis for breakdown into fuels or chemical feedstocks, or they may be incinerated for energy recovery. On the other hand, thermoplastic materials include two major classes: addition polymers, which are difficult to reverse to their monomers, and step-growth polymers like polyesters, polyamides, and polyurethanes, which are formed through condensation reactions and are more amenable to chemical recycling processes (Chanda 2021).

Although chemical recycling methods are more complex and technically demanding compared to conventional disposal or mechanical recycling techniques, they offer significant advantages. These methods are highly effective for recovering high-purity raw materials and generating reusable products, making them a preferred choice for advanced plastic waste management (Zia et al. 2007). Moreover, physical recycling methods are ineffective for crosslinked polymers like rigid and flexible polyurethanes. However, since polyurethane formation involves reversible reactions, its urethane, ether, and ester bonds can be broken down through chemical depolymerization processes (Yang et al. 2012).

Considering the currently available recycling technologies, chemical recycling emerges as one of the best options available for recycling of synthetic and biobased plastics. In order to undergo degradation through chemical route the polymer needs to have linkages that are susceptible to chemical attack. The number of polymers is endless but in order to ease the classification, we can divide them into polymers containing clevageable linkages, i.e., esters, urethanes, urea, carbonyl and amide, and the other class of polyolefins. The polyolefins are usually produced by free radical polymerization and the main linkage present are (C-C) (Sauter et al. 2017). The bond energy of C-C linkage is quite high which makes it very difficult to break at normal operating conditions even with the use of catalyst as well as gives the olefins their chemically inert nature making them difficult to undergo cleavage (Wang et al. 2023b).

There are different methods for chemical recycling; the common ones are listed as glycolysis (Simón et al. 2018), hydrolysis (Mahoney et al. 1974), methanolysis (Asahi et al. 2004), aminolysis (Elidrissi et al. 2007), ammonolysis (Lentz and Mormann 1992), acidolysis (Gama et al. 2020), and phosphorolysis (Grancharov et al. 2007) (Fig. 10.2).

#### 10.3.2.1 Glycolysis (Fig. 10.3)

Glycolysis is a depolymerization reaction by glycols. It can be performed with or without a catalyst, but mostly a catalyst is preferred. Amines, inorganic salts, inorganic acids, phosphorous compounds, etc. are used as catalysts (Heiran et al. 2021). Glycolysis, due to its advantages in terms of operating temperature, mild reaction conditions, yield, and scalability, is one of the most promising depolymerization processes (Shirazimoghaddam et al. 2023).

Most common glycols that are used include Diethylene glycol (DEG), Ethylene glycol, propylene glycol, dipropylene glycol,1,4-Butanediol,1,6-Hexanediol, 1,3-Propanediol, 1,5-pentanediol, Glycerol, Sorbitol, etc. (Rossignolo et al. 2024) (Table 10.3).

The reaction time can depend on five major factors: (Datta and Rohn 2007)

- Temperature for glycolysis
- Presence and quantity of the catalyst

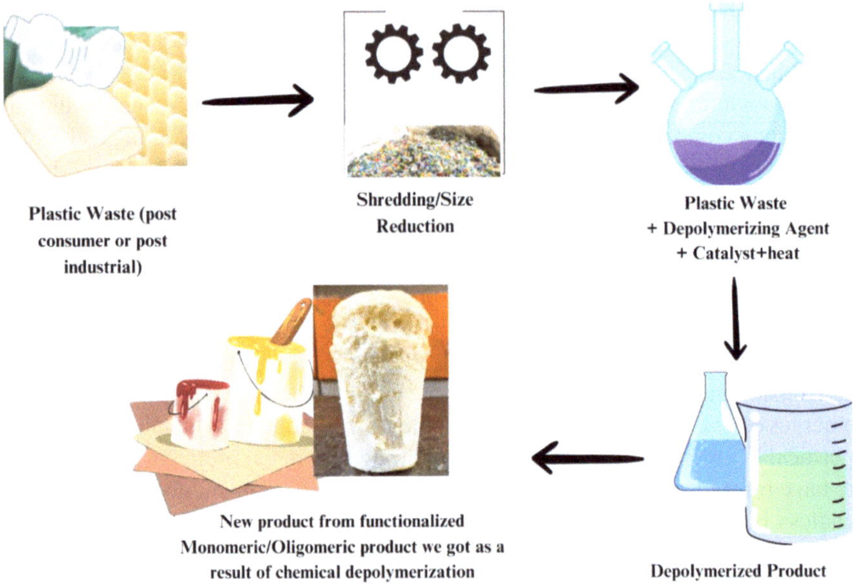


**Fig. 10.2** Schematic representation of chemical recycling of polymer

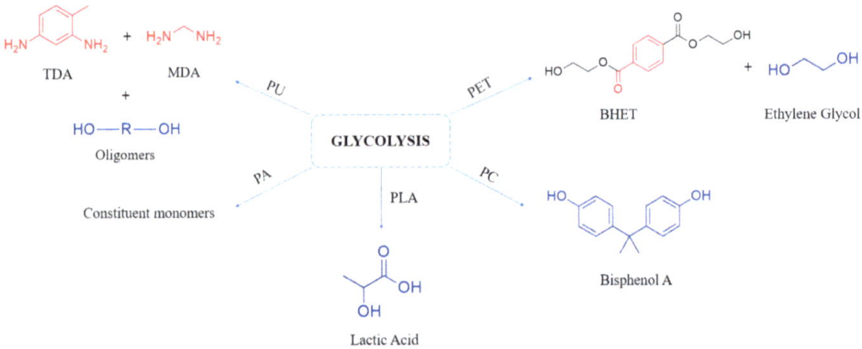


**Fig. 10.3** Reaction scheme of polymer recycling by glycolysis

- Molar/mass ratio of waste to glycol
- Rate of addition of waste in the reactor
- The degree of size reduction of polymer waste

#### 10.3.2.2 Acidolysis (Fig. 10.4)

Acidolysis of waste is an emerging technology and is gaining importance in the recent times. Till date only fewer literature on polyurethane and polyethylene terephthalates is available. More and more research needs to be conducted in the coming times in order to explore its potential in commercialization. There is an exchange

**Table 10.3** Some examples of recycling by glycolysis process performed on different polymers

| Polymer | Catalyst | Glycol Used | Reaction Conditions | Products | Reference |
|---|---|---|---|---|---|
| Poly (ethylene terephthalate) (PET) | Zinc acetate (heterogeneous) | Ethylene glycol (EG) | 165–185 °C | Oligomers (2000–10,000 Daltons) | Vaidya and Nadkarni (1989) |
| PET | Antimony (III) oxide, | Ethylene glycol (EG) | 240 °C, 400 W microwave power | Oligomers (DP ~37) | Mohammadi et al. (2023) |
| PET | Bamboo leaf ash as a bio-waste derived recyclable heterogeneous catalyst | Ethylene glycol (EG) | Stirred tank reactor | BHET, MHET, TPA | Lalhmangaihzuala et al. (2020) |
| Polycarbonate (PC) | NaOH or DBU | Ethylene glycol (EG), 1,2-propanediol (1,2-PD), glycerol | 180 °C | BPA and cyclic carbonate derivatives | Quaranta et al. (2018) |
| Polyurethane (PU) | Stannous Octate | Glycerol, diethylene glycol (DEG) | 179 °C to 189 °C | Polyols, urethane moieties | Simón et al. (2013) |
| PU Adhesives in Multilayer Films | – | Ethylene glycol | 150 °C | Delaminated layers, ethylene glycol oligomer | O'Rourke et al. (2022) |
| Rigid Polyurethane Foam (RPUF) | NaOH or sodium acetate | Ethylene glycol (EG), DEG | 180 °C to 198 °C | Green polyols | Miguel-Fernández et al. (2022) |
| PU (Flexible Foam) | NaOH, Na acetate, and Diethanolamine | Glycerol, DEG | Single or split-phase glycolysis | Recovered polyols | Jašek et al. (2024) |
| Polyamides (e.g., Nylon-6, Nylon-66) | Diammonium hydrogen phosphate | Ethylene glycol (EG) | 190 °C | Monomers (e.g., caprolactam) | Datta et al. (2018). |

(continued)

**Table 10.3** (continued)

| Polymer | Catalyst | Glycol Used | Reaction Conditions | Products | Reference |
|---|---|---|---|---|---|
| PET | 1,3-dimethylimidazolium-2-carboxylate | Ethylene glycol | 160–200 °C for 0.5–4 h | 60% of bis (2-hydroxyethyl terephthalate) (BHET) recovered | Wang et al. (2020) |
| PET | DBN-based DES catalysts | EG | 150 rpm, unreacted PET is hot-filtered under reduced pressure and washed with hot water to prevent the precipitation of oligomers. | 100% conversion rate of PET and 78.7% yield of BHET | Wei et al. (2025) |
| PET water bottles | Zn $(OAc)_2$ | EG | 200 °C for 3 h. | Recycled BHET were used for synthesizing high-performance PU elastomers | Wang et al. (2025) |
| PET | Zinc acetate with dimethyl aniline as co-catalyst | EG, DEG | 180–230 °C | Preparation of unsaturated polyester resins | Salem et al. (2025) |
| PET | 1-ethyl-3-methylimidazolium acetate ([Emim] [OAc]) | Ethylene glycol | 196 °C, and the reaction time was about 2.5 h | Waterborne polyurethane (WPU) was prepared | Xu et al. (2025) |
| PET | Titanium isophthalate (Ti-IPA) | Ethylene glycol | 3.15 h at 194 °C | BHET yield of 84.86% | Wen et al. (2024) |
| Rigid polyurethane (PU) foam | NaOH | Crude glycerol | 150 ~ 230 °C | Bio polyol can be synthesized by glycolysis product | Chang et al. (2025) |
| Rigid polyurethane (PU) waste | Sodium acetate | Diethylene glycol | 200 °C | Synthesis of new PU products. | Ivashchuk et al. (2025) |
| Polyurethane (PUR) foam | $K_2CO_3$ | 2-hydroxypropyl ricinoleate (2-HPR) from castor oil | Microwave,500 W 151 °C | REC used in the standard PUR foam | Jašek et al. (2024) |

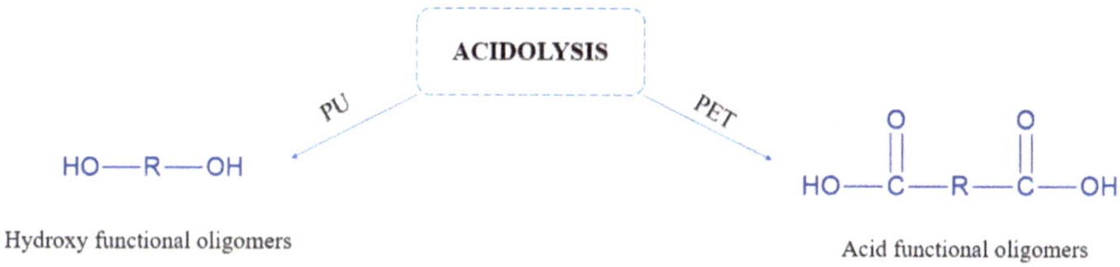


**Fig. 10.4** Reaction scheme of polymer recycling by acidolysis

of molecules taking place between the ester and the carboxylic acid group. The reactivity of the ester depends on the order of alcohol (Hoang et al. 2022a). Initial study of acid alcoholysis included combination of dicarboxylic acids (adipic, succinic, and sebacic) and combination with butanediol. This makes the system more complex making it difficult to comment of actual acidolysis reaction (Kárpáti et al. 2018). A study was also conducted using acetic and propanoic acid, the use of these cleavage agents increased the efficiency of TPA generation. In principle acidolysis reaction is considered reversible in nature due to similarity in its reactants and products. So, if one of the products is a volatile material, it would shift the equilibrium in the forward direction so esters of acetate would be more favourable (Pereira et al. 2024). Succinic acid and adipic acid was also studied for acidolysis of PET without the use of glycol or alcohol and solvents under microwave irradiation. These methods showed that acidolysis could be a viable approach for PET recycling and the obtained products can be further utilized (Hoang et al. 2022b). Acidolysis of polyurethane acts by attacking the carbamate linkage and producing polyols.

Acidolysis is a chemical depolymerization technique that breaks down urethane linkages in polyurethane (PU) using organic or inorganic acids. While initial research on PU acidolysis dates back to 1982, it has gained significant attention in recent years due to its potential as a sustainable recycling method. This process decomposes PU into recyclable polyols, offering an alternative to conventional petroleum-based raw materials. A key advantage of acidolysis, as highlighted by Soltysinski et al. (2018), is that it does not lead to the formation of carcinogenic primary aromatic amines such as toluene diamine (TDA) and methylene diamine (MDA), unlike other chemolysis techniques. Gadhave et al. (2019) reported various chemical depolymerization routes, including acidolysis, but noted that commercial-scale implementation remains limited. However, research suggests that acidolysis using dicarboxylic acids, rather than inorganic acids like HCl, is a more promising approach due to its efficiency and reduced toxicity. The research group at the University of Aveiro identified succinic acid as the most effective reagent for PU acidolysis, demonstrating its ability to produce rigid polyurethane foams containing up to 30% recycled polyol (Gama et al. 2020; Godinho et al. 2021a). Additionally, Grdadolnik et al. explored the use of microwave-assisted acidolysis with adipic acid as a depolymerizing agent, yielding polyols that were successfully utilized in flexible foam application (Grdadolnik et al. 2022). Literature also suggests the use of maleic acid (Liu et al. 2024a), vapour phase dicarboxylic acid, anhydrides (Liu

**Table 10.4** Some examples of recycling by acidolysis process performed on different polymers

| Polymer | Catalyst | Acid used | Reaction condition | Products | Ref |
|---|---|---|---|---|---|
| Polyurethane | NA | Succinic acid | 200 °C | Polyols and oligomers | Gama et al. (2020) |
| Polyurethane | NA | Maleic acid | 175-220 °C | Polyurethane foams | Liu et al. (2024a) |
| Polyurethane | NA | Succinic acid | 200 °C | Polyurethane coatings | Godinho et al. (2021a) |
| Polyurethane | NA | Succinic acid | 205 °C | Polyurethane adhesive | Donadini et al. (2024) |
| Polyurethane | Zinc acetate | Succinic acid | 200 °C | Polyurethane foams | He et al. (2023) |
| Polyurethane | NA | Succinic and phthalic acid | 140 °C | Polyols and oligomers | Liu et al. (2024b) |
| PET | NA | Acetic acid | 280 °C | Terephthalic acid | Peng et al. (2023) |
| PET | NA | Adipic acid | 180 W microwave irradiation | Epoxy Coatings | Hoang et al. (2022b) |
| PET | NA | Succinic acid | 180 W microwave irradiation | Epoxy Coatings | Hoang et al. (2022a) |

et al. 2024b) as well as studies on catalysis (He et al. 2023) and structure property relationships on depolymerization (Westman et al. 2024). The growing industrial interest in acidolysis, with companies like Rampf Group and H&S Anlagentechnik investing in its development, highlights its potential for commercialization (Table 10.4).

#### 10.3.2.3 Solvolysis (Fig. 10.5)

Solvolysis is use of solvents particularly low molecular weight alcohols like butanol, ethanol. Methanol for the degradation of thermoplastic and thermoset polymers having C-O or C-N linkages in their backbone, i.e., polyesters, polyurethanes, and polyamides (Xanthos and Patel 1998). Solvolysis is a transesterification reaction carried out at an elevated temperature of about 200–300 °C (Petrus et al. 2016). Use of methanol for recycling of PET usually is a better way to recover ethylene glycol, terephthalic acid, dimethyl terephthalate or bis(hydroxyethyl) terephthalate (BHET) (Kárpáti et al. 2018). However, the toxicity of methanol and the byproducts generated such as glycols and phthalate derivatives making the separation difficult and in turn reducing its financial viability (Bharadwaj et al. 2024). Polyurethane solvolysis generally results into glycols with mono and dicarbamates. The carbamate compounds generated when monoatomic alcohols have no reactive OH end group which makes it difficult for further use and lessens the interest of this particular method in PU recycling (Suprun et al. 2022). Hydrocarbon solvents are chemically inert towards polyamides and the main role of solvents is to bring the waste to a high temperature so that it undergoes intermolecular cyclization (Tonsi et al. 2023).

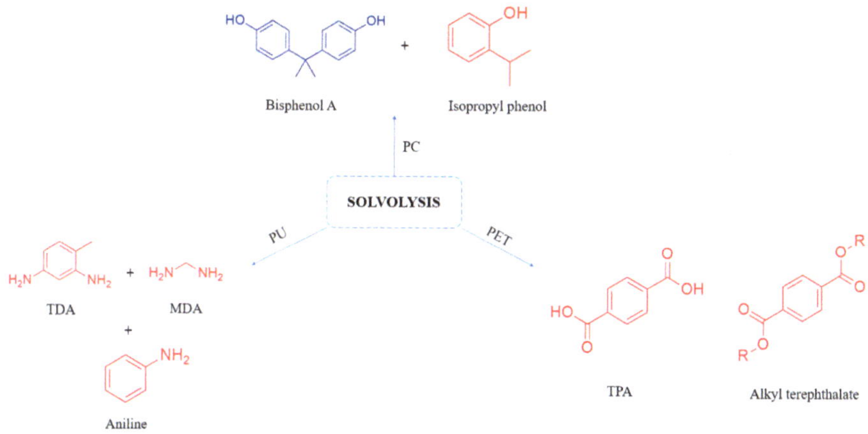


**Fig. 10.5** Reaction scheme of polymer recycling by solvolysis

Under sub and supercritical conditions, less nucleophilic solvents such as secondary and tertiary alcohols give higher yields (Kamimura et al. 2008). Polycarbonate mainly bisphenol A based are largely studied for recycling giving BPA as the main product after recycling. Many researchers have studied different solvents such as isopropanol (Knight and Fieser 2024), tetralin, decalin (Sato et al. 2005) and cyclohexanol (Kawai 2002). Ethanol and methanol are ideal for recovering the pure BPA monomer but the low solubility of PC has forced the researchers to introduce sub and supercritical conditions in the process (Antonakou and Achilias 2013) (Table 10.5).

#### 10.3.2.4 Hydrolysis (Fig. 10.6)

Hydrolysis involves the reaction of polymers having hydrolysable groups with water in order to obtain the monomers utilized for their synthesis or their derivatives. Hydrolysis is of three types, neutral, acid, and alkaline depending on the pH of the system (Abedsoltan 2023). All the methods require very harsh operating conditions making them difficult to be carried out on the large scale as well as increasing the cost of the process. In case of PET the chain scission results into ethylene glycol and terephthalic acid with temperature and pressure in the range of 100–400 °C and 15-20 bar respectively (Sinha et al. 2010). For acid hydrolysis of polyamides, usually organic acids are used with normal pressure and around 100-200 °C temperature. Polyamides being a class of polymers there are different raw materials used for their synthesis and hence each polyamide has different degradation speeds in different media (Brette et al. 2023). Alkaline hydrolysis of polyamide is faster in comparison with acid hydrolysis but requires very high concentration of alkali solutions which not only increases the cost but also increases environmental pollution (Zheng et al. 2024). Biopolymers like polyhydroxy acids including polylactic acid, polyglycolic acids easily undergo hydrolysis due to presence of ester linkages and the degradation reaction is autocatalyzed due to presence

**Table 10.5** Some examples of recycling by solvolysis process performed on different polymers

| Polymer | Catalyst | Solvent | Reaction condition | Products | Reference |
|---|---|---|---|---|---|
| PET | NaOH | Ethanol and Propanol | 100 °C | Terephthalic acid (TPA) | Sanda et al. (2016) |
| PET | Cu/ZnO | Methanol | 65 °C | Dimethyl terephthalate | Lin et al. (2024) |
| PET | – | Ethanol | 150 bar/275 °C | Diethyl terepthalate, TPA, ethylene glycol | Lozano-Martinez et al. (2021) |
| PET | NaOH | Butanol | 150 °C | Disodium terepthalate | Sanda et al. (2017) |
| PU | NaOAc, NaOH, KOH, KOAc | t-amyl alcohol | 225–250 °C | Toluene diamine and Aniline | Johansen et al. (2022) |
| PU | KOH | t-amyl alcohol | 225 °C | Methylene diamine | Johansen et al. (2023) |
| Polycarbonate | Alkali-earth, alkali carbonate | Tetralin, cyclohexanol | 450 °C/2 MPa | Isopropyl phenol, Bisphenol-A | Sato et al. (2005) |
| Polycarbonate | Ammonium carbonate | Methanol | 120 °C | Bisphenol-A, Urea | Shaikh et al. (2024) |

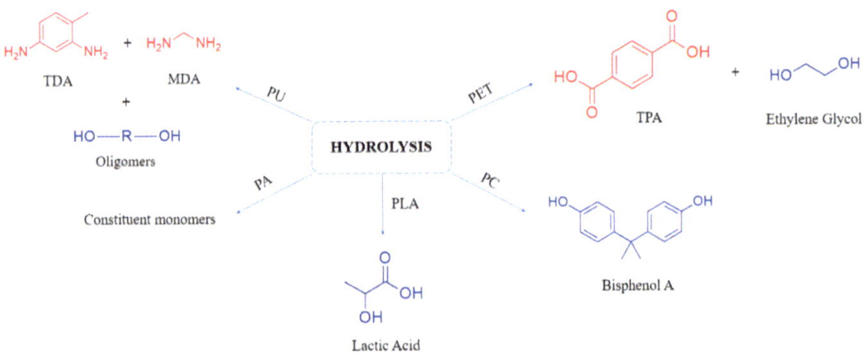


**Fig. 10.6** Reaction scheme of polymer recycling by hydrolysis

of carboxylic acid groups getting generated as the reaction proceeds (Elsawy et al. 2017). Polyurethanes also undergo hydrolysis but is the least preferred method to recycle them as it always would result in formation of 4,4′-diaminodiphenylmehane (MDA) or 2,4- and 2,6-toluene diamines (TDA) which are classified as carcinogens. Literature suggests that hydrolysis by use of supercritical and subcritical water for recycling of plastics has potential to be scaled up from lab scale to commercial scale

(Meng et al. 2004; Goto 2009; Čolnik et al. 2021). Numerous studies are published on hydrolysis of polycarbonate using different catalysts which mainly yields BPA and its degradation products such as 4-isopropenyl phenol (Grause et al. 2010; Tsintzou and Achilias 2013). Anhydride cured epoxies have ester linkages which makes them easily cleavable through hydrolysis resulting into an acid and the epoxy initially used (Zhang et al. 2022) (Table 10.6).

#### 10.3.2.5 Aminolysis (Fig. 10.7)

Aminolysis is a chemical recycling method that involves the cleavage of bonds in polymers using amines. This process can be applied to various polymers, including polyesters (PET), Polyamide and polyurethane, to break them down into valuable chemicals or recover monomers. Aminolysis involves a nucleophilic attack of an amine on the carbonyl carbon of an ester bond, leading to chain scission and the formation of amide bonds. PET can be depolymerized using aminolysis with reagents like ethanolamine, ethylene diamine, or other amines. This process yields diamides of terephthalic acid, such as bis(2-hydroxyethyl) terephthalamide (BHETA), which can be further converted into value-added products like unsaturated polyesters and polyurethanes ('zhang-et-al-2014-from-aminolysis-product-of-pet-waste-to-value-added-products-of-polymer-and-assistants', n.d.). Polyurethane can be chemically recycled through microwave-assisted aminolysis using amine reagents with primary and tertiary amino groups. This method efficiently degrades the urethane groups, recovering valuable chemicals (Rossignolo et al. 2024) (Table 10.7).

### 10.3.3 Enzymatic Recycling

#### 10.3.3.1 Mechanism of Enzymes

Several enzymes, such as reductases, lipases, laccases and peroxidases from microbial sources, have been studied for their ability to degrade plastic (Mohanan et al. 2020). The utility of enzymes as powerful heterogeneous catalysts acting at the solid-liquid interface was demonstrated as early as 1981, when nylon was degraded by a hydrolase enzyme isolated from *Flavobacterium* sp. KI72 by a group of Japanese researchers (Kinoshita et al. 1981).

Enzymes work by adding $H_2O$, $O_2$ or $H_2O_2$ across the linkages connecting monomers to cause oxidation or hydrolysis of the polymer. The mechanisms vary depending on the type of enzymes and bonds involved; for example, esters are cleaved by esterases, while polypeptides are cleaved by peptidases (Kaushal et al. 2021).

Polymers such as PE, PET, polystyrene, and nylon have also been subjected to microbial degradation. Microbes are able to utilize these polymers as their sole carbon source to generate $CO_2$, $H_2O$ and energy, or are able to produce economically valuable products such as biosurfactants and organic acids (Ru et al. 2020).

Microbial enzymes involved in plastic degradation can be broadly classified into extracellular and intracellular enzymes. Extracellular enzymes are essential for the de-polymerization of polymers into their corresponding oligomers and monomers,

**Table 10.6** Some examples of recycling by hydrolysis process performed on different polymers

| Polymer | Catalyst/ reaction mixture | Reaction conditions | Products | Reference |
|---|---|---|---|---|
| Polyurethanes (PUs) | Carbonic acid generated from $CO_2$ acted as an acid catalyst | 190 °C for 24 h at 8.0 MPa $CO_2$ | 1,4-butanediol and diamines | Motokucho et al. (2018) |
| Polyurethane | NaOH, water and ethanol | 80 °C, 2 h | Polyols and Benzene derivative | Liu et al. (2025) |
| Polyether-based polyurethane foam | Dry steam | 190 and 230 °C | Polyol, isomeric toluenediamines, and carbon dioxide as primary products. | Gerlock et al. (1980) |
| PET | $NaHCO_3$ and $KHCO_3$ used together as catalysts,<br>marine water, stainless-steel batch reactor with stirrer. | 195 °C<br>34.5 atm for 2 h | TPA and EG | Stanica-Ezeanu and Matei (2021) |
| PET | High-temperature liquid | 573 K | TPA and EG | Sato et al. (2010) |
| PET | Humi-cola insolenscutinase (HiC) as catalyst, reactor tube. | 250 °C and 39.5 atm | TPA and EG | Quartinello et al. (2017) |
| PET | Plasma treatment to enhance the wettability of PET. | 205 °C and 16 atm | TPA and EG | Mancini et al. (2013) |
| PET | Zinc acetate as catalyst, hot compressed water, autoclave reactor. | 220–280 °C<br>21–61 atm | TPA and EG | Liu et al. (2012) |
| PET | Hexadecyltrimethylammonium chloride as cationic surfactant (accelerator) along with 1.5 mol. mL−1 NaOH as catalyst | 100 °C | TPA and EG | Čorak et al. (2022) |
| PET | Stainless steel batch reactor heated by the silicon oil bath, γ-valerolactone (GVL) as both swelling and dissolution agent for PET, 15 wt% NaOH as catalyst. | 90 °C | TPA and EG | Chen et al. (2021) |

| | | | | |
|---|---|---|---|---|
| PET | Batch Teflon sealed reactor with agitator, 0.5 wt% $[C_{16}H_{33}N(CH_3)_3]_3PW_{12}O_{40}$ (known as (CTA)3PW) as PTA and 4 wt% NaOH as catalyst. | 110 °C | TPA and EG | Wang et al. (2022a) |
| PET | Pressure-tight reactor with stirrer, 0.1 g.ml – 1 of TPA in water (0.6 M TPA) as the solid catalyst. | 220 °C | TPA and EG | Yang et al. (2021) |
| PET | Microwave irradiation with the power range of 60–940 Watt, glass reactor sealed with Teflon-silicon septa, H + @ZSM-5-25 as Brønsted–Lewis acid catalyst (1.25 mmol. $g^{-1}$). | 121 °C<br>3–28 atm | TPA and EG | Kang et al. (2020) |
| PA66 | Supercritical $H_2O$ | 380 °C, 28 MPa | Monomers, oligomers and Hypo products from decomposition of monomers such as cyclopentanone and cycloheptylamine. | Meng et al. (2004) |
| PA6 | Subcritical $H_2O$, 3% Phosphotungstic heteropoly acid (HPA) | 85 min at 300 °C | ε-caprolactam | Chen et al. (2010) |
| PA6.6 | $H_2O$ + 1.25 HCl/amide mol ratio | 200 °C<br>Microwave heating | 90% adipic acid and 86% hexamethylene diamine | Češarek et al. (2020) |
| PA6 | $H_2O$ and $H_3PO_4$ | 200 W/ 100 mL water<br>Microwave heating | ε-amino-caproic acid | Klun (2000) |
| PA6 | 5 mol% DMAP, 5 equiv. $Ac_2O$ | 250 °C<br>microwave heating | N-acetyl caprolactam | Alberti et al. (2019) |
| Aliphatic polyamides (PA 66, PA 1010, PA 11, and PA 12) | HCl as an acid catalyst | 200 °C<br>microwave reactor | Constituent Monomers | Češarek et al. (2020) |

(continued)

**Table 10.6** (continued)

| Polymer | Catalyst/ reaction mixture | Reaction conditions | Products | Reference |
|---|---|---|---|---|
| Polycarbonate | Alkaline (NaOH), phase transfer catalyst (1-hexadecyl trimethyl ammonium bromide) | 160 °C, after 40mins | Bisphenol-A | Tsintzou and Achilias (2013) |
| Polycarbonate | Hydrolysis in a steam atmosphere, semibatch reactor in the presence of MgO and CaO | 300 and 500 °C | Bisphenol A (BPA) | Grause et al. (2010) |
| Polycarbonate | NaOH as catalyst, different amount of water as main solvent and constant amount of different kind of co-solvent (tetrahydrofuran and 1,4-dioxane) | Microwave irradiated | Bisphenol A (BPA) | Deirram and Rahmat (2012) |
| Polycarbonate | Concentrated sulfuric acid ($H_2SO_4$) as the catalyst | 110 to 150 °C | Bisphenol A (BPA) | Abedsoltan (2024) |
| Polycarbonate | 1,8-diazabicyclo [5.4.0] undec-7-ene (DBU) as catalyst and solvent | 140 °C for 3.0 h | Bisphenol A (BPA) | Liu et al. (2018) |
| Poly (lactic acid) (PLA) | Water/PLA ratio up to 50% of PLA by weight | 140 °C to 180 °C,120 mins | Water-soluble lactic acid | Piemonte and Gironi (2013) |
| Poly (lactic acid) (PLA) | Hexadecyltrimethylammonium bromide as phase transfer catalyst in a 10% w/v NaOH | Microwave irradiation, 100 °C for 10 min | Lactic acid | Siddiqui et al. (2020) |
| Poly (lactic acid) (PLA) | Water-ethanol solutions | 60–90 °C | Water-soluble lactic acid | Iñiguez-Franco et al. (2018) |

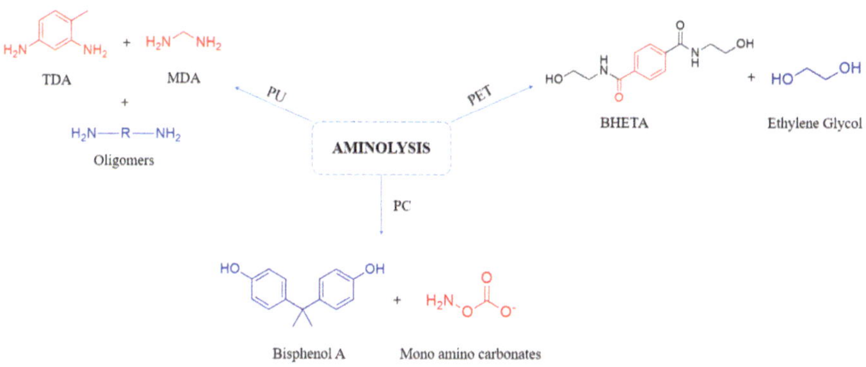


**Fig. 10.7** Reaction scheme of polymer recycling by aminolysis

as the entire polymer cannot enter the cytoplasm due to its high molecular weight. The degradation products are then transported inside the cytoplasm and are further metabolized by intracellular enzymes (Mohanan et al. 2020). The further metabolism can either be anaerobic or aerobic; in the former, nitrates and sulphates act as the terminal electron acceptor, while in the latter, $O_2$ acts as the terminal electron acceptor (Ma et al. 2022).

An example of microbial enzymes that can degrade polymers is the CYP 450 mono-oxygenase class of enzymes, which are predominantly found inside fungal hyphae bound to the endoplasmic reticulum. These enzymes are responsible for the oxidative degradation of plastic monomers via phase I and phase II oxidation and conjugation reactions (Srikanth et al. 2022). Lipase secreted extracellularly by *Aspergillus niger* is capable of hydrolysing PS, PET, and PE (Safdar et al. 2024). Enzymes identified from microbial sources that can depolymerize plastics are listed below (Table 10.8):

#### 10.3.3.2 PETase and MHETase

The discovery of *Ideonella sakaiensis* in 2016 and its role in the degradation of PET proved to be a breakthrough in enzymatic degradation of plastic waste, as a 58% reduction in PET weight was obtained after 40 days of incubation. Since then, the enzymes PETase and MHETase have been studied extensively to improve their thermal stability and degradation efficiency (Yang 2016).

PETase is secreted extracellularly where it catalyses the conversion of PET into MHET (mono(2-hydroxyethyl) terephthalate), which then enters the bacterial cell via porins present on the outer cell membrane. MHETase is present in the periplasmic space. It breaks MHET into EG and TPA. TPA enters through TPA transporters after it binds to TPA binding protein as shown in Fig. 10.8. PETase and MHETase belong to the alpha/beta hydrolase family which works via the aspartate-histidine-serine catalytic triad mechanism (as depicted in Fig. 10.9) (Choi et al. 2024; Sagong et al. 2020; Han et al. 2017). Serine from the triad attacks the carbonyl carbon of PET, which results in acylation at the site of attack. The final step is the deacylation of the substrate and regeneration of serine to yield acid group at the substrate site.

**Table 10.7** Some examples of recycling by aminolysis process performed on different polymers

| Polymer | Catalyst | Amine used | Reaction conditions | Products | Reference |
|---|---|---|---|---|---|
| Polyurethane foams (PUFs) | Triethylenediamine (DABCO) | Tris(2-aminoethyl) amine (TREN), hyperbranched poly(ethylenimines) | Microwave reactor Monowave 400 175 °C–210 °C | The flexible PUFs made from recovered polyols have comparable mechanical properties to those made from virgin polyols. | Grdadolnik et al. (2023) |
| Rigid Polyurethane Foam | Sodium hydroxide | Diethylenetriamine | 180°C for 70 minutes | Depolymerization increased as the amount of NaOH increased. | Chuayjuljit et al. (2002) |
| Flexible polyurethane foams | – | TEPA | 160 °C FOR 5 hrs | Depolymerized product was used to assist in the devulcanization of waste rubbers | Wang et al. (2023a) |
| Polyurethane shoe sole | NaOH | 4,7,10-trioxatridecan-1,13-diamine (1,13-diamine) and 1,3-propanediamine (1,3-diamine) | 140 °C, 3 hrs | The products depolymerized were further used as hardeners for epoxy coatings | Guchait et al. (2024) |
| PET | Sodium acetate | Ethanolamine | 8 h | 85 wt% Bis (2-hydroxy ethylene) terephthalamide (BHETA) and 15 wt% dimer | Mir Mohamad Sadeghi et al. (2011) |
| PET | 1,5,7-triazabicyclo [4.4.0] dec-5-ene (TBD) | Ethylenediamine,1,3-diaminopropane | 180 °C, 4 Hrs | A broad range of crystalline terephthalamides | Nica et al. (2018) |
| Polyimide | – | Diethylamine and N-butylamine | 8 h, atmospheric pressure | Completely degraded in ethylenediamine and MEA, and the yield of 4,4′-ODA recovered from both reaction systems is high | Chen et al. (2025) |
| Polycarbonate | $ZnX_2$ | Chiral amino alcohols | Mild condition | Tansformation of BPA-PC plastic waste to BPA and high value-added chiral chemicals. | Wang et al. (2022b) |

| | | | | | |
|---|---|---|---|---|---|
| Polycarbonate | $Na_2CO_3$ | N, N′-dimethyl-1,2-diaminoethane | A short period at 100 °C | Produced DMI and BPA, | Hata et al. (2002) |
| Polycarbonate | – | m-methoxy benzylamines | 80 °C | Urea derivatives by reacting with primary amines | [62] |
| Poly (bisphenol A carbonate) (BPA-PC) | – | Ethanolamine | 120 °C, 46 Hrs | BPA (>99%) and 2-oxazolidone (>99%) were obtained | Singh et al. (2015) |
| Polyethersulfone (PES) | – | Bis(4-(tert-butyl) phenyl) amine, carbazole, diphenylamine | 120 °C, dissolved in DMF | Functionalized oligomers and/or small molecules with amine and phenol products useful for the synthesis of other polycondensates, | Liotier et al. (2024) |
| Polycarbonate | – | Amoxapine and desloratadine | Additive/ catalyst free conditions | BPA, mono amino carbamate | Padhi et al. (2025) |

**Table 10.8** Microbial enzymes demonstrated to degrade plastics

| Enzyme | Source | Reaction scheme | Polymers degraded | References |
|---|---|---|---|---|
| Laccases | *Cochliobolus sp*, *Penicillium simplicissimum* | 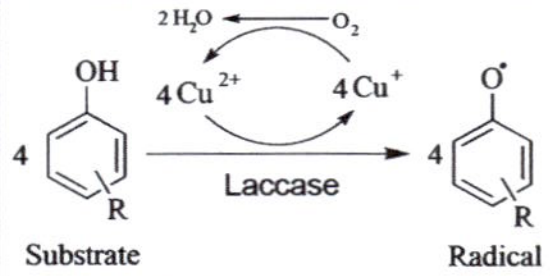 | PVC, PE | (Sumathi et al. 2016) (Sowmya et al. 2015) |
| Beta Di Hydroxyalse | *Pseudomonas sp* | 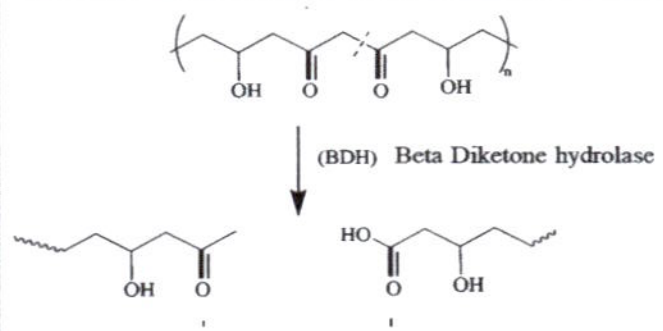 | PVA | (Wilkes and Aristilde 2017) |
| Lignin peroxidases | *Phanerocheate chrysosporium*, *Aspergillus niger*, *Aspergillus flavus* | 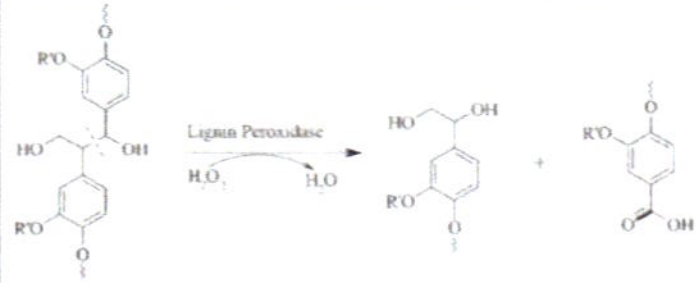 | PVC, PE | (Khatoon et al. 2019), (Srikanth et al. 2022) |
| Alkane Mono oxygenase (Alkb) | *Rhodococcus* sp. | 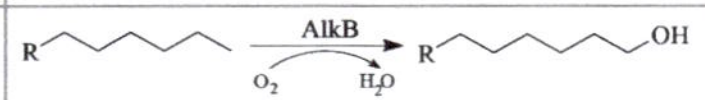 | PE | (Rong et al. 2024) |
| Lipases | *Aspergillus niger*, *Candida antarctica* | Triacyl glycerol $\xrightarrow[\text{Ester hydrolysis}]{H_2O}$ Diacyl glycerol + Carboxylic acid | PLA, PET | (Safdar et al. 2024), (Srikanth et al. 2022) |
| Cutinases | *Thermobifida fusca*, *Cryptococcus sp* | 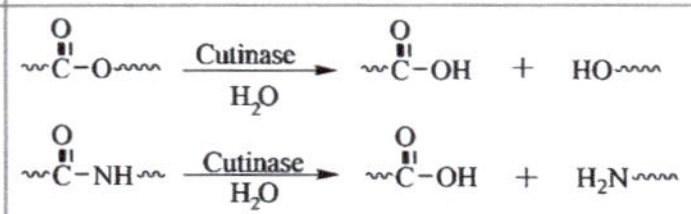 | PET, Polyurethane, PLA | (Flores-Castañón et al. 2022), (Srikanth et al. 2022) |
| PETase, MHETase | *Ideonella sakaiensis* | PET $\xrightarrow[H_2O]{\text{PETase}}$ MHET $\xrightarrow[H_2O]{\text{MHETase}}$ TPA + EG | PET | (Mohanan et al. 2020) |
| Protease | *Bacillus* sp. | 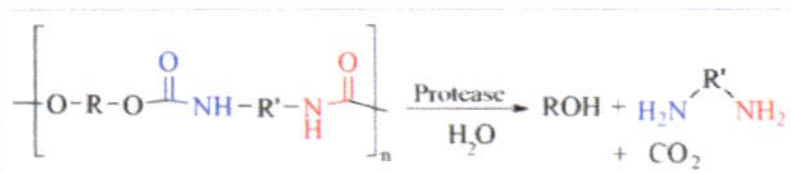 | Polyurethane | (Katnic et al. 2024) |

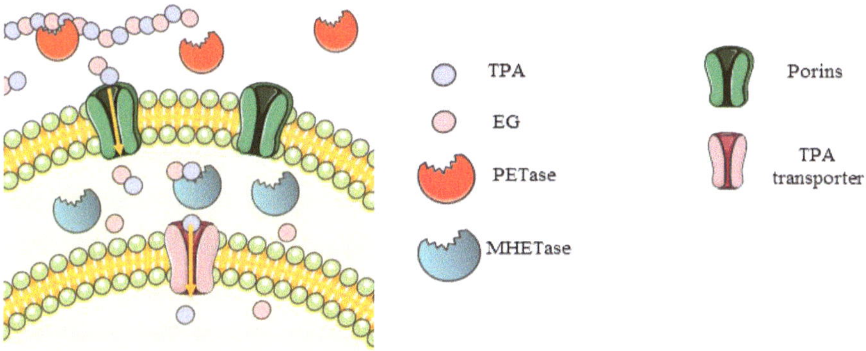


**Fig. 10.8** Biodegradation of PET by *Ideonella sakiensis*

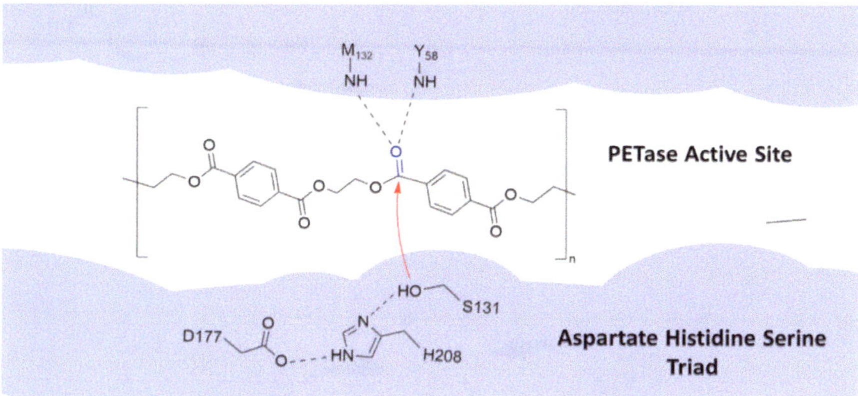


**Fig. 10.9** PETase mechanism (aspartate histidine serine catalytic triad)

The same mechanism of action is seen in hydrolysis by MHETase (Pinto et al. 2021) (Fig. 10.10).

#### 10.3.3.3 Strategies to Improve Efficiency of Enzymatic Degradation of Polymers

Three commonly used strategies to improve efficiency (see Fig. 10.11) of enzymatic degradation of polymers are (1) use of pretreatment methods, (2) process optimization, and (3) genetic modification of enzymes and microbes (Crystal Thew et al. 2024).

1. Pretreatment

Hydrophobic surface of the plastic polymers acts as a major hindrance in its degradation. PE is such a polymer that requires a longer incubation period with the

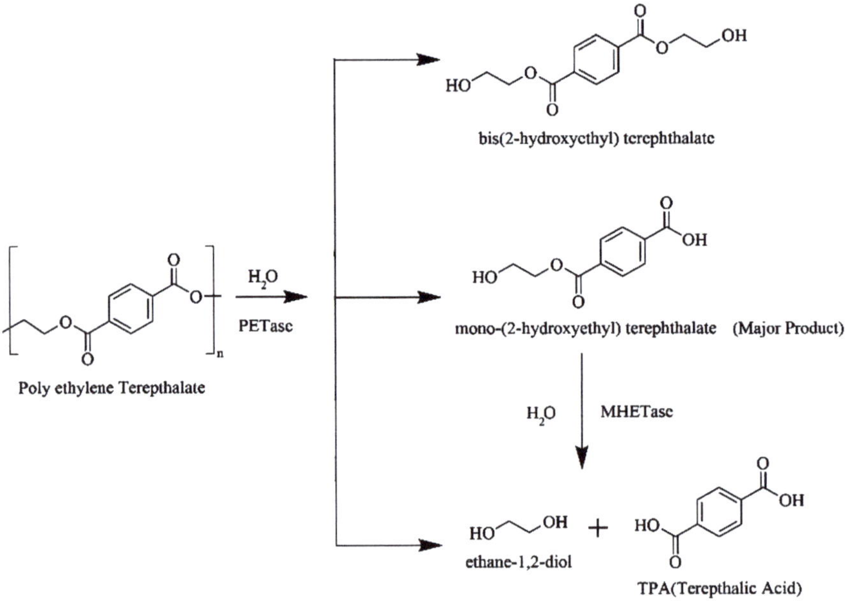


**Fig. 10.10** PET hydrolysis by PETase and MHETase

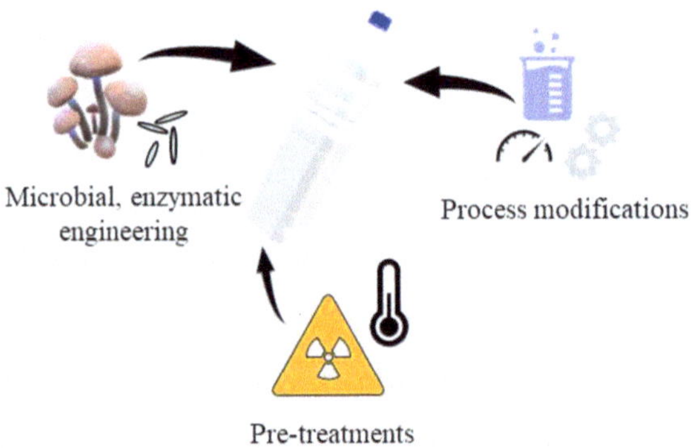


**Fig. 10.11** Strategies used to improve efficiency of plastic biodegradation

enzymes and shows poor degradability owing to its highly hydrophobic nature. The pretreatment process aims to produce surface roughness, improve the total exposed surface area, and generate more hydrophilic functional groups on the polymer surface, thus hastening the bio degradation process (Ciuffi et al. 2024).

UV exposure for 3–15 days results in photo-oxidation of polymers (Taghavi et al. 2021c). Heating polymers at 60–70 °C generates polar groups such as ketone and aldehyde, which alters polymer structure and aids in biofilm formation (Awasthi et al. 2017). Thermal pretreatment is a highly scalable and industrially sustainable method. Chemical pretreatments involve the addition of acids and oxidising agents to generate polar functional groups on the polymer surface. Treatment with alcohol and benzene may remove plasticisers and other additives that may affect the biodegradation. Chemical based pretreatments are also scalable and industrially viable as the chemicals can be recycled for multiple batches (Crystal Thew et al. 2024). The concern, however, is the generation of microplastics and proper disposal of such microplastic-containing chemicals.

2. Process Modifications

The addition of surfactants increases the interaction between microbes and plastics thus aiding in biofilm formation. Studies have proven that the addition of Tween 20 improves biodegradation rate (Giyahchi and Moghimi 2024). The absence of other C sources in the media (carbon-starved conditions) enables the induction of plastic cleaving enzymes. However, many bacteria are unable to survive in such harsh conditions.

3. Enzyme Modifications

Enzymes can be modified to improve their biodegrading capacity. The major challenges are poor enzyme thermostability, long incubation periods, and poor degradation efficiency. A new PET hydrolase was engineered in 2020 that surpasses all preexisting enzymes in terms of incubation period, efficiency, and percent degradation (Tournier et al. 2020).

#### 10.3.3.4 Challenges and Future Perspectives

Although enzyme-based degradation is a sustainable plastic recycling method, it has a few drawbacks. For example, a long period of incubation is required to obtain any degradation. Thus, the efficiency of degradation is very poor. This is because no natural enzyme exists that can degrade plastics. Such enzymes are only synthesized in environments where microbes are forced to break down plastic polymers as their carbon source, such as landfills (Shilpa, Basak and Meena 2022). The degradation process itself generates a lot of microplastics.

The thermal stability of enzymes is also a major subject of concern. Plastic degradation is enhanced at higher temperatures, which is deleterious for enzymes. Plastic structure, hydrophilicity and polymer thickness influence the degradation process. Pretreatment processes may improve biodegradation; however, the overall cost of the process will increase significantly (Taghavi et al. 2021a).

### 10.3.4 Microbial/Fungal Degradations

Many bacteria and fungi have been isolated that are able to bring about biochemical transformation of plastic polymers to either oxidize them completely to generate $CO_2$ and $H_2O$ or utilize plastic monomers in their biochemical pathways (Srikanth et al. 2022) This is interpreted by observing the changes in plastic morphology, structure, and molecular weight post incubation with such microbes.

Microbial degradation of plastics involves four steps (see Fig. 10.12): biodeterioration, bio fragmentation, assimilation and mineralization. All these processes are executed via various enzymes produced by the microbes (Amobonye et al. 2021).

1. Biodeterioration

   The first step in biodegradation involves the firm attachment of microbes on the polymer surface, which is accelerated by virtue of hydrophilic functional groups on the polymer surface. The microbe proliferates and forms a biofilm or mycelia at the site of adsorption. This is aided by the production of exopolysaccharides by the microbes that improve adherence and penetration. Factors like temperature and pH influence this step. The molecular structure, hydrophobicity, and surface roughness also play an important role. Branched and amorphous plastics with many hydrophilic groups at the polymer surface enhance biofilm formation. The addition of plasticisers promotes initial colony formation as they are smaller in size and can be easily consumed by the microbe (Amobonye et al. 2021), (Yuan et al. 2020). However, common plastic additives like dibutyl tin dilaurate retard biofilm formation (Han and Cooney 1995).

2. Biofragmentation

   The process involves the depolymerization of large polymers into oligomers of smaller molecular weight by enzymatic hydrolysis, oxidation and peroxidation. Extracellular enzymes bring about the reduction of the polymer into smaller molecules that can easily be assimilated into the cell. This process is influenced

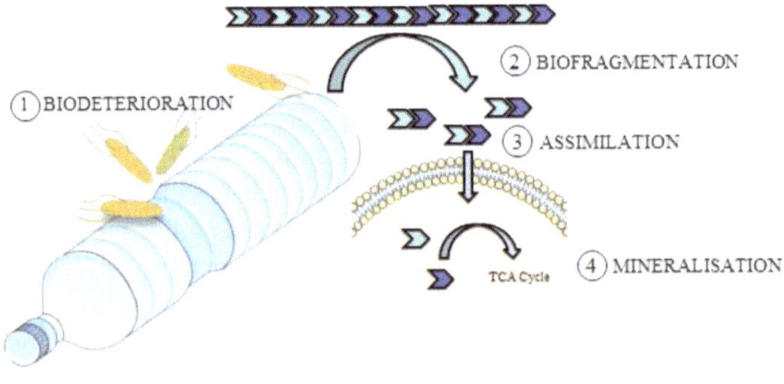


**Fig. 10.12** Mechanism of microbial deterioration

by external factors like temperature, humidity and pH, as well as the presence of organic and inorganic compounds (Amobonye et al. 2021).

3. Assimilation

   The entry of the bio-fragmented oligomers inside the cells is facilitated via passive or active transport. Porins and ATP binding cassettes have been reported to play an active part in this step, depending on the type of organism and polymer. Some transporters have a dual role; they catalyze further oxidation/hydrolysis while transporting the oligomer. *Ideonella sakaiensis* utilises porins at the cell membrane surface to assimilate MHET, the major metabolite of PET degradation (Choi et al. 2024).

4. Mineralization

   In the last stage of the biodegradation process, the oligomer is hydrolyzed to its corresponding monomers, which then enter into the metabolic pathway of the organism to either be completely oxidized as $CO_2$ and water or enter other biosynthetic pathways in the microbe. For example, acetic acid produced as a result of polyethylene degradation enters into the Krebs cycle or the lipid synthesis pathways (Wilkes and Aristilde 2017).

   Table 10.9 shows the rate and extent of microbes in degrading different plastic polymers (Taghavi et al. 2021b).

#### 10.3.4.1 Metabolic Pathways/Mechanism

Once assimilated, the monomers are incorporated into the microbial metabolic processes where they are converted into other secondary metabolites, or enter the TCA cycle to be completely oxidized into $CO_2$ and $H_2O$. For example, in PVA degradation by *Pseudomonas* spp. (shown in Fig. 10.13), acetic acid is formed, which enters lipid biosynthetic pathways, essential for microbial cell membrane synthesis or is completely oxidized to form $CO_2$ (Wilkes and Aristilde 2017). PET degradation yields various economically valuable metabolites (valorization). Metabolic and biosynthetic pathways have been manipulated for production of value-added chemicals

**Table 10.9** Plastic degradation efficiencies of different microbes [213]

| Strain | Plastic type | Time (day) | Weight loss (%) |
|---|---|---|---|
| *Bacillus subtilis* | Polystyrene | 30 | 23 |
| *Rhodococcus ruber* C208 | Polystyrene | 56 | 0.8 |
| *Bacillus* sp. YP1 | Polyethylene | 28 | 10.7 |
| *Pseudomonas aeruginosa* | Polyethylene | 120 | 20 |
| *Aspergillus niger* | Polyethylene | 30 | 4.32 |
| *Trichoderma harzianum* | Polyethylene | 60 | 2.26 |
| *Bacillus subtilis* | Polyethylene | 30 | 9.26 |
| *Aspergillus flavus* | Polyethylene | 28 | 3.9 |
| *Idionella sakaiensis* 201-f6 | PET | 42 | 58 |
| *Bacillus subtilis* UCP 999 | PET | 60 | 0.06 |
| *Bacillus cereus* | PET | 40 | 6.6 |

such as polyphenols, biosurfactants, and degradable plastics (PLA) (Ru et al. 2020).

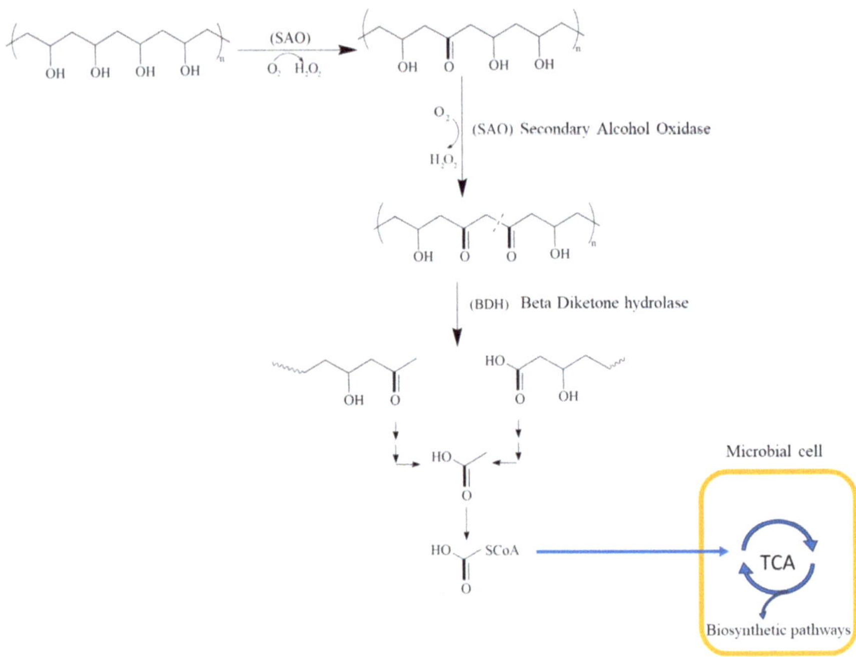


**Fig. 10.13** PVA biodegradation by *Pseudomonas* spp.

#### 10.3.4.2 Genetic Engineering to Enhance Degradation Capabilities

Several strategies have been employed to enhance the efficiency of enzymes and microbes for the degradation of plastics. Enzyme mutations, linking of synergistic enzymes, and cell surface expression of enzymes are commonly used strategies (Zhu et al. 2022).

*Enzyme mutations:* PET degradation is enhanced at 75° C in the presence of thermostable PETase developed after several mutations. Addition of proline, disulfide bonds and glycosidic linkages imparts thermodynamic and molecular stability to the enzyme, thus preventing protein aggregation at higher temperature. To overcome the challenge of poor interaction of hydrophobic plastics with hydrophilic functional groups of protein, which reduces the degradation efficiency, active site hydrophobicity has been increased, along with changing the electrostatic charges of the active site. Computational studies of polymer-enzyme interactions help in designing mutations to improve the interaction.

*Bifunctional biocatalysts:* Lipase-cutinase and PETase-MHETase fusion proteins show improved degradation of plastics (Liu et al. 2019; Li et al. 2023). Simultaneous catalysis is seen with such fusion proteins as a result of synergism. Lipase catalyses breakdown of large PCL (Polycaprolactone) polymers into smaller oligomers, while catalase hydrolyses these oligomers into smaller soluble monomers.

*Cell surface expressed enzymes:* A study was conducted using recombinant *Vibrio natriegens*, a halophile, to evaluate its efficacy in degrading PET microplastic waste in the oceans. MHETase-PETase fusion protein was expressed on the cell membrane of *V. natriegens*. Cell surface-bound enzymes proved more advantageous than the extracellularly secreted PETase of *I. sakaiensis* as multiple PETase units could act on the polymer, simultaneous hydrolysis of MHET took place, and there was no diffusive loss of enzymes as they were bound to the microbe (Li et al. 2023).

#### 10.3.4.3 Integration with Current Tech (Hybrid Approach)

Biodegradation involving bacteria and fungi is a sustainable recycling approach as it does not involve harmful chemicals, toxic gas release, and challenges associated with disposal of by-products. Moreover, biodegradation pathways can be manipulated to obtain desired value-added products; however its efficiency is very poor (Shilpa, Basak and Meena 2022).

A hybrid chemical-biological strategy offers a promising route for upcycling mixed plastic waste, enabling cost-effective processing with a reduced carbon footprint. Pre-treatment of plastic results in an increase in exposed surface area and change in surface polarity, allowing better enzymatic degradation. Mixed plastics (PLA and PET) can first be separated by use of ionic liquids and then subjected to microbial degradation for complete recycling/valorization (Dou et al. 2023).

The integration of nanotechnology with enzymatic reactions has been attempted to overcome the stability and recyclability issues of enzymes. PETase and MHETase immobilized on $CO_3(PO_4)_2$ nanoparticles have been developed recently to improve enzyme stability and recyclability. An increase in catalytic efficiency was observed when immobilized enzymes were used (Jia et al. 2021).

### 10.3.5 Energy Recovery

The European Directive on Waste Management establishes a precise hierarchy to be taken into account to determine the best end-of-life option for any waste product. This heavily depends on the type and composition of the feedstock material. Tertiary and quaternary recycling techniques are promising for the management of plastic waste that is difficult to be processed via primary or secondary recycling (e.g., non-biodegradable bioplastics like bioPET, bioPE, etc.). It involves techniques like pyrolysis and incineration. Pyrolysis, in particular, is a better alternative to "disposal," which is the least preferred approach for waste management as it results in the loss of valuable raw materials associated with bioderived molecules. In contrast, chemical recycling processes like pyrolysis can yield monomers that can be used as an alternative feedstock, thereby reducing the environmental impact of certain non-biodegradable bioplastics.

Quaternary recycling or energy recovery is particularly useful in non-recyclable and non-biodegradable fractions of plastic waste. Such approaches are ranked lower in the waste-disposal hierarchy but are still preferred over landfilling for disposal. The following sub-sections discuss the necessary conditions, challenges, and parameters affecting the yield of useful primary renewable resources (like monomers or intermediate compounds) and energy in case of pyrolysis and incineration (Fredi and Dorigato 2021).

#### 10.3.5.1 Pyrolysis

Pyrolysis is the thermochemical conversion of plastic waste into valuable chemicals or alternative energy in an oxygen-free atmosphere. It is considered to be a chemical recycling technology with considerable commercial maturity. The oxygen-free degradation of long-chain molecules in pyrolysis causes them to decompose into saturated hydrocarbon-rich vapours that can be condensed to obtain pyrolytic oil. It also yields solids (char) and non-condensable gaseous products, the relative composition of which depends on factors such as the elemental composition of feedstock material (Fredi and Dorigato 2021; Rahman et al. 2023; Dogu et al. 2021; Jiao et al. 2024).

The potential of different types of plastics to yield valuable chemicals can be understood by their ultimate (elemental analysis) and proximate analyses. The ultimate analyses (elemental analysis) reported for various non-biodegradable bioplastics indicate that PE and PP possess higher carbon content as compared to PLA and PA polymers. PE and PP were found to contain greater hydrogen content, too, when compared to PET. The elemental composition of such non-biodegradable bioplastics makes them potentially suitable for pyrolysis. However, the presence of significant oxygen content in the feedstock (in the case of PET) is unfavourable due to the generation of dioxins, polychlorinated dioxins, and oxygenated aromatics like benzaldehyde, benzoic acid, and phenols, which negatively impact the quality of the pyrolytic oil that is being produced. Therefore, polymers with high carbon content and negligible oxygen content, like bioPE, are best suited for hydrocarbon fuel production through pyrolysis (Rahman and Bhoi 2021; Dogu et al. 2021). On the other hand, polymers like PET and PLA with abundant oxygen content are preferably subjected to gasification, which converts the organic solid material into a mixture of gases like Carbon monoxide (CO), hydrogen ($H_2$), Carbon dioxide ($CO_2$), and methane ($CH_4$) by partial oxidation.

A proximate analysis is done for various plastics to determine the composition of solid char, gases (non-condensable substances), and liquids (oils and condensed vapours) produced on thermochemical degradation. This includes an assessment of moisture, fixed carbon, volatiles and ash content. Bioplastics like bioPE, bioPP, and bioPET have a higher fraction of volatile content (more than 90%), which is favourable for pyrolysis, which yields high-quality bio-oil and syngas in such cases (Rahman and Bhoi 2021). Principally, the entire volatile content of the feedstock can be converted to bio-oil. The composition of solids, condensable and non-condensable vapours can also be controlled by optimizing process parameters like residence time, temperature, catalyst and pre-treatment of the waste. The yield and its quality are also affected by plastic additives such as dyes, plasticizers, and flame-retardants. For example, brominated flame retardants like tetrabromobisphenol A (TBBA) decompose to HBr, brominated dibenzo-*p*-dioxins, and phenols, which pose environmental and human health dangers (Dogu et al. 2021). Hence, a high-carbon and low-oxygen content polymer with a higher volatile fraction and minimal fixed carbon content becomes the ideal feedstock for pyrolysis (e.g., HDPE).

A drawback of this process is the high temperatures that the plastics are usually subjected to (300–700 °C), resulting in undesired side reactions. This negatively impacts the quality of the end products. For instance, the pyrolysis of PLA at higher temperatures to obtain lactide monomer leads to undesired racemization of the monomers. This can be overcome by using catalysts to lower the reaction temperatures. Catalytic pyrolysis with metal oxides and hydroxides like CaO and $Al(OH)_3$ have been reported to lower the reaction temperature to avoid such by-products (Fredi and Dorigato 2021).

Extensive work has been done to develop various pyrolysis techniques, such as vacuum pyrolysis and fluid-catalytic cracking. Catalytic pyrolysis involves a catalyst placed in the reactor that interacts with the molten plastic feed or the use of a catalytic bed that interacts with the vapours developed, i.e., vapour phase process. Fluid Catalytic Cracking (FCC) involves using a solvent for dissolving the plastic feed followed by the pyrolysis of the solution, thereby enabling it to become a continuous process with continuous feeding of the solubilized plastic as feedstock. The use of silica-alumina, ZSM-5, zeolites, and mesoporous materials has been reported in such processes. Other techniques include hydrocracking, a combination of pyrolysis and gasification, etc. Hydrocracking requires the addition of hydrogen to the cracking process at elevated pressure of approximately 70 atm and temperatures ranging between 375 and 500 °C. The need for large amounts of hydrogen is a major limitation because of its high cost.

Pyrolysis-gasification, on the other hand, involves two phases. An initial reductive phase of pyrolysis with limited oxygen content at 600 °C, followed by the gasification stage that requires the addition of oxygen and natural gas to increase the temperature to 2000 °C to oxidize the inorganic compounds. Therefore, a mineral residue is produced at the end of the process apart from the pyrolytic oil and char. This mineral residue can either be disposed of at landfills or recycled. There are certain drawbacks to catalytic pyrolysis, including the inactivation of catalysts by carbonaceous deposits, inorganic materials, chloride, and nitrogen-containing components of the feedstock (Dogu et al. 2021).

#### 10.3.5.2 Incineration

Energy recovery involves incinerating solid plastic waste (SPW) to harness energy from the combustion of waste. Incineration is a Waste-to-Energy (WTE) generation technique that is done by the controlled combustion of the waste with the goal of volume reduction and energy recovery from the heat produced during combustion. During incineration, the waste is burned in a furnace at high temperatures ranging from 900 °C to 1200 °C in the presence of a sufficient quantity of $O_2$ for a fixed period. The three most important factors that affect the efficiency of incineration are the time of residence of flue gases in the system, the temperature of the incineration chamber, and the agitation provided (Moharir et al. 2019; Mallick et al. 2022).

Incineration ranks lower in the waste management hierarchy as compared to options like mechanical recycling. However, it is still a useful EoL treatment option for non-recyclable and non-biodegradable SPW. It is also a favourable option in the case of treating the residues obtained from plastic waste sorting plants (Dogu et al. 2021). It is a much more preferred option as compared to disposal in landfills. High-calorific value plastics with an elemental composition of C, O and H are generally accepted for incineration, which includes bioplastics. However, incineration is not the desirable EoL option when it comes to biodegradable bioplastics (Fredi and Dorigato 2021).

In the case of bioplastics, which contain only C, H, and O as their constituents and are sourced from renewable materials, the $CO_2$ emissions resulting from combustion can be considered as net-zero (Rosenboom et al. 2022). Moreover, the energy generated from the incineration of such bio-based plastic waste is considered renewable (Fredi and Dorigato 2021).

However, it should be noted that incineration is accompanied by the need to introduce costly pollution control measures mandated by environmental regulations in several countries (e.g., EU Hazardous Waste Incineration Directive). This is because the emission of toxic volatile organic compounds (VOCs) like dioxins, polychlorinated dioxins, polyaromatic hydrocarbons, furans, dibenzofurans, halogenated gases, cyano compounds, etc., depending on the type and composition of the waste, pose serious environmental and human health concerns. Hence, remediation technologies, such as thermal cracking of the pollutants in the released gas (at approximately 850 °C), are employed in modern incineration plants (Chidambarampadmavathy et al. 2017). Considering the above limitations, energy recovery might not be favourable from an environmental and economic point of view (Dogu et al. 2021). Another point of concern is a "locking-in" effect because of high investments made for setting up incineration plants. There is a continuous need for waste influx for such incineration plants, which may end up in reducing the flow of waste towards EoL options like recycling (Rosenboom et al. 2022).

## 10.4 Value Added Applications

### 10.4.1 Development of Functional Oligomers

Hydrolysis of PU waste typically generates MDA and TDA, both of which are classified as carcinogens, thereby limiting their utilization in further processes. In contrast, the hydrolysis of polyhydroxy acids, polyhydroxyalkanoates, and PET predominantly yields terephthalic acid, ethylene glycol, and carboxyl acid-terminated oligomers. The monomers obtained can be purified and reused for PET regeneration. Similarly, the carboxyl acid-terminated oligomers can be employed in the production of biodegradable polymers by reacting them with biobased polyols such as glycerol, 1,4-butanediol, 1,3-propanediol, polyols derived from transesterification (Shahdan and Hirzin 2021; Tajau et al. 2018) or epoxidation of vegetable oils (Patil et al. 2021; Kirpluks et al. 2020; Hoang et al. 2023). Additionally, some degree of biodegradability can be incorporated into polyamides synthesized using these acids. UV-curable coatings can also be developed by reacting these acids with hydroxyl-functional acrylates such as glyceryl acrylate, 2-hydroxyethyl acrylate, 3-hydroxypropyl acrylate, and 4-hydroxybutyl acrylate. The hydrolysis of polyamides and polycarbonates yields their respective monomers, although with some impurities, which can be purified and reutilized in polyamide production. While hydrolysis generates monomers and oligomers suitable for upcycling, the significant energy consumption required for their recovery makes them costlier than petroleum-based raw materials.

Aminolysis of polymers such as PET, PU, polyhydroxyalkanoates, and polyhydroxy acids results in amine-terminated oligomers. Since PU is a class of polymers with diverse compositions, predicting the exact structure of the resulting oligomers depends on the specific waste feedstock. In case of PET, the product is consistently BHETA (bis-hydroxyethyl terephthalamide). Thus, focusing on BHETA for developing value-added products is advantageous, as its structure is well-defined, enabling modifications based on customer requirements. Similarly, the oligomers obtained from PHAs and polyhydroxy acids are structurally consistent and can be effectively utilized. This level of structural predictability is not feasible with polyurethane.

Polyamides synthesized from recycled oligomers and dimer fatty acids, citric acid, tartaric acid, or itaconic acid can enhance sustainability of product while reducing reliance on petroleum-based raw materials. Recycled amines can also serve as curing agents for epoxy resins. Additionally, polyurea coatings can be developed by utilizing isocyanates as hardeners. Behrouz et al. demonstrated the use of aminolyzed PET for synthesizing non-isocyanate polyurea's (Shojaei and Najafi 2024). Chan converted PET bottle waste into hydrogels using ethylene glycol diglycidyl ether, facilitating dye adsorption and copper ion removal (Chan et al. 2023; Chan and Zinchenko 2021). Singh utilized BHETA derived from aminolysis with ethylene diamine and dimer acid to formulate polyester amide hot-melt adhesives (Singh et al. 2024). Polycarbonate aminolysis yields carbamates, which can be applied similarly to those mentioned above (Wu et al. 2018; Quaranta et al. 2021; Padhi et al. 2025) as well as bisphenol A, which can be repurposed for epoxy resin synthesis.

Acidolysis of PET produces oligomers with carboxylic acid end groups (Hoang et al. 2022b). Epoxy resins are well known to be cured with carboxylic acids (Jing et al. 2022). Hoang employed the recycled product in various ratios as a curing agent for epoxy resins (Hoang et al. 2022a). The resulting ethylene adipate or ethylene adipate-co-terephthalate oligomers possess -COOH terminations at both ends, making them suitable for further modifications such as polyester polyol or polyamide synthesis through amine reactions. Notably, no literature currently explores the modification of PET oligomers obtained via acidolysis, highlighting a significant opportunity for further research.

Acidolysis of PU yields hydroxyl-functional oligomers, composition which includes aliphatic, aromatic, and cycloaliphatic groups varies based on composition of PU waste. The hydroxyl value is also dependent on the type of waste utilized. The OH-functional groups present numerous opportunities for modifications and applications. Gama and team have utilized OH functional oligomers for polyurethane coating (Godinho et al. 2021a; De Smet and Vanneste 2024), foam (Gama et al. 2020; Gama et al. 2024; Grdadolnik et al. 2022; Aksu et al. 2025), and adhesive (Beran et al. 2021; Godinho et al. 2021b; Donadini et al. 2024) applications. More modifications can be carried out with the OH groups for preparation of polyester polyols, PU dispersions, composites and elastomers, the drawback being that with different waste the composition of the chain would change and in turn would affect the properties of the final application.

Glycolysis is one of the most extensively explored recycling technologies and the only one that has been industrially scaled. Glycolysis of polycarbonate generates bisphenol A (BPA) and ethylene carbonate, the latter being crucial for lithium-ion battery electrolytes. Several studies have investigated its application in Li-ion batteries (Chaudhari et al. 2018; Lundström et al. 2024). Sheel et al reported a method for converting PC waste to epoxies and have used them in UV curable coatings (Sheel and Pant 2023). Polyurethane glycolysis mainly yields carcinogens MDA and TDA with OH terminated oligomers as discussed above due to variation in structures of the waste used the properties of the upcycled product cannot be same every time. In the case of PET, we get ethylene glycol and BHET (bis-2 hydroxy ethyl terephthalate). BHET having OH termination at both its end makes it useful for further modification using various components. A huge amount of research has been done on upcycling BHET into value added products like PU coatings (Adamu et al. 2021; De Smet and Vanneste 2024), foams (Lee et al. 2022; Donadini et al. 2023), adhesives (Pu et al. 2024; Cevher and Sürdem 2021), elastomers (Li et al. 2024; Pu et al. 2023), alkyd resin (Ma et al. 2020; Ouyang et al. 2021), unsaturated polyesters (Rubeš et al. 2024; Jamdar et al. 2024), UV curable coatings (Liu et al. 2023; Jamdar et al. 2018), waterborne PU coatings (Xu et al. 2025), etc the applications are endless and depends upon the imagination and application of knowledge by the researcher.

#### 10.4.1.1 Cost Analysis and Economic Feasibility

In order to manage plastic waste, lessen environmental pollution, and preserve fossil fuel resources, polymer recycling has drawn a lot of attention. But the economic feasibility of recycling technologies continues to be a critical determinant of their uptake and scalability. Assessing economic feasibility is crucial to determine whether a proposed process is financially viable. This evaluation typically involves estimating both capital and operational expenditures, often based on costs from comparable processes and prevailing market prices of raw materials. Economic analysis in the context of recycling methods involves evaluating the cost-to-benefit ratio of each approach. It encompasses assessing the expenses associated with implementing a specific recycling process and determining its overall profitability and economic viability (Salahuddin et al. 2023).

Cost analysis and economic feasibility are critical determinants in the transition toward a sustainable, circular economy for polymers. With increasing regulatory pressures and market demand for sustainable products, polymer recycling both mechanical and chemical has become a focus of extensive research and development. However, establishing economically viable recycling processes requires a comprehensive evaluation of cost drivers, market dynamics, and process efficiencies (Gao et al. n.d.).

A primary factor in determining feasibility is the cost of raw materials versus recycled feedstocks. Virgin polymers benefit from well-established production processes and economies of scale, whereas recycled polymers must overcome challenges related to feedstock variability and contamination. Mechanical recycling, for example, has traditionally been the most cost-effective one due to its lower capital expenditure and energy consumption yet, its limitations in handling mixed or contaminated waste streams, along with degradation in polymer quality, can reduce the market value of the recyclate. In contrast, the chemical recycling methods such as glycolysis, hydrolysis offer a potential for recovering monomers with near-virgin quality, making them attractive for high-value applications. However, these processes typically require more intensive energy inputs, sophisticated catalysts, and complex separation steps, which collectively drive up both operating costs and capital investments (Kumar et al. 2025; Nosrati-Ghods et al. 2025).

The economic analysis of polymer recycling must take into account both fixed and variable costs. Capital expenditures (CAPEX) include investments in specialized reactors, separation equipment, and pre-treatment facilities. Operating expenditures (OPEX) are largely influenced by energy consumption, catalyst replenishment, labor, and maintenance costs (Kumar et al. 2025). For instance, studies have shown that even small improvements in catalyst efficiency or reaction conditions can significantly lower energy requirements and, consequently, operational expenses. Detailed techno-economic analyses (TEA) help identify these critical cost hotspots and guide process optimization. By coupling TEA with life cycle analysis (LCA), researchers and industry professionals can assess not only the economic viability but also the environmental impact of different recycling technologies, thereby ensuring that cost reductions do not come at the expense of increased environmental burdens(Clark and Shaver 2024).

Feedstock quality and availability are additional major determinants of process economics. High-quality, homogenous waste streams enable more predictable and efficient recycling operations, whereas heterogeneous and contaminated streams may necessitate costly pre-treatment and sorting processes. The cost-effectiveness of a recycling process is therefore often region-specific, depending on the local waste management infrastructure and the regulatory environment. Moreover, economies of scale play a vital role. Larger, integrated recycling facilities can leverage higher throughput and lower per-unit costs, which is essential for competing with the lower cost of virgin polymers. However, scaling up remains a challenge for many chemical recycling technologies that are still in the pilot phase, where uncertainties in feedstock supply, process yields, and product quality can deter large-scale investments (Mastellone 2019).

Market drivers and policy frameworks also significantly influence the economic feasibility of polymer recycling. Governments worldwide are beginning to impose stricter regulations on plastic waste disposal and offer incentives for recycling and upcycling initiatives. Subsidies, tax credits, and favourable procurement policies can lower the financial risks associated with investing in recycling infrastructure.

Furthermore, consumer demand for the sustainable products is gradually increasing the willingness to pay a premium for recycled content, which can enhance the revenue potential of recycled polymers. This trend is particularly notable in industries such as packaging and textiles, where high-quality recycled materials are increasingly sought after (Jiang and Bateer 2025; Kumar et al. 2025).

Integration of advanced technologies, such as microwave-assisted heating and novel catalytic systems, has the potential to shift the economic balance in favour of chemical recycling. For example, emerging research indicates that incorporating renewable energy sources and developing more efficient catalysts can reduce the overall energy intensity of chemical depolymerization processes (Clark and Shaver 2024).

In summary, the cost analysis and economic feasibility of polymer recycling involve a complex interplay of technological, market, and regulatory factors.

#### 10.4.1.2 Case Studies for Lab Scale to Industrial Scale

Plastic wastes represent an increasing pollution problem due to the exponential growth of their presence in the waste stream. Epoxy-based thermosets, which is used widely but lack recyclability. A groundbreaking technological advancement has been introduced that allows epoxy thermosets to be converted into thermoplastics, while also enabling the reuse of reinforcements from composites an unprecedented achievement in the thermoset sector. This recyclable epoxy solution, specifically designed for rotor blades, offers a novel approach for recovering both the matrix and fiber reinforcement in fiber-reinforced composites. Utilizing the proprietary Recyclamine® technology platform, a series of recyclable epoxy formulations have been developed for wet lay-up and infusion techniques, catering to the requirements of both onshore and offshore wind turbine blades, including those of the next generation. These systems have been assessed for processing and performance characteristics and have demonstrated notable improvements over traditional, non-recyclable epoxy systems. The composite recycling process has been successfully validated using a low-energy solvolysis method, which enables the recovery of reinforcement fibers and transformation of the epoxy matrix into a reusable thermoplastic. These reclaimed materials have been further utilized to fabricate new composites and thermoplastic components, thereby recovering value from waste and advancing sustainability in the wind energy sector through a circular economy approach (Dubey and Mahanth n.d.; Dubey et al. 2020; Ferrari et al. 2021).

In a study led by José Maria Rodrigues da Luz and colleagues, the potential of *Pleurotus ostreatus* PLO6 to utilize oxo-biodegradable plastics as a carbon and energy source was investigated. Oxo-biodegradable polymers, formulated with pro-oxidant additives, are designed to enhance their degradation under environmental conditions, thereby reducing plastic accumulation in landfills. To simulate environmental weathering, plastic bags made from these polymers were exposed to sunlight for up to 120 days. The resulting fragments were then used as substrates for fungal growth. The researchers found that sunlight exposure alone was insufficient to initiate degradation. However, mechanical alterations and a reduction in titanium dioxide ($TiO_2$) concentration induced by UV exposure played a key role in enabling microbial degradation. This process was further supported by the non-specific activity of lignocellulolytic enzymes and the presence of nitrogen-fixing endosymbiotic microorganisms. The same research team also explored the degradation of green polyethylene (GP), a bio-based plastic derived from renewable sources, aimed at lowering greenhouse gas emissions. Similar to the previous experiment, GP bags were photoaged under sunlight for 120 days and subsequently incubated with *P. ostreatus*. After a 30-day incubation period, notable physical and chemical changes in the polymer matrix were observed, indicating the progression of degradation. The study concluded that pre-exposure of GP to sunlight followed by fungal treatment significantly reduces its half-life and enhances its mineralization potential (da Luz et al. 2014, 2015).

d2w® is a biodegradable plastic technology developed by Symphony Environmental Ltd. It employs the oxo-biodegradable additive that transforms conventional plastics into biodegradable materials at the end of their useful life when exposed to oxygen. In d2w plastic technology-based plastics, the natural oxidation process is accelerated in presence of oxygen the catalyst which reduces the molecular weight of the polymer at a rapid rate, to the point where it is no longer a plastic but has become a waxy substance that can be bio-assimilated by bacteria and fungi in the natural environment (symphony environmental n.d.).

PureCycle Technologies has successfully processed post-consumer carpet waste through its Feedstock Evaluation Unit (FEU), converting it into clear, odourless, ultra-pure polypropylene (UPRP) resin using its proprietary recycling technology originally developed by Procter & Gamble. This achievement demonstrates the scalability of the process and highlights its potential to recover a wide range of waste polypropylene (PP) materials. The resulting UPRP exhibits properties comparable to virgin PP, thereby unlocking significant value from plastic waste streams. This innovation is expected to boost the demand for waste PP and offer sustainable alternatives to conventional disposal methods such as landfilling and export (PureCycle Technologies celebrates successful run of groundbreaking plastics recycling technology 2019). Garbo has implemented a medium scale unit 55–65 ton per day of BHET production, an innovative process technology able to produce chemical recycled PET at very high efficiency Chem PET.

The advantages brought by ChemPET relate to the possibility of using a heterogeneous mix of PET wastes, including dyed PET, which is otherwise unrecyclable at the state of the art, and limits therefore an effective circularization of the entire plastics value chain (*The Chempet Project can be accessed at;* https://garbo.it/en/chempet/, 2023).

RAMPF Eco Solutions, based in Pirmasens, Germany, has developed innovative chemical recycling processes that allow various materials—including PET, PSA, other polyesters such as PLA, PC, and PHB, as well as renewable or bio-based resources like vegetable oils to be utilized as feedstock for the production of high-quality recycled polyols.('RAMPF—Holistic Solutions For Polyurethane Upcycling, https://www.rampf-group.com/en/news/2024/rampf-holistic-solutions-for-polyurethane-upcycling/' 2024).

## 10.5 Sustainability Assessment

### 10.5.1 Lifecycle Assessment

The need for sustainability in the production, consumption, and overall life cycle of plastics has led to several systematic studies being done to assess the impact of plastics on the environment. Life cycle assessment (LCA) is the quantitative analytical method of evaluating the environmental aspects of such products at each stage of their life cycle. It has four phases (see Fig. 10.14: (i) defining the goal and scope of the LCA, (ii) life cycle inventory analysis (LCI), (iii) life cycle impact assessment (LCIA), and (iv) life cycle interpretation. The guidelines and recommendations for LCA are described in the ISO standards (e.g., ISO 14040:2006 and ISO 14044:2006 guidelines).

Performing LCA for bioplastics is essential as it paves the way for strategic future planning, policy making, and further research and development. It is useful especially in identifying more environmentally sustainable options, processes, and materials. This is possible because LCA encompasses the evaluation of each stage of the life cycle, spanning from the extraction of raw materials to their transportation, consumption, and disposal. The assessment begins with defining the goal and scope of the study, which includes defining the product and its function (e.g., a bottle used to hold beverages), its functional unit (e.g., each bottle has a capacity of holding 2 L of water), reference flow, and product system boundaries that determine all the unit processes associated with the product. This is followed by the life cycle inventory generation, which involves the collection of data regarding the input of materials and energy and the output (e.g., emissions, products, etc.) in each unit process. This data is normalized per unit mass for every unit process. The life cycle inventory thus sums up all the contributions of the unit processes that fall under the system boundary that quantifies the product's environmental load. Based on the environmental load of the product determined from the inventory analysis, the key

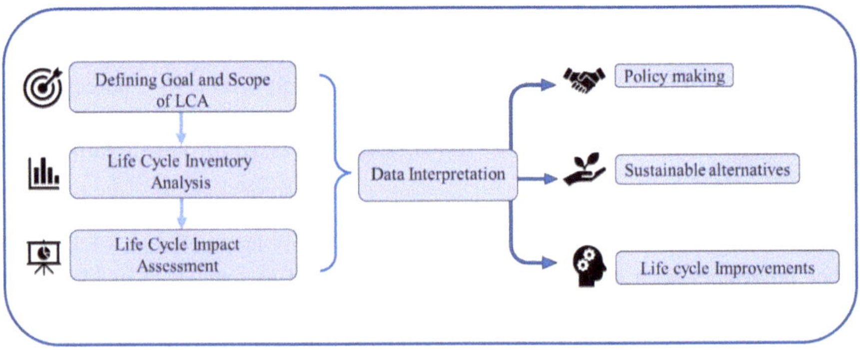


**Fig. 10.14** Summary of the methodology of Lifecycle Assessment (LCA) (Life Cycle Assessment Best Practices of ISO 14040 Series Ministry of Commerce, Industry and Energy Republic of Korea Asia-Pacific Economic Cooperation Committee on Trade and Investment 2004)

impacting categories are determined (such as greenhouse gas emissions (GHG) and ecotoxicity, for example). Finally, the impact of the product's life cycle is assessed and compared with other products. Further interpretations are done to identify areas for improvement in the product life cycle, and recommendations are made for sustainable alternatives. It also aids in policy making (Life Cycle Assessment Best Practices of ISO 14040 Series Ministry of Commerce, Industry and Energy Republic of Korea Asia-Pacific Economic Cooperation Committee on Trade and Investment 2004; Jiao et al. 2024).

For instance, GHG (Greenhouse Gas) emissions can be considered as a key impacting category, and different bioplastics and the processes involved in their lifecycle can be subjected to LCA (Van Roijen and Miller 2022). The use of biogenic resources to generate bioplastics has been estimated to reduce GHG emissions by up to 225% compared to conventional plastics. The cultivation of feedstock for such bioplastics acts as a carbon sink, owing to $CO_2$ uptake during cultivation. This results in the reduction of emissions in the cradle-to-gate LCA of such products (Van Roijen and Miller 2022). As compared to conventional plastics, the end-of-life (EoL) pathway of biodegradable bioplastics is a significant contributor to emissions due to biodegradation. This is especially significant for bioplastics that degrade completely in landfills. Therefore, ineffective disposal methods in landfills can lead to significant GHG emissions like methane.

The LCAs of various EoL pathways for bioplastics also depend on the assumptions made during the assessment. For instance, the highest GHG emissions were obtained for PLA disposal in landfills by Papong et al., assuming complete biodegradation without any provision for capture of the released methane (Papong et al. 2014). The assumed degree of biodegradation significantly impacts the GHG emissions, which ranged from 1.7 to 4.5 kg $CO_2$-eq/kg for a range of 0% to 85% extent of biodegradation. The lowest emissions for PLA disposal in landfills were obtained by Posen et al. by assuming 0% biodegradation and the emissions arising only from the transportation of waste (Posen et al. 2016).

As compared to other EoL options, LCA studies indicate that thermomechanical recycling of plastics can have 100–130% reduced GHG emissions. However, thermomechanical recycling of bioplastics offers only modest benefits as compared to conventional plastics due to the deterioration of the material during the recycling process due to ageing that happens with exposure to temperature and moisture. This ageing is more pronounced in bioplastics like PLA that have a greater biogenic carbon content.

Incineration, which is a more preferred EoL option as compared to landfills, as per the waste management hierarchy, contributes approximately 1 kg $CO_2$ -eq/kg for PLA, 1.6 kg $CO_2$ -eq/kg for polyhydroxybutyrate, and 1.4 kg $CO_2$-eq/kg for thermoplastic starch. These values of emissions reported in literature can be correlated to the stoichiometric carbon content of the respective bioplastics. For bioplastic waste that is incinerated, the $CO_2$ emissions can be considered net-neutral as compared to petroleum-based plastics. It is also imperative to analyze the impact of incineration as an EoL option in terms of human health and ecosystem damage, which are discussed in Sect. 10.3.4.2.

Composting as an EoL pathway for bioplastics generally yields higher overall GHG emissions as compared to other methods since composting emissions are usually not captured. The emissions vary with the assumptions made, e.g., the emissions from biodegradation when assumed as net zero for PLA lead to lower values per kg of PLA [243].

Different outputs like excess electricity generation, fertilizers from the digestate, etc., from the anaerobic digestion of PLA, TPS, and PHBV (poly(3-hydroxybutyrate-co-3-hydroxyvalerate)) can be considered as credits that can be removed from the overall emissions, resulting in a nearly 55% drop from 1.47 kg $CO_2$-eq/kg to 0.67 kg $CO_2$-eq/kg (Hermann et al. 2011). The potential utilization of the outputs of anaerobic digestion in meeting the energy demands of the process and other value-added products can have a major impact on their sustainability assessment.

Significant variability also arises based on the scope of the LCA being done (e.g., cradle-to-grave or cradle-to-gate analysis). A comparative study on the cradle-to-grave life cycle assessment of bio-based polyethylene terephthalate (PET) bottles and petrochemical PET bottles was done considering several environmental impact categories (Vural Gursel et al. 2021). Bio-based PET bottles produced using Brazilian sugarcane as the source of the ethylene glycol (EG) part of the polymer were considered as baseline while comparing other alternative hypothetical product systems that use European wheat straw and a mix consisting of maize, wheat, and sugar beet instead of Brazilian sugarcane. Bio-based PET is often referred to as a "drop-in" product that is chemically identical to its petrochemical-derived analogue. The difference in bioPET lies in the use of bio-MEG (monoethylene glycol) in place of petrochemical MEG. However, 70% by weight of the polymer, which comprises of PTA (purified terephthalic acid), is still sourced from petrochemicals in both bio-based PET and petrochemical PET. Therefore, both of these products have identical EoL pathways. The major difference in their environmental impacts would be owing to raw-material acquisition and processing, production, and transportation.

However, it is important to note that the scope of assessment, assumptions made for the LCA, and the key impacting categories chosen have a significant impact on the results. It is, therefore, imperative to identify the research gaps present in the existing LCAs and conduct further studies to harness the complete potential of bioplastics as a sustainable option (Vural Gursel et al. 2021; Van Roijen and Miller 2022).

### 10.5.2 Economic Viability

In recent times, bio-based plastic alternatives have become available for almost every application of conventional, petrochemical-based plastics. It has been estimated that the global market for bioplastics will increase from 2431.9 kilotons, as of 2024, to 5634.6 kilotons by 2029, with a compound annual growth rate (CAGR) of 18.3% (BCC Research Report Overview Global Markets and Technologies for Bioplastics 2025). "Drop-in plastics" such as bioPE and bioPET are anticipated to encounter the largest growth in demand and capacity, along with cost-competitive ones such as PLA blends with large-scale facilities already existing. This may be because "drop-in" plastics can be recycled in the same recycling facilities as their petrochemical counterparts (Fredi and Dorigato 2021). However, the higher cost of production of bioplastics as compared to fossil-based plastics is a barrier to its commercial and economic viability. Moreover, the lower production costs combined with cheaper fossil feedstock make the prices of conventional plastics much lower than bioplastics. The prices of oil are expected to remain competitive with renewable resources owing to technological advances like fracking and horizontal deep drilling that enable oil harvesting even from remote reservoirs. Hence, most of the alternatives to fossil-derived polymers are more expensive.

Due to the insufficient capability of recycling or biodegradation, most bioplastics also end up in landfills or incinerators. This would lead to the loss of several value-added products, monomers for the production of further bioplastics, etc., that could otherwise have been recycled or valorized. For instance, PLA recycling or biodegradation facilities are currently inadequate, leading to poorer EoL pathways (BCC Research Report Overview Global Markets and Technologies for Bioplastics 2025; Rosenboom et al. 2022).

Two interesting approaches to address the economic viability of bio-based plastics are implementing restrictive regulations against certain unfavorable plastic types (e.g., single-use plastic) and incentivizing policies in favour of bioplastics and environmentally sound EoL options like recycling. Establishing a balance between material properties desirable for specific functions and ensuring a sustainable life cycle of bioplastics would pave the way for replacing conventional plastics in potential markets such as food packaging and fast-moving consumer goods.

Just like the sustainability and environmental impacts of a product are analyzed using tools like the LCA, the economic feasibility can be inspected and validated by performing a techno-economic analysis (TEA). It helps in forming a framework that helps to estimate the commercial feasibility and cost-effectiveness of industrial plants or processes by considering economic characteristics like feedstock expenditure, operational costs, and capital. It is also employed to design protocols and simulations with the help of practical databases to evaluate mass and energy balances (Sangtani et al. 2023).

Employing stricter regulations on conventional and single-use plastics, while emphasizing carbon-footprint reduction through government policies, will drive the global bioplastics market towards a circular economy (discussed in Sect. 10.6). Corporate initiatives towards sustainability by reducing fossil fuel dependency and emphasizing renewable resources further drive global plastic manufacturers to shift towards the bioplastics market. Such an environment proves favourable for the expansion of the bioplastics market (BCC Research Report Overview Global Markets and Technologies for Bioplastics 2025).

## 10.6 Policies and Regulations

In order to tackle the growing concern of plastic waste, governments and countries across the globe have sought to find a solution by implementing various regulations and policies. The principle of circular economy has been adopted by several nations to combat plastic waste. Some global initiatives, such as the New Plastics Economy, have come to the forefront with a view to promote plastic recycling using this framework. "European Strategy for Plastics in a Circular Economy" issued by the European Commission requires plastic bottle manufacturing companies to produce bottles containing 25% recyclable plastic by the year 2025 and 90% by 2029 ('Plastic waste and recycling in the EU: facts and figures I Topics. (2024, June 25). European Parliament. Retrieved April 12, 2025, from https://www.europarl.europa.eu/topics/en/article/20181212STO21610/plastic-waste-and-recycling-in-the-eu-facts-and-figures', n.d.). Countries such as India, Germany, France, and Belgium have implemented Extended Producer Responsibility (EPR) regulations (Tumu et al. 2023). A tax of €217.85 per tonne on plastic packaging with less than 30% recycled content has been enforced by the UK (HM Revenue and Customs 2021), while the Australian Packaging Covenant Organisation (APCO) works with the Australian Government with the aim to have 70% of plastic packaging made from recycled materials (Australian Packaging Covenant Organisation n.d.). Worldwide, regulations have been made to reduce the use of plastic. The following are some noteworthy policies across the globe.

### 10.6.1 EU Circular Economy Action Plan

The "Circular Economy Action Plan" was launched by the EU in March 2020 with the aim of achieving carbon neutrality goals within the next 30 years. It prioritizes safe disposal and recycling of plastics and chemicals. The plan aims to encourage sustainable consumption by the public and industries to reduce usage of natural resources and prevent the loss of biodiversity. This plan encompasses global agreements as well as legislative and non-legislative provisions in order to curb the vicious life cycle of plastics. CEAP has implemented various policies with an approach to reduce usage of plastic bags, single use plastics, and microplastics generated using plastic products (THE EUROPEAN COMMISSION 2020).

The CEAP has adopted several methods to curb the generation of microplastics. For example, tyres produce TWPs (tyre wear particles) that are formed as a result of friction and abrasion in poorly maintained tyres or at high speeds. By controlling chemical composition of tyres, regulating lower speed limits and encouraging use of devices that capture TWPs, CEAP aims to reduce the contribution of tyres in the generation of microplastics. Similarly, by strictly regulating paint compositions, accounting for pellet losses at every stage of its life cycle, the CEAP plans to eliminate these major microplastic sources (Author: Vivienne Halleux 2025).

The directives that have been laid focus on nullifying the effects caused by certain plastic products on the environment and human health. It also promotes the transition to a circular economy by proposing innovative ideas for sustainable use of plastic. The guidelines consist of reusable alternatives to single use plastics along with spreading awareness among individuals. It also lays down penalties and brings a substantial reduction in use of oxo degradable plastic ('DIRECTIVE (EU) 2019/904 OF THE EUROPEAN PARLIAMENT AND OF THE COUNCIL of 5

June 2019 on the reduction of the impact of certain plastic products on the environment' 2019).

### 10.6.2 Basel Convention for Plastic Trade

The Basel Convention is the only global legal agreement that deals with the transboundary movement of hazardous waste, including plastic waste. Till date, 191 countries are a part of this Convention, which aims to improve the management of plastic disposal in a sustainable manner and thereby significantly reducing the generation of microplastics in the environment (*Basel Convention on the Control of Transboundary Movements of Hazardous Wastes and their Disposal* n.d.). The collaborative effort of parties is to facilitate changes in the disposal manner taking place at regional, national, and global levels. Some initiatives taken at the regional level include schools involved in beach cleaning activities and repairing old fishing nets instead of discarding them. It also ensures that the country receiving the waste has an environmentally sound management of hazardous waste.

In order to export wastes falling under the categories of Annex II (plastics which are difficult to recycle or which require special considerations) and Annex VIII (hazardous plastic waste), a Prior Informed Consent (PIC) has to be granted by the importing country granting a permit. Waste categorized under Annex IX (which includes plastic waste that is nearly contamination-free and non-hazardous if recycled in an environmentally responsible way, such as low-density polyethylene and other plastics used in packaging) can be moved among parties without a PIC for recycling (*Basel Convention on the Control of Transboundary Movements of Hazardous Wastes and their Disposal* n.d.).

The stages for issuing a PIC include:

1. Notification by the exporting country
2. Consent and issuance of a movement document
3. Tran boundary movement
4. Confirmation of disposal

The implementation of this framework is expected to have a positive impact on the plastic disposal and recycling.

The Basel Convention aims to prevent and minimize waste by designing sustainable alternatives and eliminating the use of toxic additives. It also ensures transparency in the transboundary movements by enforcing a legally binding framework. The development of infrastructures for collecting, recycling, and disposing plastic waste through this legislation is another positive impact of this Convention.

With a view of enhancing the Parties' capabilities in managing hazardous waste and other waste, the Basel Convention provides various tools and resources such as a framework for Environmentally Sound Management (ESM) that highlights the strategies for an effective implementation of ESM and ESM toolkits which offers a concise guidance on management of hazardous waste and other types of waste(*Basel*

*Convention on the Control of Transboundary Movements of Hazardous Wastes and their Disposal* n.d.).

### 10.6.3 High Ambition Coalition to End Plastic Pollution (Aug 2022)

Seventy countries have taken part in this initiative headed by Rwanda and Norway. The HAC aims to end plastic pollution by 2040 by employing three strategies:

1. *Confining the manufacture and consumption of plastics to sustainable levels*
2. *Facilitate the development of a circular economy for plastics that safeguards the environment and human health*
3. *Manage and recycle plastic wastes without harming the environment.*

The above is achieved by a set of directives issued by the HAC. They include the complete banning of certain types of plastics while restricting the use of some, controlling plastic compositions and ensuring complete transparency in what goes in the plastics, complete documentation of every stage of plastic lifecycle to effectively eliminate the generation of waste at each stage, and definition of global standards and goals for sustainability across the entire plastic lifecycle (*High Ambition Coalition to End Plastic Pollution* n.d.; International Cooperation on Plastic Pollution I Plastics and the Environment Series 2025).

### 10.6.4 National Action Plan for Marine Plastic Waste (2024–26)

Microplastic of terrestrial origin often end up in water bodies along with rain drainage and sewage wastewater. Indiscriminate use of plastic and its improper disposal is the major cause of this growing concern (Microplastics 2023). Abandoned, lost or discarded fishing gear (ALDFG), which includes fishing nets and lines, form a major percentage of marine plastic waste (Macfadyen n.d.).

The objectives of the National Action Plan (NAP) are to prevent, reduce, and control sea-based marine plastic litter (SBMPL) in the country, especially in the shipping and fisheries sectors in India ('NATIONAL ACTION PLAN ON MARINE PLASTIC LITTER FROM SEA-BASED SOURCES' 2024).

Some important actions of NAP ('NATIONAL ACTION PLAN ON MARINE PLASTIC LITTER FROM SEA-BASED SOURCES' 2024):

1. Legal policies enforcement reforms: effective and stringent enforcement of existing rules and regulations, stringent marking and tracking of fishing gears and development of databases to monitor the same.
2. The plan provides guidelines associated with waste collection and processing.
3. The National Task Force, composed of government officials, scientists and volunteers, ensures the effective implementation of the plan.

4. As per NAP 2024, Government-sponsored incentive programs are administered through the State Fisheries Department for the collection and return of damaged, irreparable nets and ALDFG as well as the recycling of marine debris.
5. R&D: Plastic alternatives for fishing gears and conducting hackathons in schools and colleges to encourage innovative solutions.
6. Conduct training on the laid guidelines and on methods to implement effective collection of marine waste.
7. The plan also encourages private organizations and startups that specialize in marine waste collection and its recycling.

### 10.6.5 Plastic Waste Management Rules, 2016 in India

Plastic Waste Management Rules, 2016 and its amendments provide guidelines for import, manufacturing, sale, and distribution of plastic sheets and carry bags. It provides guidelines on the addition of pigments to single use plastic products, allowed thickness, etc. It also mandates the need of valid registration for purchase and sale of plastic as raw material ('Plastic Waste Management (Amendment) Rules, 2021' 2021). Non-recyclable waste should be used for energy recovery, while recyclable plastics should be recycled only by registered plastic waste recyclers.

Local bodies are responsible for setting up infrastructure collection, segregation, processing, and disposal of plastic waste, generating public participation in the form of cleanliness drives, and creating awareness. Waste generating organizations must ensure segregation and storage of waste at source itself in accordance with the Solid Waste Management Rules, 2000. The Rules mandate that such organizations must minimize waste generation, and hand over segregated waste to waste processing facilities. Individuals organizing an event in open spaces, which involves service of food stuff in plastic or multi-layered packaging, shall segregate and manage the waste generated and hand it to the concerned authority. Retailers and vendors can face penalties if they sell plastic sheets and carry bags which are not manufactured or labelled as per guidelines.

### 10.6.6 Extended Producer Responsibility (EPR) in India

Extended Producer Responsibility is applied to the organizations that produce or import plastics or are brand owners of plastic products. It mandates the collection and recycling of plastic wastes by such organizations. Plastic producers are required to minimise production of plastics and implement methods of segregation and recycling of plastics. Local bodies, Central and State Pollution Control Boards are responsible for mandating EPR on plastic waste generators as well as on retailers and street vendors to curb plastic waste (Guidelines on Extended Producer Responsibility for Plastic Packaging 2022).

Promoting environmentally friendly packaging practices and gradually replacing plastic packaging are the primary goals of EPR. The obligation to EPR by an entity ends once it uses 100% biodegradable material without having any adverse health effects and environmental hazards and produces no chemical residue or microplastics.

### 10.6.7 Incentives for Sustainable Recycling

By using incentives, governments can promote the reuse and recycling of plastic waste. Many countries use strategies wherein an individual may receive money or rewards for recycling plastic waste(Recycling Incentive Scheme 2023). Vouchers and/or reduction in waste collection fees on waste segregation by housing societies empower sustainable recycling at a local level. Governments may also support private companies, non-governmental organizations and local bodies in the collection, segregation, treatment, and disposal of solid and liquid waste through income tax exemptions and financial assistance.

Initiatives have been taken to support the establishment of recycling plants and startups that specialize in treating discarded fishing gear and coastal plastic waste by providing one-time grants.

In India, the project "RePLAN" (Reducing Plastic in Nature) was undertaken in rural areas by the Khadi and Village Industries Commission to upcycle single-use plastics in the form of handicrafts, which further generated employment and business opportunities (*KVIC's Unique Plastic-Mixed Handmade Paper Gets Patent for Invention to Recycle Waste Plastic* 2021).

In Hong Kong, individuals can submit their recyclable waste in specific recycling facilities. 9680 tonnes of plastic waste was collected in 2023. With the GREEN$ mobile app, on submitting waste, the individual receives GREEN$ points. These points can be exchanged for rewards, groceries, and other incentives (*GREEN@ COMMUNITY* n.d.).

Operating in developing countries, the Plastic Bank collects plastics and sends them to processing centres where plastic waste is processed and upcycled. The bank operates through membership by organizations/companies who need to fulfil their EPR requirements. The volunteers of the Bank collect and recycle the amount of plastic required by the organizations (*About Plastic Bank* n.d.).

### 10.6.8 Barriers to Implementation

Lack of awareness and sense of responsibility among the public is a major barrier to implementation of plastic waste management regulations. The lack of infrastructure for plastic recycling and processing burdens organizations that wish to recycle and process waste but are unable to (Kibria et al. 2023). Developing countries face a major challenge in meeting sustainability and recycling goals. Collection and processing of waste from coastlines and oceans is a laborious procedure (Debnath et al. 2023).

Mixed plastics and additives make recycling complicated. The fear of being penalized may result in companies manipulating records, as traceability of plastic waste is poor in developing countries, and indulging in illegal waste dumping.

A gap in knowledge and technical skills exists and the logistics of such training at local levels is poor. High energy and resource consumption make recycling processes costlier than manufacturing new/virgin plastic products. Greenwashing is a major strategy used by companies in manipulating the public to increase their plastic consumption (Zhu and Wang 2020). Recycling plastics is not the solution; rather, stopping its consumption and reducing its manufacture may be the key for mitigating the problem of plastic waste [22].

## 10.7 Future Perspectives

The future of plastic recycling would go through significant advancements driven by technological innovations, regulatory frameworks, and the growing global commitment to sustainability. Several key areas promise to improve recycling efficiency is to expand the scope of recyclable materials and integrate sustainable alternatives into mainstream applications.

The lack of a universal recycling systems which makes it necessary to segregate the polymer in order to undergo recycling through a specific method. Automation coupled with artificial intelligence (AI) can revolutionize waste sorting and recycling. AI-powered robotic sorting systems can identify and separate different types of plastics with high accuracy, reducing contamination and increasing recycling efficiency. Machine learning algorithms can optimize sorting by analysing waste patterns and improving classification over time. Integrating sensor-based technologies, such as near-infrared (NIR) spectroscopy and hyperspectral imaging, can enhance the precision of material identification, enabling better quality control in recycled products obtained.

It is necessary to study raw materials from agricultural produces and waste for their potential in polymerization as India being an agricultural nation. Biopolymers derived from renewable sources offer a promising alternative to conventional synthetic plastics. Materials such as polylactic acid (PLA), polyhydroxyalkanoates (PHA), and starch-based polymers exhibit biodegradability and reduced environmental impact. However, challenges remain in their mechanical performance, scalability, and cost competitiveness. Research is focused on enhancing the properties of biopolymers through blending, copolymerization, and nanocomposite development to make them viable substitutes for conventional synthetic plastics.

Scaling up innovative recycling technologies from laboratory to industrial levels remains a significant challenge. High initial investment costs, lack of standardized processing techniques, and varying feedstock quality hinder the widespread adoption of advanced recycling methods. Addressing these issues requires industry collaboration, investment in pilot-scale facilities, and government incentives to support the transition towards large-scale, sustainable recycling systems.

For enhancing the efficiency and economic feasibility of recycling, process optimization plays a crucial role. Advanced catalyst development, reaction condition optimization, and integration of hybrid recycling approaches can improve the yield and quality of recycled materials. For instance, combining chemical and enzymatic recycling can increase the breakdown efficiency of complex polymers while minimizing energy consumption. Data-driven approaches, including computational modelling and machine learning, can further refine recycling processes and enhance material recovery rates.

A universal recycling technology capable of processing diverse plastic types without requiring extensive sorting is a long-term goal for the industry. Innovations such as solvent-based recycling and depolymerization techniques are being explored to handle mixed plastic waste streams efficiently. Additionally, modular recycling units adaptable to different waste compositions could enable decentralized recycling, reducing transportation costs and carbon footprint. Future developments in universal recycling solutions will require interdisciplinary research, investment in infrastructure, and collaboration across stakeholders to achieve a truly circular economy.

The future of plastic recycling will be shaped by technological advancements, policy interventions, and societal commitment to sustainability. By addressing existing challenges and leveraging innovative approaches, the transition towards a circular bioeconomy can be accelerated, minimizing plastic waste and maximizing resource efficiency. Blockchain technology can improve traceability and transparency in plastic waste management. By maintaining immutable records of waste collection, sorting, and recycling, blockchain can help track recycled content in products, verify sustainability claims, and incentivize responsible waste disposal through tokenized reward systems. A hybrid approach that combines mechanical and chemical recycling could enhance overall recycling efficiency. Mechanical recycling can be used for less degraded plastics, while chemical recycling can handle contaminated or complex polymer structures, increasing the overall circularity of plastic materials. The development of smart packaging with embedded indicators for degradation, freshness, or recyclability can improve waste management. Additionally, advancements in biodegradable polymers tailored for specific environments (marine, soil, compost) could help mitigate long-term plastic pollution. Despite technological advancements, consumer participation remains crucial for effective recycling. Large-scale awareness campaigns, educational programs, and behavioural nudges such as deposit-return schemes can encourage responsible disposal and increase recycling rates.

## References

Abedsoltan H (2023) A focused review on recycling and hydrolysis techniques of polyethylene terephthalate. Polym Eng Sci 63(9):2651–2674. https://doi.org/10.1002/pen.26406

Abedsoltan H (2024) Concentrated sulfuric acid as a catalyst for chemical recycling of polycarbonate in water. Waste Biomass Valorization 15(5):2793–2806. https://doi.org/10.1007/s12649-023-02326-x

About Plastic Bank (n.d.)
Adamu AA, Muhamad Sarih N, Gan SN (2021) Thermal and anticorrosion properties of polyurethane coatings derived from recycled polyethylene terephthalate and palm olein-based polyols. Royal Soc Open Sci 8(4):rsos.201087. https://doi.org/10.1098/rsos.201087
Aksu Y, Haykiri-Acma H, Yaman S (2025) Recycle of flexible polyurethane foam by acidolysis and reuse of recovered polyol. J Polym Environ 33(2):1147–1158. https://doi.org/10.1007/s10924-024-03467-3
Alassali A et al (2021) Towards higher quality of recycled plastics: limitations from the material's perspective. Sustainability 13(23):13266. https://doi.org/10.3390/su132313266
Alberti C et al (2019) Chemical recycling of end-of-life polyamide 6 via ring closing depolymerization. ChemistrySelect 4(43):12638–12642. https://doi.org/10.1002/slct.201903970
Ali SS et al (2023) Bioplastic production in terms of life cycle assessment: a state-of-the-art review. Environ Sci Ecotechnol 15:100254. https://doi.org/10.1016/j.ese.2023.100254
Al-Salem SM, Lettieri P, Baeyens J (2009) Recycling and recovery routes of plastic solid waste (PSW): a review. Waste Manag 29(10):2625–2643. https://doi.org/10.1016/j.wasman.2009.06.004
Amobonye A et al (2021) Plastic biodegradation: frontline microbes and their enzymes. Sci Total Environ 759:143536. https://doi.org/10.1016/j.scitotenv.2020.143536
Andrady AL (2011) Microplastics in the marine environment. Mar Pollut Bull 62(8):1596–1605. https://doi.org/10.1016/j.marpolbul.2011.05.030
Antonakou EV, Achilias DS (2013) Recent advances in polycarbonate recycling: a review of degradation methods and their mechanisms. Waste Biomass Valorization 4(1):9–21. https://doi.org/10.1007/s12649-012-9159-x
Asahi N et al (2004) Methanolysis investigation of commercially available polyurethane foam. Polym Degrad Stab 86(1):147–151. https://doi.org/10.1016/j.polymdegradstab.2004.04.002
Australian Packaging Covenant Organisation (n.d.) About APCO
Author: Vivienne Halleux (2025) Reducing microplastic pollution from plastic pellet losses
Awasthi S et al (2017) Biodegradation of thermally treated low density polyethylene by fungus Rhizopus oryzae NS 5. 3 Biotech 7(1):73. https://doi.org/10.1007/s13205-017-0699-4
Babu RP, O'Connor K, Seeram R (2013) 'Current progress on bio-based polymers and their future trends. Progress Biomaterials 2(1):8. https://doi.org/10.1186/2194-0517-2-8
Shilpa, Basak N, Meena SS (2022) Microbial biodegradation of plastics: challenges, opportunities, and a critical perspective. Front Environ Sci Eng 16(12):161. https://doi.org/10.1007/s11783-022-1596-6
Basel Convention on the Control of Transboundary Movements of Hazardous Wastes and their Disposal (n.d.)
BCC Research Report Overview Global Markets and Technologies for Bioplastics (2025)
Beltrán FR et al (2021) Evaluation of the technical viability of distributed mechanical recycling of PLA 3D printing wastes. Polymers 13(8):1247. https://doi.org/10.3390/polym13081247
Beran R et al (2021) Wood adhesives from waste-free recycling depolymerisation of flexible polyurethane foams. J Clean Prod 305:127142. https://doi.org/10.1016/j.jclepro.2021.127142
Bergmann M et al (2019) White and wonderful? Microplastics prevail in snow from the Alps to the Arctic. Sci Adv 5(8). https://doi.org/10.1126/sciadv.aax1157
'Beyler_Hirschler_SFPE_Handbook_3' (n.d.)
Bharadwaj C et al (2024) A review on sustainable PET recycling: strategies and trends. Mater Today Sustain 27:100936. https://doi.org/10.1016/j.mtsust.2024.100936
Brette MM et al (2023) Pure hydrolysis of polyamides: a comparative study. Chemistry 6(1):13–50. https://doi.org/10.3390/chemistry6010002
Češarek U, Pahovnik D, Žagar E (2020) Chemical recycling of aliphatic polyamides by microwave-assisted hydrolysis for efficient monomer recovery. ACS Sustain Chem Eng 8(43):16274–16282. https://doi.org/10.1021/acssuschemeng.0c05706
Cevher D, Sürdem S (2021) Polyurethane adhesive based on polyol monomers BHET and BHETA depolymerised from PET waste. Int J Adhes Adhes 105:102799. https://doi.org/10.1016/j.ijadhadh.2020.102799

Chan K, Zinchenko A (2021) Conversion of waste bottles' PET to a hydrogel adsorbent via PET aminolysis. J Environ Chem Eng 9(5):106129. https://doi.org/10.1016/j.jece.2021.106129

Chan K et al (2023) Copper ion removal using a waste-plastic-derived hydrogel adsorbent prepared via microwave-assisted PET Aminolysis. Gels 9(11):874. https://doi.org/10.3390/gels9110874

Chanda M (2021) Chemical aspects of polymer recycling. Adv Indust Eng Polym Res 4(3):133–150. https://doi.org/10.1016/j.aiepr.2021.06.002

Chang C et al (2025) Glycolysis of waste polyurethane foam for preparing recycled polymer materials: optimization and techno-economic evaluation. Chem Eng J 507:160583. https://doi.org/10.1016/j.cej.2025.160583

Chaudhari MI et al (2018) Assessment of simple models for molecular simulation of ethylene carbonate and propylene carbonate as solvents for electrolyte solutions, pp 53–77. Available at: https://doi.org/10.1007/978-3-030-00593-1_3

Chen Y, Selvinsimpson S (2022) Current trends, challenges, and opportunities for plastic recycling. In: Plastic and microplastic in the environment. Wiley, pp 205–221. https://doi.org/10.1002/9781119800897.ch11

Chen J et al (2010) Catalytic hydrolysis of waste nylon 6 to produce ε-caprolactam in sub-critical water. J Anal Appl Pyrolysis 87(1):50–55. https://doi.org/10.1016/j.jaap.2009.10.004

Chen W et al (2021) Biomass-derived γ-valerolactone: efficient dissolution and accelerated alkaline hydrolysis of polyethylene terephthalate. Green Chem 23(11):4065–4073. https://doi.org/10.1039/D1GC00665G

Chen H et al (2025) Full recovery of polyimide wastes into high-value products through break and reconstruction of imide ring. Advan Sci [preprint]. Available at: https://doi.org/10.1002/advs.202414416

Chidambarampadmavathy K, Karthikeyan OP, Heimann K (2017) Sustainable bio-plastic production through landfill methane recycling. Renew Sust Energ Rev 71:555–562. https://doi.org/10.1016/j.rser.2016.12.083

Chiemchaisri C, Visvanathan C (2008) Greenhouse gas emission potential of the municipal solid waste disposal sites in Thailand. J Air Waste Manage Assoc 58(5):629–635. https://doi.org/10.3155/1047-3289.58.5.629

Choi J et al (2024) Recent advances in microbial and enzymatic engineering for the biodegradation of micro- and nanoplastics. RSC Adv 14(14):9943–9966. https://doi.org/10.1039/D4RA00844H

Chuayjuljit S, Norakankorn C, Pimpan V (2002) Chemical recycling of rigid polyurethane foam scrap via base catalyzed aminolysis. J Metals Mater Minerals

Ciuffi B, Fratini E, Rosi L (2024) Plastic pretreatment: the key for efficient enzymatic and biodegradation processes. Polym Degrad Stab 222:110698. https://doi.org/10.1016/j.polymdegradstab.2024.110698

Clark RA, Shaver MP (2024) Depolymerization within a circular plastics system. Chem Rev 124(5):2617–2650. https://doi.org/10.1021/acs.chemrev.3c00739

Collignon A et al (2014) Annual variation in neustonic micro- and meso-plastic particles and zooplankton in the Bay of Calvi (Mediterranean–Corsica). Mar Pollut Bull 79(1–2):293–298. https://doi.org/10.1016/j.marpolbul.2013.11.023

Čolnik M, Knez Ž, Škerget M (2021) Sub- and supercritical water for chemical recycling of polyethylene terephthalate waste. Chem Eng Sci 233:116389. https://doi.org/10.1016/j.ces.2020.116389

Čorak I et al (2022) Sustainable alkaline hydrolysis of polyester fabric at low temperature. Materials 15(4):1530. https://doi.org/10.3390/ma15041530

Crystal Thew XE et al (2024) Enhancing plastic biodegradation process: strategies and opportunities. Crit Rev Biotechnol 44(3):477–494. https://doi.org/10.1080/07388551.2023.2170861

Cywar RM et al (2021) Bio-based polymers with performance-advantaged properties. Nat Rev Mater 7(2):83–103. https://doi.org/10.1038/s41578-021-00363-3

da Luz JMR et al (2014) Abiotic and biotic degradation of Oxo-biodegradable plastic bags by Pleurotus ostreatus. PLoS One 9(11):e107438. https://doi.org/10.1371/journal.pone.0107438

da Luz JMR et al (2015) Degradation of green polyethylene by Pleurotus ostreatus. PLoS One 10(6):e0126047. https://doi.org/10.1371/journal.pone.0126047
Datta J, Rohn M (2007) Glycolysis of polyurethane wastes. Part I. Glycolysis agents and catalysts. Polimery 52(07/08):579–582. https://doi.org/10.14314/polimery.2007.579
Datta J et al (2018) A new approach to chemical recycling of polyamide 6.6 and synthesis of polyurethanes with recovered intermediates. J Polym Environ 26(12):4415–4429. https://doi.org/10.1007/s10924-018-1314-4
De Smet D, Vanneste M (2024) Design of Debondable PU coating for degradation on demand. Coatings 14(6):731. https://doi.org/10.3390/coatings14060731
Debnath B et al (2023) Modelling the barriers to sustainable waste management in the plastic-manufacturing industry: an emerging economy perspective. Sustain Analyt Model 3:100017. https://doi.org/10.1016/j.samod.2023.100017
Deirram N, Rahmat AR (2012) Hydrolysis degradation of polycarbonate using different co–solvent under microwave irradiation. APCBEE Procedia 3:172–176. https://doi.org/10.1016/j.apcbee.2012.06.065
Desidery L, Lanotte M (2022) Polymers and plastics: types, properties, and manufacturing. In: Plastic waste for sustainable asphalt roads. Elsevier, pp 3–28. https://doi.org/10.1016/B978-0-323-85789-5.00001-0
DIRECTIVE (EU) 2019/904 OF THE EUROPEAN PARLIAMENT AND OF THE COUNCIL of 5 June 2019 on the reduction of the impact of certain plastic products on the environment (2019) THE EUROPEAN PARLIAMENT AND THE COUNCIL OF THE EUROPEAN UNION [preprint]. Official Journal of the European Union
Dogu O et al. (2021) The chemistry of chemical recycling of solid plastic waste via pyrolysis and gasification: state-of-the-art, challenges, and future directions. In: Progress in Energy and Combustion Science. Elsevier Ltd. https://doi.org/10.1016/j.pecs.2020.100901
Donadini R et al (2023) Chemical recycling of polyurethane waste via a microwave-assisted glycolysis process. ACS Omega 8(5):4655–4666. https://doi.org/10.1021/acsomega.2c06297
Donadini R et al (2024) Closing the loop of polyurethane adhesives: acidolysis process optimization. Int J Adhes Adhes 135:103843. https://doi.org/10.1016/j.ijadhadh.2024.103843
Dou C et al (2023) A hybrid chemical-biological approach can upcycle mixed plastic waste with reduced cost and carbon footprint. One Earth 6(11):1576–1590. https://doi.org/10.1016/j.oneear.2023.10.015
Dubey PK, Mahanth SK (n.d.) Recyclamine ®-Novel Amine Building Blocks for a Sustainable World. Available at: www.epotec.info
Dubey PK et al (2020) Recyclable epoxy systems for rotor blades. IOP Conf Ser: Mater Sci Eng 942(1):012014. https://doi.org/10.1088/1757-899X/942/1/012014
Elidrissi A et al (2007) Synthesis, characterisation, and chemical degradation of segmented polyurethanes with butylamine for chemical recycling. J Appl Polym Sci 105(3):1623–1631. https://doi.org/10.1002/app.26194
Elsawy MA et al (2017) Hydrolytic degradation of polylactic acid (PLA) and its composites. Renew Sust Energ Rev 79:1346–1352. https://doi.org/10.1016/j.rser.2017.05.143
Ferrari F et al (2021) Fully recyclable bio-based epoxy formulations using epoxidized precursors from waste flour: thermal and mechanical characterization. Polymers 13(16):2768. https://doi.org/10.3390/polym13162768
Flores-Castañón N, Sarkar S, Banerjee A (2022) Structural, functional, and molecular docking analyses of microbial cutinase enzymes against polyurethane monomers. J Hazad Mater Lett 3:100063. https://doi.org/10.1016/j.hazl.2022.100063
Fredi G, Dorigato A (2021) Recycling of bioplastic waste: a review. In: Advanced Industrial and Engineering Polymer Research. KeAi Communications Co, pp 159–177. https://doi.org/10.1016/j.aiepr.2021.06.006
Gadhave RV et al (2019) Recycling and disposal methods for polyurethane wastes: a review. Open J Polym Chem 09(02):39–51. https://doi.org/10.4236/ojpchem.2019.92004
Gama N et al (2020) Recycling of polyurethane scraps via acidolysis. Chem Eng J 395:125102. https://doi.org/10.1016/j.cej.2020.125102

Gama N et al (2024) Polyurethane recycling through acidolysis: current status and prospects for the future. J Polym Environ 32(10):4777–4793. https://doi.org/10.1007/s10924-024-03278-6
Gao W et al. (n.d.) Plastics recycling: Using an economic-feasibility lens to select the next moves
Gautam M, Agrawal M (2021) Greenhouse gas emissions from municipal solid waste management: a review of global scenario, pp 123–160. Available at: https://doi.org/10.1007/978-981-15-9577-6_5
Gecol H et al (2004) Use of surfactants to remove water-based inks from plastic film: effect of calcium ion concentration and length of surfactant hydrophobe. Colloid Polym Sci 283(2):154–163. https://doi.org/10.1007/s00396-004-1082-7
Gerlock JL et al (1980) Reaction of polyurethane foam with dry steam: kinetics and mechanism of reactions. J Polym Sci: Polym Chem Ed 18(2):541–557. https://doi.org/10.1002/pol.1980.170180215
Geyer R (2020) Production, use, and fate of synthetic polymers'. In: Plastic waste and recycling. Elsevier, pp 13–32. https://doi.org/10.1016/B978-0-12-817880-5.00002-5
Giyahchi M, Moghimi H (2024) Acceleration a yeast-based biodegradation process of polyethylene terephthalate microplastics by Tween 20: efficiency, by-product analysis, and metabolic pathway prediction. Environ Pollut 351:124106. https://doi.org/10.1016/j.envpol.2024.124106
Global production capacities of bioplastics 2018–2024 Global production capacities of bioplastics (2019) (by material type) Global production capacities of bioplastics 2024 (by material type) (n.d.). Available at: http://www.european-bioplastics.org/news/publications/
Godinho B et al (2021a) Recycling of different types of polyurethane foam wastes via acidolysis to produce polyurethane coatings. Sustainable Mater Technol 29:e00330. https://doi.org/10.1016/j.susmat.2021.e00330
Godinho B et al (2021b) Recycling of polyurethane wastes using different carboxylic acids via acidolysis to produce wood adhesives. J Polym Sci 59(8):697–705. https://doi.org/10.1002/pol.20210066
Goring PD, Priestley RD (2023) Polymer recycling and upcycling: recent developments toward a circular economy. JACS Au 3(10):2609–2611. https://doi.org/10.1021/jacsau.3c00544
Goto M (2009) Chemical recycling of plastics using sub- and supercritical fluids. J Supercrit Fluids 47(3):500–507. https://doi.org/10.1016/j.supflu.2008.10.011
Grancharov G et al (2007) Smart polymer recycling: synthesis of novel rigid polyurethanes using phosphorus-containing oligomers formed by controlled degradation of microporous polyurethane elastomer. J Appl Polym Sci 105(2):302–308. https://doi.org/10.1002/app.25676
Grause G et al (2010) High-value products from the catalytic hydrolysis of polycarbonate waste. Polym J 42(6):438–442. https://doi.org/10.1038/pj.2010.21
Grdadolnik M et al (2022) Insight into chemical recycling of flexible polyurethane foams by acidolysis. ACS Sustain Chem Eng 10(3):1323–1332. https://doi.org/10.1021/acssuschemeng.1c07911
Grdadolnik M et al (2023) Chemical recycling of flexible polyurethane foams by aminolysis to recover high-quality polyols. ACS Sustain Chem Eng 11(29):10864–10873. https://doi.org/10.1021/acssuschemeng.3c02311
GREEN@COMMUNITY (n.d.) Hing Kong Waste Reduction Website
Guchait A, Maity D, Sabnis AS (2024) Recycling of waste polyurethane shoe sole via aminolysis route. J Mater Cycles Waste Manag 26(2):860–871. https://doi.org/10.1007/s10163-023-01866-7
Guerrero AM et al (2021) What influences and inhibits reduction of deforestation in the soy supply chain? A mental model perspective. Environ Sci Pol 115:125–132. https://doi.org/10.1016/j.envsci.2020.10.016
Guidelines on Extended Producer Responsibility for Plastic Packaging (2022) Ministry of Environment, Forest and Climate Change
Guo X, Wang J (2019) The chemical behaviors of microplastics in marine environment: a review. Mar Pollut Bull 142:1–14. https://doi.org/10.1016/j.marpolbul.2019.03.019
Han G, Cooney JJ (1995) Effects of butyltins and inorganic tin on chemotaxis of aquatic bacteria. J Ind Microbiol 14(3–4):293–299. https://doi.org/10.1007/BF01569942

Han X et al (2017) Structural insight into catalytic mechanism of PET hydrolase. Nat Commun 8(1):2106. https://doi.org/10.1038/s41467-017-02255-z
Hata S et al (2002) Chemical conversion of poly(carbonate) to 1,3-dimethyl-2-imidazolidinone (DMI) and bisphenol a: a practical approach to the chemical recycling of plastic wastes. Polymer 43(7):2109–2116. https://doi.org/10.1016/S0032-3861(01)00800-X
He H et al (2023) Chemical recycling of waste polyurethane foams: efficient acidolysis under the catalysis of zinc acetate. ACS Sustain Chem Eng 11(14):5515–5523. https://doi.org/10.1021/acssuschemeng.2c07260
Heiran R et al (2021) Glycolysis: an efficient route for recycling of end of life polyurethane foams. J Polym Res 28(1):22. https://doi.org/10.1007/s10965-020-02383-z
Hermann BG et al (2011) To compost or not to compost: carbon and energy footprints of biodegradable materials' waste treatment. Polym Degrad Stab 96(6):1159–1171. https://doi.org/10.1016/j.polymdegradstab.2010.12.026
High Ambition Coalition to End Plastic Pollution (n.d.)
HM Revenue & Customs (2021) Plastic Packaging Tax: steps to take
Hoang CN et al (2022a) Acidolysis of poly(ethylene terephthalate) waste using succinic acid under microwave irradiation as a new chemical upcycling method. ACS Omega 7(50):47285–47295. https://doi.org/10.1021/acsomega.2c06642
Hoang CN et al (2022b) Novel efficient method of chemical upcycling of waste poly(ethylene terephthalate) bottles by acidolysis with adipic acid under microwave irradiation. Polym Degrad Stab 206:110175. https://doi.org/10.1016/j.polymdegradstab.2022.110175
Hoang PH, Dat NM, Hung BQ (2023) One-pot synthesis of bio-polyol from vegetable oil for preparation of bio-polyurethane based porous sorbent with high oil adsorption capacity. J Appl Polym Sci 140(35). https://doi.org/10.1002/app.54332
Hopewell J, Dvorak R, Kosior E (2009) Plastics recycling: challenges and opportunities. Philos Trans R Soc Lond B Biol Sci 364(1526):2115–2126. https://doi.org/10.1098/rstb.2008.0311
Ignatyev IA, Thielemans W, Vander Beke B (2014) Recycling of polymers: a review. ChemSusChem 7(6):1579–1593. https://doi.org/10.1002/cssc.201300898
Iñiguez-Franco F et al (2018) Chemical recycling of poly(lactic acid) by water-ethanol solutions. Polym Degrad Stab 149:28–38. https://doi.org/10.1016/j.polymdegradstab.2018.01.016
International Cooperation on Plastic Pollution | Plastics and the Environment Series (2025) Geneva Environment Network
Ivashchuk OS et al (2025) Depolymerization of rigid polyurethane waste by catalytic glycolysis with diethylene glycol. South Afr J Chem Eng 52:61–67. https://doi.org/10.1016/j.sajce.2025.01.009
Jamdar V, Kathalewar M, Sabnis A (2018) Glycolytic depolymerization of PET waste using MP-diol and utilization of recycled product for UV-curable wood coating. J Coat Technol Res 15(2):259–270. https://doi.org/10.1007/s11998-017-9992-8
Jamdar V et al (2024) Chemical recycling of PET waste and utilization of recycled product for synthesis of hybrid unsaturated polyester–urethane coatings. Polym Bull 82(5):1625–1642. https://doi.org/10.1007/s00289-024-05587-1
Janajreh I, Alshrah M, Zamzam S (2015) Mechanical recycling of PVC plastic waste streams from cable industry: a case study. Sustain Cities Soc 18:13–20. https://doi.org/10.1016/j.scs.2015.05.003
Jašek V et al (2024) Chemically recycled commercial polyurethane (PUR) foam using 2-hydroxypropyl ricinoleate as a glycolysis reactant for flexibility-enhanced automotive applications. RSC Adv 14(41):29966–29978. https://doi.org/10.1039/D4RA04972A
Jia Y et al (2021) Nano-immobilization of PETase enzyme for enhanced polyethylene terephthalate biodegradation. Biochem Eng J 176:108205. https://doi.org/10.1016/j.bej.2021.108205
Jiang X, Bateer B (2025) A systematic review of plastic recycling: technology, environmental impact and economic evaluation. Waste Manag Res. https://doi.org/10.1177/0734242X241310658
Jiao H et al (2024) A critical review on plastic waste life cycle assessment and management: challenges, research gaps, and future perspectives. In: Ecotoxicology and environmental safety. Academic Press. https://doi.org/10.1016/j.ecoenv.2024.115942

Jing F et al (2022) Influence of the epoxy/acid stoichiometry on the cure behavior and mechanical properties of epoxy Vitrimers. Molecules 27(19):6335. https://doi.org/10.3390/molecules27196335
Johansen MB et al (2022) Tert -amyl alcohol-mediated deconstruction of polyurethane for polyol and aniline recovery. ACS Sustain Chem Eng 10(34):11191–11202. https://doi.org/10.1021/acssuschemeng.2c02797
Johansen MB et al (2023) Selective chemical disassembly of elastane fibres and polyurethane coatings in textiles. Green Chem 25(24):10622–10629. https://doi.org/10.1039/D3GC02994H
Kamimura A et al (2008) Supercritical secondary alcohols as useful media to convert polyamide into monomeric lactams. ChemSusChem 1(1–2):82–84. https://doi.org/10.1002/cssc.200700024
Kang MJ et al (2020) Depolymerization of PET into terephthalic acid in neutral media catalyzed by the ZSM-5 acidic catalyst. Chem Eng J 398:125655. https://doi.org/10.1016/j.cej.2020.125655
Karan H et al (2019) Green bioplastics as part of a circular bioeconomy. Trends Plant Sci 24(3):237–249. https://doi.org/10.1016/j.tplants.2018.11.010
Kárpáti L et al (2018) Oligoester and polyester production via Acido-alcoholysis of PET waste. Periodica Polytechnica Chem Eng 62(3):336–344. https://doi.org/10.3311/PPch.11513
Katnic SP, de Souza FM, Gupta RK (2024) Recent progress in enzymatic degradation and recycling of polyurethanes. Biochem Eng J 208:109363. https://doi.org/10.1016/j.bej.2024.109363
Kaushal J, Khatri M, Arya SK (2021) Recent insight into enzymatic degradation of plastics prevalent in the environment: a mini - review. Cleaner Eng Technol 2:100083. https://doi.org/10.1016/j.clet.2021.100083
Kawai N (2002) Chemical recovery of bisphenol-a from polycarbonate resin and waste. In: Proceedings of ISFR'2002, the 2nd international symposium on feedstock recycling of plastics and other innovative recycling techniques, September 8–11, Ostend, Belgium
Khan B et al (2017) Thermoplastic starch: a possible biodegradable food packaging material—a review. J Food Proc Eng 40(3). https://doi.org/10.1111/jfpe.12447
Khatoon N, Jamal A, Ali MI (2019) Lignin peroxidase isoenzyme: a novel approach to biodegrade the toxic synthetic polymer waste. Environ Technol 40(11):1366–1375. https://doi.org/10.1080/09593330.2017.1422550
Kibria MG et al (2023) Plastic waste: challenges and opportunities to mitigate pollution and effective management. Int J Environ Res 17(1):20. https://doi.org/10.1007/s41742-023-00507-z
Kinoshita, S. et al (1981) Purification and characterization of 6-Aminohexanoic-acid-oligomer hydrolase of *Flavobacterium* sp. KI72. Eur J Biochem 116(3):547–551. https://doi.org/10.1111/j.1432-1033.1981.tb05371.x
Kirpluks M et al (2020) High Functionality Bio-Polyols from Tall Oil and Rigid Polyurethane Foams Formulated Solely Using Bio-Polyols'. Materials 13(8):1985. https://doi.org/10.3390/ma13081985
Klun U (2000) Rapid microwave induced depolymerization of polyamide-6. Polymer 41(11):4361–4365. https://doi.org/10.1016/S0032-3861(99)00658-8
Knight KD, Fieser ME (2024) Divergent methods for polyester and polycarbonate depolymerization with a cobalt catalyst. Inorg Chem Front 11(1):298–309. https://doi.org/10.1039/D3QI01833D
Korhonen M-R, Dahlbo H (n.d.) Ympäristön-Suojelu Environmental Protection Reducing Greenhouse Gas Emissions by Recycling Plastics and Textiles into Products Reducing Greenhouse Gas Emissions by Recycling Plastics and Textiles into Products
Kumar S, Krishnan S (2020) Recycling of carbon fiber with epoxy composites by chemical recycling for future perspective: a review. Chem Pap 74(11):3785–3807. https://doi.org/10.1007/s11696-020-01198-y
Kumar A et al (2018) Estimation of the generation rate of different types of plastic wastes and possible revenue recovery from informal recycling. Waste Manag 79:781–790. https://doi.org/10.1016/j.wasman.2018.08.045
Kumar B et al (2025) Designing cost-effective supply chains for plastics at the end-of-life. J Clean Prod:145227. https://doi.org/10.1016/j.jclepro.2025.145227

KVIC's Unique Plastic-Mixed Handmade Paper gets Patent for Invention to Recycle Waste Plastic (2021) PIB Delhi
La Mantia FP (2000) Recycling of plastic materials. Chemtec Publishing
Lalhmangaihzuala S et al (2020) Glycolysis of poly(ethylene terephthalate) using biomass-waste derived recyclable heterogeneous catalyst. Polymers 13(1):37. https://doi.org/10.3390/polym13010037
Lee PS et al (2022) One-pot reaction of waste PET to flame retardant polyurethane foam, via deep eutectic solvents-based conversion technology. J Polym Environ 30(1):333–343. https://doi.org/10.1007/s10924-021-02202-6
Lentz H, Mormann W (1992) Chemical recycling of polyurethanes and separation of the components by supercritical ammonia. Makromolekulare Chemie. Macromol Symp 57(1):305–310. https://doi.org/10.1002/masy.19920570127
Li H et al (2022) Expanding plastics recycling technologies: chemical aspects, technology status and challenges. Green Chem 24(23):8899–9002. https://doi.org/10.1039/D2GC02588D
Li T, Menegatti S, Crook N (2023) Breakdown of <scp>polyethylene therepthalate</scp> microplastics under saltwater conditions using engineered Vibrio natriegens. AIChE J 69(12). Available at: https://doi.org/10.1002/aic.18228
Li F et al (2024) Directional glycolysis of waste PET using deep eutectic solvents for preparation of aromatic-based polyurethane elastomers. Green Chem 26(18):9802–9813. https://doi.org/10.1039/D4GC00292J
Life Cycle Assessment Best Practices of ISO 14040 Series Ministry of Commerce, Industry and Energy Republic of Korea Asia-Pacific Economic Cooperation Committee on Trade and Investment (2004)
Lin MM, Tay Zheng J, Yu W-Y (2024) One-pot methanolysis of poly(ethylene terephthalate) enabled by isopropanol-assisted CO2 hydrogenation. J Taiwan Inst Chem Eng 158:105069. https://doi.org/10.1016/j.jtice.2023.105069
Liotier J et al (2024) Polyethersulfones upcycling to luminescent materials by aminolysis. Available at: https://doi.org/10.26434/chemrxiv-2024-vfkz0
Liu Y, Wang M, Pan Z (2012) Catalytic depolymerization of polyethylene terephthalate in hot compressed water. J Supercrit Fluids 62:226–231. https://doi.org/10.1016/j.supflu.2011.11.001
Liu M et al (2018) Degradation of waste polycarbonate via hydrolytic strategy to recover monomer (bisphenol A) catalyzed by DBU-based ionic liquids under metal- and solvent-free conditions. Polym Degrad Stab 157:9–14. https://doi.org/10.1016/j.polymdegradstab.2018.09.018
Liu M et al (2019) Efficient enzymatic degradation of poly (ε-caprolactone) by an engineered bifunctional lipase-cutinase. Polym Degrad Stab 160:120–125. https://doi.org/10.1016/j.polymdegradstab.2018.12.020
Liu Z et al (2023) Bio-based upcycling of poly(ethylene terephthalate) waste to UV-curable polyurethane acrylate. Polym Chem 14(10):1110–1116. https://doi.org/10.1039/D2PY01506D
Liu B, Westman Z, Richardson K, Lim D, Stottlemyer AL, Farmer T et al (2024a) Polyurethane foam chemical recycling: fast acidolysis with maleic acid and full recovery of polyol. ACS Sustain Chem Eng 12(11):4435–4443. https://doi.org/10.1021/acssuschemeng.3c07040
Liu B, Westman Z, Richardson K, Lim D, Stottlemyer AL, Gillis P et al (2024b) Vapor-phase dicarboxylic acids and anhydrides drive depolymerization of polyurethanes. ACS Macro Lett 13(4):435–439. https://doi.org/10.1021/acsmacrolett.4c00008
Liu L-J et al (2025) Rapid hydrolysis of waste polyurethane and facile system separation. Sep Purif Technol 354:129068. https://doi.org/10.1016/j.seppur.2024.129068
Lozano-Martinez P, Torres-Zapata T, Martin-Sanchez N (2021) Directing Depolymerization of PET with subcritical and supercritical ethanol to different monomers through changes in operation conditions. ACS Sustain Chem Eng 9(29):9846–9853. https://doi.org/10.1021/acssuschemeng.1c02489
Lundström R et al (2024) Unveiling reaction pathways of ethylene carbonate and Vinylene carbonate in Li-ion batteries. J Phys Chem C 128(20):8147–8153. https://doi.org/10.1021/acs.jpcc.4c00927

Ma Y et al (2020) Eco-friendly waterborne alkyd resin from polyethylene terephthalate waste. J Polym Environ 28(3):1083–1094. https://doi.org/10.1007/s10924-020-01666-2

Ma J et al (2022) Elucidating degradation properties, microbial community, and mechanism of microplastics in sewage sludge under different terminal electron acceptors conditions. Bioresour Technol 346:126624. https://doi.org/10.1016/j.biortech.2021.126624

Macfadyen G, HT, CR (n.d.) Abandoned, lost or otherwise discarded fishing gear, UNEP Regional Seas Reports and Studies No.185; FAO Fisheries and Aquaculture Technical Paper, No. 523. Rome, UNEP/FAO. 2009. 115p. [Preprint]

Maharana T, Negi YS, Mohanty B (2007) Review article: recycling of polystyrene. Polym-Plast Technol Eng 46(7):729–736. https://doi.org/10.1080/03602550701273963

Mahoney LR, Weiner SA, Ferris FC (1974) Hydrolysis of polyurethane foam waste. Environ Sci Technol 8(2):135–139. https://doi.org/10.1021/es60087a010

Mallick D et al (2022) Emerging commercial opportunities for conversion of waste to energy: aspect of gasification technology. In: Inorganic chemistry frontiers-to-energy approaches towards zero waste. Elsevier, pp 105–127. https://doi.org/10.1016/B978-0-323-85387-3.00012-4

Mancini SD et al (2013) Solid-state hydrolysis of postconsumer polyethylene terephthalate after plasma treatment. J Appl Polym Sci 127(3):1989–1996. https://doi.org/10.1002/app.37591

Mastellone ML (2019) A feasibility assessment of an integrated plastic waste system adopting mechanical and thermochemical conversion processes. Resour Conserv Recycl: X 4:100017. https://doi.org/10.1016/j.rcrx.2019.100017

Meng L et al (2004) Studies on the decomposition behavior of nylon-66 in supercritical water. Polym Degrad Stab 83(3):389–393. https://doi.org/10.1016/j.polymdegradstab.2003.08.001

Microplastics (2023) Marine Debris Program Office of Response and Restoration National Oceanic and Atmospheric Administration

Miguel-Fernández R et al (2022) Recovery of green polyols from rigid polyurethane waste by catalytic depolymerization. Polymers 14(14):2936. https://doi.org/10.3390/polym14142936

Mir Mohamad Sadeghi G, Shamsi R, Sayaf M (2011) From aminolysis product of PET waste to novel biodegradable polyurethanes. J Polym Environ 19(2):522–534. https://doi.org/10.1007/s10924-011-0283-7

Mohammadi S, Bouldo MG, Enayati M (2023) Controlled glycolysis of poly(ethylene terephthalate) to oligomers under microwave irradiation using antimony(III) oxide. ACS Appl Polym Mater 5(8):6574–6584. https://doi.org/10.1021/acsapm.3c01071

Mohanan N et al (2020) Microbial and enzymatic degradation of synthetic plastics. Front Microbiol 11. https://doi.org/10.3389/fmicb.2020.580709

Moharir RV, Gautam P, Kumar S (2019) Waste treatment processes/technologies for energy recovery'. In: Current developments in biotechnology and bioengineering. Elsevier, pp 53–77. https://doi.org/10.1016/B978-0-444-64083-3.00004-X

Motokucho S et al (2018) Environment-friendly chemical recycling of aliphatic polyurethanes by hydrolysis in a <scp>CO</scp> $_2$ -water system. J Appl Polym Sci 135(8). https://doi.org/10.1002/app.45897

Narancic T et al (2018) Biodegradable plastic blends create new possibilities for end-of-life management of plastics but they are not a panacea for plastic pollution. Environ Sci Technol 52(18):10441–10452. https://doi.org/10.1021/acs.est.8b02963

Nasrollahi M et al. (2020) Plastic packaging, recycling, and sustainable development, pp 544–551. Available at: https://doi.org/10.1007/978-3-319-95726-5_110

'NATIONAL ACTION PLAN ON MARINE PLASTIC LITTER FROM SEA-BASED SOURCES' (2024)

Nica S et al (2018) Functionalized 1,5,7-triazabicyclo [4.4.0] dec-5-ene (TBD) as novel organocatalyst for efficient depolymerization of polyethylene terephthalate (PET) wastes. Rev Chim 69(10):2613–2616. https://doi.org/10.37358/RC.18.10.6591

Nosrati-Ghods N et al (2025) Techno-economic assessment of waste poly(lactic acid) enzymatic hydrolysis and recycling for poly(lactic acid) production. Biochem Eng J 215:109615. https://doi.org/10.1016/j.bej.2024.109615

Núñez SS et al (2023) Decontamination of recycled LDPE using different washing methods. Resour Conserv Recycl 195:107017. https://doi.org/10.1016/j.resconrec.2023.107017

O'Rourke G et al (2022) Delamination of polyamide/polyolefin multilayer films by selective glycolysis of polyurethane adhesive. Green Chem 24(18):6867–6878. https://doi.org/10.1039/D2GC01531E

Ouyang S et al (2021) Preparation of autoxidative water-reducible alkyd resins from waste polyethylene terephthalate. Royal Soc Open Sci 8(6):202375. https://doi.org/10.1098/rsos.202375

Padhi G et al (2025) Depolymerization of waste polycarbonates to value-added products. ChemSusChem 18(2). https://doi.org/10.1002/cssc.202400756

Pan D et al (2020) Research progress for plastic waste management and manufacture of value-added products. Adv Compos Hybrid Mater 3(4):443–461. https://doi.org/10.1007/s42114-020-00190-0

Papong S et al (2014) Comparative assessment of the environmental profile of PLA and PET drinking water bottles from a life cycle perspective. J Clean Prod 65:539–550. https://doi.org/10.1016/j.jclepro.2013.09.030

Patil CK et al (2021) Nonedible vegetable oil-based polyols in anticorrosive and antimicrobial polyurethane coatings. Polymers 13(18):3149. https://doi.org/10.3390/polym13183149

Peng Y et al (2023) Acetolysis of waste polyethylene terephthalate for upcycling and life-cycle assessment study. Nat Commun 14(1):3249. https://doi.org/10.1038/s41467-023-38998-1

Pereira P, Savage PE, Pester CW (2024) 'Acid catalyst screening for hydrolysis of post-consumer PET waste and exploration of acidolysis. Green Chem [preprint]. Available at: https://doi.org/10.1039/D3GC03906D

Petrus R, Bykowski D, Sobota P (2016) Solvothermal alcoholysis routes for recycling polylactide waste as lactic acid esters. ACS Catal 6(8):5222–5235. https://doi.org/10.1021/acscatal.6b01009

Picuno C et al (2021) Flows of post-consumer plastic packaging in Germany: an MFA-aided case study. Resour Conserv Recycl 169:105515. https://doi.org/10.1016/j.resconrec.2021.105515

Piemonte V, Gironi F (2013) Kinetics of hydrolytic degradation of PLA. J Polym Environ 21(2):313–318. https://doi.org/10.1007/s10924-012-0547-x

Pinto AV et al (2021) Reaction mechanism of MHETase, a PET degrading enzyme. ACS Catal 11(16):10416–10428. https://doi.org/10.1021/acscatal.1c02444

Plastic waste and recycling in the EU: facts and figures | Topics. (2024, June 25). European Parliament. Retrieved April 12, 2025, from https://www.europarl.europa.eu/topics/en/article/20181212STO21610/plastic-waste-and-recycling-in-the-eu-facts-and-figures' (no date)

Plastic Waste Management (Amendment) Rules, 2021' (2021) MINISTRY OF ENVIRONMENT, FOREST AND CLIMATE CHANGE NOTIFICATION New Delhi [Preprint]

Posen ID, Jaramillo P, Griffin WM (2016) Uncertainty in the life cycle greenhouse gas emissions from U.S. production of three biobased polymer families. Environ Sci Technol 50(6):2846–2858. https://doi.org/10.1021/acs.est.5b05589

Pu M et al (2023) A facile, alternative and sustainable feedstock for transparent polyurethane elastomers from chemical recycling waste PET in high-efficient way. Waste Manag 155:137–145. https://doi.org/10.1016/j.wasman.2022.10.032

Pu M et al (2024) Sustainable and renewable microcross-linked polyurethane pressure-sensitive adhesive from waste polyethylene terephthalate. ACS Sustain Chem Eng 12(32):12226–12235. https://doi.org/10.1021/acssuschemeng.4c04411

PureCycle Technologies celebrates successful run of groundbreaking plastics recycling technology (2019)

Quaranta E, Minischetti CC, Tartaro G (2018) Chemical recycling of poly(bisphenol a carbonate) by glycolysis under 1,8-Diazabicyclo[5.4.0]undec-7-ene catalysis. ACS Omega 3(7):7261–7268. https://doi.org/10.1021/acsomega.8b01123

Quaranta E et al (2021) Chemical recycling of poly-(bisphenol A carbonate) by diaminolysis: a new carbon-saving synthetic entry into non-isocyanate polyureas (NIPUreas). J Hazard Mater 403:123957. https://doi.org/10.1016/j.jhazmat.2020.123957

Quartinello F et al (2017) Synergistic chemo-enzymatic hydrolysis of poly(ethylene terephthalate) from textile waste. Microb Biotechnol 10(6):1376–1383. https://doi.org/10.1111/1751-7915.12734

Rafey A, Siddiqui FZ (2023) A review of plastic waste management in India—challenges and opportunities. Int J Environ Anal Chem 103(16):3971–3987. https://doi.org/10.1080/03067319.2021.1917560

Ragaert K, Delva L, Van Geem K (2017) Mechanical and chemical recycling of solid plastic waste. Waste Manag 69:24–58. https://doi.org/10.1016/j.wasman.2017.07.044

Rahman MH, Bhoi PR (2021) An overview of non-biodegradable bioplastics. J Clean Prod. Elsevier Ltd. Available at: https://doi.org/10.1016/j.jclepro.2021.126218

Rahman MH, Bhoi PR, Menezes PL (2023) Pyrolysis of waste plastics into fuels and chemicals: a review. Renew Sust Energ Rev 188:113799. https://doi.org/10.1016/j.rser.2023.113799

Rajvanshi J et al (2023) Biomaterials: a sustainable solution for a circular economy. In: RAiSE-2023. MDPI, Basel Switzerland, p 133. https://doi.org/10.3390/engproc2023059133

RAMPF—Holistic Solutions For Polyurethane Upcycling, https://www.rampf-group.com/en/news/2024/rampf-holistic-solutions-for-polyurethane-upcycling/' (2024)

Ravve A (2000) Degradation of polymers. In: Principles of polymer chemistry. Springer US, Boston, MA, pp 581–616. https://doi.org/10.1007/978-1-4615-4227-8_9

Recycling Incentive Scheme (2023) Plastic smart cities

Rong Z et al (2024) Degradation of low-density polyethylene by the bacterium Rhodococcus sp. C-2 isolated from seawater. Sci Total Environ 907:167993. https://doi.org/10.1016/j.scitotenv.2023.167993

Rosenboom JG, Langer R, Traverso G (2022) Bioplastics for a circular economy. Nat Rev Mater Nat Res:117–137. https://doi.org/10.1038/s41578-021-00407-8

Rossignolo G, Malucelli G, Lorenzetti A (2024) Recycling of polyurethanes: where we are and where we are going. Green Chem 26(3):1132–1152. https://doi.org/10.1039/D3GC02091F

Ru J, Huo Y, Yang Y (2020) Microbial degradation and valorization of plastic wastes. Front Microbiol 11. Available at: https://doi.org/10.3389/fmicb.2020.00442

Rubeš D et al (2024) Bio-based unsaturated polyester resin from post-consumer PET. RSC Advances 14(12):8536–8547. https://doi.org/10.1039/D3RA08500G

Safdar A, Ismail F, Imran M (2024) Biodegradation of synthetic plastics by the extracellular lipase of Aspergillus niger. Environ Adv 17:100563. https://doi.org/10.1016/j.envadv.2024.100563

Sagong H-Y et al (2020) Decomposition of the PET film by MHETase using Exo-PETase function. ACS Catal 10(8):4805–4812. https://doi.org/10.1021/acscatal.9b05604

Salahuddin U et al. (2023) Plastic recycling: a review on life cycle, methods, misconceptions, and techno-economic analysis. Adv Sustain Syst 7(7). https://doi.org/10.1002/adsu.202200471

Salem T et al (2025) Chemical recycling of poly(ethylene terephthalate) waste using a new co-catalytic system. Iranian Polym J. https://doi.org/10.1007/s13726-024-01429-z

Sambyal P et al (2024) Plastic recycling: challenges and opportunities. Cad J Chem Eng [preprint]. Available at: https://doi.org/10.1002/cjce.25531

Sanda O et al (2016) A Study of the Alkaline Solvolysis of Postconsumer Polyethylene Terephthalate in Primary C1—C3 Aliphatic Alcohols. Am Chem Sci J 16(1):1–14. https://doi.org/10.9734/ACSJ/2016/27697

Sanda O, Taiwo E, Osinkolu G (2017) Alkaline Solvolysis of poly(ethylene terephthalate) in Butan–1–ol media: kinetics and optimization studies. Br J Appl Sci echnol 21(5):1–12. https://doi.org/10.9734/BJAST/2017/32597

Sangtani R et al (2023) Systematizing microbial bioplastic production for developing sustainable bioeconomy: metabolic nexus modeling, economic and environmental technologies assessment. J Polym Environ Springer:2741–2760. https://doi.org/10.1007/s10924-023-02787-0

Sato Y et al (2005) Degradation behaviour and recovery of bisphenol-A from epoxy resin and polycarbonate resin by liquid-phase chemical recycling. Polym Degrad Stab 89(2):317–326. https://doi.org/10.1016/j.polymdegradstab.2005.01.015

Sato O et al (2010) Chemical recycling process of poly(ethylene terephthalate) in high-temperature liquid water. J Cheml Eng 43(3):313–317. https://doi.org/10.1252/jcej.09we137

Sauter D, Taoufik M, Boisson C (2017) Polyolefins, a success story. Polymers 9(6):185. https://doi.org/10.3390/polym9060185
Shahdan SN, Hirzin RSFN (2021) Pre-treatment of used cooking oil followed by transesterification reaction in the production of used cooking oil-based polyol. Sci Res J 18(2):129–146. https://doi.org/10.24191/srj.v18i2.13749
Shaikh TM et al (2024) Chemical recycling of polycarbonates via ammonolysis and polycaprolactones by KOH-catalyzed methanolysis. Eur J Organ Chem 27(10). https://doi.org/10.1002/ejoc.202301308
Sheel A, Pant D (2023) Epoxy paint from waste polycarbonate (PC). Int J Environ Sci Technol 20(7):7393–7402. https://doi.org/10.1007/s13762-022-04365-8
Shen M et al (2020) (Micro)plastic crisis: un-ignorable contribution to global greenhouse gas emissions and climate change. J Clean Prod 254:120138. https://doi.org/10.1016/j.jclepro.2020.120138
Shirazimoghaddam S et al (2023) Chemical recycling of used PET by glycolysis using Niobia-based catalysts. ACS Eng Au 3(1):37–44. https://doi.org/10.1021/acsengineeringau.2c00029
Shojaei B, Najafi M (2024) Eco-friendly synthesis of non-isocyanate polyurea using PET and $CO_2$ upcycling: a double green approach. Polym Adv Technol 35(9). https://doi.org/10.1002/pat.6587
Siddiqui MN et al (2020) Depolymerization of PLA by phase transfer catalysed alkaline hydrolysis in a microwave reactor. J Polym Environ 28(6):1664–1672. https://doi.org/10.1007/s10924-020-01716-9
Simón D et al (2013) Glycolysis of flexible polyurethane wastes using stannous octoate as the catalyst: study on the influence of reaction parameters. Polym Degrad Stab 98(1):144–149. https://doi.org/10.1016/j.polymdegradstab.2012.10.017
Simón D et al (2018) Glycolysis of advanced polyurethanes composites containing thermoregulating microcapsules. Chem Eng J 350:300–311. https://doi.org/10.1016/j.cej.2018.05.158
Singh S, Lei Y, Schober A (2015) Direct extraction of carbonyl from waste polycarbonate with amines under environmentally friendly conditions: scope of waste polycarbonate as a carbonylating agent in organic synthesis. RSC Adv 5(5):3454–3460. https://doi.org/10.1039/C4RA14319A
Singh N et al (2017) Recycling of plastic solid waste: a state of art review and future applications. Compos Part B 115:409–422. https://doi.org/10.1016/j.compositesb.2016.09.013
Singh AK, Chugh K, Mhaske ST (2024) Improvement in BHETA yield by aminolysis and its application in hot melt adhesive. Int J Adhes Adhes 130:103610. https://doi.org/10.1016/j.ijadhadh.2023.103610
Sinha V, Patel MR, Patel JV (2010) Pet waste management by chemical recycling: a review. J Polym Environ 18(1):8–25. https://doi.org/10.1007/s10924-008-0106-7
Soltysinski M et al (2018) Conversion of polyurethane technological foam waste and post-consumer polyurethane mattresses into polyols—industrial applications. Polimery 63(03):234–238. https://doi.org/10.14314/polimery.2018.3.8
Sowmya HV et al (2015) Degradation of polyethylene by Penicillium simplicissimum isolated from local dumpsite of Shivamogga district. Environ Dev Sustain 17(4):731–745. https://doi.org/10.1007/s10668-014-9571-4
Spierling S et al (2018) Bio-based plastics – a review of environmental, social and economic impact assessments. J Clean Prod 185:476–491. https://doi.org/10.1016/j.jclepro.2018.03.014
Srikanth M et al (2022) Biodegradation of plastic polymers by fungi: a brief review. Bioresour Bioprocess 9(1):42. https://doi.org/10.1186/s40643-022-00532-4
Stanica-Ezeanu D, Matei D (2021) Natural depolymerization of waste poly(ethylene terephthalate) by neutral hydrolysis in marine water. Sci Rep 11(1):4431. https://doi.org/10.1038/s41598-021-83659-2
Sumathi T et al (2016) Production of laccase by *Cochliobolus* sp. isolated from plastic dumped soils and their ability to degrade low molecular weight PVC. Biochem Res Int 2016:1–10. https://doi.org/10.1155/2016/9519527

Suprun VY, Marukha VI, Sylovaniuk VP (2022) Recycling Technologies for Polyurethane Wastes (a survey). Mater Sci 57(6):755–764. https://doi.org/10.1007/s11003-022-00605-x

Symphony environmental (n.d.) Biodegradable Plastic technology—an environmentally responsible solution for your plastic product, film, or packaging needs. https://www.symphonyenvironmental.com/technologies/biodegradable-plastic/'

Taghavi N, Udugama IA et al (2021a) Challenges in biodegradation of non-degradable thermoplastic waste: from environmental impact to operational readiness. Biotechnol Adv 49:107731. https://doi.org/10.1016/j.biotechadv.2021.107731

Taghavi N, Singhal N et al (2021b) Degradation of plastic waste using stimulated and naturally occurring microbial strains. Chemosphere 263:127975. https://doi.org/10.1016/j.chemosphere.2020.127975

Taghavi N, Zhuang W-Q, Baroutian S (2021c) Enhanced biodegradation of non-biodegradable plastics by UV radiation: part 1. J Environ Chem Eng 9(6):106464. https://doi.org/10.1016/j.jece.2021.106464

Tajau R et al (2018) Development of new bio-based polyol ester from palm oil for potential polymeric drug carrier. Adv Polym Technol 37(8):3552–3560. https://doi.org/10.1002/adv.22139

Tang Y et al (2021) A review: research progress on microplastic pollutants in aquatic environments. Sci Total Environ 766:142572. https://doi.org/10.1016/j.scitotenv.2020.142572

Thakur S et al (2018) Recent developments in recycling of polystyrene based plastics. Curr Opin Green and Sustainable Chemistry 13:32–38. https://doi.org/10.1016/j.cogsc.2018.03.011

The Chempet Project can be accessed at; https://garbo.it/en/chempet/ (2023)

THE EUROPEAN COMMISSION (2020) A new Circular Economy Action Plan For a cleaner and more competitive Europe

Tonsi G et al (2023) Nylon recycling processes: a brief overview. Chemical Eng Trans 100(December 2022):727–732. https://doi.org/10.3303/CET23100122

Tournier V et al (2020) An engineered PET depolymerase to break down and recycle plastic bottles. Nature 580(7802):216–219. https://doi.org/10.1038/s41586-020-2149-4

Tsintzou GP, Achilias DS (2013) Chemical recycling of polycarbonate based wastes using alkaline hydrolysis under microwave irradiation. Waste Biomass Valorization 4(1):3–7. https://doi.org/10.1007/s12649-012-9125-7

Tumu K, Vorst K, Curtzwiler G (2023) Global plastic waste recycling and extended producer responsibility laws. J Environ Manag 348:119242. https://doi.org/10.1016/j.jenvman.2023.119242

Ügdüler S et al (2023) Understanding the complexity of deinking plastic waste: an assessment of the efficiency of different treatments to remove ink resins from printed plastic film. J Hazard Mater 452:131239. https://doi.org/10.1016/j.jhazmat.2023.131239

Ulutan S (2003) Influence of additional thermal stabilizers on the reprocessing of postconsumer poly(vinyl chloride) bottles. J Appl Polym Sci 90(14):3994–3999. https://doi.org/10.1002/app.13133

Vaidya UR, Nadkarni VM (1989) Polyester polyols from glycolyzed PET waste: effect of glycol type on kinetics of polyesterification. J Appl Polym Sci:1179–1190. https://doi.org/10.1002/app.1989.070380615

van den Tempel P, Picchioni F (2024) Polymer recycling: a comprehensive overview and future outlook. Recycling 10(1):1. https://doi.org/10.3390/recycling10010001

Van Roijen EC, Miller SA (2022) A review of bioplastics at end-of-life: linking experimental biodegradation studies and life cycle impact assessments. Resour Conserv Recycl. Elsevier B.V. https://doi.org/10.1016/j.resconrec.2022.106236

Vu DH et al (2020) Recycling strategies for polyhydroxyalkanoate-based waste materials: an overview. Bioresour Technol 298:122393. https://doi.org/10.1016/j.biortech.2019.122393

Vural Gursel I et al (2021) Comparative cradle-to-grave life cycle assessment of bio-based and petrochemical PET bottles. Sci Total Environ 793. https://doi.org/10.1016/j.scitotenv.2021.148642

Wang L et al (2020) Glycolysis of PET using 1,3-Dimethylimidazolium-2-Carboxylate as an Organocatalyst. ACS Sustain Chem Eng 8(35):13362–13368. https://doi.org/10.1021/acssuschemeng.0c04108

Wang Y et al (2022a) Towards recycling purpose: converting PET plastic waste back to terephthalic acid using pH-responsive phase transfer catalyst. Chin J Chem Eng 51:53–60. https://doi.org/10.1016/j.cjche.2021.10.028

Wang Z et al (2022b) Chemical upcycling of poly(bisphenol a carbonate) plastic catalyzed by $ZnX_2$ via an amino-alcoholysis strategy. ACS Sustain Chem Eng 10(14):4529–4537. https://doi.org/10.1021/acssuschemeng.1c08408

Wang W et al (2023a) Mechano-chemical rubber reclamation using aminolysis products of waste flexible polyurethane foams as the devulcanizing agent. J Clean Prod 384:135421. https://doi.org/10.1016/j.jclepro.2022.135421

Wang X-Y, Gao Y, Tang Y (2023b) Sustainable developments in polyolefin chemistry: progress, challenges, and outlook. Prog Polym Sci 143:101713. https://doi.org/10.1016/j.progpolymsci.2023.101713

Wang Y et al (2025) Self-healing conducting composite electrodes derived from chemical recycling of PET plastic wastes for flexible supercapacitors. Polym Eng Sci [Preprint]. Available at: https://doi.org/10.1002/pen.27117

Wei X et al (2025) Efficient glycolysis of poly(ethylene terephthalate) to bis(hydroxyethyl) terephthalate catalyzed by non-metallic deep eutectic solvent. Chem Eng J 508:161115. https://doi.org/10.1016/j.cej.2025.161115

Wen R et al (2024) One-pot synthesis of titanium isophthalate and highly active catalytic glycolysis of waste PET. New J Chem 48(39):17254–17260. https://doi.org/10.1039/D4NJ03300K

Westman Z et al (2024) Influence of carboxylic acid structure on the kinetics of polyurethane foam acidolysis to recycled polyol. JACS Au 4(8):3194–3204. https://doi.org/10.1021/jacsau.4c00495

Wilkes RA, Aristilde L (2017) Degradation and metabolism of synthetic plastics and associated products by *Pseudomonas* sp.: capabilities and challenges. J Appl Microbiol 123(3):582–593. https://doi.org/10.1111/jam.13472

Worm B et al (2017) Plastic as a persistent marine pollutant. Annu Rev Environ Resour 42(1):1–26. https://doi.org/10.1146/annurev-environ-102016-060700

Wright SL, Kelly FJ (2017) Plastic and human health: a micro issue? Environ Sci Technol 51(12):6634–6647. https://doi.org/10.1021/acs.est.7b00423

Wu C-H et al (2018) 100% atom-economy efficiency of recycling polycarbonate into versatile intermediates. ACS Sustain Chem Eng 6(7):8964–8975. https://doi.org/10.1021/acssuschemeng.8b01326

Xanthos M, Patel SH (1998) Solvolysis. In: Frontiers in the science and technology of polymer recycling. Springer Netherlands, Dordrecht, pp 425–436. https://doi.org/10.1007/978-94-017-1626-0_20

Xiang Q et al (2002) Effects of melt reprocessing on volatile emissions and structural/rheological changes of unstabilized polypropylene. Polym Degrad Stab 77(1):93–102. https://doi.org/10.1016/S0141-3910(02)00083-6

Xu K et al (2025) One-pot preparation of flame-retardant waterborne polyurethane by glycolysis of PET with ionic liquids. Polym Degrad Stab 233:111146. https://doi.org/10.1016/j.polymdegradstab.2024.111146

Yang Y (2016) Empirical mode decomposition as a time-varying multirate signal processing system. Mech Syst Signal Process 76–77:759–770. https://doi.org/10.1016/j.ymssp.2016.02.022

Yang W et al (2012) Recycling and disposal methods for polyurethane foam wastes. Procedia Environ Sci 16:167–175. https://doi.org/10.1016/j.proenv.2012.10.023

Yang W et al (2021) Hydrolysis of waste polyethylene terephthalate catalyzed by easily recyclable terephthalic acid. Waste Manag 135:267–274. https://doi.org/10.1016/j.wasman.2021.09.009

Yin S et al (2015) Mechanical reprocessing of polyolefin waste: a review. Polym Eng Sci 55(12):2899–2909. https://doi.org/10.1002/pen.24182

Yuan J et al (2020) Microbial degradation and other environmental aspects of microplastics/plastics. Sci Total Environ 715:136968. https://doi.org/10.1016/j.scitotenv.2020.136968

Zhang N et al (2022) Efficient catalytic degradation of anhydride-cured epoxy resin by amphiphilic molecule catalysts. Green Chem 24(19):7395–7402. https://doi.org/10.1039/D2GC02248F

'Zhang-et-al-2014-from-aminolysis-product-of-pet-waste-to-value-added-products-of-polymer-and-assistants (n.d.)
Zheng L et al (2024) Recycling and Degradation of Polyamides'. Molecules 29(8):1742. https://doi.org/10.3390/molecules29081742
Zhu J, Wang C (2020) Biodegradable plastics: green hope or greenwashing? Mar Pollut Bull 161:111774. https://doi.org/10.1016/j.marpolbul.2020.111774
Zhu B, Wang D, Wei N (2022) Enzyme discovery and engineering for sustainable plastic recycling. Trends Biotechnol 40(1):22–37. https://doi.org/10.1016/j.tibtech.2021.02.008
Zia KM, Bhatti HN, Ahmad Bhatti I (2007) Methods for polyurethane and polyurethane composites, recycling and recovery: a review. React Funct Polym 67(8):675–692. https://doi.org/10.1016/j.reactfunctpolym.2007.05.004
Zou L et al (2023) Chemical recycling of polyolefins: a closed-loop cycle of waste to olefins. Natl Sci Rev 10(9). https://doi.org/10.1093/nsr/nwad207
Zúñiga DM, Aranda FL, Rivas BL (2023) Polymers recycling: Upcycling Techniques. An overview. J Chil Chem Soc 68(2):5876–5886. https://doi.org/10.4067/s0717-97072023000205876

# Sustainable Biorefinery for the Production of Biomaterials

# 11

Abas Siraj Hamda, Melkiyas Diriba Muleta, Teketel Alemu Anshebo, and Mani Jayakumar

**Abstract**

Biorefinery systems, which integrate different techniques for converting biomass to produce an enormous variety of products, including biofuels, biochemicals, and bio products, represent a promising approach to sustainable resource utilization. The fundamental principles incorporate the physical, chemical, and biological processes that convert biomass into various value added biomaterials and products. These processes typically include biomass pretreatment, conversion, separation, and purification steps to obtain desired products efficiently. Technological developments, process integration, and product diversification have increased the range of industries in which bio refineries can be applied, supporting international efforts to reduce climate-change, improve energy-security, and advance sustainable development. This chapter thoroughly summarizes the general concept of biorefinery systems, potential application, classification, and current technologies of biorefinery for biomaterial production. In addition,

A. S. Hamda · M. D. Muleta
Department of Chemical Engineering, Haramaya Institute of Technology, Haramaya University, Dire Dawa, Ethiopia

T. A. Anshebo
Department of Chemical Engineering, Wachemo University, Hossana, Ethiopia

M. Jayakumar (✉)
Department of Chemical Engineering, Haramaya Institute of Technology, Haramaya University, Dire Dawa, Ethiopia

Department of Biotechnology, Faculty of Engineering, Karpagam Academy of Higher Education, Coimbatore, Tamilnadu, India

Centre for Natural Products and Functional Foods, Karpagam Academy of Higher Education, Coimbatore, Tamilnadu, India
e-mail: drjayakumarmani@haramaya.edu.et



V. Singh (ed.), *Sustainable Waste Management Towards Circular Bioeconomy*,
https://doi.org/10.1007/978-981-95-2227-9_11

techno-economic aspects and sustainability of biorefinery in the waste management system are discussed.

**Keywords**

Biomaterials · Biorefinery · Biofuels · Biochemicals · Bio products

## 11.1 Introduction

A global shift towards sustainable solutions has been sparked by the volatility of oil prices and the approaching shortage of fuel supply. Due to upcoming restrictions on carbon dioxide emissions, a number of nations are implementing laws to encourage the adoption of bio-based alternatives (Sethumadhavan et al. 2024; (Orejuela-Escobar et al. 2021). The foundation of these initiatives is the idea of biorefineries, biomass can be processed into a variety of products using this flexible method, from specialty chemicals to bioenergy. Through the combination of different biomass conversion techniques, biorefinery systems present a viable path toward resource sustainability (Pathak and Mandavgane 2023).

To optimize the value extracted from feedstock, biorefineries take advantage of the diversity of biomass components and intermediary chemicals (Hiloidhari et al. 2020). Biomass products are marketed with the aim of maximizing their value, fossil fuel utilization should be minimized, cutting greenhouse-gas-emissions (GHGs), and fostering economic growth in rural and regional areas. Integrated biorefineries guarantee environmental sustainability as well as increased profitability by combining several conversion processes (Reshmy et al. 2022b).

Major variations in raw material and process sophistication can be seen when comparing biorefineries to traditional oil refineries (Enawgaw et al. 2023). Biomass is renewable and does not require millions of years for biodegradation, in contrast to crude oil. Furthermore, biomass is broken down into its basic components by biorefineries utilizing a various technologies (Talan et al. 2022; Enawgaw et al. 2023). These elements are then processed by physical, chemical, and biological methods to produce products with additional value.

Sustainable biorefineries play a major part in advancing the bio based circular-economy, which seeks to carbon, shut the loop on resources, minerals, and water (Mahari et al. 2022). These bioprocesses effectively use biomass assets to generate a variety of metabolites and value added end products with commercial potential, such by way of proteins, lipids, carbohydrates, bioactive chemicals, and biomaterials. Furthermore, waste biorefineries are becoming more and more popular as efficient waste management options that reduce greenhouse gas emissions from burning fossil fuels, lessening carbon footprints and safeguarding the environment (Talan et al. 2022).

Biorefineries are the epitome of circularity and sustainability; its goal is to maximize biomass value while reducing waste and negative environmental effects. These facilities are able to effectively convert a variety of feedstocks into useful products by integrating several biomass conversion processes (Mohan et al. 2016). Because of the flexibility of biorefinery processes, less reliance on finite fossil fuels can be placed on a range of biomass sources, including agricultural waste and specialized energy crops (Sut and Kataki 2020).

The capacity of biorefineries to generate a variety of biomaterials in addition to bioenergy and biochemicals is one of their unique selling points. These goods include bioplastics, bio composites, bio fibers, and biopolymers. They are used in a range of industries, from textiles and automobiles to packaging and construction (Heo et al. 2019; Saini et al. 2019). Biorefineries help create sustainable substitutes for traditional petrochemical-based materials by utilizing this abundance of biomaterials, intensifications their resource proficiency and which lessens their negative ecological effects.

Furthermore, the construction of biorefineries has the potential to spur economic growth and revitalize rural areas, especially in agricultural areas. These facilities can help reduce rural unemployment and support local communities by giving farmers new sources of income and opening up job opportunities in the biomass supply chains and biorefinery operations (Reshmy et al. 2022a). Moreover, localizing production is made possible by the decentralized structure of biorefineries, which lowers transportation costs and fosters regional independence.

To sum up, sustainable biorefineries are a revolutionary method of using biomass and provide a way forward for a more robust and sustainable bioeconomy. Biorefineries have the ability to significantly contribute to the solution of urgent environmental issues, promote economic expansion, and advance social advancement by utilizing cutting-edge technologies and the natural renewable potential of biomass. Biorefineries are positioned to become important catalysts for sustainable development, opening doors to new possibilities for resource efficiency, innovation, and wealth as the shift to a bio-based circular economy quickens.

## 11.2 General Concept of Biorefinery Systems

A flexible and sustainable method of utilizing biomass, biorefinery technologies provide a path in the direction of a bio-based and more circular economy. The term "biorefineries" is defined and interpreted differently, but in general, a range of useful products and energy sources can be produced by integrating biomass-conversion-processes (BCP) (Gutierrez-Macias et al. 2017). Biorefinery, according to Bioenergy Task 42, is the viable conversion of biomass into a variety of commercial items, including as chemicals, energy, feed, food, and biomaterials in the form of electricity, heat, and fuels (Ghatak 2011). Likewise, the American National Renewable Energy Laboratory describes a biorefinery as an integrated system for transforming biomass into chemicals, fuels, and electricity. To expound further, define biorefineries as procedures that, like contemporary petroleum refineries, utilize biomass from forestry or agriculture as a source of bio-based resources to create chemicals, energy, and bio-based products.

In general, biorefineries are establishments that process various forms of biomass, such by way of energy crops, organic residues, and aquatic biomass, to generate a variety of products, such as chemicals, biomaterials, fuels, power and heat, feed, and food. Biorefinery systems sometimes require multi-step processes comprising biomass pretreatment, conversion, separation, and purification in order to efficiently create desired products (Yan et al. 2021). Biorefineries produce a wide range of products. At lower sale prices, they can generate large quantities of bulk energy in the practice of biofuels or heat, or they can manufacture molecules specially intended for fine chemistry, cosmetics, or medicine at higher volumes and higher costs.

In order to classify biorefineries holistically, one must take into account a number of important factors, such as feedstocks, products, conversion procedures, new developments, and possible applications (Duan et al. 2022). A comprehensive overview of biorefineries is shown in Fig. 11.1, which also highlights the range of capabilities and possible contributions that these facilities can make to a sustainable bio-based economy. Fundamentally, biorefinery systems are important nodes in the shift to a future that is more resource-efficient and sustainable. They do this by utilizing biomass's natural ability to address a variety of societal demands while reducing its negative effects on the environment. Biorefinery systems have a wide range of possible advantages in biomaterial production since they use sustainable biomass use to create a variety of products. These uses support the growth of a more sustainable and circular economy and range from bioenergy and biofuels to biochemicals, biomaterials, and beyond (Igbokwe et al. 2022). Biomass materials are converted into fuels like biodiesel, bioethanol, and biogas through thermochemical and biochemical conversion processes, which decreases dependence on fossil fuels and GHGs (Budzianowski and Postawa 2016).

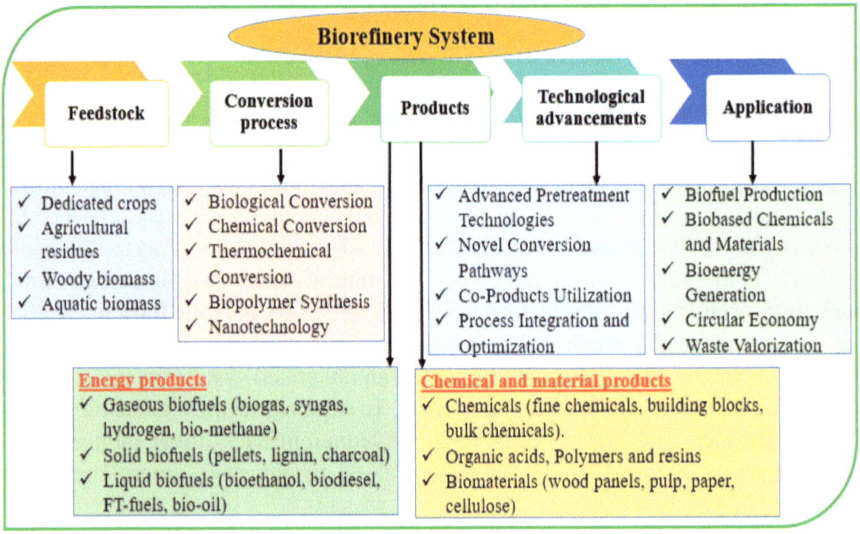


**Fig. 11.1** General overview of biorefineries system

## 11.3 Potential Applications of Biorefinery in Biomaterial Production

Biomaterials are a subset of bio products, which are materials derived from renewable resources. They are essential to the shift from an economy based on fossil fuels to one based on biofuels. As the chemical industry transitions from petrochemical-based production to bio-based production, the significance of biorefinery as a key technology platform for biomaterial production is increasing. Furthermore, the expansion of biorefinery is narrowly associated to the governmental drive to replace fossil-based industry with bio-based industry. Many nations have published roadmaps and strategies for the bioeconomy, aiming to double the industry value by 2050 and reduce reliance on non-renewable resources (Gardossi et al. 2023).

Biorefineries also produce biochemicals, such as organic acids, vitamins, and enzymes, through fermentation and enzymatic hydrolysis. These biochemicals are essential components for a number of industries, including cosmetics, food and beverage, pharmaceuticals, and agriculture (Tsarpali et al. 2021). Sustainable substitutes for conventional materials are created by turning bio-based polymers like polyhydroxyalkanoates (PHAs) or polylactic acid (PLA) into biomaterials like bioplastics, which reduce environmental impact (Chen et al. 2023). Moreover, the waste materials and byproducts generated during the conversion process are transformed into beneficial products like lignin by biorefineries, which can be used as a sustainable feedstock for chemicals, adhesives, and composites (Bilal and Iqbal 2022). Bioactive compounds, flavors, perfumes, and nutraceuticals are among the high-value chemicals and specialty goods manufactured, with a variety of applications in pharmaceuticals, cosmetics, and fine chemicals.

Biorefineries contribute greatly to carbon capture and utilization programs by converting biomass feedstocks into bio-based products, lowering greenhouse gas emissions, and advancing carbon neutrality (Chen et al. 2023). Biorefineries are essential in the process of creating a circular economy because they close the loop on biomass feedstock, waste streams, and by-products while optimizing resource use and minimizing environmental effect.

## 11.4 Raw Materials for Biorefinery in Biomaterial Production

Biorefineries can use a range of platforms, depending on the feedstocks used and the technological processes involved, such as carbon-rich chains, sugar (biochemical), syngas (thermochemical), or biogas (Adu et al. 2018). To maximize resource usage, these platforms can integrate several processing routes and procedures from many categories. The main feedstock sources for biorefineries can be classified as lignocellulosic biomass or crops, which combine cereals with residues as a backup feedstock supply (Hagman et al. 2018). An important consideration in deciding the sustainability and profitability of biorefinery operations is the selection of feedstock generation as shown in Table 11.1.

**Table 11.1** Classification of biorefinery feedstock

| Classification | Feedstock's | Source | Name | Reference |
|---|---|---|---|---|
| First generation | Dedicated crops | Oil crops | Soybean, sunflower, rape, palm, Jatropha, | Goswami et al. (2022) |
| | | Sugar crop | Sugarcane, beet | |
| | | Starch crop | Potato, corn, wheat, cassava, sorghum | |
| | | Wood plantations | Softwoods, hardwoods | |
| Second generation | Residues/ wastes | Lignocellulose | Paper mill waste sludge, black liquor, bagasse, rice straw, timber house waste, wheat straw. | Patel and Shah (2021) |
| | | Fats and oils | Animal farm wastes, fat from slaughterhouse, municipal wastes, Slaughterhouse trimmings and bones, vegetables and fruit processing wastes, waste cooking oil. | |
| Third generation | Microalgae | Aquatic plants, algae and cyanobacteria | Algae (micro or macro) | Lara et al. (2020) |
| Fourth generation | $CO_2$ | Non-edible crops | *Jatropha or Karanja* | Esteghamat et al. (2024) |

First-generation feedstock's usually come from specially planted food harvests such by way of corn and sugarcane. However, because they compete with food production and land use, these feedstocks provide serious social, economic, and environmental issues (Moncada et al. 2016; Hamda et al. 2023). As a result, there has been a move toward second-generation feedstocks, which are more sustainable options. The disadvantages of first-generation feedstocks can be mitigated by using second-generation raw materials, such by way of ligno-cellulosic biomass derived from forestry and agricultural waste. Crop leftovers and byproducts from agriculture are excellent sources of feedstock that don't conflict with the supply of food. These lignocellulosic components, which include lignin, cellulose, and hemicelluloses, be able to remain effectively handled to create an extensive range of distinct produces (FitzPatrick et al. 2010)

Woody biomass from forests, while less seasonal than agricultural wastes, nevertheless needs to be carefully considered in relations of sustainability of the economy and environment, depending on the nature and location of the forest (Yang and Yu 2013). The usage of agricultural leftovers as raw materials for bio-refineries is a topic of ongoing controversy because it raises several concerns. Despite this, each tonne of farming leftovers applied in a biorefinery system has the potential to produce multiple times as much renewable energy as non-renewable energy while also drastically reducing greenhouse gas emissions.

Additionally, third-generation feedstocks use biomass sources other than food, like aquatic plants and algae, to produce high-value chemicals, medicines, and biofuels. Particularly in light of their high biomass production, sustainability, and potential for carbon capture, algae-based feedstocks have drawn interest (Mohammad et al. 2023). As overall, choosing the right raw materials for biorefinery operations is essential to maintaining sustainability, reducing negative environmental effects, and maximizing the use of available resources. Biorefineries have the potential to make a substantial contribution to the shift towards a bio-based economy that is more efficient with resources and sustainable by employing a variety of feedstock sources.

## 11.5 Classification of Biorefinery

Feedstocks, products, and conversion procedures can all be used to categories biorefineries.

### 11.5.1 Feed Stock Based Classification

The raw materials that are employed in a biorefinery are referred to as feedstock. A vast range of plant species, each with a unique physical structure and chemical makeup, are included in the general term biomass. Biomass may be divided into four primary categories, regardless of phenotype: lipids, proteins, carbohydrates (including cellulose, hemicelluloses, and starch), and lignin. Biomass, which is created by photosynthetic reactions between atmospheric carbon dioxide and water, yields sugars (Chaturvedi et al. 2020; Hu et al. 2020). This sugar is used by plants to create complex compounds that provide them energy (Temesgen et al. 2024). The process by which proteins are made in biomass is largely similar to that of other eukaryotic creatures, as amino acids are the components that make up proteins. Nitrates, a kind of nitrogen, are extracted by plants from their roots and transformed into amino acids. An elongating peptide chain is created by ribosomes accumulating amino acids. Then, in order to create proteins, chemical groups such as phosphates or other compounds are usually attached to this chain. Feedstock is an essential requirement for a biorefinery, and it must be consistent, reliable, and renewable; in fact, this feature sets several different biorefinery designs apart (Cherubini 2010). Feed stock based bio-refineries remain delivered from five different sources.

A. *Agricultural Biorefineries:* Produce biofuels, biochemicals, and other bio products by using crops like corn, sugarcane, wheat, and so on.
B. *Forestry Biorefineries:* To create biofuels, biochemicals, and other bio products, utilize the wood biomass found in forests.
C. *Aquatic Biorefineries:* Produce high-value chemicals, biofuels, and other products by using the biomass from algae.
D. *Municipal Solid Waste (MSW) Biorefineries:* For the synthesis of biochemicals and biofuels, utilize the organic portion of municipal garbage.

E. *Industrial biorefineries:* Utilize process residues and leftovers.

The three components that make up the main biomass are carbohydrates, lignin, mixed organic residues, proteins, and triglycerides (Cherubini 2010).

#### 11.5.1.1 Lignin and Carbohydrates

Carbohydrates (found in cellulose, hemicellulose, and starch) are the utmost frequent biomass component in plant feedstocks; they are molecules of oxygen, hydrogen, and carbon. Arabinose and xylose are the utmost prevalent 5-carbon sugars ($C_5H_{10}O_5$), while mannose, galactose, and glucose are the 6-carbon, 1-molecule "monosaccharide" sugars ($C_6H_{12}O_6$). Sugar cane and sugar beetroot are the two utmost significant sugar harvests and contribute almost all of the ethanol generated in the world today, together with corn, a commodity used to make carbohydrates (Tsai et al. 2007; Rajagopal and Zilberman 2008).

The relatively starch ($C_6H_{10}O_5$) is large polymer which made up of hundreds or thousands of glucose molecules, or polysaccharides. Prior to fermentation, the starch molecule must be broken down into one or two pieces. The two most commonly produced starch crops are wheat and maize (Koutinas et al. 2005; Deswarte et al. 2007; Humbird 2011). Sugars may be effortlessly fermented to harvest bioethanol or utilized as a feedstocks for a range of chemical processes that produce an extensive spectrum of organic compounds when they are extracted (for sugar crops) or depolymerized (for starch crops).

The three primary constituents of lignocellulosic biomass materials are generally composed of lignin, hemicellulose, and cellulose (Gari et al. 2024). Made from lengthy chains of glucose molecules ($C_6$ sugar), cellulose ($C_6H_{10}O_6$) has a sturdy molecular arrangement (Mosier et al. 2005). Starch differs from other substances due to the arrangement of bonds transversely the oxygen molecule that connects double hexose divisions. While the separate glucose monomers in starch be able to freely hydrolyzed by acidic assaults or enzymes, it is considerable supplementary challenging to hydrolyze and liberate individual glucose monomers from cellulose, which makes about 30–50% of all lignocellulosic dry matter (Hamda et al. 2024). Hemicellulose ($C_5H_8O_5$) is a more amorphous ingredient than cellulose, breaking down more quickly in the presence of heat and/or chemicals due to its mixing of $C_6$ and $C_5$ sugars.

Constituting 20–40% of the feedstock's dry material, it is the subsequent principal constituent of ligno-cellulosic bio-mass. Lignin ($C_9H_{10}O_2$ ($OCH_3$) n), which is composed of phenolic-polymers, is essentially the glue that gives the general stiffness of plant and tree structures. Unlike hemicellulose and cellulose, which are polysaccharides that be able to hydrolyzed to sugars and then further fermented to bioethanol, lignocellitate cannot be utilized in fermentation practices, while it may be advantageous for other uses (energy generation or chemical extraction). Lignin, derived primarily from lignocellulose, makes up 15–25% of the dry matter feedstock in total, the mainstream of which is non-carbohydrate. Figure 11.2 demonstrates lignin structure.

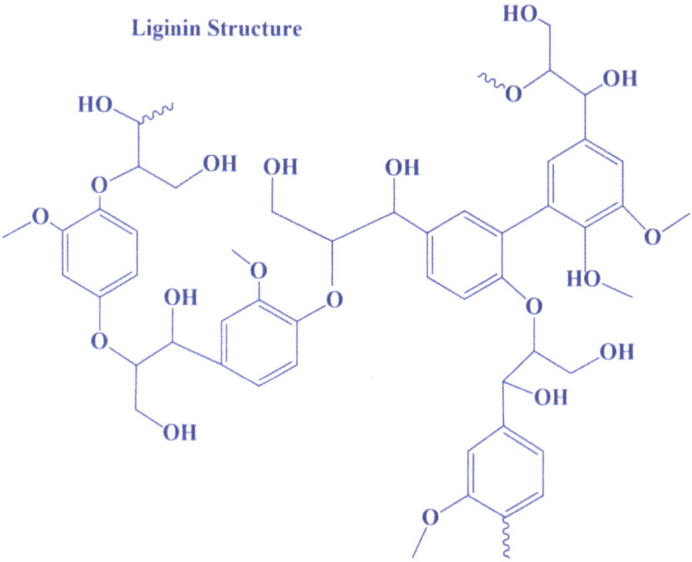


**Fig. 11.2** Lignin structure

Lignocellulosic biomass can be supplied in two ways: as a crop and as residue. Cellulosic biomass can be produced in considerable amounts by specialized harvests, such as shorter rotation woody harvests or perennial herbaceous plant species. Some more sources of lignocellulosic biomass and waste are forestry residue. In the pulp and paper industry, wood waste is generated, and agricultural straw (Jayakumar et al. 2023a). Using waste biomass offers a way to replace fossil fuels with material that would typically decay, bringing value to society without requiring new land for production (Chaturvedi et al. 2020). Figure 11.3 shows an illustration of a typical lignocellulosic structure.

#### 11.5.1.2 Triglycerides

Saturated and unsaturated fatty acids, glycerol, and other triglycerides make up oils and fats. They range in chain length from $C_8$ to $C_{20}$, with the most common numbers of carbons being 16, 18, and 20. Oils and fats are derived from animal and vegetable raw resources, yet the most common feedstock for oil production worldwide is sunflower, soybean, palm, rapeseed, and oil (Demirbas 2010). These feedstocks are also the most significant in terms of globally consumed oil. Nowadays, vegetable oils are used to treat low molecular weight alcohols like methanol or ethanol in the existence of a base catalyst like KOH or NaOH. This process produces glycerin and esters, which are subsequently used to make biodiesel. If methanol is the alcohol utilized, biodiesel is stated to chemically as a methyl-ester; if ethanol is used, the term is an ethyl ester. Oils derived from vegetables include dual reactive sites: the double bond of the unsaturated fat chain and the acid group of the fatty chain (Gizaw et al. 2024). Vegetable oil can be a good alternative for producing biodiesel because

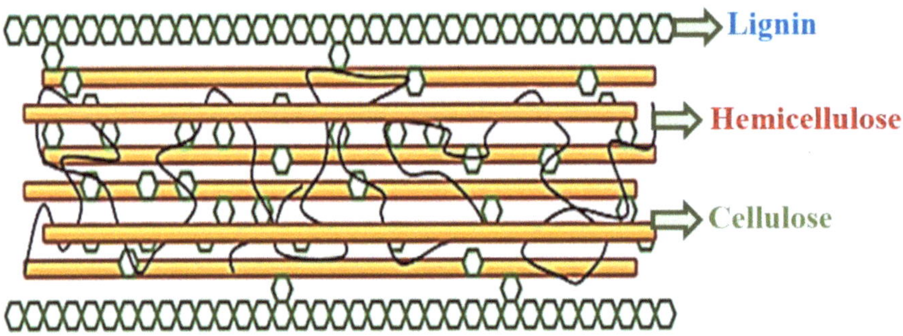


**Fig. 11.3** Typical lignocellulosic structure

of these two reactive sites (Sims et al. 2010). The high input utilization and low yield of oil seed crops with the highest glucose recoveries of 75% and 88% are comparable to those of sugar and starch crops (Pan et al. 2006). Nevertheless, it will become very contentious in the future to use food crops, notably oil crops, to produce fuel. Subsequently, non-food crops like *Salicornia bigelovii* and *Jatropha curcas* will proliferate as the most popular oil crops for biorefinery applications because they are more suitable to marginal areas and require less inputs (Achten et al. 2007; Cybulska et al. 2014). Vegetable oil waste is easily obtainable from manufacturing facilities, homes, and the food sector. They are an excellent first step toward creating a circular economy and can also be useful possibilities for obtaining feedstock (Tsai et al. 2007).

#### 11.5.1.3 Diverse Organic Divisions

Proteins, leftovers, and manure from the vegetable and fresh fruit processing sectors are included in this class, in the organic fraction of municipal solid waste, this is also called OFMSW (Organic Fraction of Municipal Solid Waste). These biomass types are impossible to identify due to the wide range of physical and chemical qualities that result from their unique chemical compositions. Some streams are less desirable for high-temperature processes like gasification because of their extremely tall moisture content upto 70 percent. Waste from food production, manure obtained from cow farms, and dairy processing sectors, in addition sewage sludge are a few examples of these streams (Chaturvedi et al. 2020). High moisture content feedstock is more suitable for producing biogas through anaerobic digestion. Other streams often have a great potential for energy recovery, such as OFMSW combined with plastics, paper, and cardboard. Since the various qualities and traits of biomass waste necessitate distinct conversion methods, an integrated processing network for biorefineries appears to be the best option for this kind of feedstock.

#### 11.5.1.4 Proteins

The structural components of proteins are polypeptide chains. Combining polypeptide chains with lipids results in lipoproteins, while polysaccharides with polypeptide chains results in Mulder (2010). Proteins have molecular weights ranging from dozens to millions of daltons. Protein molecules fall into two broad classes: globular and fibrous proteins. Generally, elongated and insoluble proteins are fibrous. The majority of globular proteins have a spherical, soluble, and compact structure. Spherical components can occasionally be combined by non-covalent forces to practice higher, more well-defined structures in more complex proteins. Animals and plants both have proteins. The principal cradles of plant derived proteins for human consumption are legumes (such beans, chickpeas, green peas, and lentils) and cereals (like wheat, barley, and sorghum). The food and non-food sectors employ casein, soy, whey based proteins, and wheat which are produced commercially in response to the growing need for proteins (Mulder 2010). The use of proteins by biomasses like grass and algae has been investigated (Vanthoor-Koopmans et al. 2013; Soh et al. 2014). Apart from the aforementioned protein sources, the biorefinery technique advocates for the consumption of waste streams, including the regaining of protein from press cakes that remain after the extraction of oil from plant derived oil-seeds like rapeseeds or sunflower (Anastopoulos et al. 2015; Kachrimanidou et al. 2015).

### 11.5.2 Processes Based Classification

Biorefining uses technical processes to deoxygenate and depolymerize biomass in order to transform it into value added end produces. Many times, one or more technical procedures need to be used afterwards or in simultaneously. As they guarantee that the lipids, sugars, carbohydrates, and other components may be made accessible for additional refining, these procedures are also known as the biomass/feedstock fractionation techniques. The processes under consideration can be broadly categorized into five groups: biological, hydrothermal, chemical, mechanical/physical, and thermochemical.

#### 11.5.2.1 Processes Related to Thermochemical

The dualistic foremost thermochemical practices that release energy and change biomass into chemical compounds are pyrolysis and gasification. The process of gasification includes heating biomass to a tall temperature (>700 °C) and short $O_2$ levels in order to harvest a mixture of CO, $H_2$, $CO_2$, and $CH_4$ (Caputo et al. 2005; Jayakumar et al. 2023c). Alternatively, syngas can be used as a platform for chemical reactions to create fuels (such as ethanol, alcohols, dimethyl ether, FT-fuels, and isobutene) or compounds (such as alcohols, ammonia, methanol, and organic acids). Biomass is broken down by pyrolysis when it is between 300 and 600 °C and there is no oxygen present. After pyrolysis, three products are produced: solid charcoal, light gasses, and liquid pyrolytic oil (Du et al. 2012). Contingent on the specifics of the practice, pyrolysis practice influence to produce different product yields. Nevertheless, because it is a prerequisite for the subsequent synthesis of several chemicals and fuel, bio-oil is the most sought-after byproduct of pyrolysis (Cherubini 2010). Using biofuel as a means of transportation Biofuel has drawbacks, and its potential as a raw material for chemicals is still untapped (Ruiz et al. 2020). Practice of direct combustion, is one of the most popular and oldest thermo-chemical practice, involves burning biomass in an $O_2$ rich atmosphere principally for the purpose of producing warmness. Conversely, new and more effective processing paths have advanced as a result of this technology's shortcomings and inefficiencies.

#### 11.5.2.2 Lower Temperature Practice

Hydro-thermal treatment practice, which classically comprises heating biomass in $H_2O$ at tall pressures (up to 20 bar), and also with higher temperature range lied between 120 and 250 °C, and residence periods 10 to 50 minutes, is one of the most widely utilized lower temperature procedures (Gundupalli et al. 2022; Ruiz et al. 2020). One of the most common types of hydrothermal therapy is liquid hot water, while another is un-catalyzed steam explosion (Ruiz et al. 2020). Large-scale application of this technique typically eliminates the need for previous size reduction, which lowers operational and capital expenses (Taherzadeh and Karimi 2008). Conversely, Hosseini and Shah claimed in their study that reducing the size of wood chips can result in a 50 percent rise in energy-efficiency (Hosseini and Shah 2009).

The hydrothermal process, which also alters the lignin arrangement by coagulating, melting, and in addition depolymerizing the lignocellulosic fibers, removes lesser quantities of lignin (the acid soluble portion). Consequently, hydrothermally processed materials cannot yield lignin in its functional state (Heitz et al. 1988). However, by expanding the cellulose's accessible surface area by non-chemical swelling, the hydrothermal treatment considerably increases the number of promising sites for enzyme action. The ruthlessness of pre-treatment practice can be summarized using a single component, Ro ($Ro = t\ exp\ ((T - 100)/14.75)$, which relations the possessions of temperature, T (0C), and time, t (min) (Heitz et al. 1988). Because of its tall severity (over 3.0), steam pretreatment is unique of the furthermost extensively used approaches for producing high concentrations of products of sugar breakdown. Conversely, at high process severity levels (over 4.0), inhibitors such as 5-hydroxyl-methyl-furfural and furfural (from the dehydration of hexoses) are also generated. During the treatment, acetic acid and breakdown products were generated. Additionally, minor quantities of formic and levulinic acids were also created, which also inhibited yeast. At advanced levels, levulinic acid is added inhibitory than acetic acid, and formic acid is additional inhibitory still. The degree to which the pretreatment is harsh determines the concentrations at which these substances become inhibitory. Levulinic acid, formic acid, and acetic acid concentrations up to 100 mmol/l are reported by Larsson et al. to enhance ethanol output; however, the presence of significant amounts of these compounds reduced ethanol yield and other fermentation microorganisms (Larsson et al. 1999). The lengthy sequence polymers that are left behind after lignin breakdown are another type of by-product (Thomsen et al. 2009). Since biogas and ethanol remain the completion harvests of the process, other organic acids, levulinic, formic, and acetic acid are classified as degradation products, but in a biorefinery, they influence as healthy be the goal produces due to their tall market price. It is also possible to optimize the process kinetics to maximize their output while ignoring the development of the previously anticipated products, such as biogas and ethanol.

As little as a few seconds are required for treatments at high temperatures (up to 230 °C), while several hours are required for treatments at low temperatures (150–180 °C). The severity factor may be used to quantify the trade-off between these requirements (Garrote et al. 1999). In the literature, it is generally accepted that steam explosion is not a part of hydrothermal treatments. The primary process step that has an impact is the progressive heating and cooling that takes place prior to and following pretreatment, respectively (Ruiz et al. 2020). This step is absent in the event of a steam explosion. Due to its ease of use and lack of chemical requirements, hydrothermal treatment is becoming increasingly popular as a pretreatment technique for the ethanol sector.

#### 11.5.2.3 Biochemical Processes

Compared to thermochemical routes, biochemical activities occur at comparatively lower temperatures (below 80 °C). Anaerobic digestion, enzymatic hydrolysis, and fermentation are the three most prevalent kinds of biochemical processes (Jayakumar et al. 2023d; Chozhavendhan et al. 2023). Enzymes and/or microorganisms cause the substrate to ferment, turning it into alcohols and organic-acids. Methane and $CO_2$ are manufactured by bacteria breaking down the substrate's biodegradable organic material during anaerobic digestion, which takes place at temperatures range lied between 35 and 55 °C in the absence of oxygen. Because of the widespread need for liquid fuel for shipping and the prompt identification of the sugarcane-ethanol-pathway (and, later, the corn-ethanol pathway), ethanol is the one of the utmost widely produced fermentation creation. Nonetheless, there is a rising body of research investigations being done on the synthesis of other chemicals, including succinic acid, methanol, and hydrogen. Fermentation has been widely accepted as a process since Louis Pasteur's discovery in the 1870s that bacteria can convert carbohydrates to alcohol without the need for oxygen. Even though the process was discovered early on, the use of sugars to produce fuel has been sporadic and fairly sluggish. Genetically engineered microbes are being used in biochemical processes to manufacture specific chemicals but $C_6$ sugars remain the most often utilized fermentation substrates, even though specialized fermentation organisms are needed for the conversion of other hydrocarbons, glycerol, and pentose (sugars from hemicellulose) to other value added products and ethanol (Argyros et al. 2011). Other gases, methane, and carbon dioxide are combined to form biogas, which is the primary byproduct of anaerobic digestion. It is possible to refine anaerobic digestion methane to a 97% purity and utilize it in place of natural gas (Montgomery and Bochmann 2014).

For lignocellulosic biomasses to be converted to sugars for ethanol fermentation, pretreatment is a necessary step; however, other products require unique procedures. It entails dissolving a naturally occurring carbohydrate–lignin barrier that prevents some enzymes from accessing cellulose and hemicellulose. A stream high in monomeric sugars is produced during the physical, chemical, or enzymatic pretreatment of lignocelluloses (Chen et al. 2011). These sugars can then be fermented biologically to produce biogas and/or ethanol.

#### 11.5.2.4 Mechanical Practices

Mechanical practices simply accomplish size reduction or feedstock component separation; they do not alter the content or condition of biomass (Fitzgerald and Themelis 2009). These are usually the foremost phases in a system of biorefineries since the conversion of biomass needs lesser units to afford an enormous surface area in order for the practice to be effective (Taherzadeh and Karimi 2008). Reducing the size of biomass involves either cutting or commuting operations that modify the bulk density, shape, and size of the biomass particles. While extraction techniques take valuable molecules out of large, heterogeneous substrates and concentrate them, separation operations separate the substrate's components.

#### 11.5.2.5 Chemical Processes

When biomass combines with other substances in the presence of catalysts, chemical reactions have the ability to change the feedstock's molecular makeup. The two chemical processes that are most frequently employed to convert biomass are enzymatic hydrolysis and transesterification. Hydrolysis is the process of breaking down polysaccharides into monomeric sugars (cellulose into glucose, for example) by using acids and alkalis. Vegetable oil is transesterified to produce biodiesel, sometimes referred to as methyl or ethyl esters of fatty acids (Jayakumar et al. 2023b; Chozhavendhan et al. 2018). Glycerin, a chemical molecule with a variety of industrial applications, is also produced in tandem with this procedure (Demirbas 2010). Methanization, steam reforming, and Fisher-Tropsch synthesis are further significant and frequent chemical processes in biorefining (Conde-Mejia et al. 2012).

### 11.5.3 Product Based Classification

There are two main categories into which the output of biorefinery systems may be divided: chemicals and energy as shown in Table 11.2. Material goods are utilized for their chemical or physical qualities, whereas energy products are employed to provide transportation, heat, or power. Some products, such as bio-ethanol or bio-hydrogen, can be employed as chemical compounds in chemical synthesis or as fuels; therefore in some circumstances, a further differentiation is necessary.

The primary objective is to produce goods from a biorefinery that have qualities comparable to those that are currently provided by petrochemical refineries. In this case, biomass turned out to be a viable substitute for fossil fuels, offering noteworthy environmental advantages as well as comparable benefits farther down the line, like petrochemical refineries. The foremost raw materials utilized in the manufacture of bulk chemicals nowadays are natural gas and crude oil. Petrochemicals have grown up with oil refining, taking advantage of cheap and easily accessible oil sources to thrive. Fourteen percent of the world's daily oil consumption and around 8% of gas are attributed to the petrochemical industry (Chaturvedi et al. 2020). Because of their strong reliance on hydro-carbons, the petro-chemical productions have joined forces with gas and oil refineries to grow into huge, stretchy, and incredibly productive manufacturing facilities. These facilities fully utilize the benefits of

**Table 11.2** Classification of bio refinery products

| Classification | Products | Examples | Reference |
|---|---|---|---|
| Energy products | Gaseous biofuels | Biogas, syngas, hydrogen, bio-methane | Padilla-Rivera et al. (2019) |
| | Solid biofuels | Pellets, lignin, charcoal | Manandhar et al. (2022) |
| | Liquid biofuels | Bioethanol, biodiesel, FT-fuels, bio-oil | Chiong et al. (2018) |
| Chemical and material products | Compounds (chemicals) | Building blocks, bulk chemicals and fine chemicals | Sieber et al. (2018) |
| | Organic-acids | Lactic acid, other sugar derivatives, levulinic acid, and succinic acid | Varriale and Ulber (2023) |
| | Resins and polymers | Furan resins, phenol resins, and starch based plastics | Ortiz et al. (2019) |
| | Bio-materials | Cellulose, paper, pulp, rayon, cellophane adsorbent, dispersants, emulsifiers, adsorbents, vanillin, adhesives, wood panels | Enawgaw et al. (2023) |
| | Animal feed and food | Protein rich | Karthikeyan et al. (2017) |
| | Biopolymers | Polylactic acid | Redondo-Gómez et al. (2020) |

co-location to access a range of energy and raw material streams for the affordable manufacture of chemicals.

Aromatics, butylene, propylene, and ethylene are among the primary chemical intermediates generated by petrochemical complexes. After that, these intermediates undergo additional processing to create intermediates for the detergent, solvent, plastic, and textile industries. With approximately 75 million tones generated annually worldwide, ethylene is the most manufactured intermediate chemical by volume (Werpy and Petersen 2004). It is possible to dehydrate bioethanol to produce ethylene, which may then be transformed into propylene, butylene, and aromatics. This procedure shows how biorefinery-based models can be quickly and readily implemented in existing markets.

## 11.6 Current Technologies of Biorefinery to Produce Biomaterials

The methods used in biorefineries today to produce biomaterials cover an extensive range of procedures meant to effectively convert feedstocks of biomass into various value added products with additional value. These technologies extract, modify, and manufacture biomaterials with a variety of features and applications by utilizing advances in chemistry, engineering, and biotechnology (Mohammad et al. 2023;

Kasirajan et al. 2022). Key technologies used in biorefineries to produce biomaterials include some of the following:

### 11.6.1 Biological Conversion

Enzymes, bacteria, or entire cells can catalyze reactions that convert biomass into biomaterials. In fermentation, for example, sugars and other organic substances are broken down by microorganisms to create biomolecules including lipids, polysaccharides, and proteins. Another biological conversion technique is enzymatic hydrolysis, in which enzymes reduce complex biomolecules, such as cellulose, to simpler sugars so they can be processed further.

### 11.6.2 Chemical Conversion

Chemical conversion methods are based on the synthesis of biomaterials by the manipulation of biomass components through chemical processes. In an alkaline or acidic solution, for example, lignocellulosic biomass is hydrolyzed to yield sugars, lignin, and other compounds that can be further processed into biomaterials. Furthermore, biomolecules can be altered by chemical processes like esterification or polymerization to produce new materials with the necessary characteristics.

### 11.6.3 Thermochemical Conversion

By applying pressure and heat to biomass feedstocks, thermochemical conversion methods reduce them into smaller molecules that can be utilized to create biomaterials. For example, pyrolysis is heating biomass without oxygen to create syngas, charcoal, and bio oil, which be able to then be processed promote to create bioplastics or bio-based compounds.

### 11.6.4 Biopolymer Synthesis

The synthesis of biodegradable polymers from renewable biomass sources is made possible by biopolymer synthesis technology. This is the process of fermenting microorganisms to harvest biopolymers, like polyhydroxyalkanoates (PHAs) and polylactic acid (PLA), which can be used in place of conventional plastics in a range of applications.

### 11.6.5 Nanotechnology

Biorefinery techniques are progressively incorporating nanotechnology to generate biomaterials with improved characteristics and capabilities. For instance, nanocellulose can be separated from biomass and utilized to create biomedical devices, enhance packaging sheets, and reinforce composite materials.

### 11.6.6 Bio-based Chemicals and Fine Chemicals

Biorefineries also produce fine chemicals and bio-based substances using a range of chemical and biological processes. These materials can be combined to form a variety of biomaterials with specific properties, including solvents, adhesives, coatings, and plant-based resins.

### 11.6.7 Bioprocessing and Downstream-Processing

The purification and refinement of biomaterials manufactured in biorefineries is largely dependent on bioprocessing and downstream processing technologies. Separation techniques including chromatography, filtration, and distillation are used to separate biomolecules and eliminate impurities in order to guarantee the quality and purity of the final biomaterial products. All things considered, these technologies help biorefineries create a broad variety of biomaterials with a variety of qualities and uses, which advances the creation of biobased and sustainable substitutes for conventional materials.

## 11.7 Techno-economic Aspects of Biorefinery to Produce Biomaterials

In mandate to solve the eco-friendly issues brought on by the practice of fossil fuels, research into renewable energy sources is growing. The balance between energy consumption and performance is an important consideration in biorefinery development. A biorefinery must produce more energy in the form of electricity, heat, gaseous, or liquid fuels than it takes in from its several process units. Therefore, energy analysis is important in these processes. To ascertain whether biorefineries are feasible, technical and financial research is required. The advantages and disadvantages of each process can be determined by technical and economic assessments, which can also offer development objectives and opportunities for biorefineries. Techno-economic analyses, for instance, have been utilized lately to assess the viability of using lignocellulosic materials to produce biodiesel, bioethanol, and biogas (Rahimi et al. 2020). A facility called a biorefinery converts renewable feedstocks (biomass), comprising waste streams and resources, into a range of products with added value, including feed, food, feedstock, other bio-based products, and biofuel.

Thus, the development of biorefineries for biofuel and food, for example, has remained given priority by bio-economy policies in various nations, particularly in the Asia and European Union (EU), as a way to address global environmental concerns like the exhaustion of fossil fuels and climate change (Silalertruksa et al. 2022; Venkatesa Prabhu et al. 2023).

### 11.7.1 Recent Technological Advancement of Biorefinery System

The efficiency, adaptability, and sustainability of the biorefinery industry have all greatly benefited from recent developments in the technology. One significant area of progress is the diversification of feedstock options, as scientists are investigating unusual sources such as algae, aquatic plants, and organic waste streams (Freitas et al. 2021). Additionally, new biomass pretreatment techniques are improving the accessibility of lignocellulosic material.

In order to enhance overall process efficiency and production, biorefinery systems have integrated various conversion paths, integrating thermochemical, catalytic, and biological approaches (Arevalo-Gallegos et al. 2017). This is another significant advance. The integration of biorefinery products into existing arrangements for energy and food provides important additional dimensions that may be analyzed in terms of the role of biorefinery products in constructing and negotiating energy and bio-product pathways and the influence of biorefineries on shaping those pathways that create the demand for and supply of biomaterials and bio-products (Chaturvedi et al. 2020; Hagman et al. 2020; Casa et al. 2021). Furthermore, the translation of biorefinery concepts from laboratory tests to commercial operations has been made easier by developments in process optimization and scale-up approaches, supported by advanced modeling and simulation tools. Furthermore, tactics like cascade usage and integrated biorefinery concepts are being used to maximize the value extracted from by-products and waste streams generated during the conversion process.

Plant, machines, and processes establish the three key dimensions of technological innovation in the biorefinery. A shift to a multidisciplinary approach based on systemic analysis is necessary, while the prevailing focus on crop selection reflects the dominant role of agricultural science in defining the state of the art in traditional rural biorefineries. The goal of an ideal biorefinery process should be to demonstrate environmental viability, generate a variety of products, and utilize the entire crop. Facilities should aim to be small-scale and adaptable, offering potential opportunities for various forms of cooperation between research, business, and local communities (Dragone et al. 2020; Cavalcanti Pessôa et al. 2022; Goswami et al. 2022).

The variety of bio products has also increased, with new molecules having uses in a range of industries being produced through fermentation, metabolic engineering, and bio catalysis (Dessbesell et al. 2017). The sustainability of biorefineries is being further improved by sustainable waste management techniques, such as the incorporation of carbon capture and use plans and waste-to-energy technologies.

Finally, biorefinery operations are being revolutionized by the introduction of digitalization and data analytics tools, which are optimizing process control, resource allocation, and decision-making. All things considered, these advancements position biorefineries as pivotal players in the transition to an additional resource efficient and ecological bio-based economy.

### 11.7.2 Economic Viability of Biorefinery for Biomaterials

Access to feedstock and processes for bio-chemicals has signified peculiar traction. This can be attributed to the economic feasibility linked with abated raw material costs and demand for eco-friendly and sustainable processes in the chemical industry. The literature cited in the paper demonstrates the augury potential and profit capacity for bio-chemical production. The use of fossil fuels has attracted ascending costs due to reinforcement of the alternation of fossil resources with bio-based nominal value. Therefore, availability of feedstock and its techno-commercial aspects are critically important for the initiation and triumph of a biorefinery. Bio-chemicals are reported to have superior market potential compared to bio-fuels. This is because value of chemicals values is higher than that of substantial and their market demand is perennially persistent. Chemical purification processes anglice—far from equilibrium methods, such as distillation and liquid-liquid extrication—are metabolically energetically subsidiary processes as a result of the thermodynamic driving force inherit in rectifications. Alternatively, the deployment of bio-fuel products is periodically subject to rigorous governmental regulations and policies in terms of $CO_2$ mitigation, tax exemption, and compulsory contingency consumption (Velankar et al. 2022). For example, in the European Union, the renewable kinetic energy sector has experienced unparalleled bulge in improvements over the last decennium under the legislative and policy measures for climate accommodations and environmental risk mitigation. However, for a paramount sum of biorefineries at an incipient stage, ethanol has been the most frequently preferred bio-fuel product. The replication of such biorefineries in Europe may promise advanced self-sustainability as the year progresses. However, the desideratum of achieving a high market share of petro-independently of ethanol kinetic energy is subject to restrictive policies and the emerging developments of the second and third generations of bio-fuel products. On the converse, due to breakthroughs in irrigation efficiency, genetic engineering for advanced crop productivity, and the integration of more environmentally pellucid instrumentations, the biomass thematic area of bio-chemicals has gained the attention of the chemical industry in the utilization of renewable carbon feedstocks for syntheses (Baral et al. 2022; Narisetty et al. 2022; Velankar et al. 2022).

## 11.8 Sustainability of Biorefinery in the Waste Management System

### 11.8.1 Waste Biorefinery

An economical and ecologically friendly substitute for conventional waste treatment techniques are bio-waste treatment facilities. Technology and conversion products selected in these biorefineries are heavily influenced by the raw materials they use (Hamda et al. 2024). The majority of studies classify biorefinery waste based on their biomass waste source, which includes food, dairy, and forestry residues. Biodegradable biomass waste with a high moisture level is referred to as organic waste, whereas trash that is lignocellulosic has a tall cellulose, hemicellulose, and lignin content (Ochieng et al. 2022). A key strategy for lowering greenhouse-gas-emissions from fossil fuels whereas preserving land usage and food security is sustainable bioenergy manufacture. In order to handle upcoming waste treatment difficulties, bio-waste treatment plants can potentially be integrated into waste management systems. However, biomass waste is considered low quality due to its diverse composition and seasonal variations. Biomass waste poses value-added conversion technology challenges, requiring additional research to increase the profitability of bio-waste treatment plants. To achieve economic and environmental sustainability, bioprocesses must be linked to process diverse biomass sources and produce a variety of bioenergy products. Sorting biomass waste based on existing biological treatment processes is essential to maximize integration, reduce unnecessary processes and increase production costs of biorefineries. The suitability of waste-to-energy technologies and integrated biorefinery design based on biomass sorting are examined in this study (Nanda and Berruti 2021; Ochieng et al. 2022).

The increasing rates of waste generation and the environmental burdens have become critical societal issues and mainstream research topics. Successful and sustainable waste administration is the main challenge in urban areas, as the cities are expanding and the amount of waste keeps on increasing. The primary goals of waste management, as stated by the Environmental Protection Agency (EPA), are the recovery of resources and energy from trash in order to create a sustainable society. In this sense, a biorefinery is a competence that combines cutting-edge transformation technology to practice lignocellulosic biomass into value added biofuels, electricity, in addition higher cost biomaterials for various applications. Nowadays, many of the biorefinery concepts are being developed and tested on pilot and/or demonstration scale. The materials and energy obtained from fossil resources can be substituted with these biorefinery products, and they also performance a dynamic part in achieving the goals of the waste hierarchy, a waste management, and resource recovery technique. Additionally, valorization of waste and waste derived resources, and minimization of the land-filling were identified as the main objectives and targets of the biorefineries with respect to the waste hierarchy strategy. It is believed that waste management has a great potential to provide sustainability in the socio-economic systems and to preserve the stable environment for the next generations.

However, the achievement of sustainable waste management, in particular in urban areas, is a challenging goal, which is considered both as an ideal and a practical future objective. However, a biorefinery offers pathways for knowledge transfer, energy generation, and economic growth in addition to solutions for the creation of a sustainable circular economy and the ability to lower annual environmental impacts per person (Duque et al. 2021; Muh et al. 2021).

### 11.8.2 Importance of Sustainable Waste Management

Waste's effects on local and global communities' health and ecology are a major source of concern these days. Municipal solid waste recycling and composting are crucial because they keep the waste out of landfills and instead direct it towards more effective transformation activities. Extracting useful items from the waste stream conserves raw materials, reduces the environmental impact of mining and manufacturing, thereby leading to less pollution and more balanced ecosystems. The recovered materials can generate income or reduce the cost of waste disposal while the compost or other recovered organic matter generated from the waste can be utilized effectively as a renewable resource in a sustainable biorefinery operation. However, as much as it is essential to implement biorefinery technologies, it is vital for an overall sustainable waste management system to be in place to optimize the advantages of such an operation. This is because sustainable waste management focuses on recycling and reusing the materials in the waste which is important for a biorefinery that intends to harvest higher value produces from the feedstocks. In addition to that, a sustainable waste management system aims to reduce the reliance on landfills for the disposal of the waste as it is not an efficient or cost-effective way to manage waste. Instead, the selection of technologies in waste management should look to move up the waste hierarchy so that the value of the waste in terms of material or energy recovery can be exploited before the disposal of the waste. This is also crucial as landfills release greenhouse gases such as methane which contribute to climate change, and once the waste is landfilled, the energy content of the waste is effectively lost. By emphasizing a sustainable waste management system, landfill dependence as a means of disposing of garbage is minimized in addition to the amount of waste that is dumped there. This would be in line with the European Union's initiatives to promote a recycling society where products, resources, and materials are retained for as lengthy as promising at their fullest potential in the economy (Nanda and Berruti 2021; Tang et al. 2021).

### 11.8.3 Waste Management and Biorefineries Role

Biorefinery appears to have a pivotal role in the framework of waste management. The knowledge of a biorefinery is parallel to that of contemporary fuel refineries, which turn crude petroleum into a diversity of goods and fuels. Biorefineries objective to harvest an extensive variety of bioenergy and goods by utilizing biomass to the fullest extent possible. Using a variety of technical techniques and technologies common to biorefineries, waste streams are separated and converted into energy and products with additional value (Ochieng et al. 2022). The output of the biorefinery, for example, bioenergy and biofuels, can be used to electrify the waste management process. From this position, the solid waste will be diminished and as well minimized the reliance on non-renewable fossil fuels in the waste management machineries. On the other hand, waste is the principle driver of global environmental issues; biorefinery adds to the progress of a more sustainable and efficient global environmental management. This also leads to the overused resources by means of maximizing the use of waste through the biorefinery. Eventually, greenhouse gas emissions and carbon footprint associated with waste management can be reduced. The integration and implementation of biorefineries clearly will make a change in shifting today's fossil-based economy to a bio-based economy through sustainable resource management for both researchers and industrial partners. The potential end products and intermediate supplies from biorefinery provide several adaptable solutions. As the world is descending into the crisis of depleting fossil fuel reserves and the environmental impacts caused by the utilization of these non-renewable resources are becoming increasingly more evident, biorefineries offer a promising and alternative resolution. By changing over biomass waste from the farms, forests, gardens, greenhouses, lakes, and other places, biorefinery could end the waste reliance on dumping and searing and facilitate in creating a sustainable and economically feasible resolution for waste management. It would also help to build an advanced and innovative waste economy which has the capability to turn currently societal, financial, and environmental problems into opportunities (Thanigaivel et al. 2022; Velvizhi et al. 2022).

## 11.9 Challenges and Prospect of Biorefinery to Produce Biomaterials

Biorefineries have great potential to produce biomaterials in a sustainable manner, but there are substantial obstacles in their way. The financial sustainability of biorefinery processes is a major obstacle because biomass feedstocks, processing techniques, and product development can be expensive. Furthermore, biorefineries frequently need large financial investments and compete with more established sectors. A further difficulty is the unpredictability and scarcity of biomass feedstocks, which can be impacted by elements including climate, farming methods, and modifications to land usage, creating uncertainty in the supply chain. The intricacy of biomass composition and the requirement for effective conversion technologies also provide technical difficulties in terms of optimizing product yields and quality. The creation and application of biorefinery technologies can also be impacted by the regulatory environment and policy structure, necessitating the need for encouraging laws and incentives to spur investment and creativity.

Biorefineries present encouraging opportunities for the sustainable production of biomaterials in spite of these obstacles. The efficiency and economics of biorefinery processes are constantly being improved by developments in biotechnology, chemistry, and engineering. This allows for the synthesis of a variety of biomaterials and the use of a larger range of feedstocks. Furthermore, the market for biobased products is expanded by the rising environmental consciousness and the growing need for sustainable substitutes for traditional resources. Furthermore, by increasing resource efficiency and monetizing waste streams, biorefineries are crucial in accelerating the shift to a circular economy. Moreover, partnerships between government, business, and academic organizations promote innovation and knowledge sharing in biorefinery technology. Overall, despite ongoing difficulties, biorefineries have a bright future ahead of them thanks to market demand, scientific developments, and the necessity of a sustainable future.

## 11.10 Conclusion

Biorefinery systems represent a versatile and sustainable approach to biomass utilization, offering a pathway concerning a more circular and bio-based economy. The chapter begins with a brief introduction biorefinery systems, potential application and raw materials for biorefinery to produce biomaterials. Keeping biorefinery systems in mind, classification and current technologies of biorefinery to produce biomaterials were also discussed. In addition, techno-economic aspects and sustainability of biorefinery in the waste management system high lightly discussed. Finally, future prospects and challenges were discussed.

## References


Achten WMJ et al (2007) Jatropha biodiesel fueling sustainability? Biofuels Bioprod Biorefining: Innov Sustain Econ 1(4):283–291

Adu C, Jolly M, Thakur VK (2018) Exploring new horizons for paper recycling: a review of biomaterials and biorefinery feedstocks derived from wastepaper. Curr Opin Green Sustain Chem 13:21–26

Anastopoulos G et al (2015) Transesterification of vegetable oils with ethanol and characterization of the key fuel properties of ethyl esters. Energies 2(2):362–376

Arevalo-Gallegos A et al (2017) Lignocellulose: a sustainable material to produce value-added products with a zero waste approach—a review. Int J Biol Macromol 99:308–318

Argyros DA et al (2011) High ethanol titers from cellulose by using metabolically engineered thermophilic, anaerobic microbes. Appl Environ Microbiol 77(23):8288–8294

Baral P, Kumar V, Agrawal D (2022) Emerging trends in high-solids enzymatic saccharification of lignocellulosic feedstocks for developing an efficient and industrially deployable sugar platform. Crit Rev Biotechnol 42(6):873–891. https://doi.org/10.1080/07388551.2021.1973363

Bilal M, Iqbal HMN (2022) Nanoengineered ligninolytic enzymes for sustainable lignocellulose biorefinery. Curr Opin Green Sustain Chem 38:100697

Budzianowski WM, Postawa K (2016) Total chain integration of sustainable biorefinery systems. Appl Energy 184:1432–1446

Caputo AC et al (2005) Economics of biomass energy utilization in combustion and gasification plants: effects of logistic variables. Biomass Bioenergy 28(1):35–51

Casa M et al (2021) A brief overview on valorization of industrial tomato by-products using the biorefinery cascade approach. Detritus 15(15):31–39. https://doi.org/10.31025/2611-4135/2021.14088
Cavalcanti Pessôa L et al (2022) A review of microalgae-based biorefineries approach for produced water treatment: barriers, pretreatments, supplementation, and perspectives. J Environ Chem Eng 10(4):108096. https://doi.org/10.1016/j.jece.2022.108096
Chaturvedi T, Torres AI et al (2020) Developing process designs for biorefineries—definitions, categories, and unit operations. Energies 13(6):1493. https://doi.org/10.3390/en13061493
Chen CY et al (2011) Cultivation, photobioreactor design and harvesting of microalgae for biodiesel production: a critical review. Bioresour Technol 102(1):71–81. https://doi.org/10.1016/j.biortech.2010.06.159
Chen Z et al (2023) Exploitation of lignocellulosic-based biomass biorefinery: a critical review of renewable bioresource, sustainability and economic views. Biotechnol Adv 69:108265
Cherubini F (2010) The biorefinery concept: using biomass instead of oil for producing energy and chemicals. Energy Convers Manag 51(7):1412–1421
Chiong MC et al (2018) Liquid biofuels production and emissions performance in gas turbines: a review. Energy Convers Manag 173:640–658
Chozhavendhan S et al (2018) Utilization of crude glycerol from biodiesel industry for the production of value-added bioproducts. In: Waste to Wealth. Singapore: Springer Singapore, pp 65–82. https://doi.org/10.1007/978-981-10-7431-8_4
Chozhavendhan S et al (2023) Current and prognostic overview on the strategic exploitation of anaerobic digestion and digestate: a review. Environ Res 216:114526
Conde-Mejia C, Jimenez-Gutierrez A, El-Halwagi M (2012) A comparison of pretreatment methods for bioethanol production from lignocellulosic materials. Process Saf Environ Prot 90(3):189–202
Cybulska I et al (2014) Chemical characterization and hydrothermal pretreatment of Salicornia bigelovii straw for enhanced enzymatic hydrolysis and bioethanol potential. Bioresour Technol 153:165–172
Demirbas A (2010) Biorefinery technologies for biomass upgrading. Energy Sources, A: Recovery Util Environ Eff 32(16):1547–1558
Dessbesell L et al (2017) Forest biomass supply chain optimization for a biorefinery aiming to produce high-value bio-based materials and chemicals from lignin and forestry residues: a review of literature. Can J For Res 47(3):277–288
Deswarte FEI et al (2007) Toward an integrated straw-based biorefinery. Biofuels Bioprod Biorefining: Innov Sustain Econ 1(4):245–254
Dragone G et al (2020) Innovation and strategic orientations for the development of advanced biorefineries. Bioresour Technol 302(January):122847. https://doi.org/10.1016/j.biortech.2020.122847
Du Z et al (2012) Hydrothermal pretreatment of microalgae for production of pyrolytic bio-oil with a low nitrogen content. Bioresour Technol 120:13–18
Duan Y et al (2022) Sustainable biorefinery approaches towards circular economy for conversion of biowaste to value added materials and future perspectives. Fuel 325:124846
Duque A et al (2021) Advanced bioethanol production: from novel raw materials to integrated biorefineries. Processes 9(2):1–30. https://doi.org/10.3390/PR9020206
Enawgaw H et al (2023) Multiple utilization ways of corn by-products for biomaterial production with bio-refinery concept; a review. Mater Circular Econ 5(1):7
Esteghamat M et al (2024) Performance optimization of fourth-generation biorefinery departments: a novel mixed macroergonomics-sustainability framework. Sustain Mater Technol 39:e00822
Fitzgerald GC, Themelis NJ (2009) Technical and economic analysis of pre-shredding municipal solid wastes prior to disposal. Unpublished MS thesis Columbia University, New York
FitzPatrick M et al (2010) A biorefinery processing perspective: treatment of lignocellulosic materials for the production of value-added products. Bioresour Technol 101(23):8915–8922
Freitas LC et al (2021) From waste to sustainable industry: how can agro-industrial wastes help in the development of new products? Resour Conserv Recycl 169:105466

Gari MT et al (2024) Effective utilization of agricultural cereal grains in value-added products: a global perspective. In: Value added products from food waste. Springer, pp 41–58
Gardossi L, Philp J, Fava F, Winickoff D, D'Aprile L, Dell'Anno B, Marvik OJ, Lenzi A (2023) Bioeconomy national strategies in the G20 and OECD countries: Sharing experiences and comparing existing policies. EFB Bioeconomy J 3:100053. https://doi.org/10.1016/j.bioeco.2023.100053
Garrote G, Dominguez H, Parajó JC (1999) Hydrothermal processing of lignocellulosic materials. Holz als roh-und werkstoff 57:191–202
Ghatak HR (2011) Biorefineries from the perspective of sustainability: feedstocks, products, and processes. Renew Sust Energ Rev 15(8):4042–4052. https://doi.org/10.1016/j.rser.2011.07.034
Gizaw DG et al (2024) Commercial products derived from vegetable processing industrial wastes and their recent conversion studies. In: Value added products from food waste. Springer, pp 147–167
Goswami L et al (2022) A critical review on prospects of bio-refinery products from second and third generation biomasses. Chem Eng J 448:137677. https://doi.org/10.1016/j.cej.2022.137677
Gundupalli MP et al (2022) Hydrothermal liquefaction of lignocellulosic biomass for production of biooil and by-products: current state of the art and challenges. In: Biofuels and Bioenergy. Elsevier, pp 61–84. https://doi.org/10.1016/B978-0-323-85269-2.00001-0
Gutierrez-Macias P et al (2017) The production of biomaterials from agro-industrial waste. Fresenius Environ Dengue Bull 26:4128–4152
Hagman L et al (2018) The role of biogas solutions in sustainable biorefineries. J Clean Prod 172:3982–3989
Hagman L, Eklund M, Svensson N (2020) Assessment of by-product valorisation in a Swedish wheat-based biorefinery. Waste Biomass Valorization 11(7):3567–3577. https://doi.org/10.1007/s12649-019-00667-0
Hamda AS et al (2023) Introduction: growth of agricultural waste, its disposal, and related environmental issues. In: Agricultural waste to value-added products: technical, economic and sustainable aspects. Springer, pp 1–19
Hamda AS et al (2024) Valorization of fruit processing industry waste into value-added chemicals. In: Value added products from food waste. Springer, pp 107–126
Heitz M et al (1988) Solvent effects on liquefaction: Solubilization profiles of a tropical prototype wood, eucalyptus. In: The presence of simple alcohols, ethylene glycol, water and phenols', research in thermochemical biomass conversion, pp 429–438
Heo JB, Lee Y-S, Chung C-H (2019) Raw plant-based biorefinery: a new paradigm shift towards biotechnological approach to sustainable manufacturing of HMF. Biotechnol Adv 37(8):107422
Hiloidhari M et al (2020) Agroindustry wastes: biofuels and biomaterials feedstocks for sustainable rural development. In: Refining biomass residues for sustainable energy and bioproducts. Elsevier, pp 357–388
Hosseini SA, Shah N (2009) Multiscale modelling of hydrothermal biomass pretreatment for chip size optimization. Bioresour Technol 100(9):2621–2628
Hu Y, Bassi A, Xu CC (2020) Energy from biomass. In: Future energy. Elsevier, pp 447–471. https://doi.org/10.1016/B978-0-08-102886-5.00021-9
Humbird D (2011) Process design and economics for biochemical conversion of lignocellulosic biomass to ethanol. Technical Report:40–43
Igbokwe VC et al (2022) Biochemical biorefinery: a low-cost and non-waste concept for promoting sustainable circular bioeconomy. J Environ Manag 305:114333
Jayakumar M, Asrat L et al (2023a) Agricultural waste as a source of fine chemicals. In: Agricultural waste to value-added products: technical, economic and sustainable aspects. Springer, pp 355–375
Jayakumar M, Bizuneh Gebeyehu K, Kuppusamy Vaithilingam S (2023b) Heterogeneous catalysts from biomass resources for biodiesel production. In: Encyclopedia of Green Materials. Springer, pp 1–9
Jayakumar M, Hamda AS et al (2023c) Comprehensive review on lignocellulosic biomass derived biochar production, characterization, utilization and applications. Chemosphere 345:140515

Jayakumar M, Prabhu SV et al (2023d) Microbial conversion of agricultural residues into organic fertilizers. In: Agricultural waste to value-added products: technical, economic and sustainable aspects. Springer, pp 113–138

Kachrimanidou V et al (2015) Integrated sunflower-based biorefinery for the production of antioxidants, protein isolate and poly (3-hydroxybutyrate). Ind Crop Prod 71:106–113

Karthikeyan OP, Mehariya S, Wong JWC (2017) Bio-refining of food waste for fuel and value products. Energy Procedia 136:14–21

Kasirajan R, Hamda AS, Jayakumar M (2022) Green membrane preparation and manufacturing practices. Singapore: Springer Nature Singapore, pp 1–13

Koutinas AA et al (2005) A whole crop biorefinery system: a closed system for the manufacture of non-food products from cereals. In: Biorefineries-industrial processes and products: status quo and future directions, pp 165–191

Lara A et al (2020) Enzymes in the third generation biorefinery for macroalgae biomass. In: Biomass, biofuels, biochemicals, pp 363–396

Larsson S et al (1999) The generation of fermentation inhibitors during dilute acid hydrolysis of softwood. Enzym Microb Technol 24(3–4):151–159

Mahari WAW et al (2022) Progress in valorisation of agriculture, aquaculture and shellfish biomass into biochemicals and biomaterials towards sustainable bioeconomy. Chemosphere 291:133036

Manandhar A et al (2022) Solid biofuels. In: Biomass, biofuels, biochemicals. Elsevier, pp 343–370

Mohammad REA et al (2023) Recent technical and non-technical biorefinery development barriers and potential solutions for a sustainable environment: a mini-review. In: Case studies in chemical and environmental engineering, p 100586. https://doi.org/10.1016/j.cscee.2023.100586

Mohan SV et al (2016) Waste biorefinery models towards sustainable circular bioeconomy: critical review and future perspectives. Bioresour Technol 215:2–12

Moncada BJ, Aristizábal MV, Cardona ACA (2016) Design strategies for sustainable biorefineries. Biochem Eng J 116:122–134. https://doi.org/10.1016/j.bej.2016.06.009

Montgomery LFR, Bochmann G (2014) Pretreatment of feedstock for enhanced biogas production. IEA Bioenergy Ireland

Mosier N et al (2005) Features of promising technologies for pretreatment of lignocellulosic biomass. Bioresour Technol 96(6):673–686. https://doi.org/10.1016/j.biortech.2004.06.025

Muh E, Tabet F, Amara S (2021) Biomass conversion to fuels and value-added chemicals: a comprehensive review of the thermochemical processes. Curr Alternative Energy 4(1):3–25. https://doi.org/10.2174/2405463103666191022121648

Mulder WJ (2010) Proteins in biomass streams. Wageningen UR Food & Biobased Research

Nanda S, Berruti F (2021) Municipal solid waste management and landfilling technologies: a review. Environ Chem Lett 19(2):1433–1456. https://doi.org/10.1007/s10311-020-01100-y

Narisetty V et al (2022) Valorisation of xylose to renewable fuels and chemicals, an essential step in augmenting the commercial viability of lignocellulosic biorefineries. Sustainable Energy & Fuels 6(1):29–65. https://doi.org/10.1039/D1SE00927C

Ochieng R, Gebremedhin A, Sarker S (2022) Integration of waste to bioenergy conversion systems: a critical review. Energies 15(7):2697

Orejuela-Escobar L et al (2021) Prospects of microalgae for biomaterial production and environmental applications at biorefineries. Sustainability 13(6):3063

Ortiz P et al (2019) Bio-based epoxy resins from biorefinery by-products. Bioresources 14(2):3200–3209

Padilla-Rivera A, Paredes MG, Güereca LP (2019) A systematic review of the sustainability assessment of bioenergy: the case of gaseous biofuels. Biomass Bioenergy 125:79–94

Pan X et al (2006) Bioconversion of hybrid poplar to ethanol and co-products using an organosolv fractionation process: optimization of process yields. Biotechnol Bioeng 94(5):851–861

Patel A, Shah AR (2021) Integrated lignocellulosic biorefinery: gateway for production of second generation ethanol and value added products. J Bioresour Bioprod 6(2):108–128

Pathak PD, Mandavgane SA (2023) Biorefinery: a sustainable approach for the production of biomaterials, biochemicals and biofuels. Springer Nature

Rahimi V, Shafiei M, Karimi K (2020) Techno-economic study of castor oil crop biorefinery: production of biodiesel without fossil-based methanol and lignoethanol improved by alkali pretreatment. Agronomy 10(10). https://doi.org/10.3390/agronomy10101538
Rajagopal D, Zilberman D (2008) Environmental, economic and policy aspects of biofuels. Found Trends Microecon 4(5):353–468
Redondo-Gómez C et al (2020) Biorefinery of biomass of agro-industrial banana waste to obtain high-value biopolymers. Molecules 25(17):3829
Reshmy R, Paulose TAP et al (2022a) Updates on high value products from cellulosic biorefinery. Fuel 308:122056
Reshmy R, Philip E et al (2022b) Lignocellulose in future biorefineries: strategies for cost-effective production of biomaterials and bioenergy. Bioresour Technol 344:126241
Ruiz HA et al (2020) Engineering aspects of hydrothermal pretreatment: from batch to continuous operation, scale-up and pilot reactor under biorefinery concept. Bioresour Technol 299:122685
Saini JK et al (2019) Integrated lignocellulosic biorefinery for sustainable bio-based economy. In: Sustainable approaches for biofuels production technologies: from current status to practical implementation, pp 25–46
Sethumadhavan A et al (2024) Food process industry waste biomass as a promising alternative for green energy production. In: Value added products from food waste. Springer, pp 275–290
Sieber V et al (2018) ChiBio: an integrated bio-refinery for processing chitin-rich bio-waste to specialty chemicals. In: Grand challenges in marine biotechnology, pp 555–578
Silalertruksa T, Wirodcharuskul C, Gheewala SH (2022) Environmental sustainability of waste circulation models for sugarcane biorefinery system in Thailand. Energies 15(24). https://doi.org/10.3390/en15249515
Sims REH et al (2010) An overview of second generation biofuel technologies. Bioresour Technol 101(6):1570–1580
Soh L et al (2014) Evaluating microalgal integrated biorefinery schemes: empirical controlled growth studies and life cycle assessment. Bioresour Technol 151:19–27
Sut D, Kataki R (2020) A biorefinery based zero-waste utilization of non-edible oilseeds for biodiesel and biofuel production along with chemicals and biomaterials. In: Biorefineries: a step towards renewable and clean energy, pp 21–55
Taherzadeh MJ, Karimi K (2008) Pretreatment of lignocellulosic wastes to improve ethanol and biogas production: a review. Int J Mol Sci 9(9):1621–1651
Talan A et al (2022) Biorefinery strategies for microbial bioplastics production: sustainable pathway towards circular bioeconomy. Bioresour Technol Rep 17:100875
Tang YY et al (2021) Malaysia moving towards a sustainability municipal waste management. Industrial Domestic Waste Manag 1(1):26–40. https://doi.org/10.53623/idwm.v1i1.51
Temesgen T et al (2024) Valorization of wastes and by-products of cane-based sugar industry. In: Value added products from food waste. Springer, pp 185–204
Thanigaivel S et al (2022) Role of nanotechnology for the conversion of lignocellulosic biomass into biopotent energy: a biorefinery approach for waste to value-added products. Fuel 322:124236. https://doi.org/10.1016/j.fuel.2022.124236
Thomsen MH, Thygesen A, Thomsen AB (2009) Identification and characterization of fermentation inhibitors formed during hydrothermal treatment and following SSF of wheat straw. Appl Microbiol Biotechnol 83:447–455
Tsai W-T, Lin C-C, Yeh C-W (2007) An analysis of biodiesel fuel from waste edible oil in Taiwan. Renew Sust Energ Rev 11(5):838–857
Tsarpali M et al (2021) Lipid-extracted algae as a source of biomaterials for algae biorefineries. Algal Res 57:102354
Vanthoor-Koopmans M et al (2013) Biorefinery of microalgae for food and fuel. Bioresour Technol 135:142–149
Varriale L, Ulber R (2023) Fungal-based biorefinery: from renewable resources to organic acids. ChemBioEng Reviews 10(3):272–292
Velankar HR, Thulluri C, Mattam AJ (2022) Development of second-generation ethanol technologies in India: current status of commercialization. In: Tuli D, Kasture S, Kuila ABT

(eds) Advanced biofuel technologies. Elsevier, pp 135–149. https://doi.org/10.1016/B978-0-323-88427-3.00004-0

Velvizhi G et al (2022) Integrated biorefinery processes for conversion of lignocellulosic biomass to value added materials: paving a path towards circular economy. Bioresour Technol 343:126151. https://doi.org/10.1016/j.biortech.2021.126151

Venkatesa Prabhu S et al (2023) Production strategies and potential repercussion of sewage sludge biochar as a futuristic paradigm toward environmental beneficiation: a comprehensive review. Environ Qual Manag 33:1

Werpy T, Petersen G (2004) Top value added chemicals from biomass: volume I--results of screening for potential candidates from sugars and synthesis gas. National Renewable Energy Lab. (NREL), Golden, CO (United States)

Yan Z et al (2021) Bringing material concepts into conventional biorefineries: considerations of sources, preparations, and applications of lignin nanomaterials. ACS Sustain Chem Eng 9(31):10403–10423

Yang S, Yu M (2013) Integrated biorefinery for sustainable production of fuels, chemicals, and polymers. In: Bioprocessing technologies in biorefinery for sustainable production of fuels, chemicals, and polymers, pp 1–26. https://doi.org/10.1002/9781118642047.ch1

# Algal Biomass Feedstock Based Biorefinery 12

Bharmjeet Singh, Abeba Haile Mariamenatu, Gourav Kumar, Bijaya Nag, Makaranga Abdalah, Getachew Tafere Abrha, and Pannaga Pavan Jutur

**Abstract**

Due to increased socioeconomic awareness, algae-based biomass has attracted much interest as a sustainable substitute for petroleum fossil fuels. This transition underscores the necessity for multifaceted processes capable of producing multiple high-value products concurrently, addressing contemporary challenges in the bioeconomy. However, it is yet unclear if different processes for turning algal biomass into fuels and fine chemicals are viable from an energy and economic standpoint. Consequently, the algal biomass feedstock-based biorefinery emerges as an innovative endeavour in ecologically sustainable algae utilization for diverse biorefinery applications. This chapter extensively explores various products and applications derived from algal biomass, including pharmaceuticals, nutritional supplements, biochemicals, biofuels, and cosmetics. Such information will provide insights into the opportunities and challenges inherent in algal biomass-based biorefineries. It critically assesses the feasibility, cost-effectiveness, and long-term prospects of employing algal biomass as a pivotal component of the bio-based economy.

B. Singh · A. H. Mariamenatu · G. Kumar · B. Nag · M. Abdalah · P. P. Jutur (✉)
Omics of Algae Group, Industrial Biotechnology, International Centre for Genetic Engineering and Biotechnology (ICGEB), Aruna Asaf Ali Marg, New Delhi, India
e-mail: Pavan.Jutur@icgeb.org

G. T. Abrha
Omics of Algae Group, Industrial Biotechnology, International Centre for Genetic Engineering and Biotechnology (ICGEB), Aruna Asaf Ali Marg, New Delhi, India

Department of Biotechnology, CoDANR, Mekelle University, Mekelle, Ethiopia

 
V. Singh (ed.), *Sustainable Waste Management Towards Circular Bioeconomy*,
https://doi.org/10.1007/978-981-95-2227-9_12

**Keywords**

Algal biomass feedstock · Biorefinery · Bioeconomy · Sustainable energy · Bioproducts

## 12.1 Introduction

The world faces twin environmental sustainability challenges and an energy crisis due to fossil fuel reliance, contributing to global warming and large greenhouse gas (GHG) emissions. This drives the search for sustainable energy alternatives, with biorefineries emerging as integrated solutions, turning microalgal biomass into products with additional value and biofuels (Thomassen et al. 2018). This multifaceted approach presents a more economically viable and sustainable option than focusing solely on fuel production (Salama et al. 2018). However, current capital expenses for biofuel production from microalgae hinder economic feasibility (Chandra et al. 2019), necessitating the inclusion of high-value co-products to enhance profitability. Microalgae, rich in lipids, carbohydrates, and proteins, offer versatile feedstock for various industries (Chandra et al. 2014). In algae-based biorefineries, it is still difficult to extract these compounds sustainably while following the principles of green chemistry (Yellapu et al. 2018). The Department of Energy emphasizes cost minimization for algal biofuel production (Bakonyi et al. 2018). This chapter reviews recent advancements in efficient and sustainable algal biomass utilization through biorefinery approaches, discussing successful implementations and challenges in producing food, fuel, and medicines (Sajjadi et al. 2018). It also outlines developments in utilizing algae biomass for value-added products within an integrated biorefinery context.

## 12.2 Algal Biomass Composition

Three fundamental biological components—proteins, carbohydrates, and lipids—dominate the biochemical composition of microalgae, while the relative amounts of each component vary among algal types. Lipids are particularly notable for their high energy density, followed by proteins and carbohydrates. In addition to these primary biomolecules, microalgae also contain trace amounts of various compounds valuable to industries such as biopharmaceuticals and nutraceuticals, including carotenoids, phycobiliproteins, polysaccharides, sterols, and omega fatty acids (Sajjadi et al. 2018).

### 12.2.1 Proteins

The protein content of microalgae, measured in dry cell weight (DCW), spans a wide range from 6% to 70%, influenced by species and environmental conditions. On average, most species exhibit protein levels around 50%. Certain species like *Chlorella* and *Spirulina* surpass 70% protein content (Nethravathy et al. 2019).

Microalgae also harbor diverse amino acids, with aspartate (Asp) and glutamate (Glu) being particularly prevalent. These amino acids are essential for many biological functions, some with significant implications in pharmacology and nutrition. Specific microalgal strains, including *Nannochloropsis oculata*, *Arthrospira platensis*, *Chlorella vulgaris*, and *Chlorella ellipsoidea*, have proven to be able to create peptides that inhibit the angiotensin-converting enzyme (ACE), valuable in regulating blood pressure. Commercially available products like algal protein isolates, exemplified by oligopeptide-based *C. vulgaris* extracts, are widely utilized in products such as Dermochlorella DG®. Additionally, microalgae feature mycosporin amino acids (MAAs), primarily found in marine environments. Abundant in species like *C. nivalis* and other green algae, MAAs bind to chromophores, providing defense against UV radiation and showcasing the expanding applications of microalgal bioactive peptides in functional medicines and nutraceuticals (Rehmanji et al. 2022; Chrapusta et al. 2017).

### 12.2.2 Carbohydrates

Microalgae produce primary energy storage molecules, chiefly carbohydrates, from which starch and cellulose can be extracted. $CO_2$ is transformed into 3-phosphoglyceric acid (3-PGA), a substrate for the synthesis of glucose, during the Calvin cycle. A monomer of many polysaccharides, including glycogen, chrysolaminarin, and floridean starch, glucose has important pharmacological and nutritional properties. For instance, chrysolaminarin, found in multiple families including haptophytes, diatoms, and chrysophytes, exhibits immune-modulatory, anti-tumor, and antioxidant properties. Certain diatoms like *Odontella aurita* and *Phaeodactylum tricornutum* demonstrate notable rates of chrysolaminarin production. Genera belonging to *Porphyridium* sp., *Phaeodactylum* sp., *Chlorella* sp., *Tetraselmis* sp., *Isochrysis* sp., and *Rhodella* sp., are identified as primary sources of microalgal polysaccharides, including sulfated polysaccharides (SPS), which have gained attention for their anti-inflammatory properties (Ampofo and Abbey 2022).

### 12.2.3 Lipids

Large lipid stores found in microalgae are essential for peroxidation, preserving cell membranes' fluidity, and preventing photosynthetic pigment oxidation. Triacylglycerols (TAGs), long-chain polyunsaturated fatty acids (LC-PUFAs), saturated, monounsaturated, and polyunsaturated fatty acids make up most of these lipids. Under varied growth conditions, microalgae can elevate their lipid content to 20–50% of DCW (Rehmanji et al. 2022; Sun et al. 2019a). Although the majority of species have an oil content of about 30%, some strains, such as *Nannochloropsis* sp., *Chlorella* sp., and *Neochloris* sp., have noticeably greater amounts. On the other hand, strains that contain more lipids often develop more slowly. Remarkably, some algae, such as *Schizochytrium* sp., possess lipid contents exceeding 80% of their total mass. Biofuels may be produced by extracting these lipids. Notably, microalgae with 80% lipid content bloom thirty times slower than those with 30% lipid content, indicating a trade-off between lipid content and growth rates (Vasudevan and Briggs 2008).

### 12.2.4 Pigments

Pigments are essential for microalgae because they give them color, absorb light energy for photosynthesis, and guard against oxidative damage. Major pigment categories include phycobiliproteins, phenolics, carotenoids, and chlorophyll (Levasseur et al. 2020).

#### 12.2.4.1 Carotenoids

Microalgae represent a significant source of various carotenoid compounds associated with notable health benefits and applications in the food industry. Nearly thirty of the roughly 1100 carotenoids discovered so far are identified in microalgae (Mudliar and Shekh 2019). Carotenoids are classified into primary and secondary categories based on their function and accumulation mode. Primary carotenoids, like as lutein and β-carotene, are vital for photosynthesis and shield the photosynthetic apparatus from an excess of light. On the other hand, environmental stressors like bright light cause the production of secondary carotenoids such astaxanthin and canthaxanthin. Numerous carotenoids of substantial nutraceutical value, such as lutein, astaxanthin, fucoxanthin, and β-carotene, are produced by microalgae. *Haematococcus pluvialis's* astaxanthin and *Dunaliella salina's* β-carotene are especially notable because of their rising market demand. Microalgae are considered efficient bio-factories for producing carotenoids commercially, with Chlorophyceae species being primary producers (Sui et al. 2021).

#### 12.2.4.2 Phenolics

Phenolics, compounds containing benzene rings with hydroxyl groups, are categorized according to their carbon structure into four classes: flavonoids, tannins, phenolic acids, and stilbenes. Microalgae like *Auxenochlorella pyrenoidosa* and *Arthrospira platensis* are rich sources of phenolics. Various microalgae species exhibit phenolic contents ranging from 54 mg/g to 375 mg/g. Phenolics offer numerous health benefits and have applications in functional foods and nutraceuticals (Ampofo and Ngadi 2020; Machu et al. 2015; Goiris et al. 2012; Widowati et al. 2017).

Table 12.1 illustrates the biochemical compositions of microalgal species identified as candidates with industrial potential for biorefinery applications.

## 12.3 Cultivation of Algae

Microalgae's rapid growth rates and efficient storage of bioproducts within their cells make them promising industrial raw material candidates, owing to their high photosynthetic effectiveness (Randrianarison and Ashraf 2017). Unlike traditional crops, microalgae cultivation doesn't necessitate fertile soil, large volumes of

**Table 12.1** Biochemical compositions of potential industrial microalgal species for Biorefinery (Sajjadi et al. 2018; Rehmanji et al. 2021; Ampofo and Abbey 2022)

| Microalgal species | Protein (% DCW) | Lipid (% DCW) | Carbohydrates (% DCW) | Carotenoids (% DCW) |
|---|---|---|---|---|
| *Thalassiosira wessflogii* | 43.0 | 5.0–20.0 | 12.0 | – |
| *Phaeodactylum tricornutum* | 30.0 | 18.0–57.0 | 8.4 | 1.5–2.0% (Fucoxanthin) |
| *Skeletonema costatum* | 25.0 | 13.0–51.0 | 4.6 | – |
| *Chaetoceros muelleri* | 31.0–43.0 | 13.0–24.0 | 7.0–28.0 | – |
| *Dunaliella primolecta* | <64.0 | 23.0 | 11.0–23.0 | – |
| *Dunaliella tertiolecta* | 20.0–29.0 | 11.0–16.0 | 12.2–14.0 | – |
| *Dunaliella salina* | 57.0 | 6.0–25.0 | 32.0 | 14.0%(β-carotene) |
| *Nannochloris* sp. | 16.7 | 20.0–56.0 | – | – |
| *Nannochloropsis oculata* | 35.0 | 22.0–29.0 | 7.8 | – |
| *Scenedesmus obliquus* | 10.0–45.0 | 30.0–50.0 | 20.0–40.0 | – |
| *Scenedesmus quadricauda* | 40.0–47.0 | 1.9 | 12.0 | – |
| *Scenedesmus dimorphos* | 8.0–18.0 | 16.0–40.0 | 21.0–52.0 | – |
| *Scenedesmus* sp. | 29.0–37.0 | 17.0–24.0 | 32.7–41.0 | – |
| *Ankistrodesmus* sp. | 16.24–18.66 | 11.5–31.0 | 4.5–5.9 | – |
| *Chlamydomonas reinhardtii* | 48.0 | 21.0 | 17.0 | – |
| *Chlamydomonas* sp. | 58.8 | 22.7 | 18.5 | – |
| *Neochloris oleoabundans* | 10.0–27.0 | 35.0–65.0 | 17.0–27.0 | – |
| *Scenedesmus falcatus* | 3.4–7.8 | 6.4–9.6 | 2.7–6.8 | – |
| *Scenedesmus protuberans* | 25.4–45.1 | 17.5–29.3 | 20.9–29.2 | – |
| *Chlorella Vulgaris* | 25.0–45.0 | 28.0–53.0 | 24.0–30.0 | – |
| *Chlorella pyrenoidosa* | 57.0 | 2.0 | 26.0 | – |
| *Chlorella protothecoides* | 10.0–28.0 | 40.0–60.0 | 11.0–15.0 | 4.6 (Lutein) |
| *Chlorella emersonii* | 36.0 | 23.0–63.0 | 41.0 | – |
| *Chlorella sorokiana* | 40.5 | 22.0–24.0 | 26.8 | – |
| *Chlorella minustissima* | 47.9 | 14.0–57.0 | 8.1 | – |
| *Haematococcus pluvialis* | 48.0 | 15.0 | 27.0 | 4.0–7.0 (Astaxanthin) |

freshwater, or the use of herbicides or pesticides, thus mitigating resource competition concerns (Khan et al. 2018).

The algal culture medium comprises essential components facilitating algae growth, including growth factors, trace metals, carbon, nitrogen, and phosphorus sources (Najiha Badar et al. 2021). The type of algae used and the intended final products determine its composition. Familiar carbon sources encompass acetate, glucose, bicarbonate, and carbon dioxide. Phosphorus is typically supplied via phosphate salts, while nitrogen sources comprise nitrates, ammonium salts, or organic nitrogen compounds. Copper, iron, manganese, and zinc are examples of trace metals that are essential for photosynthesis and enzyme activity. At the same time, vitamins and growth hormones are crucial for diverse metabolic pathways and cellular processes.

Fine-tuning the algal culture medium is imperative for achieving the desired biochemical compositions and optimal biomass yield. This entails adjusting pH levels, temperature, light intensity, culture techniques, nutrient amounts, and ratios (Juneja et al. 2013). Optimization techniques, including factorial design tests, response surface methods, and mathematical modeling, aid in determining the ideal growth and productivity conditions. Continuous monitoring and adjustment of media components based on growth stage and environmental conditions are essential to sustain optimal culture conditions.

Cultivation conditions profoundly influence microalgae's physiological functions, modulating cellular composition, growth rate, lipid accumulation, and other characteristics. Nutrient availability, salinity, temperature, light intensity, and carbon source significantly affect microalgal growth and productivity. Adjustments in nitrogen and phosphorus have proven particularly effective in substantially augmenting lipid accumulation in microalgae, surpassing the effects of other nutrients. Ideally, the algal culture medium should adhere to a stoichiometric formula of C106H181O45N16P to optimize growth conditions, with a high ratio indicative of phosphorus limitation (around 30:1) and a low N to P ratio signifying nitrogen constraint (approximately 5:1) (Sajjadi et al. 2018). Table 12.2 delineates the most prevalent environmental factors and their impacts on microalgae.

## 12.4 Algal Cultivation Techniques

### 12.4.1 Open Pond Systems

Algae are frequently cultivated outside using open pond systems. These ponds are typically shallow, with depths ranging from 15 to 30 cm, facilitating uniform sunlight distribution and atmospheric carbon dioxide absorption by the algae (Song et al. 2023). Mixing the algal cultures is essential to optimize light exposure, ensuring movement from poorly illuminated lower layers to well-lit upper ones within the ponds. Various open systems with distinct shapes, sizes, construction materials, agitation methods, and slopes are utilized in industrial and research settings.

Raceway ponds, characterized by long concrete canals with oval or racetrack configurations, are commonly used to continuously supply algal cultures with nutrients and carbon dioxide (Mahmood et al. 2023). Circular pond tanks, owing to their spherical shape, offer a controlled cultivation environment suitable for algae growth. Shallow large ponds, consisting of vast water bodies with shallow depths, typically between 15 and 30 cm, are designed to maximize sunlight exposure and atmospheric carbon dioxide absorption (Sen Tan et al. 2020). Closed ponds provide a controlled environment for algae cultivation, minimizing concerns regarding contamination and external influences. Location selection for open ponds is crucial due to algae's reliance on sunlight for photosynthesis.

Cultivating algae in wastewater is a sustainable approach that aids in contaminant removals, such as phosphorus and nitrogen, while generating algal biomass.

**Table 12.2** Summary of the environmental factors and their effect on the microalgae

| Environmental and nutrient factors | Microalgal species | Cultivation condition | Effects | References |
|---|---|---|---|---|
| Temperature | *Botryccoccus braunii* | Temperature increased from 25 to 32 °C | Accumulation of polysaccharides<br>The amount of intracellular lipid in DCW dropped from 22% to 5% | Juneja et al. (2013) |
| | *Haematococcus pluvialis* | Rise in temperature from 20 to 35 °C | A threefold rise in astaxanthin | |
| | *Chlorococcum* sp. | Temperature increase with nitrogen depletion from 20 to 35 °C | The overall amount of cellular contents has doubled | |
| | *Scenedesmus obliquus* | Temperature increases from 20 to 35 °C | The protein and carbohydrate content of *S. obiquus* rose first and then decreased, and the maximum protein and carbohydrate content was shown at 25 °C.<br>Above 25 °C, the content of protein and carbohydrate decreased<br>Beyond 25 °C the efficiency of protein synthesis and starch production was reduced resulting in a decrease in protein and carbohydrate content<br>Over the course of 4 days, the lipid content increased progressively between 25 and 35 °C before gradually declining | Zhang et al. (2019) |
| Light intensity | *Dunaliella viridis* | No light | The amount of total lipids in fatty acids and sterols rose and fell. | Juneja et al. (2013) |
| | *Nannochloropsis* sp. | Low light | Total lipid content increased, and high in EPA proportion. | |
| | *Chlorella vulgaris* | Red light | A rise in the production of starch | |
| | | Blue light | A rise in the lipid fraction | |
| | *Chlamydomonas reinharditii* | Reg-orange | Increased lipid accumulation<br>Increase carbohydrate accumulation | Li et al. (2023) |
| | | Blue-light | Promoted protein synthesis | |
| Salinity | *Scenedesmus dimorphus* | 5% Salinity (NaCl) | A 34% w/w increase in lipid DCW considerably raised the level of saturated fatty acids (SFA) while lowering that of PUFAs. | Pushpakumari Kudahettige et al. (2018) |
| | *Selenastrum minutum* | 5% Salinity (NaCl) | A substantial rise in SFA and a drop in PUFA contents were observed at 39% w/w of DCW of lipid increasement. | |
| | *Tetraselmis suecica* | Effect of salinities from 10 to 60 ppt, where the normal seawater level is (30 ppt) | The overall lipid production increased by almost 22% when the salinity was raised from 30 to 50 and 60 ppt.<br>The amount of total protein, carbohydrates, and fats did not significantly alter under the various salinities.<br>Eicosapentaenoic acid (EPA) was the main LC-PUFA, while the majority of fatty acids had chain lengths between C16 and C18. | Pugkaew et al. (2019) |

(continued)

**Table 12.2** (continued)

| Environmental and nutrient factors | Microalgal species | Cultivation condition | Effects | References |
|---|---|---|---|---|
| Carbon | *Dunaliella salina* | For 1 day, the CO2 concentration increased from 2% to 10%. | A 30% w/w increase in fatty acids in DCW | Juneja et al. (2013) |
| | *Dunaliella salina* | For 7 days, the CO2 concentration increased from 25% to 10%. | A 2.7% rise in fatty acids | |
| | *Spirulina platensis* | Increased concentration of CO2<br>$CO_2$ concentration | Rise in carbohydrates and fall in pigments and protein content | |
| Nitrogen | *Nannochloropsis oculata* | 75% nitrogen depletion | Lipid synthesis increased from 7.90% to 15.31%. | |
| | *Phaeodactylum tricornutum* | Nitrogen depletion | Decrease in protein concentration and an increase in lipid synthesis | |
| | *Chlorella vulgaris* | 75% nitrogen depletion | Lipid production rose from 5.9% to 16.41%. | |
| | *Haematococcus pluvialis* | Nitrogen depletion | 13% (w/w of DCW) increase in carotenoid formation | |
| | *Desmodesmus communis* | 50% Nitrogen depletion media | Protein synthesis increased | Akgül (2020) |
| | | 75% nitrogen depletion media | Lipid accumulation increased | |
| | *Rhodomonas* | N-deprivation for 192 h of incubation | The amount of lipid in the dry cell weight rose dramatically from 9% (t = 0 h) to 30% (t = 120 h).<br>C16:0, C18:1, C18:2, and C18:3 showed the largest increases in FA concentration, with the maximum concentrations occurring after 120 h of fasting.<br>After 48 h of hunger, the highest levels of EPA and docosahexaenoic acid (DHA) were noted, and at the conclusion of the famine, the highest DHA to EPA ratio was found. | Latsos et al. (2020) |
| | *Chlorella* sp. | Nitrogen concentration (0-5 g/L) | Lipid rose from 42% of DCW under ideal growth circumstances to 73% w/w when nitrogen was lacking. | Munir et al. (2022) |
| | *Oedogonium* | Nitrogen concentration (0-5 g/L) | The lipid content of DCW rose from 40 to 69% w/w when nitrogen was lacking. | |

| | | | | |
|---|---|---|---|---|
| Other types of nutrient deprivation | *Scenedesmus dimorphus* | The quantities of phosphate, nitrate, and ammonium were less than the instrument detection limit, or less than 0.01 mg/L. | Highest lipid content (35%) | Juneja et al. (2013) |
| | *Selenastrum minutum* | | Highest lipid content (40%) | |
| Phosphorus | *Ankistrodesmus falcatus* | Phosphorus limitation | Protein and chlorophyll a levels are down, whereas fats and carbohydrates are up. | |
| Iron | *Dunaliella tertiolecta* | Iron limitation | A reduction of chlorophyll in cells | |
| | *Chlorella vulgaris* | Increased iron concentration | Rise in the amount of lipid | |
| | *Haematococcus pluvialis* | Increased iron concentration | Rise in the amount of carotenoids | |

Despite being cost-effective, open pond systems have drawbacks, including management complexity, land requirements, and significant water loss through evaporation. Consideration of weather conditions and light levels, along with pollution risks, is essential.

For continuous cultivation of a single algal species in open ponds, extremophiles capable of thriving in specific environments, such as those with extreme pH levels or salinity, are preferred. Spirulina, for instance, thrives in high-pH environments (pH 9 to 11.5) and dominates soda lakes due to its spiral structure (Ismaiel et al. 2016). Additionally, *Dunaliella salina*, a unicellular green algae species prevalent in highly saline conditions, produces an abundance of carotenoids to shield itself from intense light in shallow salt pan ponds (Shetty et al. 2019). Algae cultivation in raceway culture ponds is common in countries like China, Israel, and the USA.

### 12.4.2 Closed Photobioreactor Systems

Closed systems offer superior control over culture parameters, minimizing contamination risks and enabling optimal growth and high-value product production when cultivating photoautotrophic microorganisms for environmentally friendly products. For microalgae, cyanobacteria, and photosynthetic bacteria to be commercialized as natural chemical products and biofuels, photobioreactors (PBRs) are essential. PBRs require optimal light utilization and mixing techniques to ensure uniform lighting and exposure, promoting high photosynthesis and growth rates. Numerous PBR designs have been created for the growing of microalgae outdoors emphasizing optimal growth and contamination prevention (Vo et al. 2019; Lindblad et al. 2019; Huang et al. 2017, Seibert and Torzillo 2018)

### 12.4.3 Hybrid Cultivation Approaches

Hybrid systems integrate both open pond and closed PBR cultivation techniques. Algae are initially grown in PBRs before being transferred to open systems to produce high-density inoculants, reducing contamination risks. Despite being inappropriate for inexpensive goods like biofuels because their batch-process nature, this approach allows for continuous production in ecologically favourable conditions by regularly introducing new inoculants to open ponds (Mahmood et al. 2023). Successful implementation requires rapid-growth algae species adaptable to both inoculum scale-up and open pond production stages.

## 12.5 Harvesting Techniques

Effective microalgae harvesting is essential for product generation. Mechanical procedures are used, including filtration, flotation, sedimentation, flocculation, and electrochemical approaches (Salim et al. 2011). The benefits and limitations of each of them are mentioned in Table 12.3.

**Table 12.3** Benefits and limitations of different harvesting methods (Salim et al. 2011)

| Methods | Biomass recovery | Benefits | Limitations |
|---|---|---|---|
| Sedimentation | 60.0–97.0% | Simple, economical, and energy-efficient | Time-consuming process, species-dependent |
| Flotation | 80.0–98.7% | Diverse flotation methods with quick harvesting and affordable pricing | High cost of certain chemical surfactants<br>Suitable for specific algae types only |
| Bioflocculation | 80.0–99.6% | Environmentally friendly and devoid of toxins<br>Cost-effective, fast, and efficient<br>Potential for reuse of culture media | Risk of microbial contamination |
| Chemical flocculation | 82.0–99.0% | Suitable for large-scale commercial use<br>Cost-effective and straightforward harvesting | Chemical removal necessary from algal cells<br>Potential toxicity of chemicals to algal cells |
| Electro-flocculation | 90.0–98.0% | Chemical-free and efficient harvesting | High energy consumption<br>Expensive metal electrodes |
| Filtration | 76.0–100.0% | Easy collection within 10 min to 24 h<br>Economical and non-chemical approach<br>Ideal for industrial-scale cultivation | Accumulation and blockage issues with tiny algal cells<br>Mainly suited for macroscopic algae<br>Periodic cleaning required for filter membranes |

(a) Sedimentation—Sedimentation stands out among mechanical algal harvesting techniques for its simplicity, relying on gravity-induced settling of denser algal cells in the culture medium. It's a cost-effective method crucial for efficiently harvesting valuable products from media. During sedimentation, cells settle according to Stokes' Law until reaching terminal velocity, influenced by their density, size, and shape. While theoretically, spherical algae like *Chlorella* have a settling velocity of 0.1; real-world measurements vary from 0.4 to 2.2 due to complex morphologies (Ghazvini et al. 2022). Despite this variability, sedimentation is widely employed in the industrial harvesting of various microalgae strains. Factors like charge neutralization and sweep flocculation significantly impact sedimentation efficiency, particularly at alkaline pH levels. Additionally, chemical species like precipitates and proteins can affect sedimentation effectiveness (Leite and Daniel 2020). It is advised to use sedimentation as a pre-concentration method, particularly in large-scale processes like the manufacture of biofuel where the end product has a low value.

(b) Flotation—Flotation is a prominent technique for microalgae separation, involving the attachment of solid particles to gaseous bubbles generated from air or gas. It includes a number of techniques such as ozonation dispersed flotation, dissolved, electrolytic, and dispersed air flotation (Deepa et al. 2023). Table 12.4 lists the microalgae for which this flotation method demonstrated increased harvesting effectiveness.

(c) Flocculation—The term flocculation describes the aggregation of minute and unstable particles once flocculants are added. This is accomplished via bridging, electrostatic patching, and neutralization of surface charges. Microalgal biomass is frequently extracted using this method. Table 12.5 outlines various flocculation methods employed in microalgae harvesting.

(d) Filtration—A vital stage in the production of microalgae for a range of industrial uses, such as wastewater treatment, medicines, and biofuel, is filtration. By employing specialized filtration systems to separate microalgae cells from the culture media, this environmentally friendly and energy-efficient method does away with the need for chemical additives or labor-intensive procedures (Chen et al. 2011). Membranes play a vital role in filtration, but fouling remains a

**Table 12.4** List of flotation techniques employed in microalgae to optimize harvesting effectiveness

| Microalgae | Flotation method used | References |
|---|---|---|
| *Chromochloris zofingiensis* | Dissolved air flotation | Zhang et al. (2012) |
| *Chlorella vulgaris, Scenedesmus obliquus* | Dispersed air flotation | Kurniawati et al. (2014) |
| *Chlorella vulgaris, Scenedesmus obliquus* | Thermal Flotation | Zou et al. (2019) |
| *Nannochloropsis* sp. | Bio-Flotation | Kandasamy and Shaleh (2018) |

**Table 12.5** illustrates various techniques of flocculation utilized for microalgae harvesting

| Types of flocculation | Key features | Microalgal strains | References |
|---|---|---|---|
| Chemical flocculation | 1. Various chemical flocculants such as Mg-sericite, poly (γ-glutamic acid), calcium oxide, alum, metal coagulant ($Fe_2(SO_4)_3$), sulfuric acid, $CaCl_2$, and Zetag 8185 are utilized to improve harvesting efficiency.<br>2. The pH of the medium significantly influences the process. | *C. vulgaris;*<br>*P. tricornutum;*<br>*Scenedesmus* sp.;<br>*Chlorella* sp. | Deepa et al. (2023) |
| Electro-flocculation | 1. Aluminium electrolysis is employed.<br>2. Higher density currents result in faster harvesting rates.<br>3. Aluminium electrodes demonstrated no impact on the lipid and pigment profile of *C. vulgaris*. | *Chlorella* sp.;<br>*Dunaliella* sp.;<br>*C. vulgaris* | |
| Bio-flocculation | 1. This technology offers an appealing approach to microalgae harvesting, avoiding the adverse effects of chemical residues.<br>2. Commonly used bio-flocculants include chitosan (cationic polyelectrolyte), cationic starch, Greenfloc 120, α-Poly-l-lysine, *Aspergillus niger*, and *Strychnos potatorum* seed powder.<br>3. Bio-flocculants exhibit adaptability to a wide pH range (3.0-9.0). | *C. protothecoides;*<br>*C. vulgaris;*<br>*C. sorokiniana* | |

significant challenge, leading to reduced fluxes and costly harvesting processes. Surface-coating materials like hydrophilic polyvinyl alcohol polymer have been employed to mitigate fouling (Deepa et al. 2023). Different filtration techniques, such as microfiltration, ultrafiltration, pressure filtration, and vacuum filtration, are utilized to extract microalgal biomass, each with distinct advantages and limitations based on factors like cell size and downstream processing requirements.

(e) Electrochemical harvesting—Electrochemical methods offer a contemporary approach for microalgal biomass extraction, including electrocoagulation, electro-flocculation, and similar processes. Electrochemical techniques have shown promising results in microalgal biomass recovery, with varying efficiency rates depending on the species and electrodes used (Deepa et al. 2023). For instance, in India, electro-flotation achieved a recovery rate as high as 99% in C. vulgaris, while in France, recovery rates ranged from 36% using aluminium and iron electrodes to higher rates with different techniques for other species. Different electrodes used in electrochemical harvesting significantly impact the biochemical composition of microalgae. Carbon cloth anodes and stainless-steel cathodes have demonstrated superior flotation efficiency compared to aluminium electrodes, with minimal pollution due to their reduced corrosive nature (Pishgar et al. 2020). The hydrophobic nature of microalgae is believed to play a crucial role in the electro-flotation mechanism.

## 12.6 Biomass Conversion Technologies

### 12.6.1 Primary Metabolites

Through primary metabolism, microalgae create organic matter—which is essential for cell development and reproduction—by using inorganic components including nitrogen, carbon, phosphorus, iron, sulfur, and trace elements. Proteins, fats, and carbohydrates are examples of primary metabolites that serve as building blocks for high-value goods including industrial chemicals, medicines, and biofuels (Batista et al. 2013).

#### 12.6.1.1 Lipid Extraction Methods in Microalgae

For the production of high-value biorenewables and carbon-neutral biofuels with uses in the food, pharmaceutical, and cosmetics sectors, lipid extraction from microalgae is essential. Standard, conventional techniques such as Soxhlet, Folch, and Bligh-Dyer extractions have the drawbacks of high reagent usage, contamination, and extended extraction periods. To overcome these obstacles and improve extraction efficiency, cutting-edge techniques including supercritical fluid extraction (SFE), pulse electric field (PEF), ultrasound-assisted extraction (UAE), microwave-assisted extraction (MAE), and ionic liquids (ILs) have been developed (Luque de Castro and Garcı́a-Ayuso 1998) (Table 12.6).

- Soxhlet Extraction: Since 1879, Soxhlet extraction, a solvent-based technique, has been used for lipid extraction, yielding higher lipid yields through repeated contact between the sample and extractant. Long extraction periods and high reagent consumption are among of its disadvantages, despite its affordability and ease of use (Luque de Castro and Garcı́a-Ayuso 1998). Folch Extraction: Developed in 1957, the Folch technique extracts lipids using chloroform and methanol, followed by phase separation. While it doesn't require high temperatures or pressure, it generates ecologically hazardous waste (Fernández et al. 2011).
- Bligh-Dyer Extraction: This method enables simultaneous lipid extraction and separation but employs hazardous reagents, necessitating cost-effective alternatives (Bligh and Dyer 1959).
- Supercritical Fluid Extraction: Utilizing carbon dioxide ($CO_2$), supercritical fluid extraction (SFE) offers advantages like environmental friendliness, ease of solvent separation, and high extraction rates. SC-CO2 has been successfully employed to extract lipids from microalgae like *N. oculata* and *C. vulgaris*, despite the greater equipment expenses (Bhargavi et al. 2018).
- Pulsed Electric Fields: PEF extraction, a non-polluting method, requires control over electric field intensity to prevent adverse impacts. Lipid extraction rates are considerably increased by PEF-assisted Bligh-Dyer extraction as opposed to ultrasonic pre-treatment (Han et al. 2019).

**Table 12.6** Lipid extraction methods in microalgae

| Microalgal strains | Extraction methods | Conditions | Remarks | References |
|---|---|---|---|---|
| *N. oculata; C. vulgaris* | $SC\text{-}CO_2$ with ethanol (10%, as co-solvent) | $CO_2$ flow: 25 g. $min^{-1}$, Solvent rate: 0.35 to 1.02 g/g·$min^{-1}$, temperature: 50 °C, pressure: 250–750 bar, ethanol flow rate: 1.9 g.$min^{-1}$ | Neutral lipid extraction was by 83% in N. oculata and 97% in *C. vulgaris* | Obeid et al. (2018) |
| *Schizochytrium* sp. | $SC\text{-}CO_2$ | $CO_2$ flow: 10 kg.$h^{-1,}$ temperature: 40/55 °C, pressure: 25/40/55/70 MPa | The higher extraction rate of lipid as compared with Soxhlet | Zinnai et al. (2016) |
| *Ankistrodesmus falcatus* | PEF chloroform ethyl acetate | Frequency of pulse repetition: 240 Hz; 24 pulses were recorded. | Higher extraction rate of lipid recovery | Zbinden et al. (2013) |
| *Nanoochloropsis* sp. | UAE hexane | Intensity of ultrasonic: 20 kHz; Time: 1/3/5 min | Increase lipid recovery by more than 6.5 times | Yao et al. (2018) |
| *Chlorella PY-ZU1* | MAE chloroform: methanol (1:1 v/v) | Temperature: 80–120 °C Time: 5/10 min | Increases the amount of saturated and short-chain fatty acids. | Cheng et al. (2013) |
| *Chlamydomonas reinhardtii* | Osmotic shock | n-hexane: methanol (0:10, 3:7, 6:4 v/v) | Lipid recovery increased by three times after the osmotic shock | Yoo et al. (2012) |
| *Schizochytrium limacinum* | $SC\text{-}CO_2$ | 35 MPa, 40 °C, 95% v/v ethanol as a co-solvent | Increase in total the lipid extraction | Tang et al. (2011) |
| *Botryococcus braunii* | | 25 MPa, 50 °C | Used for thermosensitive compounds | Santana et al. (2012) |
| *Nannochloropsis salina* | 1-butyl-3-methylimidazolium hydrogen sulphate [BMIM] [$HSO_4$] | Temperature: 120 °C: Time: 10 to 60 min: Power: 800 W | Lipid extraction rate increased up to 370% on the addition of ILs | Pan et al. (2016) |

- Ultrasound-Assisted Extraction: UAE, a low-temperature, low-energy approach, exhibits high extraction rates and improved photosynthetic efficiency in microalgae like *Scenedesmus* sp. (Sivaramakrishnan and Incharoensakdi 2019).
- Microwave-Assisted Extraction: Although solvent polarity may have an impact on the process, MAE shortens extraction times and uses less energy. While it enhances lipid output, its suitability varies across microalgae species (Yao et al. 2018).
- Ionic Liquids: Despite being more expensive, ILs offer low toxicity and stability, making them ideal for lipid extraction. ILs have been successfully used to pretreat microalgae and recover lipids with significantly improved recovery rates compared to traditional methods (Shankar et al. 2019).

#### 12.6.1.2 Carbohydrate Extraction Techniques in Microalgae

The primary constituents of microalgal biomass comprise polysaccharides such as starch, cellulose, and monosaccharides encased within a rigid cell wall, thereby limiting their accessibility for direct utilization (Laurens et al. 2015). Pre-treatment methods facilitate sugar solubilization through various physical, chemical, physicochemical, and enzymatic methods, thus enhancing the efficiency of subsequent biological processes (Naresh Kumar et al. 2020). Among physical methods, thermal treatment and ultrasonication have been investigated. Research by Passos et al. (2015) demonstrated that thermal pre-treatment significantly increased carbohydrate solubilization in mixed microalgal biomass compared to untreated controls (Jeon et al. 2013). Similarly, sonication-based carbohydrate solubilization using *Scenedesmus obliquus* YSW15 showed effective results within 15 min at 40 kHz (Chen et al. 2013).

Chemical pre-treatments, particularly under acid-catalyzed conditions, have also shown promising results in carbohydrate solubilization (Chen et al. 2013). A study by (Laurens et al. 2015) investigated the impact of acid-catalyzed conditions on sugar solubilization using *Scenedesmus* and *Chlorella* biomass, showing significant glucose solubilization rates post-lipid extraction pre-treatment. Enzymatic pretreatment methods offer high sugar yields with low inhibitory compound formation compared to physical and chemical processes (Prajapati et al. 2015; Lee et al. 2015). Enzymes like cellulases, amylases, and xylanases have demonstrated effective solubilization of algal biomass (Prajapati et al. 2015).

Microalgal biomass was first studied for the manufacture of biodiesel, but because of its high protein and carbohydrate content, it is now being studied more and more as a feedstock for different biobased products (Rajesh Banu et al. 2020). Studies have demonstrated the potential of various algal strains in biomass growth and carbohydrate accumulation under specific conditions (Tibbetts et al. 2015; Illman et al. 2000). High biomass productivity contributes to adequate $CO_2$ fixation and enhances the synthesis of renewable products from algal biomass units, ensuring economic feasibility and industrial scalability (Chen et al. 2013; Sim 2009; Gruhn et al. 2016).

Overall, developing efficient carbohydrate extraction methods from microalgae holds significant promise for advancing the utilization of microalgal biomass in diverse biobased applications (Table 12.7).

**Table 12.7** Evaluation of microalgal biomass for various bio-based products

| Biomass | Culture conditions | Reactor type | Protein (% DCW) / yield | Carbohydrate (% DCW) / yield | Lipid (% DCW) / yield | References |
|---|---|---|---|---|---|---|
| *Spirulina platensis* | Mesophilic (35 °C) | – | 46.0–63.0 | 25.0–30.0 SCFA: 1.92 $g.L^{-1}$ | 4.0–7.0 | Tibbetts et al. (2015), Ghosh et al. (2020), Becker (2013), Becker (2007) |
| *Dunaliella salina* | Mesophilic (35 °C) | – | 57.0 | 32.0 | 6.0 | |
| *Chlorella vulgaris* | Thermophilic (55 °C) | Anaerobic flasks (0.2 L) | 475.0 $mg.g^{-1}$ | 204.0 $mg.g^{-1}$ | 90.6 $mg.g^{-1}$ | |
| *Scenedesmus* sp. | Mesophilic (35 °C) | Continuous stirred tank reactor | 44.0 | 45.0 CFA: 171 mg COD eq $g^{-1}$ VS | 4.0 SCFA: 88 mg COD eq $g^{-1}$ VS | Gruhn et al. (2016) |
| *Laminaria japonica* | Mesophilic (35 °C) | Anaerobic flasks (0.3 L) | 48.4 | 10.0 SCFA: 6.3 $g.L^{-1}$ | 1.5 | Pham et al. (2012) |
| *Pachymeniopsis elliptica* | Mesophilic (35 °C) | – | 12.1 | 53.8 SCFA: 4.4 $g.L^{-1}$ | 1.0 | Pham et al. (2012) |
| *Enteromorpha crinite* | Mesophilic (35 °C) | – | 17.0 | 41.2 SCFA: 3.6 $g.L^{-1}$ | 1.8 | Pham et al. (2012) |
| *Laminaria digitate* | Mesophilic (35 °C) | Anaerobic flasks (0.3 L) | 10.0 | 48.0 | 1.0 | Xia et al. (2016) |
| *Arthrospira platensis* | Mesophilic (35 °C) | Anaerobic flasks (0.3 L) | 19.0 | 76.0 | 5.0 | Xia et al. (2015) |
| *Microcystis sp* | Mesophilic (35 °C) | Anaerobic fermenter (1.0 L) | 6.9 | 56.4 SCFA: 12.4 $g.L^{-1}$ | 0.35 | Cho et al. (2017) |
| *Chlorella vulgaris* | Mesophilic (35 °C) | Continuous stirred tank reactor (9 g COD/L) | 57.9 | 21.6 SCFA: 0.4 $g.L^{-1}$ | 13.4 | Magdalena et al. (2019) |

#### 12.6.1.3 Protein Extraction Techniques in Microalgae

Proteins represent the predominant primary metabolites in microalgae. Some strains exhibit high protein levels, indicating their potential as protein sources (Table 12.8). Marine microalgae can have up to 70% protein by dry weight, depending on species, medium composition, and seasonal fluctuations. *Arthrospira maxima* is renowned for having 60–71% protein. Similarly, various microalgal species like *S. platensis* and *C. vulgaris* possess protein contents ranging from 55% to 70% (Becker 2013; Becker 2007; Matos 2019). The Kjeldahl method, often used to estimate total nitrogen content, may overestimate protein content due to non-protein nitrogen components present in microalgae (Schwenzfeier et al. 2011). Non-protein nitrogen contents have been reported in various microalgae species, averaging around 10% of the nitrogen content (Becker 2013). The nutritional value of algal protein is evaluated using metrics including the digestibility coefficient, biological value, protein efficiency ratio, and net protein utilization (Becker 2013). Additionally, Table 12.8 summarizes the total amino acid concentration in different microalgal species (% of DCW) (Bonke et al. 2020; Paula da Silva et al. 2021; Liu et al. 2021; Nateghpour et al. 2021; Sui and Harvey 2021).

### 12.6.2 Secondary Metabolites

Microalgae synthesize secondary metabolites, with approximately 15,000 newly identified through chemical analysis (Park et al. 2022). However, many compounds and their functions remain unidentified, underscoring the complexity of microalgae's biochemical repertoire (Bourgaud et al. 2001). While not vital for survival, these secondary metabolites play crucial roles in ecological interactions, including

**Table 12.8** Amino acid composition (% of DCW) in microalgal protein sources

| Amino acids | Representative species | | |
|---|---|---|---|
| | *Spirulina maxima* | *Chlorella vulgaris* | *Dunaliella bardawil* |
| Serine | 5.1 | 5.8 | 4.6 |
| Glycine | 5.7 | 6.3 | 5.5 |
| Proline | 4.2 | 5.0 | 3.3 |
| Histidine | 2.2 | 2.0 | 1.8 |
| Glutamic acid | 10.3 | 13.7 | 12.7 |
| Isoleucine | 6.7 | 3.2 | 4.2 |
| Alanine | 9.5 | 9.4 | 7.3 |
| Valine | 7.1 | 7.0 | 5.8 |
| Asparagine | 11.8 | 9.3 | 10.4 |
| Arginine | 7.3 | 6.9 | 7.3 |
| Tyrosine | 0.3 | – | 0.7 |
| Threonine | 6.2 | 5.3 | 5.4 |
| Methionine | 2.5 | 1.3 | 2.3 |
| Cysteine | 0.9 | – | 1.2 |
| Phenylalanine | 5.3 | 5.5 | 5.8 |

defense, competition, and species communication (Nilesh Hemantkumar and Ilza Rahimbhai 2020). Exploiting microalgae for non-essential compounds has garnered interest due to its potential therapeutic applications and industrial utility in addressing various medical conditions and as nutritional supplements (Rehmanji et al. 2022). Additionally, certain species accumulate valuable metabolites such as pigments, vitamins, and carotenoids, which find commercial applications in diverse industries (Paliwal et al. 2015; Skjånes et al. 2013). These compounds, derived either directly from primary metabolism or synthesized later as secondary metabolites, exhibit activities like antioxidant and antimicrobial properties, making them valuable for pharmaceutical and nutraceutical purposes. The global market for microalgae-based products is projected to exceed $3700 million by 2024, with algal products anticipated to reach $6.09 billion by 2026, indicating significant growth opportunities in various sectors.

## 12.7 Biorefinery Applications

### 12.7.1 Biofuels

Microalgae's rapid growth rate and lipid yield position them as promising alternatives to fossil fuels in biodiesel production. Microalgae offer a lipid yield of up to 70%, primarily generated through carbon fixation during photosynthesis, making them an ideal raw material for biodiesel production (Zhang et al. 2022a). Notably, species like *Parachlorella kessleri* and *Nannochloropsis oculata* have shown potential as sustainable triacylglycerol (TAG) sources for biodiesel production. *Parachlorella kessleri*, in particular, exhibits excellent biomass, lipid, and starch productivity, making it a good option for producing biodiesel (Ota et al. 2016). Similarly, *Nannochloropsis oculata*, a marine unicellular microalga, has demonstrated promise in biofuel production by extracting cellular lipids for conversion into biodiesel through transesterification reactions. Research has indicated that growing *N. oculata* in environments with high CO2 concentrations and moderate light levels improves the plant's lipid content and general usefulness as a source of biodiesel (Zhang et al. 2022b).

Microalgae are grown in closed photobioreactors and open raceway ponds. Then their biomass or isolated macromolecules are extracted to produce biofuels like bioethanol, biodiesel, crude oil, biomethane, and bio-jet fuel. This process entails several steps and techniques. A critical factor in assessing microalgae suitability as a biofuel feedstock is the quantity and ratio of saturated and unsaturated fatty acids in their lipids. The fatty acid composition significantly impacts the fuel characteristics of biodiesel produced from microalgae. Therefore, while choosing candidates for the production of biofuel, it is crucial to take into account the capacity of microalgae strains to generate large amounts of lipids with advantageous fatty acid compositions (Sajjadi et al. 2018).

### 12.7.2 Nutraceuticals

Carbohydrates: Microalgae serve as valuable sources of storage carbohydrates, including starch, cellulose, and chrysolaminarin. These storage polysaccharides exhibit economic and nutritional importance, with chrysolaminarin demonstrating anti-tumor, antioxidant, and immune-modulatory properties (Simas-Rodrigues et al. 2015).

Lipids: Microalgae lipids predominantly consist of TAGs, glycolipids, phospholipids, and various LC-PUFAs such as EPA, DHA ($\omega - 3$), and arachidonic acid ($\omega - 6$). These lipid components have economic potential in a variety of bioproducts and are essential for human nutrition. By 2026, the worldwide market for omega fatty acids is anticipated to grow to a value of USD 9 billion (Rehmanji et al. 2021).

Amino Acids and Proteins: Amino acids, especially aspartate (Asp) and glutamate (Glu), are abundant in microalgae proteins and are necessary for a variety of biological processes. It is known that some species of Spirulina and Chlorella have about 70% protein by weight. These proteins serve as potential sources of bioactive peptides with nutritional and pharmacological implications. Marine algal proteins are also gaining attention for their immunomodulatory, anticancer, antihypertensive, and antioxidant properties (Bleakley and Hayes 2017).

Vitamins: Microalgae are abundant in essential vitamins such as B12, K, D, and E, serving as precursors or co-enzymes in vital metabolic pathways while exhibiting antioxidant and photo-protective properties. Tocopherols, including α, β, γ, and δ variants, are particularly noteworthy for their roles in these functions (Rehmanji et al. 2022).

### 12.7.3 Pharmaceuticals

Numerous physiologically active substances with potential uses in medicine may be found in microalgae, including antimicrobials, antivirals, therapeutic proteins, and drugs. Various microalgal species produce commercially critical secondary metabolites such as astaxanthin, β-carotene, canthaxanthin, and phycobiliproteins (phycoerythrin and phycocyanin) (Kamath et al. 2008; Brizio et al. 2013; Esfahani-Mashhour et al. 2009). In medicine and nutrition management, extracts from algae such as *Lyngbya majuscula* are used as immune modulators (Pulz and Gross 2004). Microalgae can also be modified to make recombinant proteins or boost the expression of native genes, which would further raise their potential for use in medicinal applications (Das et al. 2011; He et al. 2007).

### 12.7.4 Cosmetics

Numerous secondary metabolites derived from microalgae benefit skin health, leading to their incorporation into cosmetics products. In line with worldwide trends toward ecologically sourced, sustainable, and healthful products, the cosmetics industry is investing more and more in the research and development of products that contain natural extracts or components. Algal extracts exhibit antioxidant properties and UV protection, making them valuable additions to various cosmetic formulations. These extracts contribute to anti-aging, skin moisturization, whitening, sunscreen, and hair care products (Ariede et al. 2017).

Microalgae extracts are known to treat seborrhea, prevent blemishes, promote skin healing, and maintain skin moisture. Red microalgae extracts are commonly found in anti-aging creams, sunscreen products, and hair care formulations. In cosmetic formulations, algae act as antioxidant, thickening, and water-binding agents, increasing their effectiveness and attractiveness (Spolaore et al. 2006). Extracts from algae like *Chlorella* sp., *Chlorococcum* sp., *Monodus* sp., and *Thalassiosira* sp. are suggested for use in hair care products to stimulate keratinocyte differentiation and melanocyte proliferation, promoting healthy hair growth (Ariede et al. 2017).

In conclusion, microalgae have a variety of uses in biorefineries, ranging from the synthesis of medicines, cosmetics, and nutraceuticals to the generation of biofuel. Their diverse biochemical composition and beneficial properties make them valuable resources for sustainable development across multiple industries.

## 12.8 Techno-Economic Feasibility and Life Cycle Assessment

### 12.8.1 Techno-Economic Analysis (TEA)

The microalgal biorefinery landscape is broadly categorized into two main segments: upstream processing, which encompasses nutrient recovery and cultivation, and downstream processing, which involves the conversion of microalgae into usable products. Installing PBRs and incurring operational costs associated with raw materials and resources are common upstream techniques. Reducing culture costs is crucial for commercial algae production systems to turn a profit (Acién et al. 2012). The cultivation of various industrial algal strains such as *Chlorella* sp., *Nannochloropsis* sp., *Spirulina* sp., and *Scenedesmus* sp. predominantly occurs in open ponds because they are simple to put up, run, and maintain (Costa and de Morais 2014). Notably, the US's Sapphire Energy Columbus Algal Biomass Farm has successfully demonstrated the use of open raceway ponds for microalgae cultivation, yielding 520 metric tonnes of dried algal biomass over 2 years. Chlorella sp. is also cultivated in Taiwan and Japan using circular open ponds, albeit growth over 10,000 $m^2$ is restricted because of homogenous mixing regulations (Maheshwari et al. 2021). Despite their popularity, open pond designs often need help with cross-contamination, seasonal variations, and uneven light distribution. Compared to open ponds, closed PBRs provide a more regulated environment for microalgae growth, increasing biomass productivity (Béchet et al. 2015). However, indoor PBRs' high energy requirements, maintenance expenses, and significant capital expenditures might occasionally prevent their widespread industrial use. According to a 2019 TEA at NREL, the minimum selling price of algal biomass varies according on the closed PBR system, ranging from 639 to 1737 \$/tonne, while it is significantly lower for open ponds at 494 \$/tonne (Clippinger and Davis 2019).

Another study indicates that open ponds represent 45.73% of the production share, with 81.17% of the overall cost of producing biofuel coming from bubble column PBRs (Rajesh Banu et al. 2020). The choice of PBR is significantly influenced by the sale price of the end product. Closed PBR setups can generate astaxanthin from *H. pluvialis* because of its higher market price, but they might not be economically viable for generating β-carotene from *D. salina* (Thomassen et al. 2016).

Numerous solutions have been developed over the past 10 years in an attempt to solve concerns including reducing the land footprint and low algal production brought on by problems with light penetration. These include the use of open system ponds with algal lawn scrubber and stacked modular raceway ponds made of acrylic (Romagnoli et al. 2020; de Souza et al. 2019). Additionally, a significant portion of production expenses is allocated to synthetic fertilizers, which provide essential nutrients for microalgal culture alongside reactors (Mallick et al. 2016). Cultivating microalgae on a large scale requires substantial amounts of nitrogen and phosphorus, primarily sourced from synthetic fertilizers, further driving up production costs (Khan et al. 2018). However, flue gas and wastewater-based carbon capture and storage techniques can help reduce these expenses (Judd et al. 2017).

Most of the expenses associated with microalgae production are incurred during post-cultivation processes, particularly harvesting and conversion into bio-based products. Harvesting, in particular, is a challenging operation due to the diluted nature of the suspension, small particle sizes, and adverse surface charge densities (Barros et al. 2015). Various chemical and physical harvesting approaches have been explored, along with VAP extraction methods, each with its associated challenges (Venkata Subhash et al. 2022). Downstream processing is crucial in minimizing costs and adverse environmental impacts, with suggested step-by-step plans for selecting extraction and dewatering techniques (Vanthoor-Koopmans et al. 2013).

The substantial expenses involved in both cultivation and downstream techniques during scale-up often result in standalone microalgae-based products being less competitive than their synthetic counterparts. Algal biorefineries are classified as either material- or energy-driven based on their products. Different scenarios have been analyzed to evaluate algal biorefineries' performance and financial viability, including biodiesel production, anaerobic digestion for biogas generation, and supercritical gasification for syngas production (Ventura et al. 2013). Combining appropriate downstream processing approaches can enhance the economics of energy-based algae biorefineries (Dong et al. 2016). For example, a research found that combining processes such as lipid extraction from residual biomass, sugar fermentation into bioethanol, and diluted acid pre-treatment greatly decreased the selling price of biofuel units (Dong et al. 2016). The economic feasibility of energy-based biorefineries can also be increased by integrating value-added biorenewables and assessing different scenarios (Rajesh Banu et al. 2020).

Optimal growth conditions, effective pre-processing, and conversion into useful items can all improve the process's economic sustainability (Beckstrom et al. 2020;

Slegers et al. 2020). Techno-economic analyses underscore the importance of cultivation and downstream processing circumstances in determining algal biorefinery products' viability and selling price.

### 12.8.2 Life Cycle Assessment

A life cycle assessment (LCA) is a thorough analysis of an algal product's viability, energy dynamics, and environmental effect. The four primary steps for doing life cycle assessment (LCA) include establishing the objective and scope, executing life cycle inventory (LCI) analysis, performing life cycle impact assessment (LCIA), and evaluating the life cycle outcomes, according to the International Organization for Standardization (ISO) 14,040/44 recommendations (Talwar and Holden 2022). Algal farming, dewatering, extraction, conversion, and producing high-value-added bio-renewables and biodiesel are scrutinized in LCA (Strazza et al. 2015).

Olguín et al. (2015) have effectively scrutinized the economic feasibility of various biofuel production techniques using LCA. LCA evaluates biofuel production techniques' environmental impact and sustainability based on algae (Chowdhury and Franchetti 2017). It assesses net energy ratios (NERs) and GHG emissions. NER represents the overall energy efficiency of an algal biorefinery, with a value less than 1 indicating a cost-effective approach (De Bhowmick et al. 2019). Growth condition improvements, procedural advancements, and co-product utilization are essential for enhancing the feasibility of the algal biorefinery process. The following is the equation for NER calculation:

$$\text{Net Energy Ratio}\left(\text{NER}\right) = \frac{\text{Energy produced}}{\text{Energy input}}$$

LCA typically examines a process's environmental impacts using integrated and conventional approaches for biofuel production (Kyriakopoulou et al. 2015). GHG emissions from the entire biorefinery process are analyzed using SimaPro software. To evaluate the production of bioenergy from microalgae, Bennion et al. (2015) used an LCA model. They found that hydrothermal liquefaction (HTL) produced a NER of 1.23 and − 11.4 g $CO_2$-eq. $(MJ)^{-1}$ GHG emission, whereas pyrolysis produced a NER of 2.27 and 210 g $CO_2$-eq. $(MJ)^{-1}$ GHG emission.

Using SimaPro software, Rajesh Banu et al. (2020) compared the environmental effects of integrated and conventional biofuel production systems. Their findings showed significant differences in GHG emissions, with the microalgae integrated technique recording 50% GHG emission compared to 94% in conventional extraction with 84% energy use. Mu et al. (2014) performed LCAs on various routes using the GREET model, determining that the least environmentally sustainable method was algal growing followed by wet lipid extraction, with GHG emissions of −6.9 kg $CO_2$-eq $kg^{-1}$.

An LCA model created using Open LCA software in research on algae-mediated biofuel generation by Sun et al. (2019b), produced NER values between 0.71 and

2.46 and GHG emissions between −173.02 and 265.26 g CO2-eq. Out of all the options examined, anaerobic digestion with pre-treatment proved to be the most economical and ecologically beneficial approach. Raw materials and any aquatic, airborne, or terrestrial discharges during synthesis are also considered in LCAs. Algal biodiesel's minimal effects on eutrophication and land use can be attributed to its high biomass output and utilization of wastelands. Additionally, its reduced use of fertilizer and pesticides results in lower impacts on soil acidification and human toxicity.

Improving growth conditions, developing new procedures, and using co-products are all necessary to make the algal biorefinery process more feasible. Typically, LCA analyzes a process's environmental effects using integrated and conventional approaches for producing biofuel (Kyriakopoulou et al. 2015). SimaPro software is used to analyze the GHG emissions from the full biorefinery process. Using the GREET Model 2013, Bennion et al. (2015) obtained an LCA model for bioenergy production from microalgae performed using HTL with NER 1.23 and −11.4 g $CO_2$-eq. $(MJ)^{-1}$ GHG emission, which is different from the NER 2.27 and 210 g $CO_2$-eq. $(MJ)^{-1}$ GHG emission for bioenergy production by pyrolysis.

## 12.9 Future Perspectives

Microalgae's high lipid content and quick development rate are drawing more attention in the future. It's becoming evident that nano-additive applications hold significant commercial potential, alongside their positive environmental impact and generation of valuable byproducts. These applications span various stages, from microalgae culture to the utilization of the end product. Nano-additives such as nanotubes, nanodroplets, nanofibers, nanosheets, nanoparticles, and other nanostructures are anticipated to enhance microalgae cultivation (Hossain et al. 2019).

There is potential for algal biodiesel to be incorporated into aviation fuel blends, potentially reducing $CO_2$ emissions from aircraft travel in the future (Bošnjaković and Sinaga 2020). The potential benefits of algae biorefineries are illustrated in Fig. 12.1. The Algal Microbial Fuel Cell (AMFC) is an alluring alternative renewable energy source that uses photosynthetic organisms and CO2 to directly generate power from synthetic water or wastewater (Chen 2018).

Efforts toward environmentally friendly energy solutions and green energy objectives can be advanced through the appropriate algal biomass processing. Biohydrogens that generate no pollution hold promise for future use as bioelectricity. Its adoption could be crucial for industrial operations due to potential cost reductions (Tsolcha et al. 2021). To achieve environmental and green energy goals, algal biomass may soon be converted into hydrogen energy; this type of hydrogen fuel may be known as bioelectricity.

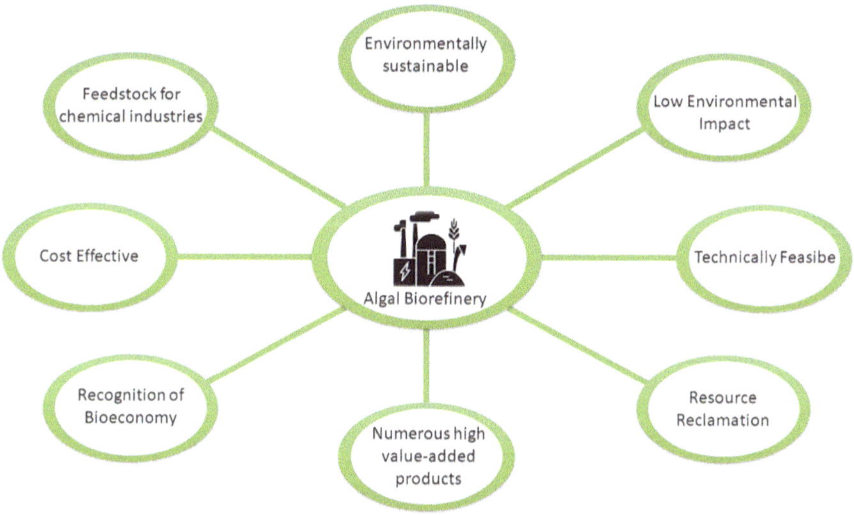


**Fig. 12.1** Potential advantages of establishing an algae biorefinery on the market

## 12.10 Conclusions

As the expansion of biorefineries progresses, we are gradually reducing our dependence on petroleum and shifting towards sustainable and renewable raw materials. The impetus behind biorefinery development stems from both economic considerations and energy concerns. Key factors crucial for the success of biorefineries include ensuring a stable supply of food and feedstock, promoting biomass diversity, minimizing ecological impact, and adapting to market dynamics. Microalgae, with their versatile metabolic capabilities, hold immense promise for the biorefinery sector. Their ability to thrive in various environments and produce essential metabolites makes them ideal candidates for sustainable bio-based production.

Microalgae serve as rich reservoirs of secondary metabolites, which find applications across numerous sectors, including food, cosmetics, pharmaceuticals, and healthcare, besides being significant sources of primary metabolites like biofuels. The significance of further research and commercialization initiatives centered on microalgal secondary metabolites is highlighted by the rising demand for these products worldwide. However, despite their potential, challenges persist in maximizing biofuel production and overcoming technological hurdles. The creation of novel strategies to overcome these obstacles and completely use the potential of microalgae-based biorefineries should be the top priority of future research initiatives.

Developing sustainable strategies for the commercialization of fuel-based and non-fuel-based goods made from microalgae requires interdisciplinary study. This all-encompassing strategy promotes a thorough comprehension of biorefinery operations, which helps the economy grow and diversify biomass byproducts for long-term, sustainable valuation. To sum up, microalgae present a viable pathway for the organic synthesis of a broad variety of metabolites with a range of biological functions, opening the door to a more ecologically friendly and sustainable future for the biorefinery industry. We can completely realize the potential of microalgae as a useful resource for bio-based industry and environmental preservation by encouraging teamwork and multidisciplinary cooperation.

## References

Acién FG, Fernández JM, Magán JJ, Molina E (2012) Production cost of a real microalgae production plant and strategies to reduce it. Biotechnol Adv 30(6):1344–1353. https://doi.org/10.1016/j.biotechadv.2012.02.005

Akgül F (2020) Effects of nitrogen concentration on growth, biomass, and biochemical composition of Desmodesmus communis (E. Hegewald) E. Hegewald. Prep Biochem Biotechnol 50(1):98–105. https://doi.org/10.1080/10826068.2019.1697884

Ampofo J, Abbey L (2022) Microalgae: bioactive composition, health benefits, safety and prospects as potential high-value ingredients for the functional food industry. Foods 11(12):1744. https://doi.org/10.3390/foods11121744

Ampofo JO, Ngadi M (2020) Ultrasonic assisted phenolic elicitation and antioxidant potential of common bean (Phaseolus vulgaris) sprouts. Ultrason Sonochem 64:104974. https://doi.org/10.1016/j.ultsonch.2020.104974

Ariede MB, Candido TM, Jacome ALM, Velasco MVR, de Carvalho JCM, Baby AR (2017) Cosmetic attributes of algae – a review. Algal Res 25:483–487. https://doi.org/10.1016/j.algal.2017.05.019

Bakonyi P et al (2018) A review of the innovative gas separation membrane bioreactor with mechanisms for integrated production and purification of biohydrogen. Bioresour Technol 270:643–655. https://doi.org/10.1016/j.biortech.2018.09.020

Barros AI, Gonçalves AL, Simões M, Pires JCM (2015) Harvesting techniques applied to microalgae: a review. Renew Sust Energ Rev 41:1489–1500. https://doi.org/10.1016/j.rser.2014.09.037

Batista AP, Gouveia L, Bandarra NM, Franco JM, Raymundo A (2013) Comparison of microalgal biomass profiles as a novel functional ingredient for food products. Algal Res 2(2):164–173. https://doi.org/10.1016/j.algal.2013.01.004

Béchet Q, Chambonnière P, Shilton A, Guizard G, Guieysse B (2015) Algal productivity modeling: a step toward accurate assessments of full-scale algal cultivation. Biotechnol Bioeng 112(5):987–996. https://doi.org/10.1002/bit.25517

Becker EW (2007) Micro-algae as a source of protein. Biotechnol Adv 25(2):207–210. https://doi.org/10.1016/j.biotechadv.2006.11.002

Becker EW (2013) Microalgae for human and animal nutrition. In: Handbook of microalgal culture. Wiley, pp 461–503. https://doi.org/10.1002/9781118567166.ch25

Beckstrom BD, Wilson MH, Crocker M, Quinn JC (2020) Bioplastic feedstock production from microalgae with fuel co-products: a techno-economic and life cycle impact assessment. Algal Res 46:101769. https://doi.org/10.1016/j.algal.2019.101769

Bennion EP, Ginosar DM, Moses J, Agblevor F, Quinn JC (2015) Lifecycle assessment of microalgae to biofuel: comparison of thermochemical processing pathways. Appl Energy 154:1062–1071. https://doi.org/10.1016/j.apenergy.2014.12.009

Bhargavi G, Nageswara Rao P, Renganathan S (2018) Review on the extraction methods of crude oil from all generation biofuels in last few decades. IOP Conf Ser Mater Sci Eng 330:012024. https://doi.org/10.1088/1757-899X/330/1/012024

Bleakley S, Hayes M (2017) Algal proteins: extraction, application, and challenges concerning production. Foods 6(5):33. https://doi.org/10.3390/foods6050033

Bligh EG, Dyer WJ (1959) A rapid method of total lipid extraction and purification. Can J Biochem Physiol 37(8):911–917. https://doi.org/10.1139/o59-099

Bonke A, Sieuwerts S, Petersen IL (2020) Amino acid composition of novel plant drinks from oat, lentil and pea. Foods 9(4):429. https://doi.org/10.3390/foods9040429

Bošnjaković M, Sinaga N (2020) The perspective of large-scale production of algae biodiesel. Appl Sci 10(22):8181. https://doi.org/10.3390/app10228181

Bourgaud F, Gravot A, Milesi S, Gontier E (2001) Production of plant secondary metabolites: a historical perspective. Plant Sci 161(5):839–851. https://doi.org/10.1016/S0168-9452(01)00490-3

Brizio P et al (2013) Astaxanthin and Canthaxanthin (Xanthophyll) as supplements in Rainbow Trout Diet: in vivo assessment of residual levels and contributions to human health. J Agric Food Chem 61(46):10954–10959. https://doi.org/10.1021/jf4012664

Chandra R, Rohit MV, Swamy YV, Venkata Mohan S (2014) Regulatory function of organic carbon supplementation on biodiesel production during growth and nutrient stress phases of mixotrophic microalgae cultivation. Bioresour Technol 165:279–287. https://doi.org/10.1016/j.biortech.2014.02.102

Chandra R, Das P, Vishal G, Nagra S (2019) Factors affecting the induction of UV protectant and lipid productivity in Lyngbya for sequential biorefinery product recovery. Bioresour Technol 278:303–310. https://doi.org/10.1016/j.biortech.2019.01.084

Chen CY (2018) Future perspective of algae: energy production and environmental cleanup. New Biotechnol 44:S127. https://doi.org/10.1016/j.nbt.2018.05.1065

Chen C-Y, Yeh K-L, Aisyah R, Lee D-J, Chang J-S (2011) Cultivation, photobioreactor design and harvesting of microalgae for biodiesel production: a critical review. Bioresour Technol 102(1):71–81. https://doi.org/10.1016/j.biortech.2010.06.159

Chen C-Y et al (2013) Microalgae-based carbohydrates for biofuel production. Biochem Eng J 78:1–10. https://doi.org/10.1016/j.bej.2013.03.006

Cheng J, Sun J, Huang Y, Feng J, Zhou J, Cen K (2013) Dynamic microstructures and fractal characterization of cell wall disruption for microwave irradiation-assisted lipid extraction from wet microalgae. Bioresour Technol 150:67–72. https://doi.org/10.1016/j.biortech.2013.09.126

Cho HU, Kim HG, Kim YM, Park JM (2017) Volatile fatty acid recovery by anaerobic fermentation from blue-green algae: effect of pretreatment. Bioresour Technol 244:1433–1438. https://doi.org/10.1016/j.biortech.2017.05.076

Chowdhury R, Franchetti M (2017) Life cycle energy demand from algal biofuel generated from nutrients present in the dairy waste. Sustain Prod Consum 9:22–27. https://doi.org/10.1016/j.spc.2016.07.001

Chrapusta E, Kaminski A, Duchnik K, Bober B, Adamski M, Bialczyk J (2017) Mycosporine-like amino acids: potential health and beauty ingredients. Mar Drugs 15(10):326. https://doi.org/10.3390/md15100326

Clippinger J, Davis R (2019) Techno-economic analysis for the production of algal biomass via closed photobioreactors: future cost potential evaluated across a range of cultivation system designs. Golden, CO (United States). https://doi.org/10.2172/1566806

Costa JAV, de Morais MG (2014) An open pond system for microalgal cultivation. In: Biofuels from Algae. Elsevier, pp 1–22. https://doi.org/10.1016/B978-0-444-59558-4.00001-2

Das P, Aziz SS, Obbard JP (2011) Two phase microalgae growth in the open system for enhanced lipid productivity. Renew Energy 36(9):2524–2528. https://doi.org/10.1016/j.renene.2011.02.002

De Bhowmick G, Sarmah AK, Sen R (2019) Zero-waste algal biorefinery for bioenergy and biochar: a green leap towards achieving energy and environmental sustainability. Sci Total Environ 650:2467–2482. https://doi.org/10.1016/j.scitotenv.2018.10.002

de Souza MP, Hoeltz M, Gressler PD, Benitez LB, Schneider RCS (2019) Potential of microalgal bioproducts: general perspectives and main challenges. Waste Biomass Valorization 10(8):2139–2156. https://doi.org/10.1007/s12649-018-0253-6

Deepa P, Sowndhararajan K, Kim S (2023) A review of the harvesting techniques of microalgae. Water (Basel) 15(17):3074. https://doi.org/10.3390/w15173074

Dong T et al (2016) Combined algal processing: a novel integrated biorefinery process to produce algal biofuels and bioproducts. Algal Res 19:316–323. https://doi.org/10.1016/j.algal.2015.12.021

Esfahani-Mashhour M, Moravej H, Mehrabani-Yeganeh H, Razavi SH (2009) Evaluation of coloring potential of Dietzia natronolimnaea biomass as source of Canthaxanthin for egg yolk pigmentation. Asian Australas J Anim Sci 22(2):254–259. https://doi.org/10.5713/ajas.2009.80202

Fernández AL, El-Diasty MM, Martínez A, Alvarez J, García-Bengochea JB (2011) A simple technique to rule out occlusion of right coronary artery after aortic valve surgery. Ann Thorac Surg 92(6):2281–2282. https://doi.org/10.1016/j.athoracsur.2011.06.016

Ghazvini M, Kavosi M, Sharma R, Kim M (2022) A review on mechanical-based microalgae harvesting methods for biofuel production. Biomass Bioenergy 158:106348. https://doi.org/10.1016/j.biombioe.2022.106348

Ghosh S, Sohn H-Y, Pyo S-J, Jensen AB, Meyer-Rochow VB, Jung C (2020) Nutritional composition of Apis mellifera drones from Korea and Denmark as a potential sustainable alternative food source: comparison between developmental stages. Foods 9(4):389. https://doi.org/10.3390/foods9040389

Goiris K, Muylaert K, Fraeye I, Foubert I, De Brabanter J, De Cooman L (2012) Antioxidant potential of microalgae in relation to their phenolic and carotenoid content. J Appl Phycol 24(6):1477–1486. https://doi.org/10.1007/s10811-012-9804-6

Gruhn M, Frigon J-C, Guiot SR (2016) Acidogenic fermentation of Scenedesmus sp.-AMDD: comparison of volatile fatty acids yields between mesophilic and thermophilic conditions. Bioresour Technol 200:624–630. https://doi.org/10.1016/j.biortech.2015.10.087

Han S-F et al (2019) Application of pulse electric field pretreatment for enhancing lipid extraction from Chlorella pyrenoidosa grown in wastewater. Renew Energy 133:233–239. https://doi.org/10.1016/j.renene.2018.10.034

He D-M et al (2007) Recombination and expression of classical swine fever virus (CSFV) structural protein E2 gene in Chlamydomonas reinhardtii chroloplasts. Colloids Surf B Biointerfaces 55(1):26–30. https://doi.org/10.1016/j.colsurfb.2006.10.042

Hossain N, Mahlia TMI, Saidur R (2019) Latest development in microalgae-biofuel production with nano-additives. Biotechnol Biofuels 12(1):125. https://doi.org/10.1186/s13068-019-1465-0

Huang Q, Jiang F, Wang L, Yang C (2017) Design of photobioreactors for mass cultivation of photosynthetic organisms. Engineering 3(3):318–329. https://doi.org/10.1016/J.ENG.2017.03.020

Illman AM, Scragg AH, Shales SW (2000) Increase in Chlorella strains calorific values when grown in low nitrogen medium. Enzym Microb Technol 27(8):631–635. https://doi.org/10.1016/S0141-0229(00)00266-0

Ismaiel MMS, El-Ayouty YM, Piercey-Normore M (2016) Role of pH on antioxidants production by Spirulina ( Arthrospira ) platensis. Braz J Microbiol 47(2):298–304. https://doi.org/10.1016/j.bjm.2016.01.003

Jeon B-H et al (2013) Ultrasonic disintegration of microalgal biomass and consequent improvement of bioaccessibility/bioavailability in microbial fermentation. Biotechnol Biofuels 6(1):37. https://doi.org/10.1186/1754-6834-6-37

Judd SJ, Al Momani FAO, Znad H, Al Ketife AMD (2017) The cost benefit of algal technology for combined CO2 mitigation and nutrient abatement. Renew Sust Energ Rev 71:379–387. https://doi.org/10.1016/j.rser.2016.12.068

Juneja A, Ceballos R, Murthy G (2013) Effects of environmental factors and nutrient availability on the biochemical composition of algae for biofuels production: a review. Energies (Basel) 6(9):4607–4638. https://doi.org/10.3390/en6094607

Kamath BS, Srikanta BM, Dharmesh SM, Sarada R, Ravishankar GA (2008) Ulcer preventive and antioxidative properties of astaxanthin from Haematococcus pluvialis. Eur J Pharmacol 590(1–3):387–395. https://doi.org/10.1016/j.ejphar.2008.06.042

Kandasamy G, Shaleh SRM (2018) Flotation removal of the microalga Nannochloropsis sp. using Moringa protein–oil emulsion: a novel green approach. Bioresour Technol 247:327–331. https://doi.org/10.1016/j.biortech.2017.08.187

Khan MI, Shin JH, Kim JD (2018) The promising future of microalgae: current status, challenges, and optimization of a sustainable and renewable industry for biofuels, feed, and other products. Microb Cell Factories 17(1):36. https://doi.org/10.1186/s12934-018-0879-x

Kurniawati HA, Ismadji S, Liu JC (2014) Microalgae harvesting by flotation using natural saponin and chitosan. Bioresour Technol 166:429–434. https://doi.org/10.1016/j.biortech.2014.05.079

Kyriakopoulou K, Papadaki S, Krokida M (2015) Life cycle analysis of β-carotene extraction techniques. J Food Eng 167:51–58. https://doi.org/10.1016/j.jfoodeng.2015.03.008

Latsos C, van Houcke J, Timmermans KR (2020) The effect of nitrogen starvation on biomass yield and biochemical constituents of Rhodomonas sp. Front Mar Sci 7. https://doi.org/10.3389/fmars.2020.563333

Laurens LML et al (2015) Acid-catalyzed algal biomass pretreatment for integrated lipid and carbohydrate-based biofuels production. Green Chem 17(2):1145–1158. https://doi.org/10.1039/C4GC01612B

Lee OK, Oh Y-K, Lee EY (2015) Bioethanol production from carbohydrate-enriched residual biomass obtained after lipid extraction of Chlorella sp. KR-1. Bioresour Technol 196:22–27. https://doi.org/10.1016/j.biortech.2015.07.040

Leite LDS, Daniel LA (2020) Optimization of microalgae harvesting by sedimentation induced by high pH. Water Sci Technol 82(6):1227–1236. https://doi.org/10.2166/wst.2020.106

Levasseur W, Perré P, Pozzobon V (2020) A review of high value-added molecules production by microalgae in light of the classification. Biotechnol Adv 41:107545. https://doi.org/10.1016/j.biotechadv.2020.107545

Li X, Huff J, Crunkleton DW, Johannes TW (2023) Light intensity and spectral quality modulation for improved growth kinetics and biochemical composition of Chlamydomonas reinhardtii. J Biotechnol 375:28–39. https://doi.org/10.1016/j.jbiotec.2023.08.007

Lindblad P et al (2019) CyanoFactory, a European consortium to develop technologies needed to advance cyanobacteria as chassis for production of chemicals and fuels. Algal Res 41:101510. https://doi.org/10.1016/j.algal.2019.101510

Liu X et al (2021) An high-throughput sequencing approach to the preliminary analysis of bacterial communities associated with changes in amino acid nitrogen, organic acid and reducing sugar contents during soy sauce fermentation. Food Chem 349:129131. https://doi.org/10.1016/j.foodchem.2021.129131

Luque de Castro MD, García-Ayuso LE (1998) Soxhlet extraction of solid materials: an outdated technique with a promising innovative future. Anal Chim Acta 369(1–2):1–10. https://doi.org/10.1016/S0003-2670(98)00233-5

Machu L et al (2015) Phenolic content and antioxidant capacity in algal food products. Molecules 20(1):1118–1133. https://doi.org/10.3390/molecules20011118

Magdalena JA, Tomás-Pejó E, González-Fernández C (2019) Volatile fatty acids production from microalgae biomass: anaerobic digester performance and population dynamics during stable conditions, starvation, and process recovery. Molecules 24(24):4544. https://doi.org/10.3390/molecules24244544

Maheshwari N, Mishra A, Thakur IS, Srivastava S (2021) Algal biofuel: a sustainable approach for fuel of future generation. In: Environmental microbiology and biotechnology. Springer Singapore, Singapore, pp 3–29. https://doi.org/10.1007/978-981-15-7493-1_1

Mahmood T, Hussain N, Shahbaz A, Mulla SI, Iqbal HMN, Bilal M (2023) Sustainable production of biofuels from the algae-derived biomass. Bioprocess Biosyst Eng 46(8):1077–1097. https://doi.org/10.1007/s00449-022-02796-8

Mallick N, Bagchi SK, Koley S, Singh AK (2016) Progress and challenges in microalgal biodiesel production. Front Microbiol 7. https://doi.org/10.3389/fmicb.2016.01019

Matos ÂP (2019) Microalgae as a Potential Source of Proteins. In: Proteins: sustainable source, processing and applications. Elsevier, pp 63–96. https://doi.org/10.1016/B978-0-12-816695-6.00003-9

Mu D, Min M, Krohn B, Mullins KA, Ruan R, Hill J (2014) Life cycle environmental impacts of wastewater-based algal biofuels. Environ Sci Technol 48(19):11696–11704. https://doi.org/10.1021/es5027689

Munir N, Hasnain M, Sarwar Z, Ali F, Hessini K, Abideen Z (2022) Changes in environmental conditions are critical factors for optimum biomass, lipid pattern and biodiesel production in algal biomass. Biologia (Bratisl) 77(11):3099–3124. https://doi.org/10.1007/s11756-022-01191-8
Najiha Badar S, Mohammad M, Emdadi Z, Yaakob Z (2021) Algae and their growth requirements for bioenergy: a review. Biofuels 12(3):307–325. https://doi.org/10.1080/17597269.2018.1472978
Naresh Kumar A et al (2020) Deoiled algal biomass-derived renewable sugars for bioethanol and biopolymer production in biorefinery framework. Bioresour Technol 296:122315. https://doi.org/10.1016/j.biortech.2019.122315
Nateghpour B, Kavoosi G, Mirakhorli N (2021) Amino acid profile of the peel of three citrus species and its effect on the combination of amino acids and fatty acids Chlorella vulgaris. J Food Compos Anal 98:103808. https://doi.org/10.1016/j.jfca.2021.103808
Nethravathy MU, Mehar JG, Mudliar SN, Shekh AY (2019) Recent advances in microalgal bioactives for food, feed, and healthcare products: commercial potential, market space, and sustainability. Compr Rev Food Sci Food Saf 18(6):1882–1897. https://doi.org/10.1111/1541-4337.12500
Nilesh Hemantkumar J, Ilza Rahimbhai M (2020) Microalgae and its use in nutraceuticals and food supplements. In: Microalgae – from physiology to application. IntechOpen. https://doi.org/10.5772/intechopen.90143
Obeid S, Beaufils N, Camy S, Takache H, Ismail A, Pontalier P-Y (2018) Supercritical carbon dioxide extraction and fractionation of lipids from freeze-dried microalgae Nannochloropsis oculata and Chlorella vulgaris. Algal Res 34:49–56. https://doi.org/10.1016/j.algal.2018.07.003
Olguín EJ, Castillo OS, Mendoza A, Tapia K, González-Portela RE, Hernández-Landa VJ (2015) Dual purpose system that treats anaerobic effluents from pig waste and produce Neochloris oleoabundans as lipid rich biomass. New Biotechnol 32(3):387–395. https://doi.org/10.1016/j.nbt.2014.12.004
Ota S et al (2016) Highly efficient lipid production in the green alga Parachlorella kessleri: draft genome and transcriptome endorsed by whole-cell 3D ultrastructure. Biotechnol Biofuels 9(1):13. https://doi.org/10.1186/s13068-016-0424-2
Paliwal C et al (2015) Selective carotenoid accumulation by varying nutrient media and salinity in Synechocystis sp. CCNM 2501. Bioresour Technol 197:363–368. https://doi.org/10.1016/j.biortech.2015.08.122
Pan J et al (2016) Microwave-assisted extraction of lipids from microalgae using an ionic liquid solvent [BMIM][HSO4]. Fuel 178:49–55. https://doi.org/10.1016/j.fuel.2016.03.037
Park YH et al (2022) Microalgal secondary metabolite productions as a component of biorefinery: a review. Bioresour Technol 344:126206. https://doi.org/10.1016/j.biortech.2021.126206
Paula da Silva S, Ferreira do Valle A, Perrone D (2021) Microencapsulated Spirulina maxima biomass as an ingredient for the production of nutritionally enriched and sensorially well-accepted vegan biscuits. LWT 142:110997. https://doi.org/10.1016/j.lwt.2021.110997
Pham TN, Nam WJ, Jeon YJ, Yoon HH (2012) Volatile fatty acids production from marine macroalgae by anaerobic fermentation. Bioresour Technol 124:500–503. https://doi.org/10.1016/j.biortech.2012.08.081
Pishgar Z, Samimi A, Mohebbi-Kalhori D, Shokrollahzadeh S (2020) Comparative study on the harvesting of marine Chlorella vulgaris microalgae from a dilute slurry using autoflocculation-sedimentation and electrocoagulation-flotation methods. Int J Environ Res 14(6):615–628. https://doi.org/10.1007/s41742-020-00277-y
Prajapati SK, Bhattacharya A, Malik A, Vijay VK (2015) Pretreatment of algal biomass using fungal crude enzymes. Algal Res 8:8–14. https://doi.org/10.1016/j.algal.2014.12.011
Pugkaew W, Meetam M, Yokthongwattana K, Leeratsuwan N, Pokethitiyook P (2019) Effects of salinity changes on growth, photosynthetic activity, biochemical composition, and lipid productivity of marine microalga Tetraselmis suecica. J Appl Phycol 31(2):969–979. https://doi.org/10.1007/s10811-018-1619-7
Pulz O, Gross W (2004) Valuable products from biotechnology of microalgae. Appl Microbiol Biotechnol 65(6):635–648. https://doi.org/10.1007/s00253-004-1647-x

Pushpakumari Kudahettige N, Pickova J, Gentili FG (2018) Stressing algae for biofuel production: biomass and biochemical composition of Scenedesmus dimorphus and Selenastrum minutum grown in municipal untreated wastewater. Front Energy Res 6. https://doi.org/10.3389/fenrg.2018.00132

Rajesh Banu J, Preethi S, Kavitha MG, Kumar G (2020) Microalgae based biorefinery promoting circular bioeconomy-techno economic and life-cycle analysis. Bioresour Technol 302:122822. https://doi.org/10.1016/j.biortech.2020.122822

Randrianarison G, Ashraf MA (2017) Microalgae: a potential plant for energy production. Geol Ecol Landscapes 1(2):104–120. https://doi.org/10.1080/24749508.2017.1332853

Rehmanji M, Suresh S, Nesamma AA, Jutur PP (2021) Microalgal cell factories, a platform for high-value-added biorenewables to improve the economics of the biorefinery. In: Microbial and Natural Macromolecules. Elsevier, pp 689–731. https://doi.org/10.1016/B978-0-12-820084-1.00027-2

Rehmanji M et al (2022) Multifaceted applications of microalgal biomass valorization to enriched biorenewables, a review of futuristic biorefinery paradigm. Bioresour Technol Rep 17:100972. https://doi.org/10.1016/j.biteb.2022.100972

Romagnoli F, Ievina B, Perera WAARP, Ferrari D (2020) Novel stacked modular open raceway ponds for microalgae biomass cultivation in biogas plants: preliminary design and modelling. Environ Clim Technol 24(2):1–19. https://doi.org/10.2478/rtuect-2020-0050

Sajjadi B, Chen W-Y, Raman AAA, Ibrahim S (2018) Microalgae lipid and biomass for biofuel production: a comprehensive review on lipid enhancement strategies and their effects on fatty acid composition. Renew Sust Energ Rev 97:200–232. https://doi.org/10.1016/j.rser.2018.07.050

Salama E-S et al (2018) Enhancement of microalgal growth and biocomponent-based transformations for improved biofuel recovery: a review. Bioresour Technol 258:365–375. https://doi.org/10.1016/j.biortech.2018.02.006

Salim S, Bosma R, Vermuë MH, Wijffels RH (2011) Harvesting of microalgae by bio-flocculation. J Appl Phycol 23(5):849–855. https://doi.org/10.1007/s10811-010-9591-x

Santana A, Jesus S, Larrayoz MA, Filho RM (2012) Supercritical carbon dioxide extraction of algal lipids for the biodiesel production. Proc Eng 42:1755–1761. https://doi.org/10.1016/j.proeng.2012.07.569

Schwenzfeier A, Wierenga PA, Gruppen H (2011) Isolation and characterization of soluble protein from the green microalgae Tetraselmis sp. Bioresour Technol 102(19):9121–9127. https://doi.org/10.1016/j.biortech.2011.07.046

Seibert M, Torzillo G (eds) (2018) Microalgal hydrogen production. Royal Society of Chemistry, Cambridge. https://doi.org/10.1039/9781849737128

Sen Tan J et al (2020) A review on microalgae cultivation and harvesting, and their biomass extraction processing using ionic liquids. Bioengineered 11(1):116–129. https://doi.org/10.1080/21655979.2020.1711626

Shankar M, Chhotaray PK, Gardas RL, Tamilarasan K, Rajesh M (2019) Application of carboxylate protic ionic liquids in simultaneous microalgal pretreatment and lipid recovery from marine Nannochloropsis sp. and Chlorella sp. Biomass Bioenergy 123:14–24. https://doi.org/10.1016/j.biombioe.2019.02.005

Shetty P, Gitau MM, Maróti G (2019) Salinity stress responses and adaptation mechanisms in eukaryotic green microalgae. Cells 8(12):1657. https://doi.org/10.3390/cells8121657

Sim SJ (2009) Hydrothermal acid pretreatment of Chlamydomonas reinhardtii biomass for ethanol production. J Microbiol Biotechnol 19(2):161–166. https://doi.org/10.4014/jmb.0810.578

Simas-Rodrigues C, Villela HDM, Martins AP, Marques LG, Colepicolo P, Tonon AP (2015) Microalgae for economic applications: advantages and perspectives for bioethanol. J Exp Bot 66(14):4097–4108. https://doi.org/10.1093/jxb/erv130

Sivaramakrishnan R, Incharoensakdi A (2019) Low power ultrasound treatment for the enhanced production of microalgae biomass and lipid content. Biocatal Agric Biotechnol 20:101230. https://doi.org/10.1016/j.bcab.2019.101230

Skjånes K, Rebours C, Lindblad P (2013) Potential for green microalgae to produce hydrogen, pharmaceuticals and other high value products in a combined process. Crit Rev Biotechnol 33(2):172–215. https://doi.org/10.3109/07388551.2012.681625

Slegers PM et al (2020) Design of value chains for microalgal biorefinery at industrial scale: process integration and techno-economic analysis. Front Bioeng Biotechnol 8. https://doi.org/10.3389/fbioe.2020.550758

Song Y, Cheng J, Yang Y, Gao Z (2023) Irradiance penetration distribution and flashing light frequency simultaneously affected with microalgal cell absorption and CO2 bubble scattering in a raceway pond. Sci Total Environ 864:160988. https://doi.org/10.1016/j.scitotenv.2022.160988

Spolaore P, Joannis-Cassan C, Duran E, Isambert A (2006) Commercial applications of microalgae. J Biosci Bioeng 101(2):87–96. https://doi.org/10.1263/jbb.101.87

Strazza C, Magrassi F, Gallo M, Del Borghi A (2015) Life cycle assessment from food to food: a case study of circular economy from cruise ships to aquaculture. Sustain Prod Consum 2:40–51. https://doi.org/10.1016/j.spc.2015.06.004

Sui Y, Harvey PJ (2021) Effect of light intensity and wavelength on biomass growth and protein and amino acid composition of Dunaliella salina. Foods 10(5):1018. https://doi.org/10.3390/foods10051018

Sui Y, Mazzucchi L, Acharya P, Xu Y, Morgan G, Harvey PJ (2021) A comparison of β-carotene, phytoene and amino acids production in Dunaliella salina DF 15 (CCAP 19/41) and Dunaliella salina CCAP 19/30 using different light wavelengths. Foods 10(11):2824. https://doi.org/10.3390/foods10112824

Sun X-M, Ren L-J, Zhao Q-Y, Ji X-J, Huang H (2019a) Enhancement of lipid accumulation in microalgae by metabolic engineering. Biochimica et Biophysica Acta 1864(4):552–566. https://doi.org/10.1016/j.bbalip.2018.10.004

Sun C-H et al (2019b) Life-cycle assessment of biofuel production from microalgae via various bioenergy conversion systems. Energy 171:1033–1045. https://doi.org/10.1016/j.energy.2019.01.074

Talwar N, Holden NM (2022) The limitations of bioeconomy LCA studies for understanding the transition to sustainable bioeconomy. Int J Life Cycle Assess 27(5):680–703. https://doi.org/10.1007/s11367-022-02053-w

Tang S, Qin C, Wang H, Li S, Tian S (2011) Study on supercritical extraction of lipids and enrichment of DHA from oil-rich microalgae. J Supercrit Fluids 57(1):44–49. https://doi.org/10.1016/j.supflu.2011.01.010

Thomassen G, Egiguren Vila U, Van Dael M, Lemmens B, Van Passel S (2016) A techno-economic assessment of an algal-based biorefinery. Clean Technol Environ Policy 18(6):1849–1862. https://doi.org/10.1007/s10098-016-1159-2

Thomassen G, Van Dael M, Van Passel S (2018) The potential of microalgae biorefineries in Belgium and India: an environmental techno-economic assessment. Bioresour Technol 267:271–280. https://doi.org/10.1016/j.biortech.2018.07.037

Tibbetts SM, Milley JE, Lall SP (2015) Chemical composition and nutritional properties of freshwater and marine microalgal biomass cultured in photobioreactors. J Appl Phycol 27(3):1109–1119. https://doi.org/10.1007/s10811-014-0428-x

Tsolcha ON, Patrinou V, Economou CN, Dourou M, Aggelis G, Tekerlekopoulou AG (2021) Utilization of biomass derived from cyanobacteria-based agro-industrial wastewater treatment and raisin residue extract for bioethanol production. Water (Basel) 13(4):486. https://doi.org/10.3390/w13040486

Vanthoor-Koopmans M, Wijffels RH, Barbosa MJ, Eppink MHM (2013) Biorefinery of microalgae for food and fuel. Bioresour Technol 135:142–149. https://doi.org/10.1016/j.biortech.2012.10.135

Vasudevan PT, Briggs M (2008) Biodiesel production—current state of the art and challenges. J Ind Microbiol Biotechnol 35(5):421. https://doi.org/10.1007/s10295-008-0312-2

Venkata Subhash G et al (2022) Challenges in microalgal biofuel production: a perspective on techno economic feasibility under biorefinery stratagem. Bioresour Technol 343:126155. https://doi.org/10.1016/j.biortech.2021.126155

Ventura J-RS, Yang B, Lee Y-W, Lee K, Jahng D (2013) Life cycle analyses of CO2, energy, and cost for four different routes of microalgal bioenergy conversion. Bioresour Technol 137:302–310. https://doi.org/10.1016/j.biortech.2013.02.104

Vo HNP et al (2019) A critical review on designs and applications of microalgae-based photobioreactors for pollutants treatment. Sci Total Environ 651:1549–1568. https://doi.org/10.1016/j.scitotenv.2018.09.282

Widowati I et al (2017) Antioxidant activity of three microalgae Dunaliella salina, Tetraselmis chuii and Isochrysis galbana clone Tahiti. IOP Conf Ser Earth Environ Sci 55:012067. https://doi.org/10.1088/1755-1315/55/1/012067

Xia A, Jacob A, Herrmann C, Tabassum MR, Murphy JD (2015) Production of hydrogen, ethanol and volatile fatty acids from the seaweed carbohydrate mannitol. Bioresour Technol 193:488–497. https://doi.org/10.1016/j.biortech.2015.06.130

Xia A, Jacob A, Tabassum MR, Herrmann C, Murphy JD (2016) Production of hydrogen, ethanol and volatile fatty acids through co-fermentation of macro- and micro-algae. Bioresour Technol 205:118–125. https://doi.org/10.1016/j.biortech.2016.01.025

Yao S, Mettu S, Law SQK, Ashokkumar M, Martin GJO (2018) The effect of high-intensity ultrasound on cell disruption and lipid extraction from high-solids viscous slurries of Nannochloropsis sp. biomass. Algal Res 35:341–348. https://doi.org/10.1016/j.algal.2018.09.004

Yellapu SK et al (2018) Recent developments of downstream processing for microbial lipids and conversion to biodiesel. Bioresour Technol 256:515–528. https://doi.org/10.1016/j.biortech.2018.01.129

Yoo G, Park W-K, Kim CW, Choi Y-E, Yang J-W (2012) Direct lipid extraction from wet Chlamydomonas reinhardtii biomass using osmotic shock. Bioresour Technol 123:717–722. https://doi.org/10.1016/j.biortech.2012.07.102

Zbinden MDA et al (2013) Pulsed electric field (PEF) as an intensification pretreatment for greener solvent lipid extraction from microalgae. Biotechnol Bioeng 110(6):1605–1615. https://doi.org/10.1002/bit.24829

Zhang X, Amendola P, Hewson JC, Sommerfeld M, Hu Q (2012) Influence of growth phase on harvesting of Chlorella zofingiensis by dissolved air flotation. Bioresour Technol 116:477–484. https://doi.org/10.1016/j.biortech.2012.04.002

Zhang Y, Ren L, Chu H, Zhou X, Yao T, Zhang Y (2019) Optimization for Scenedesmus obliquus cultivation: the effects of temperature, light intensity and pH on growth and biochemical composition. Microbiol Biotechnol Lett 47(4):614–620. https://doi.org/10.4014/mbl.1906.06005

Zhang S, Zhang L, Xu G, Li F, Li X (2022a) A review on biodiesel production from microalgae: influencing parameters and recent advanced technologies. Front Microbiol 13. https://doi.org/10.3389/fmicb.2022.970028

Zhang J, Tsai W-Y, Hsu C-H, Peng C-A (2022b) Biodiesel production from Nannochloropsis oculata cultured at stressful carbon dioxide concentration and light illumination. Biofuels 13(4):527–535. https://doi.org/10.1080/17597269.2020.1787699

Zinnai A, Sanmartin C, Taglieri I, Andrich G, Venturi F (2016) Supercritical fluid extraction from microalgae with high content of LC-PUFAs. A case of study: Sc-CO2 oil extraction from Schizochytrium sp. J Supercrit Fluids 116:126–131. https://doi.org/10.1016/j.supflu.2016.05.011

Zou X, Xu K, Wen H, Xue Y, Qu Y, Li Y (2019) Efficient microalgae harvesting using a thermal flotation method with response surface methodology. Water Sci Technol 80(3):426–436. https://doi.org/10.2166/wst.2019.287

# Active Pharmaceutical Ingredients Using Biocatalyst for Circular Bioeconomy

# 13

Shejal Ravalnath Patil, Vidya Rajendra Dighule, Kuber Kumar Bhagat, Shamlan Reshamwala, and Nitin Dnyaneshwar Arote

**Abstract**

Biocatalysis has emerged as powerful and sustainable alternative to traditional chemical synthesis techniques in pharmaceutical manufacture of Active Pharmaceutical Ingredients (APIs). In light of mounting global demand for green chemistry and circular bioeconomy, biocatalysts, especially enzymes, have been crucial in reshaping pharmaceutical production processes. They operate under mild conditions, reducing the use of hazardous chemicals, catalyst and solvents, less byproduct formation due to high regio-, chemo-, and enantio-selectivity thus increasing yields and quality with minimized environmental impact. This chapter delves into the transformative role of biocatalysis in API synthesis by presenting case studies of successful enzymatic conversions, underscoring its environmental sustainability, industrial scalability, economic and technological advantages, and contribution to the circular bioeconomy.

**Keywords**

Biocatalyst, Enzymes · Circular bioeconomy · Biosynthesis methods · Active pharmaceutical ingredients

S. R. Patil · V. R. Dighule · K. K. Bhagat · N. D. Arote (✉)
Department of Pharmaceutical Sciences and Technology, Institute of Chemical Technology (Formerly UDCT), Matunga, Mumbai, Maharashtra, India
e-mail: nd.arote@ictmumbai.edu.in

S. Reshamwala
Department of Biological Sciences & Biotechnology, Institute of Chemical Technology, Matunga, Mumbai, Maharashtra, India



V. Singh (ed.), *Sustainable Waste Management Towards Circular Bioeconomy*,
https://doi.org/10.1007/978-981-95-2227-9_13

## 13.1 Introduction

In the pharmaceutical industry, an Active Pharmaceutical Ingredient (API) refers to the primary substance in a drug that produces the intended therapeutic effect by interacting with biological targets in the body. While APIs are not consumed in their raw form, they are crucial in the formulation of medication such as capsules, injections, and topical applications (NCI 2025). APIs and drug products are closely linked, as the active ingredients in drug products are derived from APIs. While APIs provide the therapeutic effects, drug products also contain other components like excipients, binders, fillers, and coatings that help with delivery, stability, and patient adherence. Together, APIs and drug products work in harmony to treat medical conditions and improve patient outcomes. In contrast to active ingredients, excipients are inert substances added to pharmaceutical formulations to facilitate drug delivery, stability, and patient admission. While active ingredients exert the desired pharmacological effects, excipients play essential roles in ensuring proper dosage form, solubility, and bioavailability. The production of APIs involves several stages, including synthesis, purification, and formulation. APIs are typically obtained through chemical synthesis, plant extraction, or biotechnological processes. The pharmacological effectiveness of APIs is directly influenced by their strength, purity, and stability. Additionally, the efficiency of API production plays a key role in determining the cost-effectiveness and global availability of medications (Canada 2013).

The manufacturing process starts with the synthesis of the active ingredient, where chemical reactions are carefully controlled to yield the desired compound. The synthesis step involves choosing the right kind of starting material, using defined reaction conditions, and optimizing the process to high purity and yield. Following synthesis, the API is purified using techniques like chromatography, distillation, or crystallization to get rid of contaminants and produce high-quality end products (Baumann and Baxendale 2015). After purification, the API is converted into various dosage forms, such as tablets, capsules, or injections. Excipients and additives are incorporated as necessary to improve stability, solubility, and patient acceptability. The production of APIs is heavily regulated to ensure public health safety and compliance with quality standards. Regulatory organizations, such as the U.S. Food and Drug Administration (FDA) and the European Medicines Agency (EMA), set the standards and requirements for API manufacturing to ensure conformity with these guidelines (Canada 2013).

### 13.1.1 Understanding of Circular Bioeconomy

The principle of the circular economy (CE) has gained increasing global attention across academia, industry, and policy as a potentially effective way to achieve sustainable development. Nevertheless, though its popularity is great, its precise definition and usage remain very much in question. Indeed, more than a hundred definitions of the circular economy are available in the literature (Bocken et al.

2016; Geissdoerfer et al. 2017; Gladek 2017; Kirchherr et al. 2017; Korhonen et al. 2018; Emaf 2019; Schöggl et al. 2020; Alhawari et al. 2021). Generally, definitions of the circular economy are classified into two broad streams: those by the Ellen MacArthur Foundation (EMAF) and those by independent researchers.

The concept of circular bioeconomy can be grouped into two key perspectives: (1) those proposed by the EMAF and, (2) theories of the independent researchers. According to the (Emaf 2019), the circular economy is more than just a new economic model, it's a way to rethink how we produce and consume. It follows three core principles: (1) designing out waste and pollution, (2) keeping products and materials in use, and (3) regenerating natural systems. The Ellen MacArthur Foundation's strategy, in contrast to the traditional linear economy, aims to minimize overall waste while separating economic growth from the consumption of finite resources. One the other hand, theories from outside EMAF usually adopt a broader or distinct viewpoint, usually highlighting the importance of systems thinking in the manufacturing and consumption of goods. Such as, Korhonen et al. defines the circular economy as a sustainable development strategy aimed at reducing linear material and energy by including renewable and cascading energy flows. It emphasizes collaboration among producers, consumers, and societal actors in promoting high-value material cycles (Korhonen et al. 2018). Similarly, Gladek et al. describes the circular economy as a new economic model that fulfils human needs while ensuring the fair distribution of resources and maintaining ecological balance within planetary boundaries as shown in Fig. 13.1. This wide range of interpretations highlights both the richness and the complexity of the circular economy as a concept (Gladek 2017).

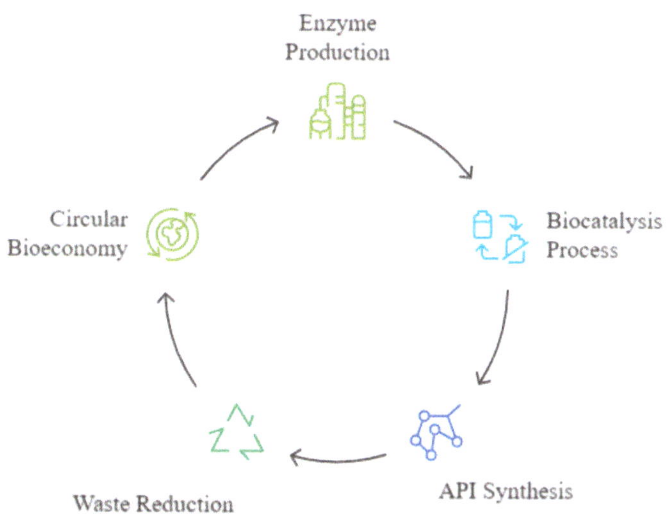


**Fig. 13.1** Schematic representation of a sustainable bio-economy; from biocatalyst-driven processes to active pharmaceutical ingredient (API) synthesis (Napkin n.d.)

## 13.1.2 Understanding Catalysts: Chemical Versus Biocatalysts

### 13.1.2.1 Catalysts

A catalyst accelerates a chemical reaction by simply reducing the activation energy, which is required to overcome the energy barrier of reaction. The catalytic reaction can also be defined as a cyclic event in which the catalyst engages in and is returned to its starting form at the end of the reaction (Chorkendorff and Niemantsverdriet 2017; Kumar et al. 2023). It accelerates the kinetics of the reaction while not altering its thermodynamics (Kumar et al. 2022). This is done by initially bonding to the reacting molecules and allowing them to react to form a product state that detaches from the catalyst and remains unchanged making it available for the next reaction (O'Connell et al. 2024).

Catalysts play a crucial role in enhancing selectivity during chemical reactions. So, reactions involving the formation of multiproduct lead to selective formation of a single product or most important one if the catalyst is involved. This is the major concern these days as the formation of by-products not only reduces the atom economy of a process but also leads to environmental contamination (Kumar et al. 2023). Thus, the use of catalysts in a chemical process or reaction is a key step toward designing sustainable technology (Kumar et al. 2022). Catalysts come in a multiple forms, ranging from atoms and molecules to large structures such as zeolites or enzymes. In addition they may be employed in various surroundings: in liquids, gases, or at the surface of solids (O'Connell et al. 2024). Catalysts can speed up chemical processes up to several orders of magnitude, typically allowing the reaction to take place at lower temperature and pressure under more favourable thermodynamic conditions. This decreases the energy requirements, increasing process efficiency and sustainability. Moreover, the design of the reactor and plant, paired with highly efficient catalysts, is a key consideration in lowering both capital and operational costs in chemical manufacturing. In industry, the overall cost-effectiveness and sustainability of chemical processes are reflected through catalysts' effects on the reaction rate and selectivity (Chorkendorff and Niemantsverdriet 2017).

Catalysts play a crucial role in the chemical industry, facilitating efficient and sustainable chemical transformations. Approximately 85–90% of chemical industry products are synthesized through catalytic processes (Topsoe 2024). Their significance extends across various domains, including:

1. *Production of Fuels*—They are essential in oil refineries (around 440 worldwide) for producing transportation fuels.
2. *Bulk and Fine Chemical Synthesis*—Used in various branches of the chemical industry to enhance reaction efficiency.
3. *Pollution Prevention*—Help minimize waste by preventing the formation of unwanted by-products.
4. *Pollution Abatement*—They aid in end-of-pipe solutions, such as automotive and industrial exhaust treatment, to reduce harmful emissions.

**Chemical Catalyst**

Chemical catalysts are the organic and inorganic materials which increase the rate of a chemical reaction by not being consumed in the reaction. Chemical catalysts are important for pharmaceutical synthesis, in the manufacture of Active Pharmaceutical Ingredients (APIs) for eventual large scale manufacturing.

There are numerous methods to categorise chemical catalysts, usually according to their composition or function. Transition-metal catalysts, which are frequently used in processes like hydrogenation, oxidation, and cross-coupling, include palladium, platinum, ruthenium, and nickel (Rahman et al. 2017). Strong acid catalysts include hydrochloric acid and sulphuric acid, while strong base catalysts include sodium hydroxide and other bases (Terry 1928). Because they promote proton transfer and hydrolysis processes in the synthesis of active pharmaceutical ingredients (APIs), acid and base catalysts are essential to drug manufacture (Yamamoto 2013). Metal-carbon bonds are present in organometallic catalysts, which include complexes like Grubbs' and Wilkinson's catalysts. These include zeolites, alumina, and silica-based catalysts, which provide surface reactions without being dissolved in the reaction medium. In pharmaceutical synthesis, organometallic catalysts are commonly employed for asymmetric hydrogenation and metathesis (Busca 2019; Zhang et al. 2022; Mahmoud et al. 2023).

**Biocatalyst**

Enzymes are natural catalysts made up of large protein structures with highly specific active sites that bind to reactant molecules (substrates) in an optimal configuration for catalysis. This precise fit helps enzymes to exhibit high specificity and catalytic efficiency (Chorkendorff and Niemantsverdriet 2017). These protein molecules help increase rates of reaction and selectivity by employing a system of active site amino acid residues that maintain the substrate precisely positioned in the binding pocket of an enzyme. The substrate gets bind to these active sites by many weak interactions such as hydrogen bonding, electrostatic interaction, dipole–dipole interaction, and van der Waals forces. These forces conspire to couple reactants, keep the substrate in the ideal conformation, and stabilize the transition state to lower its energy and accelerate the reaction (Mahmoud et al. 2023). One of the significant advantage of enzymes is their ability to maintain high levels of chemo-, regio-, and stereoselectivity in reactions. Chemeoselectivity refers to the catalyst's preference for reacting with one functional group over others, while regioselectivity determines the preferred direction of bond formation or cleavage in a molecule (Muller 1994). Stereoselectivity is the formation of one stereoisomer over the other, and also subdivided into enantioselectivity (preference for one enantiomer) and diastereoselectivity (preference for one diastereomer). These selectivities are quantitatively expressed as enantiomeric excess and diastereomeric excess, respectively (Moss 1996).

The recent developments in molecular biology, bioinformatics, enzyme engineering, and high-throughput screening have profoundly enhanced our knowledge of enzymes, which has resulted in their common use for the synthesis of active pharmaceutical ingredients (APIs) and numerous other chemicals. Biocatalysis covers a wide scope of biological systems ranging from purified enzymes to crude enzyme mixtures, whole-cell biocatalysts with particular enzymes, and hybrid systems that integrate these methods. These innovations have led to more sustainable and efficient chemical production processes (O'Connell et al. 2024).

### 13.1.3 Challenges of Chemical Catalyst in API Synthesis

Biocatalysis has become an essential tool for the synthesis of APIs as it offers numerous advantages. First, optimized enzymes used as biocatalysts offer superior regio-, stereo-, and enantioselectivity, ensuring high precision in the chemical transformations. The specificity of enzymes will eliminate the need for further purification and separation steps for by-products, which makes for a more efficient process. Biocatalysis can take out a number of steps in multistep syntheses, allowing reaction and reaction sequences that cannot be done with traditional chemical paths (Kinner et al. 2022). Moreover, biocatalysis improves atom utilization, reducing waste and increasing overall efficiency in API synthesis (Kinner et al. 2022). Another key advantage is that many enzymes function well in water-based systems, eliminating the need for organic solvents. This characteristic makes enzymatic reactions eco-friendly by reducing solvent and reagent usage, aligning with green chemistry principles. Biocatalysis also usually takes place under mild reaction conditions and does not include toxic reagents and catalysts that need extreme temperatures or pressures. This means less problematic waste and fewer side products (Domínguez De María and Hollmann 2015).

The expansion and use of non-biodegradable solvents (e.g., toluene, dichloromethane, benzene), which become both industrial waste and toxic emissions, are an additional factor that pose a serious risk of harm to human health and the environment (Poliakoff et al. 2002). The carbon footprint of chemical process—driven most specifically by energy needs and reliance on fossil fuels—is another major concern, as carbon attribution indirectly leads to emissions of greenhouse gases that directly point to the accelerated pace of global warming (Trost 1995). Given the increased environmental awareness and social responsibility, API manufacturers are encouraged to utilize greener methodologies to minimize waste output (i.e., chemical wastes and solvents). Companies continue to seek optimal ways to make production more efficient (faster and cheaper), but the only way to effectively reduce waste is to minimize the number of reaction steps in the API synthesis process (Kumar et al. 2022). To address these challenges, a more green chemistry approach suggests the substitution of harmful catalysts chemical with more environmentally friendly alternatives, such as biocatalysts or highly recyclable heterogeneous catalysts, and/or solvent-free or aqueous-phase reactions (Sheldon and Woodley 2018). For example, biocatalysis (e.g., enzyme catalysis) can operate under mild conditions and offers great selectivity, biocompatible, and biodegradable transformations, and drastically reduce reaction steps, energy, and hazardous waste generation and can also occur in solvent free or aqueous environment (Kinner et al. 2022). They hold a great potential for addressing several drawbacks of conventional chemical catalysts. All these characteristics contribute to designing more sustainable and less polluting processes in API synthesis, in line with the green chemistry principles and circular bioeconomy strategy (Pollard and Woodley 2007; Sheldon and Woodley 2018).

### 13.1.4 Advantages of Biocatalysis in API Synthesis

Biocatalysis has become an essential tool for the synthesis of APIs as it offers numerous advantages. First, optimized enzymes used as biocatalysts offer superior regio-, stereo-, and enantioselectivity, ensuring high precision in the chemical transformations. The inherent specificity of enzymes minimizes the need for additional purification and separation of by-products, leading to more efficient processes thus, biocatalysis can significantly shorten multi-step synthesis routes by enabling reactions and reaction sequences that cannot be achieved through traditional chemical methods (Kinner et al. 2022). Moreover, biocatalysis improves atom utilization, reducing waste and increasing overall efficiency in API synthesis (Kinner et al. 2022). Another key advantage is that many enzymes function well in water-based systems, eliminating the need for organic solvents. This characteristic makes enzymatic reactions eco-friendly by reducing solvent and reagent usage, aligning with green chemistry principles. Additionally, biocatalysis typically operates under mild reaction conditions, avoiding the use of toxic reagents and catalyst that require extreme temperatures or pressures. This results in reduced generation of problematic waste and fewer side products (Domínguez De María and Hollmann 2015).

Furthermore, advances in protein engineering have revolutionized biocatalysis, allowing for the conceptual design of desired chemical processes and the adaptation of biocatalysts to the required reaction conditions. Along with this the advancement of technologies such as enzyme and whole-cell immobilization enhances enzyme reusability, reducing resource consumption and waste generation, thereby aligning with the principles of the circular bioeconomy by promoting sustainable and efficient bioprocesses (O'Connell et al. 2024). Enzyme-catalyzed reactions offer many appealing characteristics for the synthesis of chiral intermediates. These reactions can be conducted at room temperature and at atmospheric pressure, which helps to limit or avoid isomerization, racemization, epimerization, and rearrangement (Rossino et al. 2022). The pharmaceutical industry has been an early adopter of biocatalysis for the synthesis of active pharmaceutical ingredients (APIs) with multiple biotransformation processes already implemented at an industrial scale. These collective benefits make biocatalysis an attractive and sustainable approach in pharmaceutical synthesis (Kinner et al. 2022).

## 13.2 General Classes of Enzymes in API Synthesis

The circular bioeconomy concept emphasizes the utilization of renewable biological materials such as agricultural residues, food waste, and other biowastes for the production of high-value products in a sustainable way with low environmental impact. Biocatalytic API synthesis revolutionizes under this regime by replacing traditional chemical processes with biocatalysts that promote reactions at mild, environmentally benign conditions, minimizing energy demands, toxic wastes, and by-product formation. Details are that these enzymes can be made from renewable biological substrates such as microorganisms (bacteria, fungi, or yeast) and plants (Intasian et al. 2021). Microbial enzymes typically are derived from microbial populations arising from biowaste—i.e., compost, fermented biomass, or residues from agro-industry—through isolation, screening, and cultivation. Illustratively, low-cost fermentation mediums such as waste cooking oil or molasses could be utilized in cultivating fungi species such as *Aspergillus niger* to produce industrially useful enzymes such as lipases (Amenaghawon et al. 2022). Similarly, bacterial strains might be enriched within organic waste systems and optimized in order to release specific enzymes of interest through the process of fermentation (Ubando et al. 2020). Plant enzymes, on the other hand, can be obtained from plant waste materials like fruit peels, seeds, or utilized pulp. Papaya peels, for example, can provide papain, which can be purified and extracted using water- or buffer-based extraction methods followed by precipitation and filtration. Pineapple stems can yield bromelain through similar processes (Muruvanda et al. 2023). By harnessing such renewable resources for enzyme production, the whole biocatalytic process is part of a closed-loop system in which waste materials are converted into useful biological tools for green synthesis, thus reflecting the fundamental principles of a circular bioeconomy (Sharma et al. 2022; Ubando et al. 2020).

### 13.2.1 Lipases

Lipases (EC 3.1.1.3), which are also referred to as triacylglycerol acylhydrolases, hydrolyze carboxylic ester bonds and are enzymes that belong to the hydrolase family. The primary role of lipases is hydrolysis of triglycerides to diglycerides, monoglycerides, fatty acids, and glycerol, and they typically are in the class of serine hydrolases. Importantly, lipases do not require cofactors, and belong to the class of serine hydrolases (Kapale et al. 2024). Beyond hydrolysis, lipases also conduct interesterification, esterification, aminolysis, and alcoholysis, so they are applied by several industries (Scheme 13.1) (Houde et al. 2004; Chandra et al. 2020).

Microbial lipases are one of the most important groups of biotechnological enzymes due to the flexibility, applied properties, and practicality of being produced on the large scale for commercial use. There is a wide variety of commercial production and application of lipases. Lipases are interesting to research to efficiently carry out biochemical transformations, especially when they occur in both

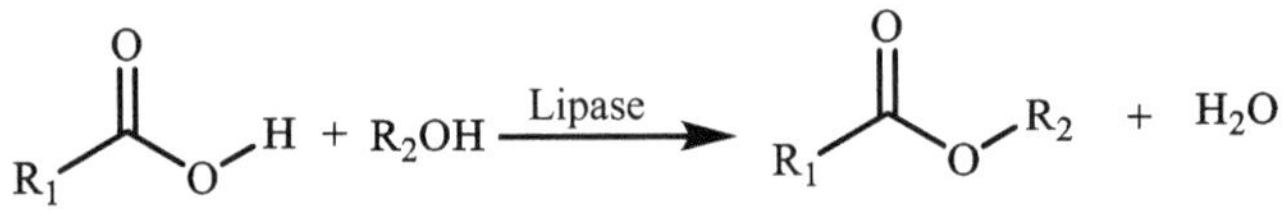


**Scheme 13.1** General mechanism of the lipase

non-aqueous and aqueous phases. The application of lipases has garnered attention in the world for a number of diverse reasons, such as ease of use, commercial availability, general safety, solvent compatibility, and substrate diversity. There are already a number of reactions that can be performed with this enzyme, such as dihydropyridines and dihydropyrimidinone pyrazole synthesis (Kapale et al. 2024).

Bacterial strains like *Pseudomonas alcaligenes*, *P. aeruginosa*, *P. fragi*, *P. fluorescens*, *Bacillus subtilis*, and *B. nealsonii* have been identified as prolific lipase producers (Ben Bacha et al. 2022). Fungal species such as *Penicillium expansum*, *Trichoderma,* and *Penicillium chrysogenum* also make important contributions to lipase production. Lipase B from *Candida antarctica* (CalB) is employed widely in personal care products, pharmaceuticals, and food ingredients since it has good regio-, chemo-, and enantioselectivity; it has broad substrate specificity, high tolerance of organic solvents, and significant temperature tolerance for many industrial processes, and in most cases is used in its immobilized form. Merck employed CalB (immobilized) in 2011 for the manufacture of odanacatib, a potent cathepsin K inhibitor developed to treat postmenopausal osteoporosis, but research and development was terminated for safety reasons (Chandra et al. 2020).

### 13.2.2 Ketoreductase

Ketoreductases (KREDs) (EC 1.1.1.) are enzymes that reduce ketones to alcohols, an important reaction in numerous biochemical and industrial processes. As members of the Aldo-Keto Reductase (AKR) superfamily, KREDs usually employ cofactors such as NADH or NADPH to donate electrons and protons to the carbonyl carbon of a ketone, converting it into a secondary alcohol. The process is extensively enantioselective and stereoselective, so KREDs find applications in synthesizing chiral alcohols from prochiral ketones (Barski et al. 2008; Bernhardt and O'Connor 2009). Among the synthesis processes for Active Pharmaceutical Ingredient (API), KREDs are quite central because, being capable of selectively reducing ketones to form chiral alcohols; it is indispensable in the process of manufacturing several drug compounds (Qiao et al. 2023). The general mechanism of action of the ketoreductases is shown in Scheme 13.2.

KREDs have different advantages in the drug manufacture process, including excellent chemoselectivity, regioselectivity, and enantioselectivity (Hu et al. 2020). For example, in montelukast synthesis, which is a drug against asthma, KREDs are used to reduce a ketone to a chiral alcohol in place of traditional chemical reducing agents with a lesser toxic and more cost-effective strategy (Sunjic and Parnham 2011). Furthermore, KREDs are steroselective in the reduction of sterically

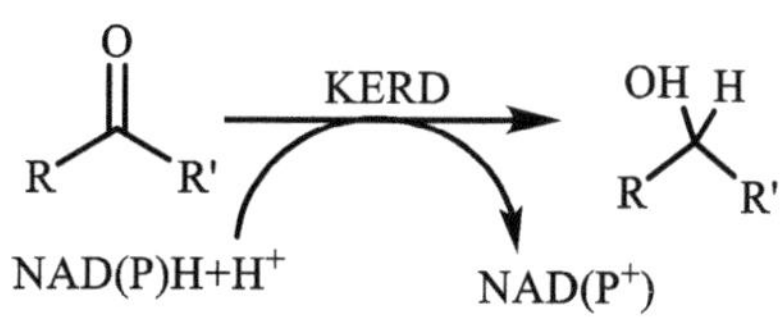


**Scheme 13.2** General mechanism of the ketoreductases enzyme

constrained ketones, a common issue in the synthesis of steroids and statins, where their targeted action ensures the formation of the right stereoisomer of the product (Hoyos et al. 2019). Another key advantage of using KREDs in API synthesis is that they can function under mild reaction conditions. These enzymes tend to work in a water environment at room temperature, meaning that harmful organic solvents and high-temperature or high-pressure reactions are not used (France et al. 2023). This is in accordance with green chemistry principles, hence biocatalysis using KREDs a greener and more environmentally friendly process than the conventional chemical means. Moreover, KREDs can reduce multi-step synthetic pathways by carrying out highly selective reductions in one step, minimizing multiple chemical conversions and enhancing efficiency in drug production. Their ability to selectively eliminate ketones in complex molecules, such as steroids and statins, also captures how critical they are in modern drug synthesis, where precise stereochemistry is a requirement (Hoyos et al. 2019). Briefly, KREDs are efficient catalysts in pharma production due to their specificity, stereoselectivity, and potency in the reduction of ketones to chiral alcohols. Application of these enzymes has enabled the preparation of key intermediates and APIs economically, in a sustainable process, and in an eco-friendly process, reducing the environmental profiles of chemical operations. Their application in the preparation of many chiral drugs like montelukast and statins underscores their significance in the pharma industry.

### 13.2.3 Alcohol Dehydrogenase

Alcohol dehydrogenases (ADHs) or ketoreductases (KREDs)/carbonyl reductases (CRs/CREDs), i.e. ADHs, are highly selective and effective biocatalysts able to carry out the stereoselective reduction of prochiral carbonyl compounds to chiral or non-chiral alcohols. Redox reactions, encompassing oxidations and reductions, are important in organic synthesis, with alcohol oxidation to aldehydes or ketones, and the reduction of carbonyl compounds to alcohols being redox processes that have a wide range of industrial applications. ADHs possess an inherent versatility, being able to catalyze both the oxidation of alcohols and the reduction of carbonyl compounds, as they are bona fide natural redox catalysts. The ability to conduct both oxidations and reductions distinguishes them as incredibly powerful, and versatile redox conversion catalysts (De Miranda et al. 2022; Abuzenah et al. 2025). Alcohol dehydrogenases bring about reversible oxidation of alcohols to their carbonyl compounds (as they are a covalent oxidoreductase). As it's a redox process, it must have a cosubstrate, at stoichiometric amounts, that balances the oxidation or reduction. Alcohol dehydrogenases require two different nicotinamide cofactors, ribose-phosphorylated [NAD(P)] and ribose-non-phosphorylated (NAD), although both forms have been proven to have the same inherent redox potentials, alcohol dehydrogenases are highly selective toward one class or the other depending upon their physiological function: alcohol dehydrogenases involved in anabolic pathways tend to use NADP and metabolically important alcohol dehydrogenases use NAD (Barski et al. 2008). Alcohol dehydrogenases employ a similar catalytic process that consists of complexation of the substrate (either alcohol or carbonyl compound) with the nicotinamide cofactor at the protein enzyme active site and hydride transfer from the conafactor to the substrate. Certain alcohol dehydrogenases also utilize metal ions (e.g., Fe) to coordinate the substrate and they are not involved in the redox reaction (Scheme 13.3) (De Miranda et al. 2022).

Apart from selectivity, an additional significant aspect of ADHs in biocatalysis is resistance toward harsh reaction conditions. Organic solvent and thermal stability are especially desirable for their efficacious application toward organic synthesis. The most important aspect of ADH-catalysis is probably reversibility, as ADHs can catalyze the NAD(P)H-directed reduction of carbonyl substrates to alcohols, and oxidize alcohols to carbonyl compounds in an NAD($P^+$)-catalyzed reaction; therefore they are far too versatile to not be utilized for various industrial and medicinal purposes. (S)-Ethyl 4-chloro-3-hydroxybutyrate ((S)-CHBE) is an important chiral intermediate in cholesterol-controlling drugs, antidepressants, and cancer chemotherapy. Conventional chemical synthesis through a ruthenium catalyst is expensive and produces only 92% optical purity. As an alternative that is greener and more selective, biocatalysis with alcohol dehydrogenase (ADH) presents a high-purity production route for (S)-CHBE under mild conditions, which is more industrially applicable (Zhu et al. 2019).

## 13.2.4 Imine Reductase

IREDs, or imine reductases, a sub family of NADPH-dependent oxidoreductases, which catalyze the asymmetrical imine-to-amine reaction require synthetic substrates, and are very useful in the pharmaceutical arena to prepare a range of chiral amines (Scheme 13.4). The IREDs were discovered as a class around 2011, and a recent surge of research interest has seen a more general advancement in employment of IREDs in pharmaceutical applications, helped also by improvements in bioinformatics and analysis of sequencing technology, meaning we can further interrogate and take advantage of the availability of many more new IREDs with varying synthetic ability and effectiveness (Mangas-Sanchez et al. 2017).

That endeavour includes the publicly available annotated sequence database over thousand putative IREDs from the fourteen superfamilies at (Bornscheuer et al.

**Scheme 13.3** General mechanism of action the ADHs

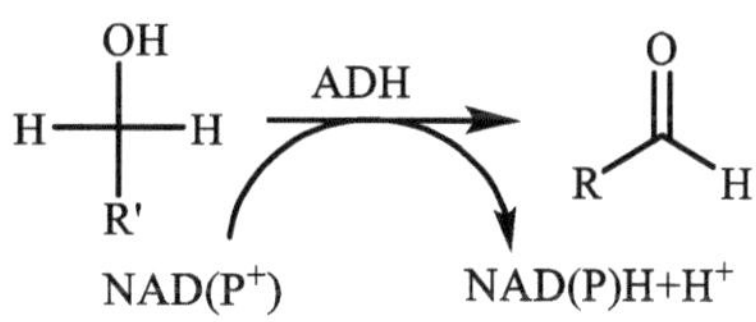


**Scheme 13.4** General mechanism of the imine reductase

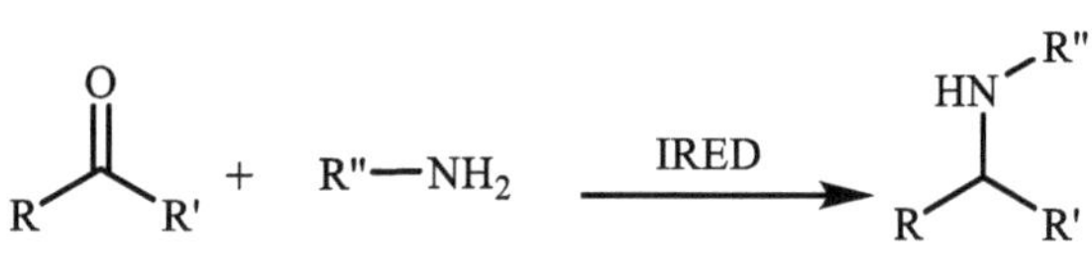

2012) a new subclass of IREDs called reductive aminases (RedAms) was established as catalysts to reductively aminate ketones to chiral amines directly in 2017 making synthesis easier by circumventing the employment of preformed imines (Farrer and Griffith 2020). One of the significant industrial applications is Pfizer's utilization of a RedAm (SpRedAm from Streptomyces purpureus) in the preparation of one of the critical intermediates for abrocitinib, a JAK inhibitor for atopic dermatitis. Although the enzyme had initial substrate tolerance limitations, it was optimized within only three rounds of protein engineering via homology modeling to fulfill the industrial need for a product titer of 100 g/L, illustrating the high-speed potential of IRED/RedAm-based biocatalysis in green and scalable pharmaceutical production (Kumar et al. 2021).

### 13.2.5 Transferase

Transferases represent a large family of enzymes (EC class 2) which catalyze the transfer of functional groups, e.g., methyl, glycosyl, acyl, phosphate, or amino groups—between a donor molecule and an acceptor. Transferases are essential in a wide variety of metabolic pathways such as amino acid biosynthesis, nucleotide metabolism, and detoxification. Among all these subtypes, aminotransferases (transaminases) play a crucial role in biocatalysis since they catalyze amino group transfers and are more commonly used in chiral amine and drug intermediate synthesis (Berg et al. 2002) (Scheme 13.5).

Transferases are considerably coveted in green chemistry and biotechnological applications because they are regioselective, stereoselective, and perform under mild reaction conditions. Their use is growing in the fine chemical and pharmaceutical industries, particularly with the development of protein engineering and enzyme evolution strategies, which provide the capability of the tailoring of these enzymes towards the needs of the specific industrial processes (Nelson et al. 2008; Bornscheuer et al. 2012).

### 13.2.6 Transaminase

Transaminase enzymes (TAms) (EC 2.6.1.) are pyridoxal phosphate-dependent enzymes that can transfer an amine from a donor to a prochiral ketone forming a chiral amine (Scheme 13.6). Transaminases have gained considerable interest due to their potential as biocatalysts for the synthesis of compounds containing chiral amine units. Chiral amines are favorable building blocks in pharmaceutical,

$$A+X \longrightarrow B+X$$

X=Functional Group

A= Doner, B= Accepter

**Scheme 13.5** General mechanism of the transferase

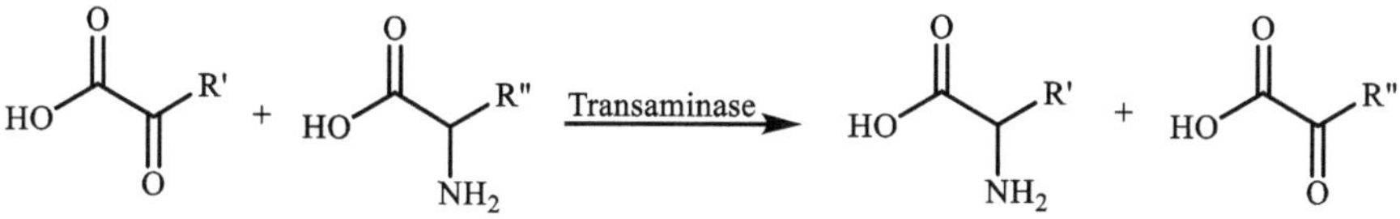


**Scheme 13.6** General mechanism of the transaminase

agrochemical, and fine chemical industries (Slabu 2017). Transaminases follow a ping-pong bi-bi mechanism, with the enzyme returning to its default state from its intermediate state and restoring the PLP cofactor to its normal state following the amine transfer mechanism (Slabu et al. 2017).

Sitagliptin is the most notable example of TAm application in the life sciences and medical products sector involving the biocatalysis of a drug substance. By employing the enzyme's large binding pocket, the two approaches of directed evolution were able to engineer the enzyme to more efficiently act upon the ketone substrate, prositagliptin, and provide better stability to the conditions necessary for the solubilization of the substrate. The third mutant resulted in a total improvement of 10–13% potency, overall productivity improvement of 53%, and waste reduction of 19% compared to the chemical routes used. The TAm reaction instead of the standard methods eliminated the requisite transition metal catalyst and expensive high-pressure hydrogenation equipment which are also positives over the standard procedures (Kelly et al. 2020).

## 13.3 Approaches for Biocatalysis

### 13.3.1 Whole Cell Biocatalyst

Whole-cell biocatalysts provide a potential substitute to purified, crude, or immobilized enzyme preparations for cofactor-dependent reactions. They naturally provide and regenerate the needed cofactors, which obviate the cost of external addition. Residual cell wall fragments also offer protection, enabling enzyme applications to remain effective under hostile conditions or non-aqueous solvents (Lin and Tao 2017). In spite of their inherent stability, whole cells may be immobilized to recover and reuse them in order to cut down operating expenses. In addition, they make it possible to conduct multi-enzyme cascade reactions with the conversion of low-cost substrates to value-added, complex products. Multi-enzyme co-expression in one cell saves reaction energy by keeping biocatalysts in close physical proximity to one another, allowing for consecutive effortless steps in the biotransformation process. In addition, such responses become economically viable owing to increased product titters (Srinivasamurthy et al. 2019; Tan et al. 2023). As a result, the combination of whole cell catalysts and non-traditional solvents is remarkably effective for economically and ecologically beneficial process designs (Pollard and Woodley 2007). The use of whole cells bypasses the need for cell lysis and enzyme purification which is associated with biocatalysis with isolated enzymes and minimises the catalyst cost (Wachtmeister and Rother 2016). Whole-cell biocatalysts do not require any additional cofactors, as none, and because zero cofactors can be supplied and regenerated by the cell, there is an added cost reduction as well. Whole-cell biocatalysts are often easier and cheaper to prepare; fermentation processes are unlikely to be excessively expensive, and the cells can usually be used more than once. Thus, whole-cell biocatalysts clearly have very good cost advantages. Biotransformation processes usually consist of two steps: cultivation of the living "catalyst" and transformation of the substrate(s). When culturing the cells, they are harvested and washed with water, or a buffer, and resuspended in a chosen buffer, for biocatalysis. The wash removes unused growth nutrients and substrates; as well as undesired metabolites generated during growth, and leads to much greater product recovery and simpler downstream processing (Lin and Tao 2017). Also, by removing growth nutrients, the cells' growth stops, and resting cells can convert more carbon feedstock into higher yields of product because almost all carbon and energy are used in product formation rather than biomass (Mu et al. 2020).

### 13.3.2 Immobilized Enzymes

Enzyme immobilization is a novel technology to improve enzymes' catalytic efficiency, stability, and reusability for pharmaceutical and industrial purposes. An immobilized enzyme, or the term immobilized enzyme, refers to an enzyme that is confined, or localized, in a specified space using physical or chemical means so that it sustains its full activity (Abedi et al. 2019). This technique mimics the natural cellular environment by attaching enzymes to solid supports or cross-linking them into stable aggregates, thereby improving operational stability and efficiency. Immobilization strategies are broadly classified into physical and chemical methods, each employing distinct preparation techniques with specific advantages.

Physical immobilization relies on non-covalent interactions such as hydrophobic, ionic, or affinity binding, preserving the enzyme's structure and catalytic properties. The commonly used physical immobilization methods include adsorption, entrapment, and encapsulation. Adsorption is one of the simplest techniques, where enzymes are weakly attached to solid carriers such as silica, activated carbon, or ion-exchange resins through Van der Waals forces, hydrogen bonding, or electrostatic interactions, providing easy enzyme recovery but lower stability due to potential desorption (Homaei et al. 2013). Entrapment involves the confinement of enzymes within porous polymeric networks, such as calcium alginate beads, polyacrylamide gels, or silica sol–gel matrices, allowing substrate diffusion while preventing enzyme loss (David et al. 2011). Encapsulation, on the other hand, encloses enzymes within a semi-permeable membrane, such as liposomes, polymersomes, or nanofibers, effectively isolating the enzyme from the external environment while enabling selective diffusion of substrates and products, thereby enhancing enzyme stability (Mohamad et al. 2015).

Chemical immobilization, in contrast, involves covalent bonding between the enzyme and a functionalized support, ensuring stronger attachment and improved operational stability. Covalent binding is widely used for industrial applications, utilizing functional groups like amines, carboxyls, and thiols to form stable enzyme-support linkages with activated carriers such as epoxy resins, functionalized silica, and chitosan (Mateo et al. 2020). Another effective chemical immobilization technique is cross-linking, where enzyme molecules are directly interconnected via bifunctional reagents such as glutaraldehyde, forming insoluble enzyme aggregates or cross-linked enzyme aggregates (CLEAs), which exhibit high stability and recyclability (Sheldon and Woodley 2018). These methods enhance enzyme rigidity and resistance to denaturation while providing operational advantages, though excessive cross-linking may lead to reduced enzyme flexibility and activity. From an environmental perspective, enzyme immobilization contributes significantly to sustainability by enabling enzyme reuse, reducing industrial waste, and eliminating the need for harsh chemical catalysts. The use of immobilized enzymes in biocatalysis promotes green chemistry by minimizing hazardous solvent use, lowering energy consumption, and decreasing toxic byproduct formation (Sheldon and Woodley 2018). However, concerns arise regarding the environmental impact of synthetic supports and immobilization reagents, which may pose disposal challenges. Research is now focused on developing eco-friendly immobilization materials, such as bio-based supports derived from chitosan, cellulose, and other biodegradable polymers, which enhance enzyme stability while reducing ecological footprints et al., (Bezerra et al. 2015). Advances in nanotechnology and hybrid organic-inorganic materials are further improving immobilization efficiency and environmental compatibility, making enzyme immobilization a key strategy for sustainable industrial and pharmaceutical applications (Zdarta et al. 2018).

### 13.3.3 Enzyme Engineering

For industrial biocatalysis needs, wild-type (WT) enzymes are usually engineered to work under unnatural reaction conditions, including pH, temperature, and cosolvent systems, or with artificial substrates. Enzymes evolved in nature are restricted from such usage, thus leading to the evolution of protein engineering methods to widen their operational range, stability, and substrate specificity. The basis of contemporary enzyme engineering is that the function of an enzyme is specified in its amino acid sequence, which is determined by its genetic code. Progress in recombinant DNA technologies like gene cloning, overexpression of the cloning in host systems, and DNA sequencing has immensely sped up the generation and inspection of enzyme variants. Initial methods included random mutagenesis using radiation, but these were not effective and lacked precision (Jackson et al. 1972).

Rational design, originally demonstrated by Smith et al. in 1978, utilizes site-directed mutagenesis to make targeted, hypothesis-based mutations. While very effective, it demands detailed information about enzyme function and structure and tends to search only a small area of the sequence (Winter et al. 1982). Directed evolution, made fashionable in the 1990s and led by Frances Arnold (Nobel Prize, 2018), replicates natural selection in the laboratory. It consists of repeated cycles of random mutagenesis (e.g., error-prone PCR), high-throughput screening, and selection of better variants. It does not depend on structural knowledge in advance, which allows us to recognize mutations in unexpected places that impact enzyme activity. Semi-rational design couples both approaches to create "smart libraries" for functionally significant residues, based on bioinformatics and structural information. Tools such as CAST (Combinatorial Active Site Saturation Test) and site-saturation mutagenesis efficiently screen for useful mutations, iterative saturation mutagenesis (ISM) going on to enhance variant performance (Mills et al. 1967). Combination of computational resources and experimental approaches has resulted in hybrid methods such as ProSAR (Protein Sequence Activity Relationship), which uses statistical models to inform mutagenesis. This method was employed by Codexis to improve halohydrin dehalogenase activity more than 4000-fold for the production of a precursor to atorvastatin (Fox et al. 2007). These engineering approaches collectively allow the design of enzymes for difficult synthetic applications, thus broadening the applicability of biocatalysis to sustainable chemical production.

### 13.3.4 Cascade Biocatalysis

An enzyme cascade refers specifically to any multi-step biochemical reaction pathway where the end product of an enzymatic reaction becomes the substrate of the subsequent enzymatic reaction, all occurring under the same reaction medium (commonly called "one-pot synthesis"). Two or more enzymes sequentially or concurrently act, frequently under the same mild conditions, to convert simple starting materials into complex target molecules efficiently (France et al. 2017).

Enzyme cascades are very beneficial in drug synthesis because of the intrinsic nature of enzymes, including high specificity, selectivity, mild aqueous condition efficiency (physiological pH and temperature), and sustainable catalysis. These cascades, which include several enzymes in a one-pot system, reduce the necessity for intermediate isolation and solvent consumption, drive equilibrium toward product formation, and enable efficient synthesis of complex molecules from simple, inexpensive starting materials. Reaction design can be tailored by enzyme engineering to enhance substrate scope and specificity. Methods such as bioretrosynthetic and chemoenzymatic strategies have been successfully utilized (France et al. 2017; Siedentop et al. 2021). In 2021, Merck & Co. created a more efficient and quicker method to produce molnupiravir with the use of enzyme cascade. They began with cytidine and utilized a designed enzyme (cytidine deaminase) to convert it into N-hydroxy-cytidine, followed by acylation with the use of Novozym® 435. The process began from simple building blocks like ribose and uracil, and included steps like acylation (Novozym® 435), phosphorylation (MTR kinase), and coupling (uridine phosphorylase) where last two enzymes were specially engineered to work much better. A smart cofactor regeneration system using three enzymes (pyruvate oxidase, acetate kinase, and catalase) made the whole process more efficient. The final chemical step formed an oxime group, resulting in molnupiravir with a 69%

overall yield—much higher and faster than the traditional 10-step method, which gave less than 10% yield (Bornscheuer et al. 2012).

### 13.3.5 Nanobiocatalyst

Nanobiocatalysts, which integrate nanotechnology with enzymatic catalysis, are gaining significant attention in the pharmaceutical industry for their role in the efficient and sustainable production of Active Pharmaceutical Ingredients (APIs). These catalysts enhance enzymatic activity by providing a high surface area for immobilization, improving reaction rates and overall process efficiency. The use of nano-biocatalysts in API production has demonstrated benefits such as improved enzyme stability, higher selectivity, and the ability to perform biotransformations under mild conditions, reducing the dependency on hazardous chemical reagent. One of the most important benefits of nano-biocatalysts is that they enable higher catalytic activity. They have a high surface area to volume, and therefore they have more enzyme immobilization site availability, leading to increased rates of reaction.

Different biomaterials are employed in the synthesis of nanobiocatalysts (NBCs), which have been reported to be of high potential in biological, medical, and pharmaceutical applications (Bilal et al. 2021; Gkantzou et al. 2021). NBCs are prepared by immobilizing enzymes on functionalized nanosupports such as nanotubes, nanocages, and nanocomposites, which are often functionalized with noble metals, metal oxides, or metal–organic frameworks for enhanced activity and stability (Gkantzou et al. 2021). Both physical and chemical immobilization techniques for enzymes are employed to maintain enzymatic activity; physically adsorbing is often used because it forms stabilizing hydrophobic or hydrophilic interactions with charge surface groups. Nanomaterial supports are also generally inorganic or organic types; inorganics provide strength and heat tolerance, while the organic compounds fullerenes and carbon nanotubes can be functionalized with improved solubility and biocompatibility (Gkantzou et al. 2021; Reshmy et al. 2021). Green nanomaterials, especially biopolymers like cellulose and chitosan, have become environmentally friendly options that can be used to increase catalytic efficiency (Bansal et al. 2018).

In addition, organic–inorganic hybrid nanoflowers (NFs) with novel flower-like morphology have shown enhanced catalytic activities compared to free enzymes (Jiang et al. 2019). Case studies have demonstrated effective uses of these materials, including the immobilization of *Candida antarctica* lipase B (CaL-B) on organic frameworks, which resulted in enhanced enzymatic activity, thermal stability, and reusability, and the immobilization of *Pseudomonas fluorescens* lipase on carbon nanofibers, which greatly improved the efficiency and stability of the biocatalyst for the synthesis of the anesthetic analog (R)-carboetomidate. These results emphasize the increasing significance of NBCs to design effective, recyclable, and sustainable biocatalytic systems (Oliveira et al. 2019).

## 13.4 Biocatalytic Synthesis of API and API Intermediates: Case Studies

### 13.4.1 Losampimod

Losmapimod is also called as GW856553. It acts as a selective p38α/β mitogen-activated protein kinase (p38 MAPK) inhibitor that helps to reduce the DUX4 expression, a key factor to facioscapulohumeral muscular dystrophy (FSHD). Fulcrum in-licensed losmapimod exclusively from GlaxoSmithKline (GSK) following the identification of its inhibitory activity against p38α/β (Willette et al. 2009; Therapeutics 2024). On January 27, 2020, the US FDA designated losmapimod as an orphan medication for the treatment of facioscapulohumeral muscular dystrophy (Fda 2022).

#### 13.4.1.1 Chemical Synthesis

Ying S. et al. have reported the losampimod through a series of subsiquential reactions. Initially, 6-chloronicotinic acid have been reacted with thionyl chloride and refluxed it for 6 h. The excess of thionyl chloride was evaporated under reduced pressure to obtain 6-chloronicotinoyl chloride. 2,2-Dimethylpropan-1-amine was dissolved in dichloromethane, also the 6-chloronicotinoyl chloride was added dropwise in dichloromethane at 0–5 °C. The reaction was carried out at room temperature for 1 h. After reaction completion, 5% hydrochloric acid and 5% sodium hydroxide solution was used to wash organic phase. Then it was the dried and then concentrated to obtain 6-chloro-N-neopentylnicotinamide, the yield was >90% (Ying Shi et al. 2020) (Scheme 13.7).

#### 13.4.1.2 Biocatalytic Synthesis

A crucial losmapimod intermediate is produced by the enzymatic coupling of 6-chloronicotinic acid with neopentylamine utilizing entire *E. coli* cells that express serotonin hydroxycinnamoyl transferase from Capsicum annuum (66CaAT) and 4-chlorobenzoate CL from Alcaligenes sp. (CBL). With a >99% conversion rate and a 10 mM substrate concentration, the reaction was conducted in LB medium,

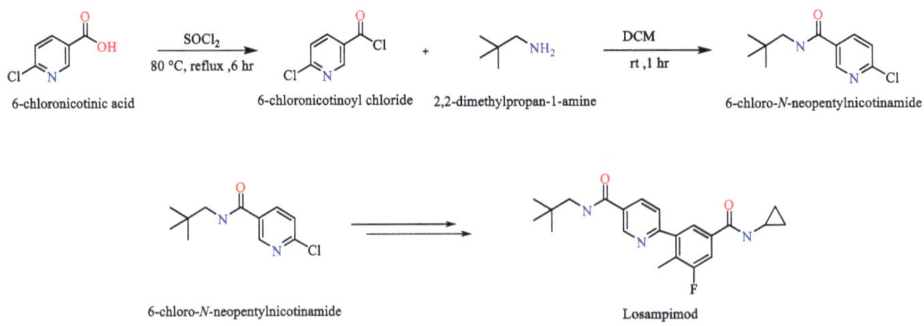


**Scheme 13.7** Chemical synthesis of the losampimod

demonstrating the effectiveness of the biocatalytic system. The conversion, however, fell to 71% when carried out in Tris buffer (pH 8.0) and to 28% when glycerol was not present. This suggests that *E. coli* is essential for the cofactor regeneration required for high reaction efficiency. The reaction was carried out at a concentration of 10 mM with a lower catalyst loading for preparative-scale synthesis. An 83% conversion rate was achieved during the transformation, and a 74% yield of the product was isolated. Chromatography was not required because the purification process only required a basic acid-base extraction and the evaporation of any remaining volatile neopentylamine (b.p. 83 °C) (Lubberink et al. 2020) (Scheme 13.8).

**Advantages**

In chemical synthesis, toxic waste and necessitates energy-intensive conditions like reflux and cooling, uses hazardous compounds like thionyl chloride and organic solvents are involved. On the other hand, biosynthesis significantly reduces energy consumption and environmental impact by using enzymes or microbes to operate under mild, ecologically friendly conditions. It also improves purity and selectivity while decreasing by-products. The successful purification procedure further demonstrates the viability of bio-catalytic amide bond production, especially for compounds having amide linkages like losmapimod. Because biosynthesis uses non-toxic, biodegradable ingredients, it is safer and more sustainable than chemical synthesis.

### 13.4.2 Tofacitinib

Tofacitinib is commercially known as XELJANZ® and developed by Pfizer. It is an oral Janus kinase (JAK) inhibitor intended to control an overactive immune system. It functions by interfering with immune cells' JAK signalling pathways, which are believed to be crucial in fostering inflammation. Tofacitinib was approved by the U.S. FDA in 2012 and is used to treat a number of inflammatory and autoimmune diseases, including as ulcerative colitis, psoriatic arthritis, polyarticular course

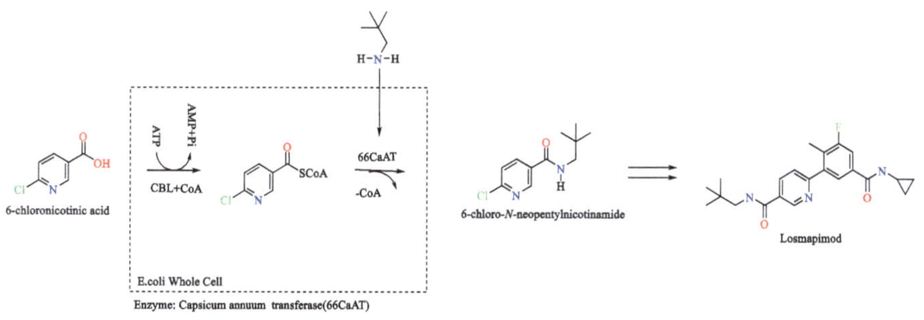


**Scheme 13.8** Biocatalytic synthesis for preparation of the key intermediate of losmapimod

juvenile idiopathic arthritis (pcJIA), and rheumatoid arthritis (Pfizer 2025; XELJANZ 2025; Fda 2018).

#### 13.4.2.1 Chemical Synthesis

Verzijl G. K. et al. have synthesized the (R,R)-amine through asymmetric reductive amination (ARA) of 1-benzyl-4-methylpiperidin-3-one under dynamic kinetic resolution (DKR) conditions. The reaction was performed in an Ir/BiPheP catalyst system and methylammonium acetate ($MeNH_2 \cdot OAc$) in isopropanol (IPA), under high temperature and hydrogen pressure for a long period. The process produced the product of interest with good stereoselectivity. Cis-1·bis-HCl was obtained as a white solid upon crystallization in good yield (Scheme 13.9) (Verzijl et al. 2018).

#### 13.4.2.2 Biocatalytic Synthesis

The biocatalytic synthesis of a tofacitinib intermediate was carried out using PocIRED-M6 which is an engineered imine reductase, demonstrating a highly efficient and stereoselective approach for intermediate production. The reaction was conducted utilizing lyophilized cell-free extract of PocIRED-M6. To maintain optimal conditions, the pH was controlled at 9.5 through titration with a 1.5 M aqueous methylamine solution. 1-Benzyl-4-methylpiperidin-3-one was reductively coupled with methylamine in this synthesis, which had a remarkable enantiomeric excess (>99.9% ee) and great diastereoselectivity (98:2 dr). Compared to conventional chemical synthesis methods, which can have harsher reaction conditions and worse stereoselectivity, the overall yield of 74% represented a notable improvement (Scheme 13.10) (Zhu et al. 2024).

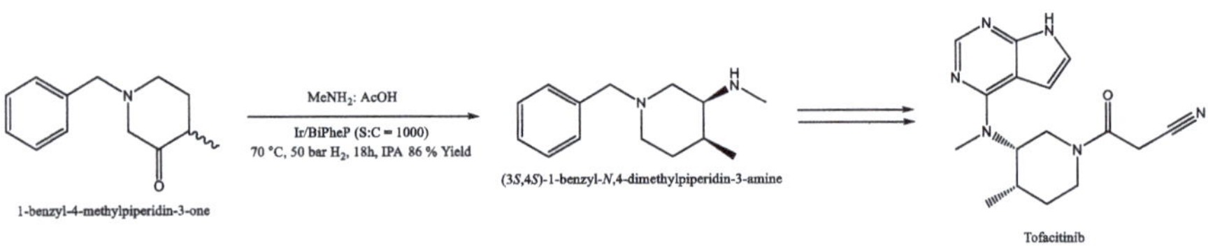


**Scheme 13.9** Chemical route of key intermediate of the tofacitinib

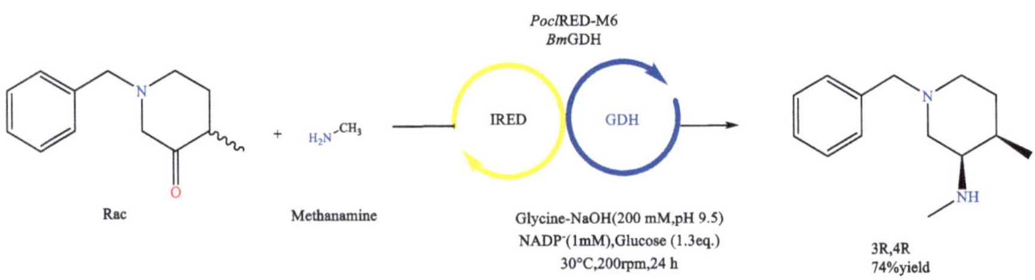


**Scheme 13.10** Biocatalytic route for synthesis of key intermediate of the tofacitinib

**Advantages**

For the production of β-branched chiral amines, asymmetric synthesis techniques such as asymmetric hydrogenation, asymmetric hydroamination, asymmetric reduction-asymmetric reductive amination, and dynamic kinetic resolution-asymmetric reductive amination have been investigated. These techniques, however, frequently rely on multi-catalyst cascades that complicate the reaction system or rare-transition-metal complexes with chiral ligands that are challenging to obtain and remove. Additionally, they usually demand severe conditions, produce a lot of waste, and exhibit inadequate diastereo- and enantioselectivity or modest turnover numbers (Zhu et al. 2024).

### 13.4.3 Esomeprazole

Esomeprazole Magnesium, sold as NEXIUM® by AstraZeneca LP, is a US FDA-approved proton pump inhibitor introduced in 2001. It is approved for the treatment of pathological hypersecretory diseases like Zollinger-Ellison (ZE) Syndrome, prevention of gastric ulcers, and treatment of gastroesophageal reflux disease (GERD). It is also prescribed for the protection against gastrointestinal injury caused by long-term NSAID therapy. Nexium Delayed-Release Granules are also indicated for short-term treatment of erosive esophagitis (EE) and GERD in pediatric patients (Fda 2001, 2007; DrugBank 2024).

#### 13.4.3.1 Chemical Synthesis

**Chemical Method 1**

Nishiguchi et al. (2018) created ideal conditions for pyrmetazole oxidation with superior selectivity. Ethyl acetate was used to dissolve the chiral Schiff base (ligand) and Fe (acac)3, which were then stirred for 30 min at rt. After adding lithium 4-dimethylaminobenzoate, the mixture was stirred for a next half hour. After cooling the reaction to −15 °C, pyrmetazole was added. Aqueous hydrogen peroxide was introduced gradually over the course of 6 h, which helped to keep the temperature below −6 °C. After that, the reaction was agitated at −13 to −10 °C for five more hours. The final product had 0.2% of unreacted starting material, 12% over-oxidized sulfone, and an 87% yield with 99.4% enantiomeric excess (ee). By treatment with methanolic potassium hydroxide, the product was separated as its potassium salt, yielding a 77% yield and >99.9% ee (Scheme 13.11) (Nishiguchi et al. 2018).

#### 13.4.3.2 Biocatalytic Synthesis

A highly enantioselective enzymatic sulfoxidation of pyrmetazole is used to biosynthesize esomeprazole, providing a more economical and environment friendly option to conventional chemical processes such as the Kagan-Sharpless-Pitchen oxidation. In the optimized biocatalytic process, an aqueous medium containing enzymes, cofactors, a seed crystal of esomeprazole, and a small amount of alcohol in buffered solution is used. Under an oxygen blanket, the reaction is conducted at

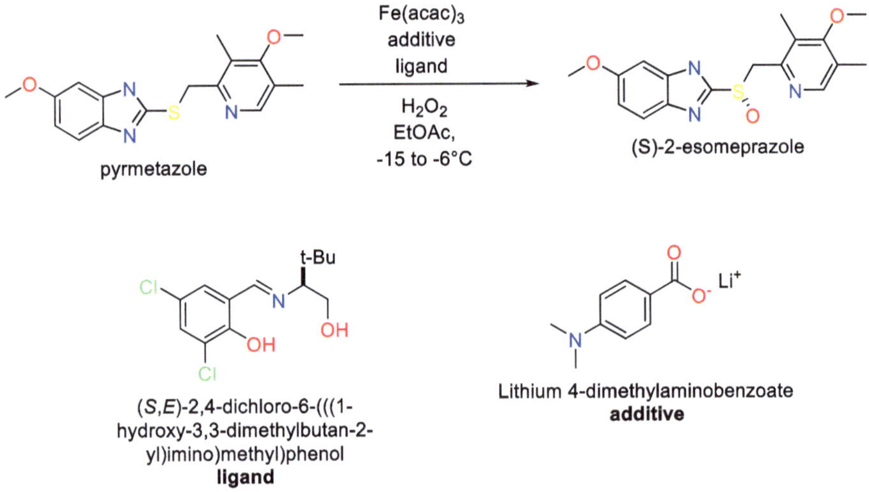


**Scheme 13.11** Chemical route of synthesis for esomeprazole

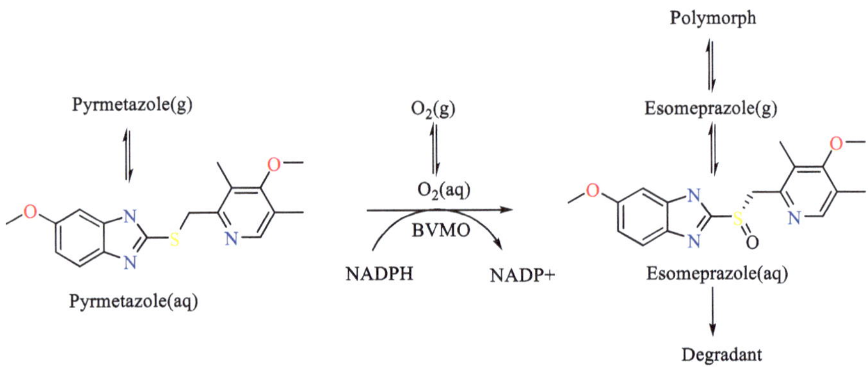


**Scheme 13.12** Biocatalytic route for synthesis of esomeprazole

25 °C and agitated for 48 h, with agitation progressively raised from 300 to 450 rpm. Key reagents, including propan-2-ol (co-solvent), NADP (cofactor), keto reductase CDX-019, Baeyer-Villiger Monooxygenase (BVMO), and catalase, are added to drive the selective oxidation while preventing unwanted by-products. After the reaction mixture is finished, it is extracted using methyl isobutyl ketone (MIBK) at 48 °C, phase separated, and filtered through Celite. Drying eliminates any remaining water, then chilling the mixture from 15 °C to 5 °C and gradually adding 240 mL of n-heptane starts the controlled crystallization process. After filtering and vacuum-drying for 72 h at 25 °C, the resultant solid had an 87% isolated yield, 99% purity (per HPLC), and 99.9% enantiomeric excess (ee). This approach is environmentally friendly due to its low usage of organic solvent and notable sulfone impurity of less than 0.1% (Bong et al. 2018) (Scheme 13.12).

**Advantages of Biocatalysis Over Improved Chemical Methods**

Compared to chemical procedures that employ huge amounts of organic solvents, this approach greatly decreases the environmental impact by using an aqueous medium and very little organic solvent (such as IPA). Additionally, this biocatalytic method produces a highly pure product with little impurity production, particularly sulfone impurities (<0.1%), and excellent enantiomeric excess (>99.9% ee), which makes the downstream workup easier. Because it uses fewer reagents and produces less waste, the biocatalytic process is very productive, yielding good results at an appealing cost. With less of an adverse effect on the environment and a simpler workup, the biosynthetic pathway provides a more economical, sustainable, and environmentally friendly option (Bong et al. 2018).

## 13.4.4 Rimegepant

The study conducted by Luo et al. reported a scalable and efficient enantioselective synthesis of a key intermediate for rimegepant, a CGRP receptor antagonist used for the treatment and prevention of migraines. The target intermediate, (R)-9-hydroxy-6,7,8,9-tetrahydro-5H-cyclohepta[b]pyridin-5-one, The synthetic target intermediate was generated using a Ru-catalyzed asymmetric transfer hydrogenation (ATH) as the key transformation. This was an appealing target since it involved the selective reduction of one of the two very similar ketone moieties in the diketone intermediate, which is a difficult transformation because their chemical environments are so similar. The synthesis starts from pyridine-2,3-dicarboxylic acid, which is esterified, cyclized, and decarboxylated to yield the diketone compound. The ATH reaction, using formic acid as the hydrogen donor, CAT05 (a ruthenium-based catalyst) at just 0.5 mol%, DIPEA as the base, and dichloromethane as the solvent, produced the desired chiral alcohol in 92.1% yield with 99.8% purity and 99.9% enantiomeric excess (ee) (Scheme 13.13). This avoids costly rhodium metal catalysts, and the cumbersome added steps for chiral resolution of the products. This is cost-effective and practical for industry. The reaction conditions are mild and favourable for kilogram-scale synthesis, and the optimised method provides a reliable, selective, and scalable access to the key intermediate of rimegepant.

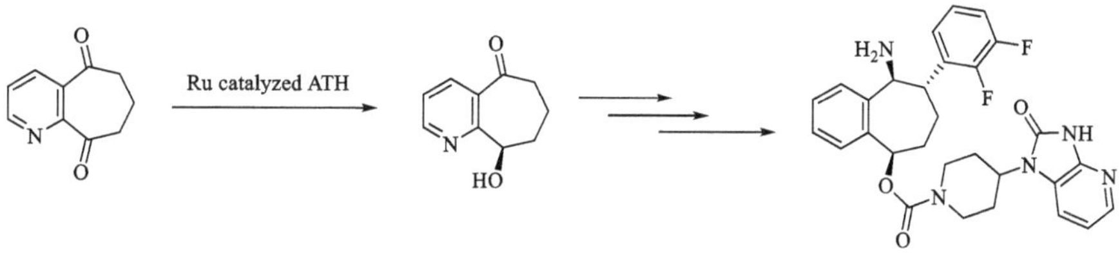


**Scheme 13.13** Chemical routes of synthesis for key intermediate of rimegepant

#### 13.4.4.1 Chemical Synthesis

**Route 1**

The study conducted by Luo et al. reported an efficient and scalable enantioselective synthesis of a key intermediate for rimegepant, a CGRP receptor antagonist used for the treatment and prevention of migraines. The target intermediate, (R)-9-hydroxy-6,7,8,9-tetrahydro-5H-cyclohepta[b]pyridin-5-one, was synthesized using a Ru-catalyzed asymmetric transfer hydrogenation (ATH) as the central transformation. This method was designed to selectively reduce one of the two similar ketone groups present in the diketone intermediate, a challenging step due to their comparable chemical environments. The synthesis starts from pyridine-2,3-dicarboxylic acid, which is esterified, cyclized, and decarboxylated to yield the diketone compound. The ATH reaction, using formic acid as the hydrogen donor, CAT05 (a ruthenium-based catalyst) at just 0.5 mol%, DIPEA as the base, and dichloromethane as the solvent, produced the desired chiral alcohol in 92.1% yield with 99.8% purity and 99.9% enantiomeric excess (ee) (Scheme 13.13). This process avoids the use of expensive rhodium catalysts and circumvents additional chiral resolution steps, making it both cost-effective and practical for industrial-scale production. The reaction conditions are mild and suitable for kilogram-scale synthesis, and the optimized method provides a robust, selective, and scalable approach to the key intermediate of rimegepant.

**Route 2**

Luo et al. have developed an efficient and scalable synthetic method for producing rimegepant, a CGRP receptor antagonist used to treat migraines. The synthesis starts with a cycloaddition between dimethyl pyridine-2, 3-dicarboxylate and dimethyl glutarate, followed by acid-induced decarboxylation to yield compound 3.

A chiral rhodium catalyst is then used for stereoselective reduction, and the resulting alcohol is protected with TIPSOTf to form compound 4. Next, a palladium-catalyzed coupling with intermediate 5 generates compound 6, which is reduced using $Li(t\text{-}BuO)_3AlH$ to yield the chiral intermediate 7. This is chlorinated with NCS and $PPh_3$ under nitrogen to produce compound 8, which is then converted to azide 10 through substitution with sodium azide. After deprotection with TBAF and condensation with NaHMDS, compound 12 is formed. The final step involves treatment with trimethylphosphine, resulting in rimegepant with high yield. The process emphasizes important transformations such as asymmetric reduction, selective protection, metal-catalyzed coupling, and azide chemistry, offering a practical route for large-scale production of the drug (Scheme 13.14) (Luo et al. 2012).

#### 13.4.4.2 Biocatalytic Synthesis

Enzymatic synthesis of a Key Intermediate for rimegepant via Transaminase Engineering Rimegepant, a calcitonin gene-related peptide (CGRP) receptor antagonist for the treatment of migraine, requires a stereochemically complex intermediate [2], in order to assemble a synthesis of the drug. Traditional chemical routes pose limitations due to the intermediate's steric complexity, necessitating the development of a biocatalytic strategy for efficient and selective synthesis. A

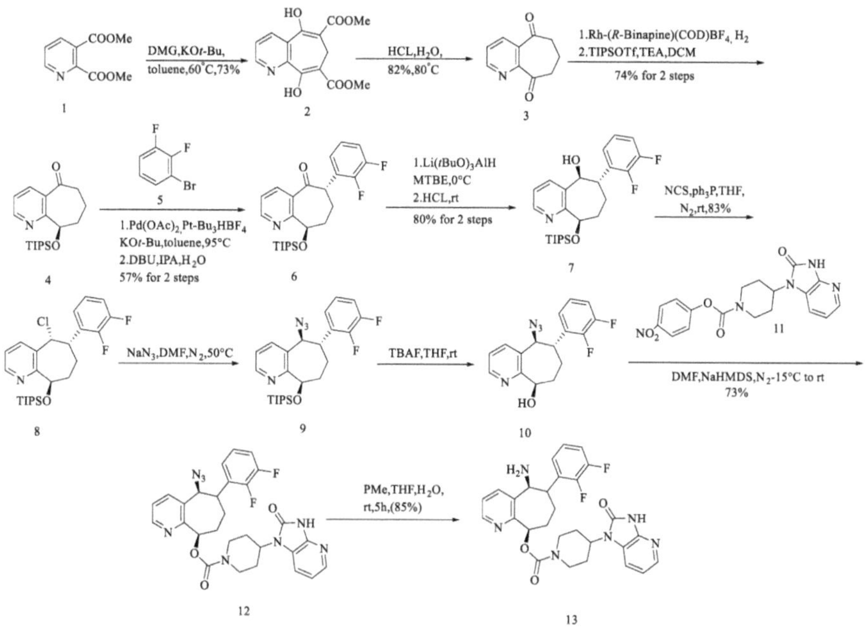

**Scheme 13.14** Chemical routes of synthesis of rimegepant

transaminase-catalyzed transformation was explored to convert the corresponding ketone precursor into the desired chiral amine. However, wild-type transaminases typically accommodate only small substituents in their active sites, resulting in negligible activity toward bulky, rigid substrates. To overcome these constraints, a comprehensive enzyme engineering approach was implemented. Initial enzyme screening and hit identification a library of over 300 transaminases—comprising wild-type and engineered variants—was screened for activity against [1]. No detectable conversion was observed due to substrate steric hindrance. A structurally simplified substrate analogue, [4], was then employed for rescreening. This enabled the identification of variant S617 from *Chromobacterium violaceum,* which demonstrated promising activity and was selected for further engineering (Scheme 13.15).

**Advantages**

The engineered and optimized transaminase variants were able to efficiently catalyze the asymmetric amination of a sterically hindered substrate in the biosynthesis of rimegepant, a calcitonin gene-related peptide (CGRP) receptor antagonist that is used for the treatment of migraine. The enzymatic process, conducted under mild aqueous conditions without transition metal catalysts, provides a more environmentally benign and sustainable alternative to traditional synthesis. By eliminating the reliance on harsh chemical reagents and minimizing the use of organic solvents, the overall reaction becomes more environmentally benign while simultaneously improving product yield. These improvements contribute not only to cleaner

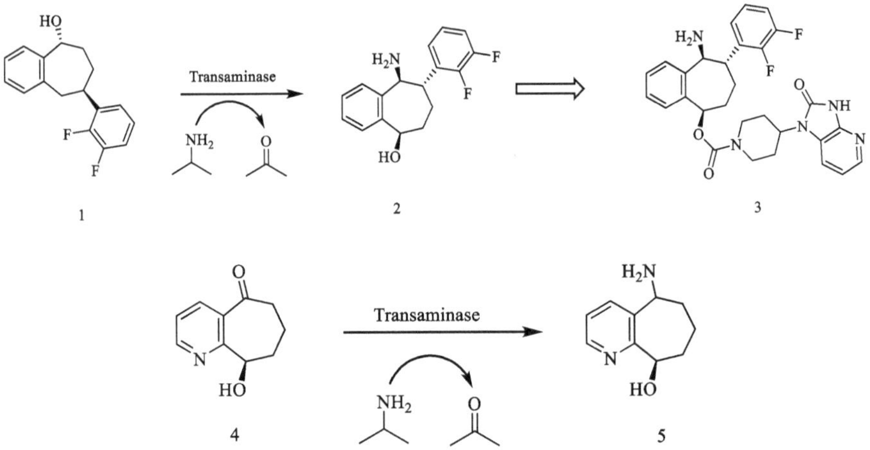

1 2 3 4 5

**Scheme 13.15** Biocatalytic route for synthesis of key intermediate of rimegepant

manufacturing practices but also to cost-efficiency and scalability, which are critical for pharmaceutical production.

### 13.4.5 Molnupiravir (Lagevrio®)

Molnupiravir (Lagevrio®) is an oral antiviral drug which limits the replication of RNA viruses by inducing viral error during their RNA replication. It is a pro-drug of the synthetic nucleoside derivative $N^4$-hydroxycytidine, that introduces errors into the viral genome during its replication, and this increases the likelihood of widespread RNA mutations due to the action of viral RNA-dependent RNA polymerase (Toots et al. 2019; Toots et al. 2020). Following the metabolism in the body, it is converted to β-D-$N^4$-hydroxycytidine 5′-triphosphate, a ribonucleoside analogue of cytidine (Amara et al. 2021; Painter et al. 2021). Molnupiravir was first developed for influenza by Drug Innovation Ventures at Emory (DRIVE) (a company completely owned by Emory University) (Jon Cohen 2020; Cully 2022). However, the drug was later developed by Merck in partnership with Ridgeback Biotherapeutics for the treatment and prevention of COVID-19. The UK granted molnupiravir the first worldwide regulatory approval on November 4, 2021, for the treatment of adults with mild to moderate COVID-19 who tested positive for SARS-CoV-2, and had at least one risk factor for severe COVID-19 illness. Although the EUA was issued it did create regulatory concern about the efficacy and the potential mutagenic nature to generate new SARS-CoV-2 variants that could escape immunity. Molnupiravir has either been granted emergency authorization or has been submitted for approval in countries such as the USA, Japan, and members of the EU. The drug achieved strong commercial success with $952 million in sales in Q4 of 2021 providing evidence of its impact during the COVID-19 pandemic (MHRA 2021, FDA 2021; Merck and Co. I 2021).

### 13.4.5.1 Chemical Synthesis

Merck Co. has partnered with Ridgeback Biotherapeutics and have claimed a patent on a 10-step pathway starting from ribose and uracil although their overall yield for the process is less than 10%. The first report of molnupiravir synthesis required five reactions with a yield of less than 17% from uridine with the majority of product resulting from the incorporation of the triazole group in what was a low yield product (Merck and Co. I 2021).

Another commercial way for molnupiravir synthesis initiates with uridine as the starting material. In the first step, the 2′,3′-diol moiety of uridine is protected by formation of an acetonide using 2,2-dimethoxypropane in the presence of acidic resin (Amberlyst-15) as a catalyst, carried out in acetone at room temperature, yielding the acetonide-protected intermediate. The second step involves esterification of the 5′-hydroxy group with isobutyric anhydride using 4-(Dimethylamino) pyridine as base in acetonitrile solvent, again conducted at room temperature, to give the 5′-O-isobutyryl ester. In the third step, the 4-position of the nucleobase is activated using 1,2,4-triazole, $POCl_3$, and DIAD (diisopropyl azodicarboxylate) in acetonitrile, resulting in formation of the triazole adduct. This reaction occurs at room temperature as well. The fourth step is a nucleophilic substitution of the triazole moiety with hydroxylamine, performed in i-PrOH (isopropanol) at 80°C, yielding the N-hydroxy cytidine derivative shown in Scheme 13.16. Finally, in the fifth step, deprotection of the acetonide group is achieved by treatment with formic acid at room temperature, affording the target molecule, molnupiravir (Hu et al. 2022) (Scheme 13.16).

### 13.4.5.2 Biocatalytic Synthesis

The enzymatic synthesis of Molnupiravir is a four-step biocatalytic cascade that demonstrates the power of enzyme engineering in pharmaceutical chemistry. The process begins with D-ribose, a naturally occurring sugar that contains multiple

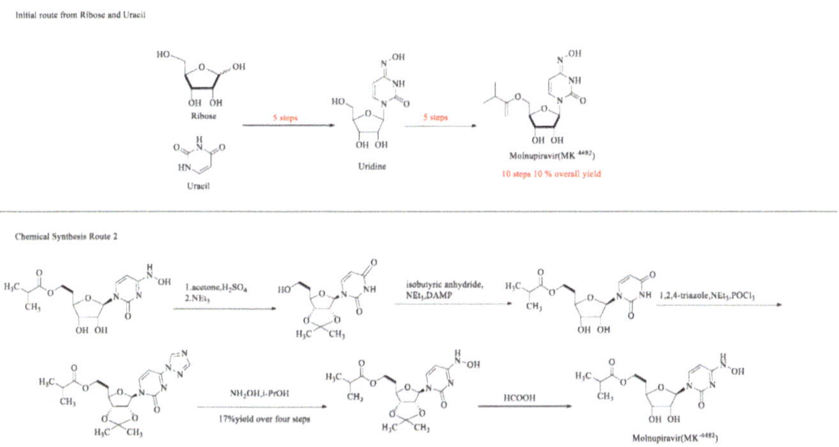


**Scheme 13.16** Chemical routes of synthesis for molnupiravir

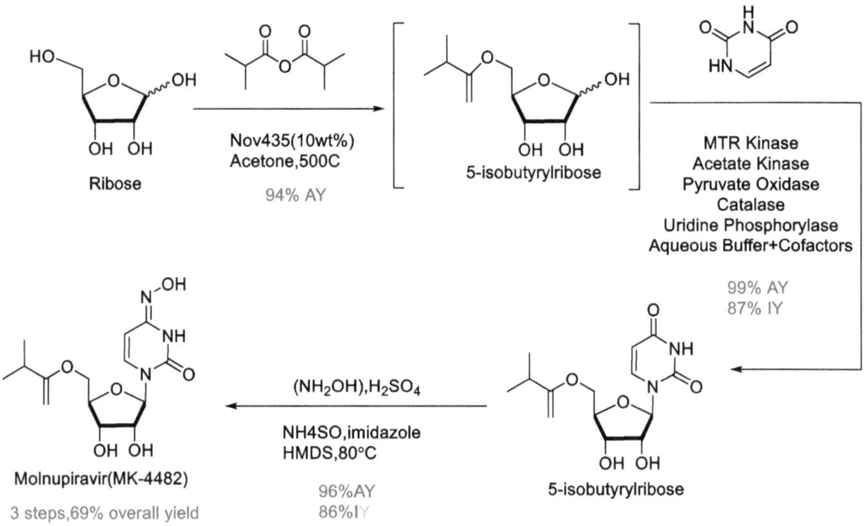


**Scheme 13.17** Biocatalytic route for synthesis of molnupiravir

hydroxyl groups. To achieve selectivity and avoid the need for protecting groups, *Candida antarctica* lipase B (CAL-B) is employed to selectively acylate the primary hydroxyl group at the 5′-position using isobutyric anhydride. This selective esterification results in the formation of 5′-O-isobutyryl ribose, setting the stage for further transformations (Scheme 13.17).

In the second step, the engineered methylthioribose (MTR) kinase catalyzes a highly regioselective phosphorylation at the anomeric (1′) position of the ribose ring, forming 5′-O-isobutyryl ribose-1-phosphate. This phosphorylation is crucial because it activates the sugar for the subsequent glycosylation step without requiring chemical activation. In the third step, uridine phosphoribosyltransferase (UPRT), another engineered enzyme, catalyzes the formation of the N-glycosidic bond between the activated sugar and the base N-hydroxycytosine. This step is particularly important as it installs the nucleobase onto the sugar backbone, yielding N-hydroxycytidine isobutyrate. Finally, the compound undergoes a mild chemical deacylation to remove the isobutyryl group, thus yielding the active pharmaceutical ingredient, Molnupiravir. This entire enzymatic route is highly efficient and sustainable, performed in aqueous media and without the use of traditional protecting group strategies, showcasing the advantages of biocatalysis in green and scalable drug synthesis (Mcintosh et al. 2021).

**Advantages**

Cost-efficient and sustainable pathway for molnupiravir has been conceived, utilizing abundant raw materials and enzymatic steps to facilitate kilogram-scale preparation. This facile three-step synthetic route, based on ribose, involves the use of one isolated intermediate alone and includes two enzymatic operations, one employing a new engineered enzyme cascade system combined with an innovative phosphate recyclization system. The final stage is a mild dehydration reaction that in aggregate affords molnupiravir in high purity and yield. Compared to the original synthesis route, this method is willing to take about 70% less time to achieve our molnupiravir, as well as have a sevenfold total yield increase. That this approach was successful indicates the impressive advance that has been made in enzyme discovery and engineering, and rapid integration of biocatalytic processes on manufacturing scale (Mcintosh et al. 2021).

## 13.5 Conclusion

The incorporation of biocatalysis in Active Pharmaceutical Ingredient (API) synthesis represents a significant step towards sustainable and efficient pharmaceutical production. Enzymes, being extremely selective and eco-friendly catalysts, provide several benefits over conventional chemical synthesis strategies-such as mild reaction conditions, less production of by-products, reduction in use of hazardous solvents and catalyst and improved process efficiency. These advantages in favour of greener chemistry also correspond with the broader goal of a circular bioeconomy, where optimal resource use and environmental responsibility are important. Through the successful implementation of enzymatic processes in various industrial case studies, it is evident that biocatalysis holds enormous potential for scalable and cost-effective API production. With increasing advancements in research, especially enzyme engineering, immobilization technologies, and process optimization, the pharmaceutical synthesis of the future will largely rely on biocatalytic approaches. In conclusion, implementing biocatalysts is not just a shift in methodology; it is a revolutionary strategy that promotes innovation while ensuring environmental and economic sustainability. This fundamental shift signifies a crucial step toward a future where pharmaceuticals are not only effective but also responsibly manufactured.

## References

Abedi D, Zhang L, Pyne M, Chou CP (2019) Enzyme biocatalysis. In: Comprehensive biotechnology. https://doi.org/10.1016/B978-0-444-64046-8.00003-3

Abuzenah H, Abdulrasheed M, Sardauna AE, Al-Qataisheh B, Musa MM (2025) Harnessing alcohol dehydrogenases in organic reaction cascades: a strategy for enhanced efficiency in synthetic organic chemistry. Catalysts 15:223

Alhawari O, Awan U, Bhutta MKS, Ülkü MA (2021) Insights from circular economy literature: a review of extant definitions and unravelling paths to future research. Sustainability 13:859

Amara A, Penchala SD, Else L, Hale C, Fitzgerald R, Walker L, Lyons R, Fletcher T, Khoo S (2021) The development and validation of a novel LC-MS/MS method for the simultaneous quantification of Molnupiravir and its metabolite ß-d-N4-hydroxycytidine in human plasma and saliva. J Pharm Biomed Anal 206:114356

Amenaghawon AN, Odika P, Aiwekhoe SE (2022) Optimization of nutrient medium composition for the production of lipase from waste cooking oil using response surface methodology and artificial neural networks. Chem Eng Commun 209:1531–1541

Bansal M, Kumar D, Chauhan GS, Kaushik A (2018) Preparation, characterization and trifluralin degradation of laccase-modified cellulose nanofibers. Mater Sci Energy Technol 1:29–37
Barski OA, Tipparaju SM, Bhatnagar A (2008) The aldo-keto reductase superfamily and its role in drug metabolism and detoxification. Drug Metab Rev 40:553–624
Baumann M, Baxendale IR (2015) The synthesis of active pharmaceutical ingredients (APIs) using continuous flow chemistry. Beilstein J Org Chem 11:1194–1219
Ben Bacha A, Alonazi M, Alharbi MG, Horchani H, Ben Abdelmalek I (2022) Biodiesel production by single and mixed immobilized lipases using waste cooking oil. Molecules 27:8736
Berg J, Tymoczko J, Stryer L (2002) Biochemistry, 5th edn. NCBI, New York
Bernhardt P, O'Connor SE (2009) Opportunities for enzyme engineering in natural product biosynthesis. Curr Opin Chem Biol 13:35–42
Bezerra CS, De Farias Lemos CMG, De Sousa M, Gonçalves LRB (2015) Enzyme immobilization onto renewable polymeric matrixes: past, present, and future trends. J Appl Polym Sci 132. https://doi.org/10.1002/app.42125
Bilal M, Qamar SA, Ashraf SS, Rodríguez-Couto S, Iqbal HM (2021) Robust nanocarriers to engineer nanobiocatalysts for bioprocessing applications. Adv Colloid Interf Sci 293:102438
Bocken NM, De Pauw I, Bakker C, Van Der Grinten B (2016) Product design and business model strategies for a circular economy. J Ind Prod Eng 33:308–320
Bong YK, Song S, Nazor J, Vogel M, Widegren M, Smith D, Collier SJ, Wilson R, Palanivel S, Narayanaswamy K (2018) Baeyer–Villiger monooxygenase-mediated synthesis of esomeprazole as an alternative for kagan sulfoxidation. J Org Chem 83:7453–7458
Bornscheuer UT, Huisman G, Kazlauskas R, Lutz S, Moore J, Robins K (2012) Engineering the third wave of biocatalysis. Nature 485:185–194
Busca G (2019) Catalytic materials based on silica and alumina: structural features and generation of surface acidity. Prog Mater Sci 104:215–249
Canada H (2013) Available: https://www.canada.ca/en/health-canada/services/drugs-health-products/compliance-enforcement/information-health-product/drugs/active-pharmaceutical-ingredients-questions-answers.html. Accessed 1 Apr 2024
Chandra P, Enespa, Singh R, Arora PK (2020) Microbial lipases and their industrial applications: a comprehensive review. Microb Cell Factories 19:1–42
Chorkendorff I, Niemantsverdriet JW (2017) Concepts of modern catalysis and kinetics. Wiley
Cully M (2022) A tale of two antiviral targets-and the COVID-19 drugs that bind them. Nat Rev Drug Discov 21:3–5
David AE, Yang AJ, Wang NS (2011) Enzyme stabilization and immobilization by sol-gel entrapment. In: Enzyme stabilization and immobilization: methods and protocols, pp 49–66
De Miranda AS, Milagre CD, Hollmann F (2022) Alcohol dehydrogenases as catalysts in organic synthesis. Front Catal 2:900554
Domínguez De María P, Hollmann F (2015) On the (un) greenness of biocatalysis: some challenging figures and some promising options. Front Microbiol 6:1257
Drugbank (2024) Esomeprazole – DB00736 [Online]. Available: https://www.drugbank.ca/drugs/DB00736. Accessed 21 Apr 2025
Emaf EMF (2019) Circular economy introduction: overview [Online]. Available: https://www.ellenmacarthurfoundation.org/topics/circular-economy-introduction/overview. Accessed 2 Apr 2025
Farrer NJ, Griffith DM (2020) Exploiting azide–alkyne click chemistry in the synthesis, tracking and targeting of platinum anticancer complexes. Curr Opin Chem Biol 55:59–68
Fda, U. S. F. A. D. A (2001) NDA 21-154 Nexium (Esomeprazole Magnesium) approval letter [Online]. Available: https://www.accessdata.fda.gov/drugsatfda_docs/nda/2001/21154_Nexium.cfm. Accessed 20 Apr 2025
Fda, U. S. F. A. D. A (2007) Esomeprazole Clinical PREA Document – NDA 022101/S-000 [Online]. Available: https://fda.report/media/71531/022101s000_Esmoprazole_Clinical_prea.pdf.pdf. Accessed 21 Apr 2025

Fda, U. S. F. A. D. A (2018) XELJANZ (tofacitinib): prescribing information [Online]. Available: https://www.accessdata.fda.gov/drugsatfda_docs/label/2018/203214s018lbl.pdf. Accessed 20 Apr 2025
Fda (2021) Coronavirus (COVID-19) update: FDA authorizes additional oral antiviral treatment for COVID-19 in certain adults [Online]. Available: https://www.fda.gov/news-events/press-announcements/coronavirus-covid-19-update-fda-authorizes-additional-oral-antiviral-treatment-covid-19-certain. Accessed 19 Apr 2025
Fda, U. S. F. A. D. A (2022) Orphan drug designation for Losmapimod (FSHD) [Online]. Available: https://www.accessdata.fda.gov/scripts/opdlisting/oopd/detailedIndex.cfm?cfgridkey=721919. Accessed 20 Apr 2025
Fox RJ, Davis SC, Mundorff EC, Newman LM, Gavrilovic V, Ma SK, Chung LM, Ching C, Tam S, Muley S (2007) Improving catalytic function by ProSAR-driven enzyme evolution. Nat Biotechnol 25:338–344
France SP, Hepworth LJ, Turner NJ, Flitsch SL (2017) Constructing biocatalytic cascades: in vitro and in vivo approaches to de novo multi-enzyme pathways. ACS Catal 7:710–724
France SP, Lewis RD, Martinez CA (2023) The evolving nature of biocatalysis in pharmaceutical research and development. JACS Au 3:715–735
Geissdoerfer M, Savaget P, Bocken NM, Hultink EJ (2017) The circular economy–a new sustainability paradigm? J Clean Prod 143:757–768
Gkantzou E, Chatzikonstantinou AV, Fotiadou R, Giannakopoulou A, Patila M, Stamatis H (2021) Trends in the development of innovative nanobiocatalysts and their application in biocatalytic transformations. Biotechnol Adv 51:107738
Gladek E (2017) The seven pillars of the circular economy. Retrieved from Metabolic website: http://www.metabolic.nl/the-seven-pillars-of-the-circular-economy
Homaei AA, Sariri R, Vianello F, Stevanato R (2013) Enzyme immobilization: an update. J Chem Biol 6:185–205
Houde A, Kademi A, Leblanc D (2004) Lipases and their industrial applications: an overview. Appl Biochem Biotechnol 118:155–170
Hoyos P, Pace V, Alcántara AR (2019) Biocatalyzed synthesis of statins: a sustainable strategy for the preparation of valuable drugs. Catalysts 9:260
Hu Y, Nie Q-Y, Pan H-X, Tang G-L (2020) 1.07-Bacterial Type II polyketide synthases. In: Liu H-W, Begley TP (eds) Comprehensive natural products Iii, pp 198–249
Hu T, Xie Y, Zhu F, Gong X, Liu Y, Xue H, Aisa HA, Shen J (2022) "One-Pot" synthesis of molnupiravir from cytidine. Org Process Res Dev 26:358–364
Intasian P, Prakinee K, Phintha A, Trisrivirat D, Weeranoppanant N, Wongnate T, Chaiyen P (2021) Enzymes, in vivo biocatalysis, and metabolic engineering for enabling a circular economy and sustainability. Chem Rev 121:10367–10451
Jackson DA, Symons RH, Berg P (1972) Biochemical method for inserting new genetic information into DNA of simian virus 40: circular SV40 DNA molecules containing lambda phage genes and the galactose operon of Escherichia coli. Proc Natl Acad Sci 69:2904–2909
Jiang T, Hou Y, Zhang T, Feng X, Li C (2019) Construction of a CaHPO 4-PGUS1 hybrid nanoflower through protein-inorganic self-assembly, and its application in glycyrrhetinic acid 3-O-mono-β-d-glucuronide preparation. Front Chem Sci Eng 13:554–562
Jon Cohen CP (2020) Emails offer look into whistleblower charges of cronyism behind potential COVID-19 drug [Online]. Available: https://www.science.org/content/article/emails-offer-look-whistleblower-charges-cronyism-behind-potential-covid-19-drug. Accessed
Kapale SS, Gaike H, Chaudhari HK (2024) Lipase as biocatalyst-for synthesis of phenol by using Box–Behnken design. Polycycl Aromat Compd 44:4261–4272
Kelly SA, Mix S, Moody TS, Gilmore BF (2020) Transaminases for industrial biocatalysis: novel enzyme discovery. Appl Microbiol Biotechnol 104:4781–4794
Kinner A, Nerke P, Siedentop R, Steinmetz T, Classen T, Rosenthal K, Nett M, Pietruszka J, Lütz S (2022) Recent advances in biocatalysis for drug synthesis. Biomedicine 10:964
Kirchherr J, Reike D, Hekkert M (2017) Conceptualizing the circular economy: an analysis of 114 definitions. Resour Conserv Recycl 127:221–232

Korhonen J, Nuur C, Feldmann A, Birkie SE (2018) Circular economy as an essentially contested concept. J Clean Prod 175:544–552

Kumar R, Karmilowicz MJ, Burke D, Burns MP, Clark LA, Connor CG, Cordi E, Do NM, Doyle KM, Hoagland S (2021) Biocatalytic reductive amination from discovery to commercial manufacturing applied to abrocitinib JAK1 inhibitor. Nat Catal 4:775–782

Kumar V, Bansal V, Madhavan A, Kumar M, Sindhu R, Awasthi MK, Binod P, Saran S (2022) Active pharmaceutical ingredient (API) chemicals: a critical review of current biotechnological approaches. Bioengineered 13:4309–4327

Kumar S, Kumar G, Saroha B, Gulati K (2023) Metal oxide heterostructures as catalysts in organic reactions. Elsevier, Metal Oxide-Based Heterostructures

Lin B, Tao Y (2017) Whole-cell biocatalysts by design. Microb Cell Factories 16:1–12

Lubberink M, Schnepel C, Citoler J, Derrington SR, Finnigan W, Hayes MA, Turner NJ, Flitsch SL (2020) Biocatalytic monoacylation of symmetrical diamines and its application to the synthesis of pharmaceutically relevant amides. ACS Catal 10:10005–10009

Luo G, Chen L, Conway CM, Denton R, Keavy D, Signor L, Kostich W, Lentz KA, Santone KS, Schartman R (2012) Discovery of (5 S, 6 S, 9 R)-5-Amino-6-(2, 3-difluorophenyl)-6, 7, 8, 9-tetrahydro-5 H-cyclohepta [b] pyridin-9-yl 4-(2-oxo-2, 3-dihydro-1 H-imidazo [4, 5-b] pyridin-1-yl) piperidine-1-carboxylate (BMS-927711): An Oral Calcitonin Gene-Related Peptide (CGRP) Antagonist in Clinical Trials for Treating Migraine. J Med Chem 55:10644–10651

Mahmoud N, Awassa J, Toufaily J, Lebeau B, Daou TJ, Cormier M, Goddard J-P (2023) Heterogeneous Photoredox catalysis based on silica mesoporous material and eosin Y: impact of material support on selectivity of radical cyclization. Molecules 28:549

Mangas-Sanchez J, France SP, Montgomery SL, Aleku GA, Man H, Sharma M, Ramsden JI, Grogan G, Turner NJ (2017) Imine reductases (IREDs). Curr Opin Chem Biol 37:19–25

Mateo C, Abian O, Fernandez-Lorente G, Pessela BC, Grazu V, Guisan JM, Fernandez-Lafuente R (2020) Multi-point covalent immobilization of enzymes on supports activated with epoxy groups: stabilization of industrial enzymes. In: Immobilization of enzymes and cells: methods and protocols, pp 109–117

Mcintosh JA, Benkovics T, Silverman SM, Huffman MA, Kong J, Maligres PE, Itoh T, Yang H, Verma D, Pan W (2021) Engineered ribosyl-1-kinase enables concise synthesis of molnupiravir, an antiviral for COVID-19. ACS Cent Sci 7:1980–1985

Merck & Co. I (2021) The Medicines Patent Pool (MPP) and Merck enter into license agreement for molnupiravir, an investigational oral antiviral COVID-19 medicine, to increase broad access in low- and middle-income countries [Online]. Available: https://www.merck.com/news/the-medicines-patent-pool-mpp-and-merck-enter-into-license-agreement-for-molnupiravir-an-investigational-oral-antiviral-covid-19-medicine-to-increase-broad-access-in-low-and-middle-income-countri/. Accessed 19 Apr 2025

(MHRA), M. A. H. P. R. A (2021) Summary of product characteristics for Lagevrio (molnupiravir) [Online]. Available: https://www.gov.uk/government/publications/regulatory-approval-of-lagevrio-molnupiravir/summary-of-product-characteristics-for-lagevrio. Accessed

Mills DR, Peterson R, Spiegelman S (1967) An extracellular Darwinian experiment with a self-duplicating nucleic acid molecule. Proc Natl Acad Sci 58:217–224

Mohamad NR, Marzuki NHC, Buang NA, Huyop F, Wahab RA (2015) An overview of technologies for immobilization of enzymes and surface analysis techniques for immobilized enzymes. Biotechnol Biotechnol Equip 29:205–220

Moss GP (1996) Basic terminology of stereochemistry (IUPAC recommendations 1996). Pure Appl Chem 68:2193–2222

Mu R, Wang Z, Wamsley MC, Duke CN, Lii PH, Epley SE, Todd LC, Roberts PJ (2020) Application of enzymes in regioselective and stereoselective organic reactions. Catalysts 10:832

Muller P (1994) Glossary of terms used in physical organic chemistry (IUPAC recommendations 1994). Pure Appl Chem 66:1077–1184

Muruvanda BU, Prakash AS, Kalapriya K, Sampath K (2023) Green enzymes from plants and microbial sources used in industries: an overview. Rasāyan J Chem 16:483–493

Napkin (n.d.) Napkin AI – AI tool for structured thinking and note-taking [Online]. Available: https://napkin.one. Accessed 21 Apr 2025

Nci (2025) Available: https://www.cancer.gov/publications/dictionaries/cancer-terms/def/active-pharmaceutical-ingredient. Accessed 21 Mar 2025

Nelson DL, Lehninger AL, Cox MM (2008) Lehninger principles of biochemistry. Macmillan

Nishiguchi S, Izumi T, Kouno T, Sukegawa J, Ilies L, Nakamura E (2018) Synthesis of esomeprazole and related proton pump inhibitors through iron-catalyzed enantioselective sulfoxidation. ACS Catal 8:9738–9743

O'Connell A, Barry A, Burke AJ, Hutton AE, Bell EL, Green AP, O'Reilly E (2024) Biocatalysis: landmark discoveries and applications in chemical synthesis. Chem Soc Rev 53:2828

Oliveira FL, De Souza SP, Bassut J, Álvarez HM, Garcia-Basabe Y, Alves De Souza RO, Esteves PM, Gonçalves RS (2019) Enzyme-decorated covalent organic frameworks as nanoporous platforms for heterogeneous biocatalysis. Chem–A Eur J 25:15863–15870

Painter WP, Holman W, Bush JA, Almazedi F, Malik H, Eraut NC, Morin MJ, Szewczyk LJ, Painter GR (2021) Human safety, tolerability, and pharmacokinetics of molnupiravir, a novel broad-spectrum oral antiviral agent with activity against SARS-CoV-2. Antimicrob Agents Chemother 65. https://doi.org/10.1128/aac.02428-20

Pfizer (2025) XELJANZ® (tofacitinib) product information [Online]. Available: https://www.pfizer.com/products/product-detail/xeljanz. Accessed 20 Apr 2025

Poliakoff M, Fitzpatrick JM, Farren TR, Anastas PT (2002) Green chemistry: science and politics of change. Science 297:807–810

Pollard DJ, Woodley JM (2007) Biocatalysis for pharmaceutical intermediates: the future is now. Trends Biotechnol 25:66–73

Qiao L, Luo Z, Chen H, Zhang P, Wang A, Sheldon RA (2023) Engineering ketoreductases for the enantioselective synthesis of chiral alcohols. Chem Commun 59:7518–7533

Rahman A, Uahengo V, Daniel L (2017) Green chemistry concept: applications of catalysis in pharmacuetical industry. Glob Drugs Ther 2:1–6

Reshmy R, Philip E, Sirohi R, Tarafdar A, Arun K, Madhavan A, Binod P, Awasthi MK, Varjani S, Szakacs G (2021) Nanobiocatalysts: advancements and applications in enzyme technology. Bioresour Technol 337:125491

Rossino G, Robescu MS, Licastro E, Tedesco C, Martello I, Maffei L, Vincenti G, Bavaro T, Collina S (2022) Biocatalysis: a smart and green tool for the preparation of chiral drugs. Chirality 34:1403–1418

Schöggl J-P, Stumpf L, Baumgartner RJ (2020) The narrative of sustainability and circular economy-a longitudinal review of two decades of research. Resour Conserv Recycl 163:105073

Sharma V, Tsai M-L, Nargotra P, Chen C-W, Kuo C-H, Sun P-P, Dong C-D (2022) Agro-industrial food waste as a low-cost substrate for sustainable production of industrial enzymes: a critical review. Catalysts 12:1373

Sheldon RA, Woodley JM (2018) Role of biocatalysis in sustainable chemistry. Chem Rev 118:801–838

Siedentop R, Claaßen C, Rother D, Lütz S, Rosenthal K (2021) Getting the most out of enzyme cascades: strategies to optimize in vitro multi-enzymatic reactions. Catalysts 11:1183

Slabu I (2017) Transaminase mediated cascade reactions. Doctor of Philosophy, The University of Manchester

Slabu I, Galman JL, Lloyd RC, Turner NJ (2017) Discovery, engineering, and synthetic application of transaminase biocatalysts. ACS Catal 7:8263–8284

Srinivasamurthy VS, Böttcher D, Bornscheuer UT (2019) A multi-enzyme cascade reaction for the production of 6-hydroxyhexanoic acid. Zeitschrift für Naturforschung C 74:71–76

Sunjic V, Parnham MJ (2011) Signposts to chiral drugs: organic synthesis in action. Springer

Tan Z, Cheng H, Chen G, Ju F, Fernández-Lucas J, Zdarta J, Jesionowski T, Bilal M (2023) Designing multifunctional biocatalytic cascade system by multi-enzyme co-immobilization on biopolymers and nanostructured materials. Int J Biol Macromol 227:535–550

Terry EM (1928) Catalytic activity of hydrochloric acid and of potassium and of sodium hydroxide in aqueous solution. J Am Chem Soc 50:[1239]-1250

Therapeutics F (2024) Fulcrum Therapeutics Enters into a Collaboration and License Agreement with Sanofi for the Development and Commercialization of Losmapimod in Facioscapulohumeral Muscular Dystrophy [Online]. Available: https://ir.fulcrumtx.com/news-releases/news-release-details/fulcrum-therapeutics-enters-collaboration-and-license-agreement. Accessed 20 Apr 2025

Toots M, Yoon J-J, Cox RM, Hart M, Sticher ZM, Makhsous N, Plesker R, Barrena AH, Reddy PG, Mitchell DG (2019) Characterization of orally efficacious influenza drug with high resistance barrier in ferrets and human airway epithelia. Sci Transl Med 11:eaax5866

Toots M, Yoon J-J, Hart M, Natchus MG, Painter GR, Plemper RK (2020) Quantitative efficacy paradigms of the influenza clinical drug candidate EIDD-2801 in the ferret model. Transl Res 218:16–28

Topsoe H (2024) Catalysts: at the core of efficient industrial processes [Online]. Available: https://www.topsoe.com/blog/catalysts-at-the-core-of-efficient-industrial-processes. Accessed 23 Mar 2025

Trost BM (1995) Atom economy—a challenge for organic synthesis: homogeneous catalysis leads the way. Angew Chem Int Ed Engl 34:259–281

Ubando AT, Felix CB, Chen W-H (2020) Biorefineries in circular bioeconomy: a comprehensive review. Bioresour Technol 299:122585

Verzijl GK, Schuster C, Dax T, De Vries AH, Lefort L (2018) Asymmetric synthesis of a key intermediate for Tofacitinib via a dynamic kinetic resolution-reductive amination protocol. Org Process Res Dev 22:1817–1822

Wachtmeister J, Rother D (2016) Recent advances in whole cell biocatalysis techniques bridging from investigative to industrial scale. Curr Opin Biotechnol 42:169–177

Willette RN, Eybye ME, Olzinski AR, Behm DJ, Aiyar N, Maniscalco K, Bentley RG, Coatney RW, Zhao S, Westfall TD (2009) Differential effects of p38 mitogen-activated protein kinase and cyclooxygenase 2 inhibitors in a model of cardiovascular disease. J Pharmacol Exp Ther 330:964–970

Winter G, Fersht AR, Wilkinson AJ, Zoller M, Smith M (1982) Redesigning enzyme structure by site-directed mutagenesis: tyrosyl tRNA synthetase and ATP binding. Nature 299:756–758

Xeljanz (2025) How XELJANZ works [Online]. Available: https://www.xeljanz.com/#how-xeljanz-works. Accessed 20 Apr 2025

Yamamoto H (2013) Acid catalysis in organic synthesis. In: Inventing reactions, pp 315–334

Ying Shi YM, Yang H, Wang X, Gong D, Bai M, Chen X, Chen Y, Zhang X, Ma Y, Gao Q (2020) Benzamide derivative. United States patent application

Zdarta J, Meyer AS, Jesionowski T, Pinelo M (2018) A general overview of support materials for enzyme immobilization: characteristics, properties, practical utility. Catalysts 8:92

Zhang H, Bin Samsudin I, Jaenicke S, Chuah G-K (2022) Zeolites in catalysis: sustainable synthesis and its impact on properties and applications. Cat Sci Technol 12:6024–6039

Zhu Y-H, Liu C-Y, Cai S, Guo L-B, Kim I-W, Kalia VC, Lee J-K, Zhang Y-W (2019) Cloning, expression and characterization of a highly active alcohol dehydrogenase for production of ethyl (S)-4-chloro-3-hydroxybutyrate. Indian J Microbiol 59:225–233

Zhu ZY, Shi M, Li CL, Gao YF, Shen XY, Ding XW, Chen FF, Xu JH, Chen Q, Zheng GW (2024) An engineered imine reductase for highly Diastereo-and enantioselective synthesis of β-branched amines with contiguous Stereocenters. Angew Chem 136:e202408686

# The Challenges, Opportunities, and Future of Sustainable Biorefineries

14

Workisa Bacha, Tesfaye Kassaw Bedru, Mani Jayakumar, and Magesh Rangaraju

**Abstract**

Developing biorefineries increasingly stands out as a response to climate change and the need for economically viable alternatives to traditional oil refineries (i.e., chemicals, fuels, and products). The rapid increase in energy demand is making conventional resources, mostly fossil fuels, scarcer. Given the imbalance between energy demand and supply in today's world, mankind is searching for sustainable energy resources not just to fulfill consumer demands but also to protect the environment. Biomass can produce many useful chemicals and fuels, offering an environmentally friendly renewable resource. A biorefinery operates by producing chemicals and fuels from biomass, similar to how a petroleum refinery works. In the context of the emerging bioeconomy, this chapter presents an overview of

W. Bacha
Department of Chemical Engineering, Haramaya Institute of Technology, Haramaya University, Haramaya, Dire Dawa, Ethiopia

T. K. Bedru
Department of Chemical Engineering, Kombolcha Institute of Technology, Wollo University, Kombolcha, Ethiopia

M. Jayakumar (✉)
Department of Chemical Engineering, Haramaya Institute of Technology, Haramaya University, Haramaya, Dire Dawa, Ethiopia

Department of Biotechnology, Faculty of Engineering, Karpagam Academy of Higher Education, Coimbatore, Tamil Nadu, India

Centre for Natural Products and Functional Foods, Karpagam Academy of Higher Education, Coimbatore, Tamil Nadu, India
e-mail: drjayakumarmani@haramaya.edu.et

M. Rangaraju
Department of Chemical Engineering, Adhiparasakthi Engineering College, Melmaruvathur, Tamil Nadu, India



V. Singh (ed.), *Sustainable Waste Management Towards Circular Bioeconomy*,
https://doi.org/10.1007/978-981-95-2227-9_14

the developments in biorefineries today and in the future. It also focuses on biorefineries' concepts, the sensibility framework, the key factors involved in the techno-economic aspects of biorefinery, commercialization, market outlook issues, prospects, and challenges.

**Keywords**

Biorefineries · Challenges · Techno-economic analysis · Prospects · Key factors

## 14.1 Introduction

Sustainable biorefineries have emerged as a vital component in the quest for a greener and more sustainable future. These innovative facilities hold great promise in addressing the pressing challenges of resource scarcity, climate change, and environmental degradation (Edenhofer et al. 2011; Hariram et al. 2023; Sandaka and Kumar 2023). By harnessing the power of renewable biomass, biorefineries have the potential to transform the way we produce energy, chemicals, and materials while minimizing our reliance on fossil fuels (Ben-Iwo et al. 2016; Benti et al. 2021).

However, any transformative effort like sustainable biorefineries has complex challenges to deal with. From technological limits and available raw materials to economic feasibility and policy frameworks, these hurdles must be crossed to unlock the full potential of biorefinery systems (Mohammad et al. 2024; Solarte-Toro and Cardona Alzate 2023). Additionally, integrating sustainability principles throughout the entire biorefinery supply chain, from obtaining raw materials to waste management, presents a multidimensional challenge that demands innovative solutions (Blasi et al. 2023; Ding et al. 2024). Despite facing difficulties, biorefineries focusing on sustainability provide many chances. Moving towards a bio-based economy opens doors to new industries, more jobs, and helping rural areas grow (Chen and Zhang 2015; Gawel et al. 2019; Ingrao et al. 2018). By using different sources like farm waste, algae, and rubbish, biorefineries can help a circular economy where waste is reused to make valuable things. Also, making biofuels, bioplastics, and chemicals can help lower greenhouse gases and deal with climate change (Chojnacka 2023; Pattnaik et al. 2024).

Advancements show sustainable biorefineries have great potential ahead. Biotech improvements, optimized processes, and distributed systems spread are clearing a path for sharper, more financially feasible biorefinery versions. Integrating digital tech and artificial intelligence further boosts how ecofriendly and lucrative these places run (Venkatesh 2022; Wei et al. 2022) Plus, the idea of biorefinery communities and nearby teamwork offers thrilling chances to share things, swap ideas, and expand together in unison ways (Bezama et al. 2019; Szarka et al. 2023).

This chapter goes into the difficulties, possibilities, and future expectations of sustainable bio-refineries. We investigate the technological, financial, and ecological obstacles that must be handled to accomplish a genuinely maintainable biorefinery industry. We examine the capability of biorefineries in driving the change towards a bio-based economy and talk about the approaches and structures important to support their progression. By understanding and addressing these viewpoints, we can open the total capability of maintainable biorefineries and clear a path for a more maintainable and affluent future.

## 14.2 Concepts of Biorefineries

Biorefineries aim to change plant and waste materials into many useful things. This includes biofuels, chemicals from plants and animals, bioplastics, and other natural products. They use renewable things like leftover crops, special energy crops, algae, and waste products (Capodaglio 2023; Gullón et al. 2014) They make different products and don't hurt the environment much. Like oil refineries change oil into products, biorefineries use plant materials instead of oil. This lets them make many products without relying on fossils fuels as much (de Jong and Jungmeier 2015). It helps move to an economy based more on natural things. It also reduces the gases that cause climate change.

There are two main categories of biorefineries: biochemical and thermochemical. Biochemical biorefineries use biological processes like fermentation and enzyme reactions to change biomass into products like biofuels and unique chemicals. Thermochemical biorefineries, however, employ high heat and chemical reactions to change biomass into biofuels, bioenergy, and other worthwhile products (Edenhofer et al. 2011; Wang et al. 2021a; Zuiderveen et al. 2023).

Breaking down plants and other organic materials is important for biorefineries. They try to use every part of the raw materials. This includes sugars, fats, fibers, and other parts. Advanced methods help separate and change these parts (Calvo-Flores and Martin-Martinez 2022; Yogalakshmi et al. 2023). This lets biorefineries use resources well with little waste. They get maximum value from each raw material (Devi et al. 2022; Haq et al. 2021).

Sugarcane biorefineries follow the idea of using biomass resources fully, without wasting anything. They extract different products from biomass in steps. For example, when making bioethanol from sugarcane, the leftover plant material can then be used to create biogas or chemicals. This makes sure every part of the sugarcane plant is put to good use (Production et al. 2019; Shibukawa et al. 2023).

Sustainable biorefineries aim to consider environmental, social, and economic factors in their work. They try to lessen their climate impact and use less water and energy. They also try to manage waste responsibly (Silva 2023; Yang et al. 2023). Biorefineries can help local areas by making jobs, supporting rural economies, and encouraging innovation and teamwork for the bioeconomy.

Biorefineries offer an environmentally friendly way to use resources and make products. By using renewable plant and animal materials, called biomass, they can replace methods that rely on oil. This reduces dependency on fossil fuels and helps create a system where waste becomes a resource. Biomass is transformed efficiently into useful things. Biorefineries are important for moving to a greener future by powering our economy with renewable sources. Their approach supports a cycle where one process feeds into another.

## 14.3 Sensibility Framework for Biorefineries

Biorefineries play an important role in transitioning to an economy that uses nature and biology instead of oil and gas. They take plants and convert them into useful products while benefiting the environment. For biorefineries to succeed long term and be sustainable, they require a strategy considering environmental, social, and financial aspects (Barragán-Ocaña et al. 2023; Chojnacka 2023). This strategy is called the sensibility framework and its classes are illustrated in Fig. 14.1. It provides biorefineries with a structure to expand and operate in a manner respecting nature over the long run.

### 14.3.1 Environmental Sensibility

According to Bhatia et al. study's, the part of the system about caring for the environment focuses on reducing the effect biorefineries have on the environment and promoting good use of resources (Bhatia et al. 2023). This includes choosing feedstocks that can last, applying practices for energy and water management, and adopting effective waste management strategies. Biorefineries should choose feedstocks that protect the environment in their work. This involves selecting feedstocks that don't harm the environment much, such as leftover plants from farming, plants grown for energy, or waste materials (Aristizábal-Marulanda and Cardona Alzate 2019; Serna-Loaiza et al. 2019). By choosing feedstocks that don't compete with making food and have a low carbon footprint, biorefineries can be sure of a lasting supply chain (Haq et al. 2021).

Being environmentally friendly with energy and water is important. Biorefineries should use energy-smart ways of working to use less energy. Using renewable power sources like solar or biomass can help biorefineries harm the environment less (Pérez-Almada et al. 2023; Solarte-Toro and Cardona Alzate 2023). Also, good ways to reuse and recycle water can cut down how much fresh water biorefineries

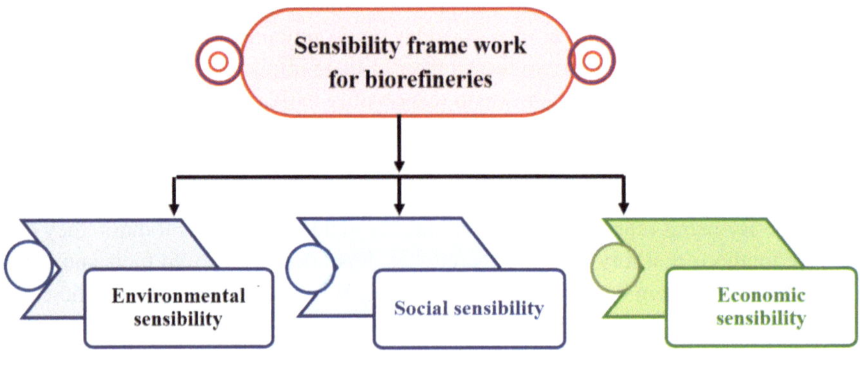


**Fig. 14.1** Sensibility framework for biorefineries

use and help the environment. Taking care of trash is important for caring for the environment. Biomass factories should use good waste plans to make less trash and to correctly take care of and throw out leftovers and scraps (Mozhiarasi and Natarajan 2022). Using trash as ingredients for other work or changing it into new products can help with a circular economy way of doing things. This cuts down on harm to the earth and uses resources in the best way possible (Vogiantzi and Tserpes 2023; Vural Gursel et al. 2022).

Doing a complete assessment of the effect products have on the environment throughout their entire lifetime (from start to finish) is important for biorefineries. This life cycle assessment or LCA measures the impacts that making and distributing a product has on the planet. Using an LCA helps biorefineries find ways to reduce harm and use resources well (Abu-Bakar et al. 2023). This ensures they work in the most eco-friendly way possible, from growing crops to getting the final product to customers.

### 14.3.2 Social Sensibility

The social considerations portion of the framework focuses on how biorefineries impact communities and society. This involves jobs, talking to interested parties, wellness and protection, and giving social advantages. Biorefineries ought to give top need to neighbourhood occupations and ability advancement, particularly in rustic territories. By encouraging nearby work openings, biorefineries can add to provincial turn of events and diminish joblessness rates (Sandaka and Kumar 2023). Preparing and limit building programs can assist with making a gifted workforce that is basic for the development and achievement of biorefineries (Vidaurre et al. 2020). It is important for biorefineries to involve local people, indigenous communities and other interested parties throughout the project. This will help make the biorefinery acceptable to those it affects and allow it to last. Biorefineries should clearly communicate with these groups, ask their opinions, and include them in decision making. This addresses concerns, builds trust, and considers different views (Bal et al. 2013).

The health and safety of workers and nearby communities must come first. Biorefineries should put strong safety rules in place, train people properly, and follow workplace health standards. This keeps the environment safe. Regular checks and risk analyses can find and deal with possible dangers, protecting employees and surrounding areas (Kaiko et al. 2013). Biorefineries should also strive to provide social benefits beyond job creation. This can include supporting local infrastructure development, contributing to educational programs, and promoting social well-being in the surrounding communities. By actively engaging with the community and addressing their needs, biorefineries can become positive agents of change and contribute to societal development.

### 14.3.3 Economic Sensibility

The economic sensibility aspect of the framework focuses on the economic viability and competitiveness of biorefineries. This includes market development, value chain integration, financial viability, and policy support. Biorefineries should identify and develop markets for their products to ensure economic viability (Stellingwerf et al. 2022). This involves conducting market research, collaborating with industry partners, and implementing targeted marketing strategies. By creating a demand for bio-based alternatives, biorefineries can contribute to the growth of the bioeconomy and enhance their market position. Value chain integration is crucial for optimizing the economic viability of biorefineries. Biorefineries should seek opportunities for collaboration and resource sharing with other industries and stakeholders. By leveraging synergies and optimizing the use of resources, biorefineries can enhance their operations and improve economic efficiency (Gajula and Reddy 2021).

Establishing sound financial models and securing access to financing mechanisms are essential for the economic sensibility of biorefineries. This includes exploring public-private partnerships, accessing grants or subsidies, and attracting private investment to support research, development, and commercialization efforts. Financial viability is crucial for the long-term sustainability and growth of biorefineries. Supportive policy frameworks play a significant role in the economic sensibility of biorefineries. Governments should develop policies that promote the growth of the biorefinery industry, such as feed-in tariffs, tax incentives, and renewable energy targets. These policies can provide a level playing field for biorefineries, incentivize sustainable practices, and encourage investment in the sector.

## 14.4 Key Factors Involved in Techno-economic Aspects of Biorefinery

### 14.4.1 Challenges of Biorefinery

Bio-refineries face many challenges that affect their performance and longevity (Calvo-Flores and Martin-Martinez 2022). These challenges include the availability and fluctuation of food sources, from different sources, posing a problem in maintaining a steady supply of biomass. Transport of biomass is complicated by the diversity and abundance of feedstock sources from different regions, complicated by seasonal changes in availability. Methods such as biomass storage period length and conversion of feed into stable intermediates can help to overcome these challenges (Mola-Yudego et al. 2024). Furthermore, competition for land resources and their impact on food supply require biorefineries to use surplus agricultural residues and sustainable sources of biomass to reduce conflicts in food production (Ekman et al. 2013; Herrmann et al. 2018). Economic viability and cost competitiveness are critical, thus requiring new budgets and continuous improvements in supply chain management and processing technologies. Regulatory compliance and standards add additional challenges, and biorefineries must adapt to changing environmental regulations to ensure their economic viability (Dragone et al. 2020; Solarte-Toro and Cardona Alzate 2023; Zetterholm et al. 2020). Public perception and social acceptance are critical, requiring community engagement, transparent communication, and proactive strategies to reduce environmental and social impacts (Bezama et al. 2019; Fytili and Zabaniotou 2017; Ganzevles et al. 2015; Peck et al. 2009). R&D gaps should be addressed through ongoing efforts to address knowledge gaps, meet technological challenges, and increase system productivity (Guddaraddi et al. 2023; Mossberg et al. 2021; Rivas-Cantu et al. 2013). There are many different aspects of funding and investment, and getting support requires strategies including clear risk disclosure, updated budgets, and focused planning (Maity 2015). To overcome these challenges, collaboration between stakeholders is essential to diversify the biorefining industry, scale up, improve public perception, address R&D gaps, and release finance of barriers to sustainable adoption of bio-based technologies, Numerous key factors involved in techno-economic aspects of biorefineries are illustrated in Fig. 14.2.

### 14.4.2 Availability of Feedstock

A variety of biomasses can be used as feedstock, including agricultural residues (such as sugarcane bagasse, corn stover, wheat straw, and rice husks), energy crops (such as switch grass and Miscanthus), forestry residues (such as sawdust and wood chips), and dedicated energy crops (such as particle conversion). tree crops) (Ahmed et al. 2022; Jayakumar et al. 2022; Ullah et al. 2015).

According to the latest estimates, about one billion tons of agricultural waste is generated worldwide each year, and agriculture is responsible for one-fifth of greenhouse gas emissions (Peng et al. 2023). Biomass production in the plant is about $200 \times 10^9$ tons/year tons per annum (Al-Battashi et al. 2019). Lignocellulose corresponds to plant cells containing cellulose, hemicellulose, and the aromatic polymer lignin (Gari et al. 2024). It also contains fewer organic ingredients, extracts, proteins, and pectin (Broxterman and Schols 2018; Hemati et al. 2022; Isikgor and Becer 2015). The proportion of cellulose in the total lignocellulosic materials is 30–50%; lignin, which contributes 20–40% and provides structural rigidity, 15–25%, and the remaining 15% is hemicellulose (Menon and Rao 2012; Vohra et al. 2014; Wang et al. 2021b). Cellulose is a lignocellulosic polymer with the highest degree of polymerization. Hemicellulose containing various monosaccharide subunits consisting of Xylan, xyloglucan, mannan, and glucomannans among others is less polymeric than cellulose and is less viscous and insoluble in pharmaceutical solutions (Hemati et al. 2022; Zoghlami and Paës 2019). Microalgae are a promising feedstock for biorefineries due to their rapid growth and ability to produce valuable chemicals. They are an environmentally friendly and renewable alternative because they can produce fats that can be converted into biodiesel (Chhandama et al. 2023; Khan et al. 2018; Kumar 2021). Several species of microalgae have been reported to be lipid-rich and can be cultivated on a large scale to produce biodiesel (Udayan et al. 2023). In addition, microalgae are high in protein and can be used as

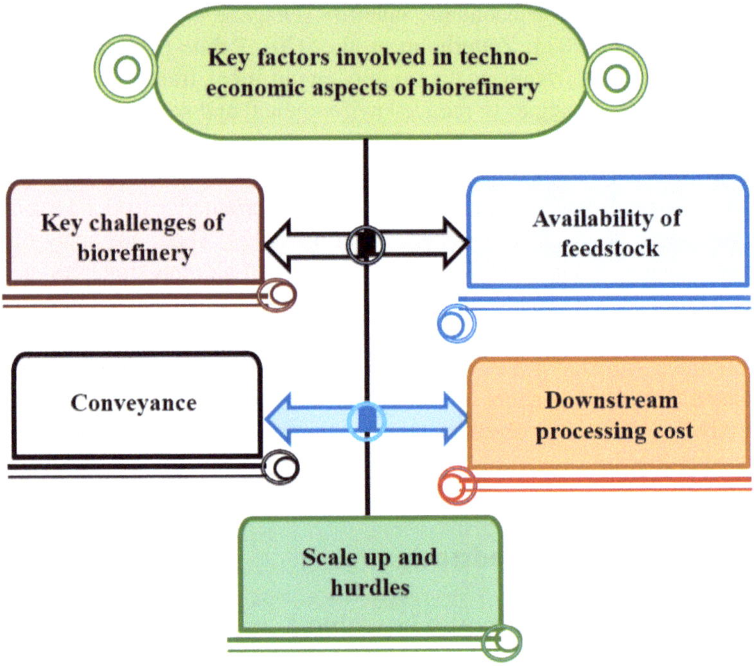


**Fig. 14.2** Numerous key factors involved in techno-economic aspects of biorefineries

animal feed, providing essential nutrients for livestock and aquatic life (Wang et al. 2021c). In addition, microalgae can produce carbohydrates, which can be converted into biofuels such as butanol and ethanol (Markou et al. 2012). In addition, microalgae can produce a variety of useful bio-chemicals for use in the pharmaceutical and cosmetic industries (Dolganyuk et al. 2020; Udayan et al. 2021).

Current research efforts are focused on the optimization of cultivation methods, enhancement of lipid and protein content, and economic strategies for converting microalgae biomass into value-added products Sunflower seed, rape seed, and soybeans production examples of oilseed that an excellent feedstock for biorefineries (Jayakumar et al. 2023b, c, 2024; Woodfield and Harwood 2016). Their vegetable oil can be extracted into biodiesel or processed for a variety of applications (Ennetta et al. 2022). Vegetable oil is obtained mechanically and can be directly converted into biodiesel (Rathore et al. 2016). It can also be processed to remove impurities and used in the food, cosmetic, and pharmaceutical industries. Oilseed production provides high-protein by-products for animal feed (Koki et al. 2023; Singh et al. 2022). Sustainable farming practices are important, and this oilseed is widely grown for its high oil content and adaptability. Using oilseeds as feed promotes renewable energy, reduces dependence on fossil fuels, and supports a sustainable biomass-based economy. Sugar cane and sugar beets are examples of sugar crops that produce sucrose which is valuable for biochemical reactions and the production of biofuels (Akram 2021). Sucrose can be fermented into ethanol, making it a sustainable fuel source. In addition, it can be turned into products used in the pharmaceutical and cosmetic industries (Kang and Lee 2015).

Sugarcane and sugar beet production requires a certain climate and requires sustainable methods. The use of sugar crops promotes renewable energy and supports biochemical performance. There are many challenges when it comes to managing municipal solid waste (MSW) in cities, but also opportunities for resource recovery and renewable energy. Anaerobic digestion is a practical way to treat MSW, especially organic waste, to produce biogas, a valuable source of energy. Biogas can be used as fuel for vehicles for energy production or converted into bio-methane (Uddin et al. 2021). These changes provide benefits on waste, reduce reliance on fossil fuels, and contribute to a more sustainable energy mix. Proper waste segregation, processing systems, and infrastructure are essential to manage MSW effectively and maximize biogas production.

### 14.4.3 Conveyance

Transportation is critical to the biorefining industry, transporting raw materials to refineries and affecting biofuel production costs (Hill 2007; Stephen et al. 2013). Costs include variable (fuel, labor, maintenance) and fixed costs (weight, insurance, and depreciation). Farm size and quantity incur costs through wages and administrative expenses. Based on biomass types, technologies, and storage methods, biomass supply includes methods of processing, collection, transportation, and storage (Nunes and Silva 2023; Roni et al. 2016). Various processes such as pelleting, dehydration, and shredding convert biomass into a useful feedstock. A reliable supply depends on feedstock management, availability, and efficient transportation.

Manufacturing, processing, and transportation account for approximately half of total project costs (Galbe and Wallberg 2019; Sikarwar et al. 2016). The increased distance between the bio-refinery and the collection site increases transportation costs, which affect the annual production costs. Effective logistics planning is essential to reduce these costs. Thickness, seasonality, abundance, and biomass distribution directly influence transportation costs. Lower-density biomass requires more material to be transported, increasing costs. Water content also affects transportation costs (Daugaard et al. 2015; Mirkouei et al. 2016; Roni 2014). Biomass transport can take various forms such as loose, chopped, baled, and pelleted, with processing requirements affecting biomass density. Carefully considered transportation and logistics systems ensure cost-effective transportation and successful biorefinery operations (Lin et al. 2016; Morgan et al. 2019).

### 14.4.4 Downstream Processing Cost

The costs incurred in a biorefinery in the process of converting organic components into a single homogeneous product are called downstream processing costs. This process includes crystallization, concentration, purification, and distillation s of the process (Blasi et al. 2023; Kiss et al. 2016; Takkellapati et al. 2018). System integration and adaptation techniques are used to ensure sustainability. The cost of downstream processing for a single type of feedstock is higher compared to using a combination of feedstock with mixed chemical compositions. Using multiple biomass sources makes the whole process self-sufficient low-cost cost and environmentally friendly. Studies show that only 20% or less of lignocellulosic biomass is digested in its original form, emphasizing the importance of combining different biomass to optimize downstream water resources (Ochieng et al. 2022; Oke et al. 2016; Purnama et al. 2024; Shukla et al. 2023).

Downstream processing in biorefineries yields a wide range of valuable products that are important for many industries. Formic acid, the essential element, passes through catalytic hydrothermal reactions to produce hydrogen and carbon dioxide, which benefits from exploring the production and storage of hydrogen. Vinegar or acetic acid is used in a wide variety of industries. Another important by-product is Levulinic acid, which is rapidly synthesized from lignocellulosic biomass and used in polymers, pharmaceuticals, and fuel additives. In addition, downstream processing yields sugar alcohols, which are valuable by-products widely used in the food industry. These factors emphasize the flexibility and importance of downstream operations to optimize the value of biomass resources and provide sustainable solutions (Capodaglio 2023; Ghoreishi et al. 2024; Oke et al. 2016; Prestigiacomo et al. 2021). All things considered; downstream production of lignocellulosic biomass enables the production of value-added chemicals at a reduced production cost. By utilizing process efficiencies and biomass sources, biorefineries can help increase productivity, reduce costs, and produce end products that are economically and environmentally sustainable permanently.

### 14.4.5 Scale Up and Hurdles

There are several challenges and barriers to scaling up a biorefinery from a laboratory or pilot scale operation to a commercial scale facility. These limitations can significantly affect biorefinery operational efficiency and economic returns. The main obstacle is the complexity of product scaling and measurement. Strategies that work well at small scales may not be easily reproduced or adapted at large scales. Factors such as heat and material transfer, reaction rates, and mixing become more difficult to control as the process size increases. Considerable technical expertise and careful consideration of design parameters are required to ensure efficient and effective scaling of biorefining processes (Pattnaik et al. 2024; Rosales-Calderon and Arantes 2019; Sanford et al. 2016). Another constraint is feedstock availability and stability. Scaling up a biorefinery requires a reliable and sustainable supply of biomass or other raw materials. However, feedstock availability can vary geographically and seasonally, creating challenges in maintaining a constant supply throughout the year. Issues such as logistics storage and feed quality must be addressed to ensure the continuity and reliability of the feed supply for the bio-refinery

(Calvo-Flores and Martin-Martinez 2022; Ray et al. 2017; Yue et al. 2014; Zetterholm et al. 2020).

Capital investment is a major barrier to large-scale biorefinery development. Building and operating a commercial facility requires a lot of capital. Finding finance for larger projects can be challenging, as investors often need adequate proof of technical feasibility and market demand. High capital costs associated with growth highlight the need for a business model further emphasizing difficulty and a clear path to profitability (Zetterholm et al. 2020). There are also technical barriers to scale-up of biorefineries. A technology developed on a small scale may not adapt quickly or cost-effectively to widespread use. Proper design, equipment selection, and integration of unit operations are critical to achieving efficient and reliable operation. Technology transfer and collaboration between research institutes, industry partners, and technical institutes will play an important role in overcoming these technical barriers (Balan 2014; Jack 2009; Krassnitzer et al. 2023).

The regulatory framework also creates challenges in the scale-up process. Biorefineries must comply with environmental, safety, and quality regulations. Going through the regulatory process, obtaining permits, and meeting compliance requirements can be time-consuming and expensive. Strong and supportive policies that encourage the development and operation of bio-refineries can help mitigate these challenges (Rashidi et al. 2022). In conclusion, some obstacles and barriers are overcome in scaling up biorefinery. Complex infrastructure, feedstock supply, capital investment, technical barriers, and regulatory compliance are among the major challenges to overcome. To overcome these barriers, researchers, and the concerted efforts of industry stakeholders, policymakers, and financial institutions are needed to realize the sustainable and economic potential of bio-refineries.

## 14.5 Commercialization and Market Outlook Issues

Commercialization is a crucial step in the development of any technology or product, including those in the biorefinery industry. It involves taking a concept or innovation from the research and development stage and transforming it into a marketable product or service (Calvo-Flores and Martin-Martinez 2022; Meramo et al. 2022). However, commercializing biorefinery technologies can present several challenges, including market viability, scaling up production, regulatory hurdles, and competition. In this article, we will explore the key issues related to commercialization and market outlook in the biorefinery sector. Commercialization and market outlook issues are illustrated in Fig. 14.3.

### 14.5.1 Market Viability

One of the primary challenges in commercializing biorefinery technologies is ensuring market viability. Biorefineries produce a wide range of bio-based products, such as biofuels, biochemicals, and biomaterials. However, the market demand for these

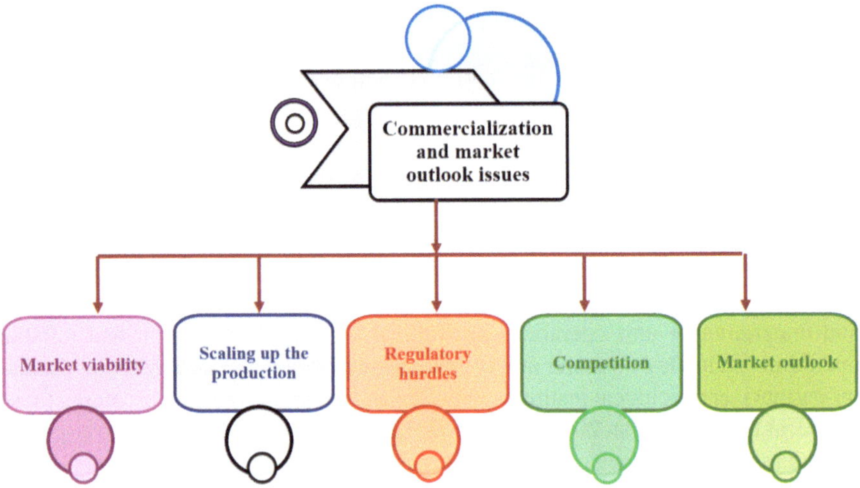


**Fig. 14.3** Commercialization and market outlook issues in biorefinery sector

products can vary, and it is essential to identify viable market segments and assess their potential size and growth. Market research and analysis play a critical role in understanding customer needs, industry trends, and competitive landscapes. This information helps biorefineries identify market gaps, develop targeted marketing strategies, and position their products effectively. They need to demonstrate the value proposition of their bio-based products, such as their environmental benefits, cost-effectiveness, and performance compared to conventional alternatives. Furthermore, developing strategic partnerships and collaborations with existing industry players and supply chain stakeholders can help biorefineries access existing markets and distribution networks, improving their market viability (Aristizábal-Marulanda and Cardona Alzate 2019; Schmidt et al. 2019).

### 14.5.2 Scaling Up the Production

Another significant challenge in commercializing biorefinery technologies is scaling up production to meet market demand. Many biorefinery technologies start at the laboratory or pilot scale, which may not be sufficient for large-scale commercial production. The transition from small-scale production to full-scale production involves addressing technical, logistical, and financial considerations. Technical challenges include optimizing and integrating various unit operations, ensuring consistent product quality, and increasing overall production efficiency. This often requires process optimization, equipment scale-up, and fine-tuning of operating parameters.

Logistical challenges involve securing a reliable and sustainable feedstock supply, developing efficient transportation and logistics systems, and establishing robust supply chain management practices. Ensuring a steady and cost-effective supply of feedstock is crucial for the economic viability of biorefineries. Financial considerations include securing sufficient funding and investment for scaling up production. This can be a significant hurdle, as scaling up requires substantial capital investment in equipment, infrastructure, and operational costs. Biorefineries need to attract investors, secure project financing, or explore public-private partnerships to overcome these financial barriers.

### 14.5.3 Regulatory Hurdles

Regulatory frameworks and policy environments can significantly impact the commercialization of biorefinery technologies. Biorefineries need to navigate complex regulatory requirements, obtain necessary permits and certifications, and comply with environmental and safety regulations. Regulatory hurdles can include obtaining permits for feedstock sourcing, waste management, emissions control, and product certifications. Compliance with sustainability certifications, such as those related to greenhouse gas emissions, land use, and biodiversity, may also be necessary to access certain markets or secure funding from investors. Additionally, biorefineries need to stay abreast of evolving regulations and policy frameworks related to bio-based products, renewable energy, and sustainability. Engaging with policymakers, industry associations, and regulatory bodies is essential to influence policy development and ensure a favourable regulatory environment for biorefinery commercialization (Jayakumar et al. 2023c).

### 14.5.4 Competition

The biorefinery sector is becoming increasingly competitive, with numerous players entering the market. Competition can arise from both conventional industries and other biorefineries. Biorefineries need to differentiate themselves from competitors by offering unique value propositions, such as novel products, improved cost-effectiveness, or superior environmental performance. Developing and protecting intellectual property (IP) is crucial for gaining a competitive edge. Biorefineries should invest in research and development to generate innovative technologies and products and secure IP rights through patents or other forms of protection. This can provide them with a competitive advantage and create barriers to entry for potential competitors. Collaborations and partnerships with industry players, research institutions, and technology providers can also help biorefineries access complementary expertise, resources, and market opportunities, enhancing their competitive position (Calvo-Flores and Martin-Martinez 2022).

### 14.5.5 Market Outlook

Despite the challenges, the market outlook for biorefineries is generally positive. The increasing focus on sustainability, climate change mitigation, and the transition to a bio-based economy creates significant opportunities for biorefinery technologies. The growing demand for biofuels, biochemicals, and biomaterials presents a vast potential market. Government policies and incentives, such as renewable energy targets, carbon pricing, and bio-based product mandates, can drive market growth and create a favourable business environment for biorefineries. Consumers are also increasingly aware of environmental issues and are willing to choose bio-based products that offer environmental benefits (Jayakumar et al. 2023c). Technological advancements and ongoing research and development efforts continue to improve the efficiency and cost-effectiveness of biorefinery processes. This, in turn, enhances the competitiveness and commercial viability of biorefineries (Bennich and Belyazid 2017; Calvo-Flores and Martin-Martinez 2022).

## 14.6 Challenges and Prospects

Global consumption of fossil fuels rises as a result of an expanding population and expanding economies. Although biorefineries have been taking shape for the last decade, they still have a long way to go before they are established. Researchers, academicians, and world leaders are concerned that the environment is being deteriorated. In today's world, sustainable solutions are only possible by introducing alternative fuels, particularly the production of biofuels through biorefineries, in place of fossil fuels (Gundupalli et al. 2022).

It has been determined that biofuel generated by biorefineries is a potentially useful energy source that is economically, ecologically, and technologically feasible. Because of the widespread use of biorefineries in several nations, producing biofuel from them is now the most advantageous alternative to conventional fuels. However, the high cost, raw material requirements, marketing, and priority given to biofuel production through biorefinery technology indicate that it will be challenging, at least for a few years, to significantly shift the world's energy use from fossil fuels to biofuels (Hamda et al. 2024; Mohanasundaram et al. 2024).

In contrast to biogas, bioethanol, biooil, and biodiesel, the generation of biofuel from biorefineries is thought to be the finest fossil fuel substitute; nevertheless, there are a few obstacles and restrictions that must be overcome before the method can be commercialized. In light of this, additional research is particularly necessary to develop current technologies and other areas in order to develop biofuels through biorefineries as a sustainable alternative to fossil fuels and address global challenges (Jayakumar et al. 2023a).

The integration of various biorefinery platforms, with the waste from these platforms being utilized to produce biofuels, is one potential future approach for biorefineries. If there is excess biofuel, it can be sent back into the grid or used for on-site energy needs. The issue with the mechanisms in place for producing biofuels now is that they are solely seen as a way to reduce organic waste and provide electricity. However, given the volatility of energy costs right now, it would be wise to take advantage of other revenue streams produced by waste lignocellulosic feedstocks (Sethumadhavan et al. 2024).

Right now, additional attention is needed to keep the nutrient levels more constant. Programs for biofuel certification and application may also persuade customers to purchase the product for use as fertilizer. Getting the useful compounds, amino acids, etc. back in the biorefinery before sending the feedstock to the biofuel production process is one of the other goals. Prior to the food processing industry's waste being treated through a variety of biofuel generation techniques, nutrients and possibly valuable compounds can be recovered from the food production and processing industries. Additionally, a thorough techno-economic study of biorefineries that takes local circumstances into account is required.

## 14.7 Summary and Conclusions

An examination of the challenges, opportunities, and future directions of sustainable biorefineries highlights the vital role they play in guiding us towards a greener and more economically viable future. Despite contending with obstacles such as technical barriers, unforeseen economic barriers, and regulatory conditions, sustainable biorefineries offer promising strategies to mitigate climate change, reduce dependency targeting fossil fuels and promoting circular economic models. The opportunities presented by advances in biotechnology, the availability of renewable feedstocks, and new processes are immense, leading to the development of efficient and cost-effective biorefinery systems which includes the integration of biorefineries into the existing industrial infrastructure and contributes to their long-term sustainability and competitiveness in the market. Looking ahead, continued research, investment, and collaboration between stakeholders is essential to overcome existing challenges and unlock the potential of sustainable biorefineries the whole of the mouth. By leveraging emerging technologies, optimizing resources, and addressing environmental concerns, biorefineries have the potential to transform many industries from multiple sectors, ranging from energy and chemicals to agriculture and waste management. Ultimately, the success of sustainable biorefineries depends on their ability to balance economic benefits with environmental stewardship, paving the way for a greener and more resilient future.

## References

Abu-Bakar NA, Roslan AM, Hassan MA, Rahman MHA, Ibrahim KN, Abd Rahman MD, Mohamad R (2023) Environmental impact assessment of rice mill waste valorisation to glucose through biorefinery platform. Sci Rep 13:1–12. https://doi.org/10.1038/s41598-023-28487-2

Ahmed E, Zeitoun A, Hamad G, Zeitoun MAM, Taha A, Korma SA, Esatbeyoglu T (2022) Lignocellulosic biomasses from agricultural wastes improved the quality and physicochemical properties of frying oils. Foods 11:1–16. https://doi.org/10.3390/foods11193149

Akram A (2021) Getting fuel from sucrose (Table sugar). J Sustain Bioenergy Syst 11:272–284. https://doi.org/10.4236/jsbs.2021.114017

Al-Battashi HS, Annamalai N, Sivakumar N, Al-Bahry S, Tripathi BN, Nguyen QD, Gupta VK (2019) Lignocellulosic biomass (LCB): a potential alternative biorefinery feedstock for polyhydroxyalkanoates production. Rev Environ Sci Biotechnol 18:183–205. https://doi.org/10.1007/s11157-018-09488-4

Aristizábal-Marulanda V, Cardona Alzate CA (2019) Methods for designing and assessing biorefineries: review. Biofuels Bioprod Biorefin 13:789–808. https://doi.org/10.1002/bbb.1961

Bal M, Bryde D, Fearon D, Ochieng E (2013) Stakeholder engagement: achieving sustainability in the construction sector. Sustainability (Switzerland) 5:695–710. https://doi.org/10.3390/su5020695

Balan V (2014) Current challenges in commercially producing biofuels from Lignocellulosic biomass. ISRN Biotechnol 2014:1–31. https://doi.org/10.1155/2014/463074

Barragán-Ocaña A, Merritt H, Sánchez-Estrada OE, Méndez-Becerril JL, del Pilar Longar-Blanco M (2023) Biorefinery and sustainability for the production of biofuels and value-added products: a trends analysis based on network and patent analysis. PLoS One 18:1–18. https://doi.org/10.1371/journal.pone.0279659

Ben-Iwo J, Manovic V, Longhurst P (2016) Biomass resources and biofuels potential for the production of transportation fuels in Nigeria. Renew Sust Energ Rev 63:172–192. https://doi.org/10.1016/j.rser.2016.05.050

Bennich T, Belyazid S (2017) The route to sustainability-prospects and challenges of the bio-based economy. Sustainability (Switzerland). https://doi.org/10.3390/su9060887

Benti NE, Gurmesa GS, Argaw T, Aneseyee AB, Gunta S, Kassahun GB, Aga GS, Asfaw AA (2021) The current status, challenges and prospects of using biomass energy in Ethiopia. Biotechnol Biofuels 14:1–24. https://doi.org/10.1186/s13068-021-02060-3

Bezama A, Ingrao C, O'Keeffe S, Thrän D (2019) Resources, collaborators, and neighbors: the three-pronged challenge in the implementation of bioeconomy regions. Sustainability (Switzerland) 11. https://doi.org/10.3390/su11247235

Bhatia L, Jha H, Sarkar T, Sarangi PK (2023) Food waste utilization for reducing carbon footprints towards sustainable and cleaner environment: a review. Int J Environ Res Public Health 20. https://doi.org/10.3390/ijerph20032318

Blasi A, Verardi A, Lopresto CG, Siciliano S, Sangiorgio P (2023) Lignocellulosic agricultural waste valorization to obtain valuable products: an overview. Recycling 8:1–46. https://doi.org/10.3390/recycling8040061

Broxterman SE, Schols HA (2018) Interactions between pectin and cellulose in primary plant cell walls. Carbohydr Polym 192:263–272. https://doi.org/10.1016/j.carbpol.2018.03.070

Calvo-Flores FG, Martin-Martinez FJ (2022) Biorefineries: achievements and challenges for a bio-based economy. Front Chem. https://doi.org/10.3389/fchem.2022.973417

Capodaglio AG (2023) Biorefinery of sewage sludge: overview of possible value-added products and applicable process technologies. Water (Switzerland) 15. https://doi.org/10.3390/w15061195

Chen HG, Zhang YHP (2015) New biorefineries and sustainable agriculture: increased food, biofuels, and ecosystem security. Renew Sust Energ Rev 47:117–132. https://doi.org/10.1016/j.rser.2015.02.048

Chhandama MVL, Ruatpuia JVL, Ao S, Chetia AC, Satyan KB, Rokhum SL (2023) Microalgae as a sustainable feedstock for biodiesel and other production industries: prospects and challenges. Energy Nexus 12:100255. https://doi.org/10.1016/j.nexus.2023.100255

Chojnacka K (2023) Valorization of biorefinery residues for sustainable fertilizer production: a comprehensive review. Biomass Convers Biorefinery 13:14359–14388. https://doi.org/10.1007/s13399-023-04639-2

Daugaard T, Mutti LA, Wright MM, Brown RC, Componation P (2015) Learning rates and their impacts on the optimal capacities and production costs of biorefineries. Biofuels Bioprod Biorefin 9:82–94. https://doi.org/10.1002/bbb.1513

de Jong E, Jungmeier G (2015) Biorefinery concepts in comparison to petrochemical refineries. In: Industrial biorefineries and white biotechnology. https://doi.org/10.1016/B978-0-444-63453-5.00001-X

Devi A, Bajar S, Kour H, Kothari R, Pant D, Singh A (2022) Lignocellulosic biomass valorization for bioethanol Production: a circular bioeconomy approach. Bioenergy Res 15:1820–1841. https://doi.org/10.1007/s12155-022-10401-9

Ding Z, Hamann KT, Grundmann P (2024) Enhancing circular bioeconomy in Europe: sustainable valorization of residual grassland biomass for emerging bio-based value chains. Sustain Prod Consum 45:265–280. https://doi.org/10.1016/j.spc.2024.01.008

Dolganyuk V, Belova D, Babich O, Prosekov A, Ivanova S, Katserov D, Patyukov N, Sukhikh S (2020) Microalgae: a promising source of valuable bioproducts. Biomolecules 10:1–24. https://doi.org/10.3390/biom10081153

Dragone G, Kerssemakers AAJ, Driessen JLSP, Yamakawa CK, Brumano LP, Mussatto SI (2020) Innovation and strategic orientations for the development of advanced biorefineries. Bioresour Technol. https://doi.org/10.1016/j.biortech.2020.122847

Edenhofer O, Madruga RP, Sokona Y, Seyboth K, Matschoss P, Kadner S, Zwickel T, Eickemeier P, Hansen G, Schlömer S, von Stechow C (2011) Renewable energy sources and climate change mitigation: special report of the intergovernmental panel on climate change. In: Renewable energy sources and climate change mitigation: special report of the intergovernmental panel on climate change. https://doi.org/10.1017/CBO9781139151153

Ekman A, Wallberg O, Joelsson E, Börjesson P (2013) Possibilities for sustainable biorefineries based on agricultural residues – a case study of potential straw-based ethanol production in Sweden. Appl Energy 102:299–308. https://doi.org/10.1016/j.apenergy.2012.07.016

Ennetta R, Soyhan HS, Koyunoğlu C, Demir VG (2022) Current technologies and future trends for biodiesel Production: a review. Arab J Sci Eng 47:15133–15151. https://doi.org/10.1007/s13369-022-07121-9

Fytili D, Zabaniotou A (2017) Social acceptance of bioenergy in the context of climate change and sustainability – a review. Curr Opin Green Sustain Chem. https://doi.org/10.1016/j.cogsc.2017.07.006

Gajula S, Reddy CRK (2021) More sustainable biomass production and biorefining to boost the bioeconomy. Biofuels Bioprod Biorefin 15:1221–1232. https://doi.org/10.1002/bbb.2227

Galbe M, Wallberg O (2019) Pretreatment for biorefineries: a review of common methods for efficient utilisation of lignocellulosic materials. Biotechnol Biofuels. https://doi.org/10.1186/s13068-019-1634-1

Ganzevles J, Asveld L, Osseweijer P (2015) Extending bioenergy towards smart biomass use Issues of social acceptance at Park Cuijk, The Netherlands. Energy Sustain Soc 5. https://doi.org/10.1186/s13705-015-0053-9

Gari MT, Asfaw BT, Abo LD, Jayakumar M, Kefalew G (2024) Effective utilization of agricultural cereal grains in value-added products: a global perspective. In: Value added products from food waste. Springer Nature Switzerland, Cham, pp 41–58. https://doi.org/10.1007/978-3-031-48143-7_3

Gawel E, Pannicke N, Hagemann N (2019) A path transition towards a bioeconomy-The crucial role of sustainability. Sustainability (Switzerland) 11:1–23. https://doi.org/10.3390/su11113005

Ghoreishi S, Løhre C, Hermundsgård DH, Molnes JL, Tanase-Opedal M, Brusletto R, Barth T (2024) Identification and quantification of valuable platform chemicals in aqueous product streams from a preliminary study of a large pilot-scale steam explosion of woody biomass using quantitative nuclear magnetic resonance spectroscopy. Biomass Convers Biorefinery 14:3331–3349. https://doi.org/10.1007/s13399-022-02712-w

Guddaraddi A, Singh A, Amrutha G, Saikanth DRK, Kurmi R, Singh G, Chowdhury M, Singh BV (2023) Sustainable biofuel Production from agricultural residues an eco-friendly approach: a review. Int J Environ Clim Change 13:2905–2914. https://doi.org/10.9734/ijecc/2023/v13i102956

Gullón P, Conde E, Moure A, Domínguez H, Parajó JC (2014) Selected process alternatives for biomass refining: a review. Open Agric J 4:135–144. https://doi.org/10.2174/1874331501004010135

Gundupalli MP, Gundupalli SP, Somavarapu Thomas AS, Jayakumar M, Bhattacharyya D, Gurunathan B (2022) Hydrothermal liquefaction of lignocellulosic biomass for production of biooil and by-products: current state of the art and challenges. In: Biofuels and bioenergy: opportunities and challenges. Elsevier, pp 61–84. https://doi.org/10.1016/B978-0-323-85269-2.00001-0

Hamda AS, Muleta MD, Jayakumar M, Periyasamy S, Gurunathan B (2024) Valorization of fruit processing industry waste into value-added chemicals. In: Value added products from food waste. Springer Nature Switzerland, Cham, pp 107–126. https://doi.org/10.1007/978-3-031-48143-7_6

Haq I, Qaisar K, Nawaz A, Akram F, Mukhtar H, Zohu X, Xu Y, Mumtaz MW, Rashid U, Azlina W, Ab W, Ghani K, Shean T, Choong Y (2021) Advances in valorization of Lignocellulosic biomass towards energy generation. Catalysts 11:1–25

Hariram NP, Mekha KB, Suganthan V, Sudhakar K (2023) Sustainalism: an integrated socio-economic-environmental model to address sustainable development and sustainability. Sustainability (Switzerland) 15. https://doi.org/10.3390/su151310682

Hemati A, Nazari M, Asgari Lajayer B, Smith DL, Astatkie T (2022) Lignocellulosics in plant cell wall and their potential biological degradation. Folia Microbiol (Praha) 67:671–681. https://doi.org/10.1007/s12223-022-00974-5

Herrmann R, Jumbe C, Bruentrup M, Osabuohien E (2018) Competition between biofuel feedstock and food production: empirical evidence from sugarcane outgrower settings in Malawi. Biomass Bioenergy 114:100–111. https://doi.org/10.1016/j.biombioe.2017.09.002

Hill J (2007) Environmental costs and benefits of transportation biofuel production from food- and lignocellulose-based energy crops. A review. Agron Sustain Dev. https://doi.org/10.1051/agro:2007006

Ingrao C, Bacenetti J, Bezama A, Blok V, Goglio P, Koukios EG, Lindner M, Nemecek T, Siracusa V, Zabaniotou A, Huisingh D (2018) The potential roles of bio-economy in the transition to equitable, sustainable, post fossil-carbon societies: findings from this virtual special issue. J Clean Prod 204:471–488. https://doi.org/10.1016/j.jclepro.2018.09.068

Isikgor FH, Becer CR (2015) Lignocellulosic biomass: a sustainable platform for the production of bio-based chemicals and polymers. Polym Chem 6:4497–4559. https://doi.org/10.1039/c5py00263j

Jack MW (2009) Scaling laws and technology development strategies for biorefineries and bioenergy plants. Bioresour Technol 100:6324–6330. https://doi.org/10.1016/j.biortech.2009.06.109

Jayakumar M, Kuppusamy Vaithilingam S, Karmegam N, Jabesa A (2022) Bioethanol Production from green biomass resources: emerging technologies. In: Encyclopedia of green materials. Springer Nature Singapore, pp 1–12. https://doi.org/10.1007/978-981-16-4921-9_110-1

Jayakumar M, Bizuneh Gebeyehu K, Deso Abo L, Wondimu Tadesse A, Vivekanandan B, Prabhu Sundramurthy V, Bacha W, Ashokkumar V, Baskar G (2023a) A comprehensive outlook on topical processing methods for biofuel production and its thermal applications: current advances, sustainability and challenges. Fuel. https://doi.org/10.1016/j.fuel.2023.128690

Jayakumar M, Gindaba GT, Gebeyehu KB, Periyasamy S, Jabesa A, Baskar G, John BI, Pugazhendhi A (2023b) Bioethanol production from agricultural residues as lignocellulosic biomass feedstock's waste valorization approach: a comprehensive review. Sci Total Environ. https://doi.org/10.1016/j.scitotenv.2023.163158

Jayakumar M, Hamda AS, Abo LD, Daba BJ, Venkatesa Prabhu S, Rangaraju M, Jabesa A, Periyasamy S, Suresh S, Baskar G (2023c) Comprehensive review on lignocellulosic biomass derived biochar production, characterization, utilization and applications. Chemosphere 345. https://doi.org/10.1016/j.chemosphere.2023.140515

Jayakumar M, Thiyagar T, Abo LD, Arumugasamy SK, Jabesa A (2024) Paddy straw as a biomass feedstock for the manufacturing of bioethanol using acid hydrolysis and parametric optimization through response surface methodology and an artificial neural network. Biomass Convers Biorefinery. https://doi.org/10.1007/s13399-024-05371-1

Kaiko M, Mills T, Adinyira E, Obeng MD, Mubita K, Mutambo J, Kahale C (2013) An Examination of occupational health and safety management practices in selected construction sites of Lusaka City. In: An examination of occupational health and safety management practices in selected construction sites of Lusaka City, vol 23, pp 32–42

Kang A, Lee TS (2015) Converting sugars to biofuels: ethanol and beyond. Bioengineering 2:184–203. https://doi.org/10.3390/bioengineering2040184

Khan MI, Shin JH, Kim JD (2018) The promising future of microalgae: current status, challenges, and optimization of a sustainable and renewable industry for biofuels, feed, and other products. Microb Cell Factories 17:1–21. https://doi.org/10.1186/s12934-018-0879-x

Kiss AA, Lange JP, Schuur B, Brilman DWF, van der Ham AGJ, Kersten SRA (2016) Separation technology–making a difference in biorefineries. Biomass Bioenergy 95:296–309. https://doi.org/10.1016/j.biombioe.2016.05.021

Koki B, Cheli F, Pinotti L, Ottoboni M (2023) Cold-pressed oilseed cakes as alternative and sustainable feed ingredients: a review. Foods 12:432

Krassnitzer P, Wenger J, Stern T (2023) Techno-economic learning in biorefinery research; a meta-level perspective of three exemplary cases. Biofuels Bioprod Biorefin 17:1736–1756. https://doi.org/10.1002/bbb.2543

Kumar A (2021) Current and future perspective of microalgae for simultaneous wastewater treatment and feedstock for biofuels Production. Chem Afr 4:249–275. https://doi.org/10.1007/s42250-020-00221-9

Lin T, Rodríguez LF, Davis S, Khanna M, Shastri Y, Grift T, Long S, Ting KC (2016) Biomass feedstock preprocessing and long-distance transportation logistics. GCB Bioenergy 8:160–170. https://doi.org/10.1111/gcbb.12241

Maity SK (2015) Opportunities, recent trends and challenges of integrated biorefinery: Part I. Renew Sust Energ Rev. https://doi.org/10.1016/j.rser.2014.11.092

Markou G, Angelidaki I, Georgakakis D (2012) Microalgal carbohydrates: an overview of the factors influencing carbohydrates production, and of main bioconversion technologies for production of biofuels. Appl Microbiol Biotechnol 96:631–645. https://doi.org/10.1007/s00253-012-4398-0

Menon V, Rao M (2012) Trends in bioconversion of lignocellulose: biofuels, platform chemicals & biorefinery concept. Prog Energy Combust Sci 38:522–550. https://doi.org/10.1016/j.pecs.2012.02.002

Meramo S, Fantke P, Sukumara S (2022) Advances and opportunities in integrating economic and environmental performance of renewable products. Biotechnol Biofuels Bioprod 15:1–18. https://doi.org/10.1186/s13068-022-02239-2

Mirkouei A, Mirzaie P, Haapala KR, Sessions J, Murthy GS (2016) Reducing the cost and environmental impact of integrated fixed and mobile bio-oil refinery supply chains. J Clean Prod 113:495–507. https://doi.org/10.1016/j.jclepro.2015.11.023

Mohammad REA, Abdullahi SS, Ad Muhammed HA, Musa H, Habibu S, Jagaba AH, Birniwa AH (2024) Recent technical and non-technical biorefinery development barriers and potential solutions for a sustainable environment: a mini review. Case Stud Chem Environ Eng 9:100586. https://doi.org/10.1016/j.cscee.2023.100586

Mohanasundaram S, Varadharajan V, Selvaraju M, Manikandan S, Ramasamy S, Jayakumar M, Prabhu Sundramurthy V, Baskar G, Pugazhendhi A (2024) Green ammonia as peerless entity for realm of clean-energy carrier toward zero carbon emission: purviews, neoteric tendencies, potentialities and downsides. Fuel 365. https://doi.org/10.1016/j.fuel.2024.131118

Mola-Yudego B, Dimitriou I, Gagnon B, Schweinle J, Kulišić B (2024) Priorities for the sustainability criteria of biomass supply chains for energy. J Clean Prod 434. https://doi.org/10.1016/j.jclepro.2023.140075

Morgan TJ, Youkhana A, Turn SQ, Ogoshi R, Garcia-Pérez M (2019) Review of biomass resources and conversion Technologies for Alternative jet Fuel Production in Hawai'i and tropical regions. Energy Fuels. https://doi.org/10.1021/acs.energyfuels.8b03001

Mossberg J, Söderholm P, Frishammar J (2021) Challenges of sustainable industrial transformation: Swedish biorefinery development and incumbents in the emerging biofuels industry. Biofuels Bioprod Biorefin 15:1264–1280. https://doi.org/10.1002/bbb.2249

Mozhiarasi V, Natarajan TS (2022) Slaughterhouse and poultry wastes: management practices, feedstocks for renewable energy production, and recovery of value added products. Biomass Convers Biorefinery. https://doi.org/10.1007/s13399-022-02352-0

Nunes LJR, Silva S (2023) Optimization of the residual biomass supply chain: process characterization and cost analysis. Logistics 7. https://doi.org/10.3390/logistics7030048

Ochieng R, Gebremedhin A, Sarker S (2022) Integration of waste to bioenergy conversion systems: a critical review. Energies (Basel). https://doi.org/10.3390/en15072697

Oke MA, Annuar MSM, Simarani K (2016) Mixed feedstock approach to Lignocellulosic ethanol production—prospects and limitations. Bioenergy Res 9:1189–1203. https://doi.org/10.1007/s12155-016-9765-8

Pattnaik F, Patra BR, Nanda S, Mohanty MK, Dalai AK, Rawat J (2024) Drivers and barriers in the Production and utilization of second-generation bioethanol in India. Recycling 9:19. https://doi.org/10.3390/recycling9010019

Peck P, Bennett SJ, Bissett-Amess R, Lenhart J, Mozaffarian H (2009) Examining understanding, acceptance, and support for the biorefinery concept among EU policy-makers. Biofuels Bioprod Biorefin 3:361–383. https://doi.org/10.1002/bbb.154

Peng X, Jiang Y, Chen Z, Osman AI, Farghali M, Rooney DW, Yap PS (2023) Recycling municipal, agricultural and industrial waste into energy, fertilizers, food and construction materials, and economic feasibility: a review. Environ Chem Lett. https://doi.org/10.1007/s10311-022-01551-5

Pérez-Almada D, Galán-Martín Á, del Contreras MM, Castro E (2023) Integrated techno-economic and environmental assessment of biorefineries: review and future research directions. Sustain Energy Fuels 7:4031–4050. https://doi.org/10.1039/d3se00405h

Prestigiacomo C, Proietto F, Laudicina VA, Siragusa A, Scialdone O, Galia A (2021) Catalytic hydrothermal liquefaction of municipal sludge assisted by formic acid for the production of next-generation fuels. Energy 232. https://doi.org/10.1016/j.energy.2021.121086

Production B, Kotarska K, Dziemianowicz W (2019) Study on the sequential combination of bioethanol and biogas production from corn straw. Molecules 24:6–11

Purnama I, Trisunaryanti W, Wijaya K, Mutamima A, Oh WC, Boukherroub R, Aziz M (2024) Multi-pathways for sustainable fuel Production from biomass using zirconium-based catalysts: a comprehensive review. Energ Technol. https://doi.org/10.1002/ente.202300901

Rashidi NA, Chai YH, Yusup S (2022) Biomass energy in Malaysia: current scenario, policies, and implementation challenges. Bioenergy Res. https://doi.org/10.1007/s12155-022-10392-7

Rathore V, Newalkar BL, Badoni RP (2016) Energy for sustainable development processing of vegetable oil for biofuel production through conventional and non-conventional routes. Energy Sustain Dev 31:24–49. https://doi.org/10.1016/j.esd.2015.11.003

Ray AE, Li C, Thompson VS, Daubaras DL, Nagle NJ, Hartley DS (2017) Biomass blending and densification: impacts on feedstock supply and biochemical conversion performance. In: Biomass volume estimation and valorization for energy. InTech. https://doi.org/10.5772/67207

Rivas-Cantu RC, Jones KD, Mills PL (2013) A citrus waste-based biorefinery as a source of renewable energy: technical advances and analysis of engineering challenges. In: Waste management and research, pp 413–420. https://doi.org/10.1177/0734242X13479432

Roni M (2014) Feedstock supply system design and analysis "The feedstock logistics design case for multiple conversion pathways". https://doi.org/10.13140/RG.2.1.4823.4961

Roni MS, Cafferty KG, Hess JR, Jacobson JJ, Kenney KL, Searcy E, Tumuluru JS (2016) Lignocellulosic crop supply chains (eg, Miscanthus, switchgrass, reed canary grass, rye, giant reed, etc.). In: Biomass supply chains for bioenergy and biorefining. Elsevier Inc, pp 271–291. https://doi.org/10.1016/B978-1-78242-366-9.00012-5

Rosales-Calderon O, Arantes V (2019) A review on commercial-scale high-value products that can be produced alongside cellulosic ethanol. Biotechnol Biofuels. https://doi.org/10.1186/s13068-019-1529-1

Sandaka BP, Kumar J (2023) Alternative vehicular fuels for environmental decarbonization: a critical review of challenges in using electricity, hydrogen, and biofuels as a sustainable vehicular fuel. Chem Eng J Adv 14:100442. https://doi.org/10.1016/j.ceja.2022.100442

Sanford K, Chotani G, Danielson N, Zahn JA (2016) Scaling up of renewable chemicals. Curr Opin Biotechnol. https://doi.org/10.1016/j.copbio.2016.01.008

Schmidt LM, Andersen LF, Dieckmann C, Lamp A, Kaltschmitt M (2019) The biorefinery approach. In: Energy from organic materials (biomass). Springer, New York, pp 1383–1412. https://doi.org/10.1007/978-1-4939-7813-7_1050

Serna-Loaiza S, Miltner A, Miltner M, Friedl A (2019) A review on the feedstocks for the sustainable production of bioactive compounds in biorefineries. Sustainability (Switzerland) 11. https://doi.org/10.3390/su11236765

Sethumadhavan A, Shah SR, Jayakumar M, Nirmala G, Rangaraju M (2024) Food process industry waste biomass as a promising alternative for green energy Production. In: Value added products from food waste. Springer Nature Switzerland, Cham, pp 275–290. https://doi.org/10.1007/978-3-031-48143-7_14

Shibukawa VP, Ramos L, Cruz-Santos MM, Prado CA, Jofre FM, de Arruda GL, da Silva SS, Mussatto SI, dos Santos JC (2023) Impact of product diversification on the economic sustainability of second-generation ethanol Biorefineries: a critical review. Energies (Basel) 16. https://doi.org/10.3390/en16176384

Shukla A, Kumar D, Girdhar M, Kumar A, Goyal A, Malik T, Mohan A (2023) Strategies of pretreatment of feedstocks for optimized bioethanol production: distinct and integrated approaches. Biotechnol Biofuels Bioprod. https://doi.org/10.1186/s13068-023-02295-2

Sikarwar VS, Zhao M, Clough P, Yao J, Zhong X, Memon MZ, Shah N, Anthony EJ, Fennell PS (2016) An overview of advances in biomass gasification. Energy Environ Sci. https://doi.org/10.1039/c6ee00935b

Silva JA (2023) Water supply and wastewater treatment and reuse in future cities: a systematic literature review. Water (Switzerland) 15. https://doi.org/10.3390/w15173064

Singh R, Langyan S, Sangwan S, Rohtagi B, Khandelwal A (2022) Protein for human consumption from oilseed cakes: a review. Front Sustain Food Syst 6:1–12. https://doi.org/10.3389/fsufs.2022.856401

Solarte-Toro JC, Cardona Alzate CA (2023) Sustainability of biorefineries: challenges and perspectives. Energies (Basel) 16. https://doi.org/10.3390/en16093786

Stellingwerf HM, Guo X, Annevelink E, Behdani B (2022) Logistics and supply chain modelling for the biobased economy: a systematic literature review and research agenda. Front Chem Eng 4:1–15. https://doi.org/10.3389/fceng.2022.778315

Stephen R, William R, Bryan R, Joseph OJ (2013) Sustainable multipurpose Biorefineries for third-generation biofuels and value-added co-products. In: Biofuels – economy, environment and sustainability. InTech. https://doi.org/10.5772/54804

Szarka N, García Laverde L, Thrän D, Kiyko O, Ilkiv M, Moravčíková D, Cudlínová E, Lapka M, Hatvani N, Koós Á, Luks A, Martín Jimenez I (2023) Stakeholder engagement in the co-design of regional bioeconomy strategies. Sustainability (Switzerland) 15. https://doi.org/10.3390/su15086967

Takkellapati S, Li T, Gonzalez MA (2018) An overview of biorefinery-derived platform chemicals from a cellulose and hemicellulose biorefinery. Clean Techn Environ Policy 20:1615–1630. https://doi.org/10.1007/s10098-018-1568-5

Udayan A, Pandey AK, Sharma P, Sreekumar N, Kumar S (2021) Emerging industrial applications of microalgae: challenges and future perspectives. Syst Microbiol Biomanuf 1:411–431. https://doi.org/10.1007/s43393-021-00038-8

Udayan A, Pandey AK, Sirohi R, Sreekumar N, Sang BI, Sim SJ, Kim SH, Pandey A (2023) Production of microalgae with high lipid content and their potential as sources of nutraceuticals. Phytochem Rev 22:833–860. https://doi.org/10.1007/s11101-021-09784-y

Uddin MN, Yasir S, Siddiki A, Mo M, Djavanroodi F, Hazrat MA (2021) Prospects of bioenergy production from organic waste using anaerobic digestion technology: a mini review. Front Energy Res 9:627093. https://doi.org/10.3389/fenrg.2021.627093

Ullah K, Kumar Sharma V, Dhingra S, Braccio G, Ahmad M, Sofia S (2015) Assessing the lignocellulosic biomass resources potential in developing countries: a critical review. Renew Sust Energ Rev 51:682–698. https://doi.org/10.1016/j.rser.2015.06.044

Venkatesh G (2022) Circular bio-economy—paradigm for the future: systematic review of scientific journal publications from 2015 to 2021. Circ Econ Sustain. https://doi.org/10.1007/s43615-021-00084-3

Vidaurre NAM, Vargas-Carpintero R, Wagner M, Lask J, Lewandowski I (2020) Social aspects in the assessment of biobased value chains. Sustainability (Switzerland) 12:1–27. https://doi.org/10.3390/su12239843

Vogiantzi C, Tserpes K (2023) On the definition, assessment, and enhancement of circular economy across various industrial sectors: a literature review and recent findings. Sustainability 15:16532. https://doi.org/10.3390/su152316532

Vohra M, Manwar J, Manmode R, Padgilwar S, Patil S (2014) Bioethanol production: feedstock and current technologies. J Environ Chem Eng 2:573–584. https://doi.org/10.1016/j.jece.2013.10.013

Vural Gursel I, Elbersen B, Meesters KPH, van Leeuwen M (2022) Defining circular economy principles for biobased products. Sustainability (Switzerland) 14:1–21. https://doi.org/10.3390/su141912780

Wang F, Harindintwali JD, Yuan Z, Wang M, Wang F, Li S, Yin Z, Huang L, Fu Y, Li L, Chang SX, Zhang L, Rinklebe J, Yuan Z, Zhu Q, Xiang L, Tsang DCW, Xu L, Jiang X, Liu J, Wei N, Kästner M, Zou Y, Ok YS, Shen J, Peng D, Zhang W, Barceló D, Zhou Y, Bai Z, Li B, Zhang B, Wei K, Cao H, Tan Z, Zhao L, bin He X, Zheng J, Bolan N, Liu X, Huang C, Dietmann S, Luo M, Sun N, Gong J, Gong Y, Brahushi F, Zhang T, Xiao C, Li X, Chen W, Jiao N, Lehmann J, Zhu YG, Jin H, Schäffer A, Tiedje JM, Chen JM (2021a) Technologies and perspectives for achieving carbon neutrality. Innovation 2. https://doi.org/10.1016/j.xinn.2021.100180

Wang F, Ouyang D, Zhou Z, Page SJ, Liu D, Zhao X (2021b) Lignocellulosic biomass as sustainable feedstock and materials for power generation and energy storage. J Energy Chem 57:247–280. https://doi.org/10.1016/j.jechem.2020.08.060

Wang Y, Tibbetts SM, McGinn PJ (2021c) Microalgae as sources of high-quality protein for human food and protein supplements. Foods 10:1–18. https://doi.org/10.3390/foods10123002

Wei X, Luo J, Pu A, Liu Q, Zhang L, Wu S, Long Y, Leng Y, Dong Z, Wan X (2022) From biotechnology to bioeconomy: a review of development dynamics and pathways. Sustainability (Switzerland) 14:1–17. https://doi.org/10.3390/su141610413

Woodfield HK, Harwood JL (2016) Oilseed crops: linseed, rapeseed, soybean, and sunflower. In: Encyclopedia of applied plant sciences, 2nd edn. Elsevier. https://doi.org/10.1016/B978-0-12-394807-6.00212-4

Yang M, Chen L, Wang J, Msigwa G, Osman AI, Fawzy S, Rooney DW, Yap PS (2023) Circular economy strategies for combating climate change and other environmental issues. Environ Chem Lett 21:55–80. https://doi.org/10.1007/s10311-022-01499-6

Yogalakshmi KN, Mohamed Usman TM, Kavitha S, Sachdeva S, Thakur S, Adish Kumar S, Rajesh Banu J (2023) Lignocellulosic biorefinery technologies: a perception into recent advances in biomass fractionation, Biorefineries, economic hurdles and market outlook. Fermentation. https://doi.org/10.3390/fermentation9030238

Yue D, You F, Snyder SW (2014) Biomass-to-bioenergy and biofuel supply chain optimization: overview, key issues and challenges. Comput Chem Eng 66:36–56. https://doi.org/10.1016/j.compchemeng.2013.11.016

Zetterholm J, Bryngemark E, Ahlström J, Söderholm P, Harvey S, Wetterlund E (2020) Economic evaluation of large-scale biorefinery deployment: a framework integrating dynamic biomass market and techno-economic models. Sustainability (Switzerland) 12. https://doi.org/10.3390/su12177126

Zoghlami A, Paës G (2019) Lignocellulosic biomass: understanding recalcitrance and predicting hydrolysis. Front Chem 7:1–11. https://doi.org/10.3389/fchem.2019.00874

Zuiderveen EAR, Kuipers KJJ, Caldeira C, Hanssen SV, van der Hulst MK, de Jonge MMJ, Vlysidis A, van Zelm R, Sala S, Huijbregts MAJ (2023) The potential of emerging bio-based products to reduce environmental impacts. Nat Commun 14. https://doi.org/10.1038/s41467-023-43797-9

MIX
Papier aus verantwortungsvollen Quellen
Paper from responsible sources
FSC® C105338

If you have any concerns about our products,
you can contact us on
**ProductSafety@springernature.com**

In case Publisher is established outside the EU,
the EU authorized representative is:
**Springer Nature Customer Service Center GmbH**
**Europaplatz 3, 69115 Heidelberg, Germany**

Printed by Libri Plureos GmbH
in Hamburg, Germany